Tributes

Volume 23

Infinity, Computability, and Metamathematics

Festschrift celebrating the 60th birthdays of

Peter Koepke and Philip Welch

Volume 13
Proofs, Categories and Computations. Essays in Honour of Grigori Mints
Solomon Feferman and Wilfried Sieg, eds.

Volume 14
Construction. Festschrift for Gerhard Heinzmann
Solomon Feferman and Wilfried Sieg, eds.

Volume 15
Hues of Philosophy. Essays in Memory of Ruth Manor
Anat Biletzki, ed.

Volume 16
Knowing, Reasoning and Acting. Essays in Honour of Hector J. Levesque.
Gerhard Lakemeyer and Sheila A. McIlraith, eds.

Volume 17
Logic without Frontiers. Festschrift for Walter Alexandre Carnielli on the occasion of his 60th birthday
Jean-Yves Beziau and Marcelo Esteban Coniglio, eds.

Volume 18
Insolubles and Consequences. Essays in Honour of Stephen Read.
Catarina Dutilh Novaes and Ole Thomassen Hjortland, eds.

Volume 19
From Quantification to Conversation. Festschrift for Robin Cooper on the occasion of his 65th birthday
Staffan Larsson and Lars Borin, eds.

Volume 20
The Goals of Cognition. Essays in Honour of Cristiano Castelfranchi
Fabio Paglieri, Luca Tummolini, Rino Falcone and Maria Miceli, eds.

Volume 21
From Knowledge Representation to Argumentation in AI, Law and Policy Making. A Festschrift in Honour of Trevor Bench-Capon on the Occasion of his 60th Birthday
Katie Atkinson, Henry Prakken and Adam Wyner, eds.

Volume 22
Foundational Adventures. Essays in Honour of Harvey M. Friedman
Neil Tennant, ed.

Volume 23
Infinity, Computability, and Metamathematics. Festschrift celebrating the 60th birthdays of Peter Koepke and Philip Welch
Stefan Geschke, Benedikt Löwe and Philipp Schlicht, eds.

Tributes Series Editor
Dov Gabbay dov.gabbay@kcl.ac.uk

Infinity, Computability, and Metamathematics

Festschrift celebrating the 60th birthdays of Peter Koepke and Philip Welch

edited by

Stefan Geschke

Benedikt Löwe

Philipp Schlicht

ISBN 978-1-84890-130-8

College Publications
Scientific Director: Dov Gabbay
Managing Director: Jane Spurr

http://www.collegepublications.co.uk

Originally published online by Templeton Press
http://www.foundationaladventures.com

Cover design by Laraine Welch

Printed by Lightning Source, Milton Keynes, UK

Table of Contents

Preface

In the year 2014, Peter Koepke and Philip Welch turned sixty and we celebrated this festive occasion with a scientific workshop in their honour. The workshop speakers included students, collaborators, colleagues, and friends of Peter and Philip whose research was influenced by them.

Peter and Philip have more in common than their age. Both were professors at the University of Bonn; Peter for many years and Philip as a Mercator Professor during the academic year 2002/3. Both were strongly influenced by Ronald Jensen early in their scientific careers. Philip got his D.Phil. in 1979 from the University of Oxford under the supervision of Robin Gandy with a thesis on combinatorial principles in Jensen's core model, with Jensen being in Oxford from 1978 to 1979. Peter received his doctoral degree from the University of Freiburg under Jensen's supervision with a dissertation titled "A Theory of Short Core Models and Some Applications". Peter and Philip continued working in set theory, but they also got interested in other subjects in logic: infinite time computation is another of these interests that they shared. In the last two decades, their interests became increasingly interdisciplinary: Philip started working with philosophers on theories of truth and Peter collaborated with linguists on natural language proofs. Three joint publications witness their common interests.

This Festschrift was handed over to Peter and Philip on 23 May 2014 during the Colloquium and Workshop "Infinity, computability, and metamathematics: Celebrating the 60th birthdays of Peter Koepke and Philip Welch", which took place from 23 May to 25 May 2014 at the Hausdorff Center for Mathematics (HCM) at the *Rheinische Friedrich-Wilhelms-Universität*

Bonn. The workshop and the production of the book were funded by the HCM. We received nineteen submissions for this Festschrift which were thoroughly refereed according to the standards of our field; nineteen referees helped us with this in very short time (we received the first submissions in late September 2013 and needed to submit the volume for production in April 2014; readers who know the usual glacial pace of mathematical publishing can appreciate how fast this is) and in the end, seventeen papers were accepted for publication in the Festschrift. As is traditional, we included lists of the publications and doctoral students of Peter and Philip in this preface to document their important role in our community. We are very happy to see five of the doctoral students among the authors of papers in this Festschrift. We should like to thank all authors and referees for their collaboration in producing this interesting book and hope that Peter and Philip will enjoy it.

April 2014, Hamburg, Amsterdam, and Bonn S. G. B. L. P. S.

Publications by Peter Koepke

[1] Peter Koepke, Karen Räsch, and Philipp Schlicht. A minimal Prikry-type forcing for singularizing a measurable cardinal. *Journal of Symbolic Logic*, 78(1):85–100, 2013.

[2] Peter Koepke and Benjamin Seyfferth. Towards a theory of infinite time Blum-Shub-Smale machines. In S. Barry Cooper, Anuj Dawar, and Benedikt Löwe, editors, *How the World Computes, Turing Centenary Conference and 8th Conference on Computability in Europe, CiE 2012, Cambridge, UK, June 18-23, 2012, Proceedings*, volume 7318 of *Lecture Notes in Computer Science*, pages 310–318, Heidelberg, 2012. Springer-Verlag.

[3] Moti Gitik and Peter Koepke. Violating the singular cardinals hypothesis without large cardinals. *Israel Journal of Mathematics*, 191(2):901–922, 2012.

[4] Peter Koepke and Julian J. Schlöder. Transition of consistency and satisfiability under language extensions. *Formalized Mathematics*, 20(3):193–197, 2012.

[5] Peter Koepke and Julian J. Schlöder. The Gödel completeness theorem for uncountable languages. *Formalized Mathematics*, 20(3):199–203, 2012.

[6] Peter Koepke and Philip D. Welch. A generalised dynamical system, infinite time register machines, and Π^1_1-CA_0. In Benedikt Löwe, Dag Normann, Ivan Soskov, and Alexandra A. Soskova, editors, *Models of Computation in Context, 7th Conference on Computability in Europe, CiE 2011, Sofia, Bulgaria, June 27–July 2, 2011, Proceedings*, volume 6735 of *Lecture Notes in Computer Science*, pages 152–159. Springer-Verlag, 2011.

[7] Marcos Cramer, Peter Koepke, and Bernhard Schröder. Parsing and disambiguation of symbolic mathematics in the Naproche system. In James H. Davenport, William M. Farmer, Josef Urban, and Florian Rabe, editors, *Intelligent Computer Mathematics, 18th Symposium, Calculemus 2011, and 10th International Conference, MKM 2011, Bertinoro, Italy, July 18-23, 2011. Proceedings*, volume 6824 of *Lecture Notes in Computer Science*, pages 180–195. Springer-Verlag, 2011.

[8] Peter Koepke and Philip D. Welch. Global square and mutual stationarity at the $\aleph_n$. *Annals of Pure and Applied Logic*, 162(10):787–806, 2011.

[9] Arthur W. Apter and Peter Koepke. The consistency strength of choiceless failures of SCH. *Journal of Symbolic Logic*, 75(3):1066–1080, 2010.

[10] Merlin Carl, Tim Fischbach, Peter Koepke, Russell Miller, Miriam Nasfi, and Gregor Weckbecker. The basic theory of infinite time register machines. *Archive for Mathematical Logic*, 49(2):249–273, 2010.

[11] Marcos Cramer, Peter Koepke, Daniel Kühlwein, and Bernhard Schröder. Premise selection in the Naproche system. In Jürgen Giesl and Reiner Hähnle, editors, *Automated Reasoning, 5th International Joint Conference, IJCAR 2010, Edinburgh, UK, July 16-19, 2010, Proceedings*, volume 6173 of *Lecture Notes in Computer Science*, pages 434–440. Springer-Verlag, 2010.

[12] Marcos Cramer, Bernhard Fisseni, Peter Koepke, Daniel Kühlwein, Bernhard Schröder, and Jip Veldman. The Naproche project, controlled natural language proof checking of mathematical texts. In Norbert E. Fuchs, editor, *Controlled Natural Language, Workshop on Controlled Natural Language, CNL 2009, Marettimo Island, Italy, June 8-10, 2009, Revised Papers*, volume 5972 of *Lecture Notes in Computer Science*, pages 170–186. Springer-Verlag, 2010.

[13] Merlin Carl and Peter Koepke. Interpreting Naproche—An algorithmic approach to the derivation-indicator view. In Alison Pease, Markus Guhe, and Alan Smaill, editors, *Proceedings of the International Symposium on Mathematical Practice and Cognition: A symposium at the AISB 2010 Convention*, pages 7–10. Society for the Study of Artificial Intelligence and the Simulation of Behaviour, 2010.

[14] Peter Koepke. Ordinal computability. In Klaus Ambos-Spies, Benedikt Löwe, and Wolfgang Merkle, editors, *Mathematical Theory and Computational Practice, 5th Conference on Computability in Europe, CiE 2009, Heidelberg, Germany, July 19-24, 2009. Proceedings*, volume 5635 of *Lecture Notes in Computer Science*, pages 280–289. Springer-Verlag, 2009.

[15] Peter Koepke and Benjamin Seyfferth. Ordinal machines and admissible recursion theory. *Annals of Pure and Applied Logic*, 160(3):310–318, 2009.

[16] Jip Veldman, Bernhard Fisseni, Bernhard Schröder, and Peter Koepke. From proof texts to logic. In Christian Chiarcos, Richard Eckart de Castilho, and Manfred Stede, editors, *Von der Form zur Bedeutung: Texte automatisch verarbeiten, From Form to Meaning: Processing Texts Automatically, Proceedings of the Biennial GSCL Conference 2009*, pages 137–146. Gunter Narr, 2009.

[17] Peter Koepke and Russell Miller. An enhanced theory of infinite time register machines. In Arnold Beckmann, Costas Dimitracopoulos, and Benedikt Löwe, editors, *Logic and Theory of Algorithms, 4th Conference on Computability in Europe, CiE 2008, Athens, Greece, June 15-20, 2008, Proceedings*, volume 5028 of *Lecture Notes in Computer Science*, pages 306–315, Heidelberg, 2008. Springer-Verlag.

[18] Arthur W. Apter and Peter Koepke. Making all cardinals almost Ramsey. *Archive for Mathematical Logic*, 47(7-8):769–783, 2008.

[19] Peter Koepke and Ryan Siders. Register computations on ordinals. *Archive for Mathematical Logic*, 47(6):529–548, 2008.

[20] Peter Koepke and Ryan Siders. Minimality considerations for ordinal computers modeling constructibility. *Theoretical Computer Science*, 394(3):197–207, 2008.

[21] Peter Koepke. Forcing a mutual stationarity property in cofinality ω_1. *Proceedings of the American Mathematical Society*, 135(5):1523–1533, 2007.

[22] Peter Koepke. Gödel's completeness theorem with natural language formulas. In Thomas Müller and Albert Newen, editors, *Logik, Begriffe, Prinzipien des Handelns, Logic, Concepts, Principles of Action*, pages 49–63. mentis, 2007.

[23] Peter Koepke. Ordinals, computations, and models of set theory. In Reinhard Kahle and Isabel Oitavem, editors, *Days in Logic '06, Two Tutorials*, volume 38 of *Texts on Mathematics*, pages 43–78. Universidade de Coimbra, 2006.

[24] Peter Koepke and Philip Welch. On the strength of mutual stationarity. In Joan Bagaria and Stevo Todorcevic, editors, *Set theory, Centre de Recerca Matemàtica, Barcelona, 2003–2004*, Trends in Mathematics, pages 309–320. Birkhäuser Verlag, 2006.

[25] Peter Koepke and Martin Koerwien. Ordinal computations. *Mathematical Structures in Computer Science*, 16(5):867–884, 2006.

[26] Arthur W. Apter and Peter Koepke. The consistency strength of $\aleph_\omega$ and $\aleph_{\omega_1}$ being Rowbottom cardinals without the axiom of choice. *Archive for Mathematical Logic*, 45(6):721–737, 2006.

[27] Sy-David Friedman, Peter Koepke, and Boris Piwinger. Hyperfine structure theory and gap 1 morasses. *Journal of Symbolic Logic*, 71(2):480–490, 2006.

[28] Peter Koepke. Computing a model of set theory. In S. Barry Cooper, Benedikt Löwe, and Leen Torenvliet, editors, *New Computational Paradigms, First Conference on Computability in Europe, CiE 2005, Amsterdam, The Netherlands, June 8-12, 2005, Proceedings*, volume 3526 of *Lecture Notes in Computer Science*, pages 223–232, Heidelberg, 2005. Springer-Verlag.

[29] Peter Koepke. Infinite time register machines. In Arnold Beckmann, Ulrich Berger, Benedikt Löwe, and John V. Tucker, editors, *Logical Approaches to Computational Barriers, Second Conference on Computability in Europe, CiE 2006, Swansea, UK, June 30-July 5, 2006, Proceedings*, volume 3988 of *Lecture Notes in Computer Science*, pages 257–266, Heidelberg, 2006. Springer-Verlag.

[30] Peter Koepke and Ralf Schindler. Homogeneously Souslin sets in small inner models. *Archive for Mathematical Logic*, 45(1):53–61, 2006.

[31] Peter Koepke. Turing computations on ordinals. *Bulletin of Symbolic Logic*, 11:377–397, 2005.

[32] Peter Koepke. *Simplified constructibility theory: minicourse Helsinki, March 2005*, volume 5 of *Graduate texts in mathematics*. University of Helsinki, 2005.

[33] Patrick Braselmann and Peter Koepke. A formal proof of Gödel's completeness theorem. *Formalized Mathematics*, 13:5–53, 2005.

[34] Peter Koepke and Marc van Eijmeren. A refinement of Jensen's constructible hierarchy. In Benedikt Löwe, Boris Piwinger, and Thoralf Räsch, editors, *Classical and new paradigms of computation and their complexity hierarchies*, volume 23 of *Trends in Logic*, pages 159–169. Kluwer Academic Publishers, 2004.

[35] Peter Koepke and Bernhard Schröder. ProofML—eine Annotationssprache für natürliche Beweise. *LDV-Forum*, 18(3), 2003.

[36] Vladimir Kanovei and Peter Koepke. Deskriptive Mengenlehre in Hausdorffs *Grundzügen der Mengenlehre*. In Egbert Brieskorn, Srishti D.

Chatterji, Moritz Epple, Ulrich Felgner, Horst Herrlich, Mirek Hušek, Vladimir Kanovei, Peter Koepke, Gerhard Preuß, Walter Purkert, and Erhard Scholz, editors, *Felix Hausdorff, gesammelte Werke. Band II. "Grundzüge der Mengenlehre"*, pages 773–787. Springer-Verlag, 2002.

[37] Peter Koepke. The category of inner models. *Synthese*, 133(1-2):275–303, 2002.

[38] Peter Koepke and Bernhard Schröder. Natürlich formal. In Gerd Willée, Bernhard Schröder, and Hans-Christian Schmitz, editors, *Computerlinguistik—Was geht, was kommt? Computational Linguistics—Achievements and Perspectives*, volume 4 of *Sprachwissenschaft, Computerlinguistik und Neue Medien*, pages 186–191. Gardez!-Verlag, 2003.

[39] Peter Koepke. An iteration model violating the singular cardinals hypothesis. In S. Barry Cooper and John K. Truss, editors, *Sets and proofs. Invited papers from Logic Colloquium '97 (European Meeting of the Association for Symbolic Logic) held in Leeds, July 6-13, 1997*, volume 258 of *London Mathematical Society Lecture Note Series*, pages 95–102. Cambridge University Press, 1999.

[40] Peter Koepke. Extenders, embedding normal forms, and the Martin-Steel theorem. *Journal of Symbolic Logic*, 63(3):1137–1176, 1998.

[41] Sy D. Friedman and Peter Koepke. An elementary approach to the fine structure of **L**. *Bulletin of Symbolic Logic*, 3(4):453–468, 1997.

[42] Peter Koepke. Embedding normal forms and $\mathbf{\Pi}^1_1$-determinacy. In Wilfrid Hodges, Martin Hyland, Charles Steinhorn, and John K. Truss, editors, *Logic: from foundations to applications. Proceedings of the Logic Colloquium held as a part of the European Meeting of the Association for Symbolic Logic at the University of Keele, Staffordshire, July 2029, 1993*, Oxford Science Publications, pages 215–224. Oxford University Press, 1996.

[43] Peter Koepke. Metamathematische Aspekte der Hausdorffschen Mengenlehre. In Egbert Brieskorn, editor, *Felix Hausdorff zum Gedächtnis, Band I. Aspekte seines Werkes*, pages 71–106. Vieweg, 1996.

[44] Peter Koepke and Juan Carlos Martínez. Superatomic Boolean algebras constructed from morasses. *Journal of Symbolic Logic*, 60(3):940–951, 1995.

[45] Peter Koepke. An introduction to extenders and core models for extender sequences. In H.-D. Ebbinghaus, J. Fernandez-Prida, M. Garrido, D. Lascar, and M. Rodríquez Artalejo, editors, *Logic Colloquium '87. Proceedings of the colloquium held at the University of Granada, Granada, July 2025, 1987*, volume 129 of *Studies in Logic and the Foundations of Mathematics*, pages 137–182. North-Holland, 1989.

[46] Peter Koepke. On the elimination of Malitz quantifiers over Archimedian real closed fields. *Archive for Mathematical Logic*, 28(3):167–171, 1989.

[47] Peter Koepke. On the free subset property at singular cardinals. *Archive for Mathematical Logic*, 28(1):43–55, 1989.

[48] Hans-Dieter Donder, Peter Koepke, and Jean-Pierre Levinski. Some stationary subsets of $\wp(\lambda)$. *Proceedings of the American Mathematical Society*, 102(4):1000–1004, 1988.

[49] Peter Koepke. Some applications of short core models. *Annals of Pure and Applied Logic*, 37(2):179–204, 1988.

[50] Peter Koepke. The consistency strength of the free-subset property for ω_ω. *Journal of Symbolic Logic*, 49(4):1198–1204, 1984.

[51] Hans-Dieter Donder and Peter Koepke. On the consistency strength of "accessible" Jónsson cardinals and of the weak Chang conjecture. *Annals of Pure and Applied Logic*, 25(3):233–261, 1983.

Publications by Philip Welch

[1] Philip D. Welch. Transfinite machine models. In Rod Downey, editor, *Turing's Legacy, Developments from Turing's Ideas in Logic*, volume 42 of *Lecture Notes in Logic*, pages 493–529. Association for Symbolic Logic, 2014.

[2] Philip D. Welch. Turing's mathematical work. In Rafał Latała, Andrzej Ruciński, Paweł Strzelecki, Paweł Świątkowski, Dariusz Wrzosek, and Piotr Zakrezewski, editors, *European Congress of Mathematics, Kraków, 2-7 July, 2012*, pages 763–777. EMS Publishing House, 2014.

[3] Philip Welch. Some observations on truth hierarchies. *Review of Symbolic Logic*, 7(1):1–30, 2014.

[4] Philip D. Welch. Truth and Turing. In S. Barry Cooper and Jan van Leeuwen, editors, *Alan Turing: his work and impact*, pages 202–205. Elsevier, 2013.

[5] Philip D. Welch. Towards the unknown region: On computing infinite numbers. In S. Barry Cooper and Jan van Leeuwen, editors, *Alan Turing: his work and impact*, pages 109–116. Elsevier, 2013.

[6] Leon Horsten and Philip Welch. The aftermath. *Mathematical Intelligencer*, 35(1):16–20, 2013.

[7] Leon Horsten, Graham E. Leigh, Hannes Leitgeb, and Philip Welch. Revision revisited. *Review of Symbolic Logic*, 5(4):642–665, 2012.

[8] Philip D. Welch. Some reflections on Alan Turing's centenary. *European Mathematical Society Newsletter*, 85:32–38, 2012.

[9] Peter Koepke and Philip D. Welch. A generalised dynamical system, infinite time register machines, and Π^1_1-CA_0. In Benedikt Löwe, Dag Normann, Ivan Soskov, and Alexandra A. Soskova, editors, *Models of Computation in Context, 7th Conference on Computability in Europe, CiE 2011, Sofia, Bulgaria, June 27–July 2, 2011, Proceedings*, volume 6735 of *Lecture Notes in Computer Science*, pages 152–159. Springer-Verlag, 2011.

[10] Philip D. Welch. Truth, logical validity and determinateness: a commentary on Field's *Saving truth from paradox*. *Review of Symbolic Logic*, 4(3):348–359, 2011.

[11] Philip D. Welch. Discrete transfinite computation models. In S. Barry Cooper and Andrea Sorbi, editors, *Computability in Context, Computation and Logic in the Real World*, pages 371–410. World Scientific, 2011.

[12] Ian Sharpe and Philip D. Welch. Greatly Erdős cardinals with some generalizations to the Chang and Ramsey properties. *Annals of Pure and Applied Logic*, 162(11):863–902, 2011.

[13] Victoria Gitman and Philip D. Welch. Ramsey-like cardinals II. *Journal of Symbolic Logic*, 76(2):541–560, 2011.

[14] Philip D. Welch. Determinacy in strong cardinal models. *Journal of Symbolic Logic*, 76(2):719–728, 2011.

[15] Sy-David Friedman and Philip D. Welch. Hypermachines. *Journal of Symbolic Logic*, 76(2):620–636, 2011.

[16] Philip D. Welch. Weak systems of determinacy and arithmetical quasi-inductive definitions. *Journal of Symbolic Logic*, 76(2):418–436, 2011.

[17] P. Koepke and Philip D. Welch. Global square and mutual stationarity at the $\aleph_n$. *Annals of Pure and Applied Logic*, 162(10):787–806, 2011.

[18] Philip D. Welch. Σ^* fine structure. In Matthew Foreman and Akihiro Kanamori, editors, *Handbook of set theory. Vol. 1*, pages 657–736. Springer-Verlag, 2010.

[19] Philip D. Welch. Relativistic computers and transfinite computation. In Cristian S. Calude, José Félix Costa, Nachum Dershowitz, Elisabete Freire, and Grzegorz Rozenberg, editors, *Unconventional Computation, 8th International Conference, UC 2009, Ponta Delgada, Azores, Portugal, September 7-11, 2009. Proceedings*, volume 5715 of *Lecture Notes in Computer Science*, pages 37–41. Springer-Verlag, 2009.

[20] Volker Halbach and Philip Welch. Necessities and necessary truths: a prolegomenon to the use of modal logic in the analysis of intensional notions. *Mind*, 118(469):96–100, 2009.

[21] Philip D. Welch. Games for truth. *Bulletin of Symbolic Logic*, 15(4):410–427, 2009.

[22] Leon Horsten and Philip Welch. Erratum: The undecidability of propositional adaptive logic. *Synthese*, 169(1):217–218, 2009.

[23] Philip D. Welch. Characteristics of discrete transfinite time Turing machine models: halting times, stabilization times, and normal form theorems. *Theoretical Computer Science*, 410(4-5):426–442, 2009.

[24] Philip D. Welch. The extent of computation in Malament-Hogarth spacetimes. *British Journal for the Philosophy of Science*, 59(4):659–674, 2008.

[25] Sy-David Friedman, Philip Welch, and W. Hugh Woodin. On the consistency strength of the inner model hypothesis. *Journal of Symbolic Logic*, 73(2):391–400, 2008.

[26] Philip D. Welch. Bounding lemmata for non-deterministic halting times of transfinite Turing machines. *Theoretical Computer Science*, 394(3):223–228, 2008.

[27] Philip D. Welch. Ultimate truth vis à vis stable truth. *Review of Symbolic Logic*, 1(1):126–142, 2008.

[28] Agustin Rayo and Philip D. Welch. Field on revenge. In J. C. Beall, editor, *Revenge of the Liar, New Essays on the Paradox*, pages 234–249. Oxford University Press, 2008.

[29] Philip D. Welch. Turing unbound: transfinite computation. In S. Barry Cooper, Benedikt Löwe, and Andrea Sorbi, editors, *Computation and Logic in the Real World, Third Conference on Computability in Europe, CiE 2007, Siena, Italy, June 18-23, 2007, Proceedings*, volume 4497 of *Lecture Notes in Computer Science*, pages 768–780. Springer-Verlag, 2007.

[30] Leon Horsten and Philip Welch. The undecidability of propositional adaptive logic. *Synthese*, 158(1):41–60, 2007.

[31] Philip Welch. Non-deterministic halting times for hamkins-kidder turing machines. In Arnold Beckmann, Ulrich Berger, Benedikt Löwe, and John V. Tucker, editors, *Logical Approaches to Computational Barriers, Second Conference on Computability in Europe, CiE 2006, Swansea, UK, June 30-July 5, 2006, Proceedings*, volume 3988 of *Lecture Notes in Computer Science*, pages 571–574, 2006.

[32] Peter Koepke and Philip Welch. On the strength of mutual stationarity. In Joan Bagaria and Stevo Todorcevic, editors, *Set theory, Centre de Recerca Matemàtica, Barcelona, 2003–2004*, Trends in Mathematics, pages 309–320. Birkhäuser Verlag, 2006.

[33] Philip Welch. On the transfinite action of 1 tape turing machines. In S. Barry Cooper, Benedikt Löwe, and Leen Torenvliet, editors, *New Computational Paradigms, First Conference on Computability in Europe, CiE 2005, Amsterdam, The Netherlands, June 8-12, 2005, Proceedings*, volume 3526 of *Lecture Notes in Computer Science*, pages 532–539. Springer-Verlag, 2005.

[34] Volker Halbach, Hannes Leitgeb, and Philip Welch. Possible worlds semantics for predicates. In Reinhard Kahle, editor, *Intensionality*, volume 22 of *Lecture Notes in Logic*, pages 20–41. Association for Symbolic Logic, 2005.

[35] Kai-Uwe Kühnberger, Benedikt Löwe, Michael Möllerfeld, and Philip Welch. Comparing inductive and circular definitions: parameters, complexity and games. *Studia Logica*, 81(1):79–98, 2005.

[36] Philip D. Welch. Some open problems in mutual stationarity involving inner model theory: A commentary. *Notre Dame Journal of Formal Logic*, 46(3):375–379, 2005.

[37] Philip D. Welch. Post's and other problems of supertasks of higher type. In Benedikt Löwe, Boris Piwinger, and Thoralf Räsch, editors, *Classical and new paradigms of computation and their complexity hierarchies*, volume 23 of *Trends in Logic*, pages 223–237. Kluwer Academic Publishers, 2004.

[38] Philip D. Welch. On the possibility, or otherwise, of hypercomputation. *British Journal for the Philosophy of Science*, 55(4):739–746, 2004.

[39] Philip D. Welch. On unfoldable cardinals, ω-closed cardinals, and the beginning of the inner model hierarchy. *Archive for Mathematical Logic*, 43(4):443–458, 2004.

[40] Joel David Hamkins and Philip D. Welch. $\mathrm{P}^f \neq \mathrm{NP}^f$ for almost all f. *Mathematical Logic Quarterly*, 49(5):536–540, 2003.

[41] Volker Halbach, Hannes Leitgeb, and Philip Welch. Possible-worlds semantics for modal notions conceived as predicates. *Journal of Philosophical Logic*, 32(2):179–223, 2003.

[42] Philip D. Welch. On revision operators. *Journal of Symbolic Logic*, 68(2):689–711, 2003.

[43] David Asperó and Philip D. Welch. Bounded Martin's maximum, weak Erdős cardinals, and ψ_{AC}. *Journal of Symbolic Logic*, 67(3):1141–1152, 2002.

[44] Philip D. Welch. On possible non-homeomorphic substructures of the real line. *Proceedings of the American Mathematical Society*, 130(9):2771–2775, 2002.

[45] Benedikt Löwe and Philip D. Welch. Set-theoretic absoluteness and the revision theory of truth. *Studia Logica*, 68(1):21–41, 2001.

[46] Philip D. Welch. On Gupta-Belnap revision theories of truth, Kripkean fixed points, and the next stable set. *Bulletin of Symbolic Logic*, 7(3):345–360, 2001.

[47] John Vickers and Philip D. Welch. On elementary embeddings from an inner model to the universe. *Journal of Symbolic Logic*, 66(3):1090–1116, 2001.

[48] Philip D. Welch. Eventually infinite time Turing machine degrees: infinite time decidable reals. *Journal of Symbolic Logic*, 65(3):1193–1203, 2000.

[49] John Vickers and Philip D. Welch. On successors of Jónsson cardinals. *Archive for Mathematical Logic*, 39(6):465–473, 2000.

[50] Philip D. Welch. Some remarks on the maximality of inner models. In Samuel R. Buss, Petr Hájek, and Pavel Pudlák, editors, *Logic Colloquium '98. Proceedings of the Annual European Summer Meeting of the Association for Symbolic Logic held at the University of Economics, Prague, August 9-15, 1998*, volume 13 of *Lecture Notes in Logic*, pages 516–540. Association for Symbolic Logic, 2000.

[51] Philip D. Welch. The length of infinite time Turing machine computations. *Bulletin of the London Mathematical Society*, 32(2):129–136, 2000.

[52] Philip D. Welch. Minimality arguments for infinite time Turing degrees. In S. Barry Cooper and John K. Truss, editors, *Sets and proofs. Invited papers from Logic Colloquium '97 (European Meeting of the Association for Symbolic Logic) held in Leeds, July 6-13, 1997*, volume 258 of *London Mathematical Society Lecture Note Series*, pages 425–436. Cambridge University Press, 1999.

[53] John R. Steel and Philip D. Welch. Σ^1_3 absoluteness and the second uniform indiscernible. *Israel Journal of Mathematics*, 104:157–190, 1998.

[54] Philip D. Welch. Determinacy in the difference hierarchy of co-analytic sets. *Annals of Pure and Applied Logic*, 80(1):69–108, 1996.

[55] Philip D. Welch. Countable unions of simple sets in the core model. *Journal of Symbolic Logic*, 61(1):293–312, 1996.

[56] Philip D. Welch. Characterising subsets of ω_1 constructible from a real. *Journal of Symbolic Logic*, 59(4):1420–1432, 1994.

[57] Philip D. Welch. Some descriptive set theory and core models. *Annals of Pure and Applied Logic*, 39(3):273–290, 1988.

[58] Philip D. Welch. Coding that preserves Ramseyness. *Fundamenta Mathematicae*, 129(1):1–7, 1988.

[59] Philip D. Welch. Doing without determinacy-aspects of inner models. In Frank R. Drake and John K. Truss, editors, *Logic colloquium '86, Proceedings of the colloquium held at the University of Hull, Hull, July 13-19, 1986, Dedicated to the memory of R. L. Goodstein*, volume 124 of *Studies in Logic and the Foundations of Mathematics*, pages 333–342. North-Holland, 1988.

[60] Philip D. Welch. Minimality in the Δ^1_3-degrees. *Journal of Symbolic Logic*, 52(4):908–915, 1987.

[61] Philip D. Welch. The reals in core models. *Journal of Symbolic Logic*, 52(1):64–67, 1987.

[62] Philip D. Welch. The natural hierarchy and quasihierarchy of constructibility degrees. *Journal of Symbolic Logic*, 51(1):130–134, 1986.

[63] Philip D. Welch. Comparing incomparable Kleene degrees. *Journal of Symbolic Logic*, 50(1):55–58, 1985.

[64] Philip D. Welch. On $\mathbf{\Sigma}^1_3$. In Gert H. Müller and Michael M. Richter, editors, *Models and Sets, Proceedings of the Logic Colloquium held in Aachen, July 18-23, 1983, Part I*, volume 1103 of *Lecture Notes in Mathematics*, pages 473–484. Springer-Verlag, 1984.

[65] Aaron Beller, Ronald B. Jensen, and Philip D. Welch. *Coding the universe*, volume 47 of *London Mathematical Society Lecture Note Series.* Cambridge University Press, 1982.

Doctoral students

Peter Koepke's doctoral students

1. Ralf-Dieter Schindler
   Rheinische Friedrich-Wilhelms-Universität Bonn 1996

2. Florian Rudolph
   Rheinische Friedrich-Wilhelms-Universität Bonn 2000

3. Jochen Löffelmann
   Rheinische Friedrich-Wilhelms-Universität Bonn 2001

4. Sandra Quickert
   Rheinische Friedrich-Wilhelms-Universität Bonn 2002

5. Merlin Carl
   Rheinische Friedrich-Wilhelms-Universität Bonn 2011

6. Ioanna Dimitriou
   Rheinische Friedrich-Wilhelms-Universität Bonn 2011

7. Marcos Cramer
   Rheinische Friedrich-Wilhelms-Universität Bonn 2013

8. Benjamin Seyfferth
   Rheinische Friedrich-Wilhelms-Universität Bonn 2013

Philip Welch's doctoral students

1. John Vickers
   University of Bristol 1993

2. Ian Sharpe
   University of Bristol 2007

3. Richard Pettigrew
   University of Bristol 2008

4. Barnaby Dawson
   University of Bristol 2009

A note on powers of singular strong limit cardinals

Arthur W. Apter*

Department of Mathematics, Baruch College, New York NY, United States of America
Department of Mathematics, The Graduate Center of the City University of New York, New York NY, United States of America

Abstract

We show via a simple forcing argument that if $\kappa \geq \aleph_0$ is any cardinal such that $\kappa^{+\omega}$ is a strong limit cardinal, then $2^{\kappa^{+\omega}} < \kappa^{+\omega_4}$. Our proof makes use of pcf theory applied only at $\aleph_\omega$ and is generalizable to other contexts.

One of the most important results in all of set theory is Shelah's celebrated theorem [2] that if $\aleph_\omega$ is a strong limit cardinal, then $2^{\aleph_\omega} < \aleph_{\omega_4}$. This theorem is proven via pcf theory and generalizes earlier work by many people (for a more detailed discussion of the relevant history, see the survey article by Abraham and Magidor [1]).

The purpose of this note is to show via a simple forcing argument that the analogue of this ZFC provable fact is true about certain larger singular strong limit cardinals as well. In particular, we have the following theorem.

Theorem 1. If $\kappa \geq \aleph_0$ is any cardinal such that $\kappa^{+\omega}$ is a strong limit cardinal, then $2^{\kappa^{+\omega}} < \kappa^{+\omega_4}$.

Note that as Shelah says in [2, p. 359], his result about $\aleph_\omega$ holds for all cardinal non-fixed points. Consequently, Theorem 1 is not a new result. What is new, however, is that a simple forcing argument immediately implies Theorem 1 without any use of pcf theory beyond $\aleph_\omega$. We will comment on this further towards the end of the note.

Proof. Theorem 1 is proven by contradiction. Specifically, suppose that $\kappa \geq \aleph_0$ is any cardinal in some $V \models \mathsf{ZFC}$ such that $\kappa^{+\omega}$ is a strong limit cardinal, and that in addition, $2^{\kappa^{+\omega}} \geq \kappa^{+\omega_4}$. Since we are giving a proof using pcf theory only at $\aleph_\omega$, we assume without loss of generality that $\kappa \geq \aleph_\omega$.

*The author's research was partially supported by PSC-CUNY grants. This paper is dedicated to Peter Koepke and Philip Welch on the occasion of their 60th birthdays. It is truly a privilege to be able to contribute a paper to this Festschrift in their honor. Its main result was proven while the author was speaking with Koepke on March 21, 2013 late in the afternoon at the Così restaurant located at 31st Street and Park Avenue South in Manhattan.

Stefan Geschke, Benedikt Löwe, Philipp Schlicht (*eds.*).
Infinity, computability, and metamathematics: Festschrift celebrating the 60th birthdays of Peter Koepke and Philip Welch. College Publications, London, 2014. Tributes, Volume 23.

Force over V with $\mathbb{P} = \text{Coll}(\omega_5, \kappa) = \{f \mid f : \omega_5 \to \kappa$ is a function such that $|\text{dom}(f)| \le \omega_4\}$, ordered as usual by inclusion. Since $\mathbb{P}$ is ω_4-closed (i.e., every increasing chain of length ω_4 has an upper bound), forcing with $\mathbb{P}$ preserves all cardinals less than or equal to ω_5. A standard density argument shows that forcing with $\mathbb{P}$ collapses κ to an ordinal having cardinality ω_5. In addition, $|\mathbb{P}| \le |[\kappa]^{\omega_5}| \le |\kappa^\kappa| = 2^\kappa$ (because by our earlier assumption, $\kappa \ge \aleph_\omega$), so since $\kappa^{+\omega}$ is a strong limit cardinal, $|\mathbb{P}| < \kappa^{+\omega}$. Consequently, it must be the case that for some fixed $k_0 < \omega$, $|\mathbb{P}| = \kappa^{+k_0}$, which of course immediately implies that $\mathbb{P}$ is κ^{+k_0+1}-c.c. From this, we may infer that there is some $n_0 \ge 1$, $n_0 < \omega$ such that for all natural numbers $0 \le m < \omega$, $\kappa^{+n_0+m} = \omega_{6+m}$, i.e., that $\kappa^{+n_0} = \omega_6$, $\kappa^{+n_0+1} = \omega_7$, etc., and that in addition, $\kappa^{+\omega} = \aleph_\omega$. It further follows that $\kappa^{+\omega}$ remains a strong limit cardinal in $V^{\mathbb{P}}$, since for all $m \ge \max(k_0, n_0)$, $m < \omega$, $(2^{\kappa^{+m}})^{V^{\mathbb{P}}} \le (2^{\kappa^{+m} \times |\mathbb{P}|})^V = (2^{\kappa^{+m}})^V < \kappa^{+\omega}$. But now, in $V^{\mathbb{P}}$, $\aleph_\omega$ is a strong limit cardinal such that $2^{\aleph_\omega} \ge (\aleph_\omega)^{+\omega_4}$, i.e., $V^{\mathbb{P}} \models$ "$2^{\aleph_\omega} \ge \aleph_{\omega_4}$". This contradiction to Shelah's theorem proves Theorem 1. Q.E.D.

Since as we observed above, Shelah's theorem generalizes to all cardinal non-fixed points, the proof of Theorem 1 also yields results such as the following.

Theorem 2. If $\kappa \ge \aleph_0$ is any cardinal such that $\kappa^{+\omega_1}$ is a strong limit cardinal, then $2^{\kappa^{+\omega_1}} < \kappa^{+\omega_4}$.

In analogy to Theorem 1, the proof of Theorem 2 requires no use of pcf theory beyound $\aleph_{\omega_1}$. By changing each occurrence of the ordinal ω to the ordinal ω_1, literally the same proof as given for Theorem 1 remains valid. Also, for the same reasons as with Theorem 1, Theorem 2 is not new. In addition, both Theorems 1 and 2 use Shelah's work as a significant black box, and in a real sense, use pcf theory as much as a direct proof not going through either $\aleph_\omega$ or $\aleph_{\omega_1}$ would. However, what is true is that the proof of Theorem 1 shows that any new ZFC bound on the size of the power set of $\aleph_\omega$, $\aleph_{\omega_1}$, etc., assuming these are strong limit cardinals, also automatically applies to additional, arbitrarily large singular strong limit cardinals as well. As a specific example, it is widely believed that eventually, some ZFC proof will show that if $\aleph_\omega$ is a strong limit cardinal, then $2^{\aleph_\omega} < \aleph_{\omega_1}$. If this were indeed the case, then the proof of Theorem 1 would show that this bound automatically transfers to all infinite cardinals κ such that $\kappa^{+\omega}$ is a strong limit cardinal. We consequently have, speaking in slogan form but still in a very concrete sense, that "the size of power sets of singular strong limit cardinals down low controls the size of power sets of certain singular strong limit cardinals up high." Further, our methods show in addition that any ZFC proof which *prima facie* only produces a bound on the size of the power set of $\aleph_\omega$, $\aleph_{\omega_1}$, etc. and doesn't appear to generalize to larger cardinals actually does produce a bound on the size of the power sets of certain larger singular strong limit cardinals. Hence, we have the quite interesting and significant fact that ZFC shows there is actually very little freedom on the size of the power sets of certain arbitrarily high singular strong limit cardinals.

References

[1] U. Abraham and M. Magidor. Cardinal arithmetic. In M. Foreman and A. Kanamori, editors, *Handbook of set theory. Volume 2*, pages 1149–1227, Dordrecht, 2010. Springer-Verlag.

[2] S. Shelah. *Cardinal arithmetic*, volume 29 of *Oxford Logic Guides*. Clarendon Press Oxford University Press, New York, 1994. Oxford Science Publications.

The consistency of a club-guessing failure at the successor of a regular cardinal

David Asperó

School of Mathematics, University of East Anglia, Norwich, United Kingdom

Abstract

We answer a question of Shelah by showing that if κ is a regular cardinal such that $2^{<\kappa} = \kappa$, then there is a $<\kappa$-closed partial order preserving cofinalities and forcing that for every club-sequence $\langle C_\delta \mid \delta \in \kappa^+ \cap \mathrm{cf}(\kappa)\rangle$ with $\mathrm{ot}(C_\delta) = \kappa$ for all δ there is a club $D \subseteq \kappa^+$ such that $\{\alpha < \kappa \mid \{C_\delta(\alpha+1), C_\delta(\alpha+2)\} \subseteq D\}$ is bounded for every δ. This forcing is built as an iteration with $<\kappa$-supports and with symmetric systems of submodels as side conditions.

1 Introduction

The purpose of this paper is to present a consistency results at κ^+ for an arbitrarily fixed regular cardinal κ satisfying $2^{<\kappa} = \kappa$. This result is obtained by a variant of the method of iterated forcing with finite supports and (finite) symmetric systems of submodels as side conditions introduced in [3] (cf. also [2]). This is the variant of that method in which one considers supports of size less than κ, rather than just finite, and systems, also of size less than κ, consisting of κ-sized structures closed under $<\kappa$-sequences. All iterands in these constructions, as well as the resulting iteration, are $<\kappa$-closed. We shall say something more about the method used to prove the main result in this paper in a moment, but first we shall introduce the result itself.

Given a set of ordinals S, a sequence $\vec{C} = \langle C_\delta \mid \delta \in S\rangle$ is called a club-sequence if C_δ is a closed and unbounded (club) subset of δ for every $\delta \in S$. Club-guessing principles are well-studied weakenings of $\diamondsuit_\kappa$, for a cardinal κ, in which the guessing object is a club-sequence defined on (some subset of) κ and in which the relevant guessing applies to closed and unbounded, rather than arbitrary, subsets of κ. It is well-known that, whereas the truth of these principles on ω_1 is easy to manipulate by forcing, many instances of club-guessing at a regular cardinal $\kappa \geq \omega_2$ are provable in ZFC (cf. [8]). The following theorem of Shelah is such a result [9, Claim 3.3]; cf. also [12] for a nicely written proof of this theorem.

Theorem 1.1. (Shelah) Let $\kappa \geq \omega_1$ be a regular cardinal. Then for every stationary $S \subseteq \kappa^+ \cap \mathrm{cf}(\kappa)$ there is a club-sequence $\langle C_\delta \mid \delta \in S\rangle$ such that for all $\delta \in S$, $\mathrm{ot}(C_\delta) = \kappa$, and for all $\alpha < \kappa$, we have $\mathrm{cf}(C_\delta(\alpha+1)) = \kappa$,

Stefan Geschke, Benedikt Löwe, Philipp Schlicht (*eds.*).
Infinity, computability, and metamathematics: Festschrift celebrating the 60th birthdays of Peter Koepke and Philip Welch. College Publications, London, 2014. Tributes, Volume 23.

and such that for every club $D \subseteq \kappa^+$ there is some $\delta \in S$ (equivalently, stationary many $\delta \in S$) such that $\{\alpha < \kappa \mid C_\delta(\alpha+1) \in D\}$ is stationary.

In the statement of Theorem 1.1, and throughout the paper, given a set C of ordinals and an ordinal ξ, we are denoting by $C(\xi)$ the ξ-th member of the strictly increasing enumeration of C.

The following corollary is a straightforward consequence of Theorem 1.1.

Corollary 1.2. Let $\kappa \geq \omega_1$ be a regular cardinal. Then for every stationary $S \subseteq \kappa^+ \cap \operatorname{cf}(\kappa)$ and every $\varrho < \kappa$ there is a club-sequence $\langle C_\delta \mid \delta \in S\rangle$ such that for all $\delta \in S$, $\operatorname{ot}(C_\delta) = \kappa$, and for all $\alpha < \kappa$, we have $\operatorname{cf}(C_\delta(\alpha+\varrho)) = \kappa$, and such that for every club $D \subseteq \kappa^+$ there is some $\delta \in S$ (equivalently, stationary many $\delta \in S$) such that $\{\alpha < \kappa \mid C_\delta(\alpha+\varrho) \in D\}$ is stationary.

The following question appears as [10, Question 5.4] (cf. [7, Question 13]).

Question 1.3. Is it true in ZFC that for every regular cardinal $\kappa \geq \omega_1$ there is a club-sequence $\vec{C} = \langle C_\delta \mid \delta \in \kappa^+ \cap \operatorname{cf}(\kappa)\rangle$ with $\operatorname{ot}(C_\delta) = \kappa$ for all δ and such that for every club $D \subseteq \kappa^+$ there is some δ such that

$$\{\alpha < \kappa \mid \{C_\delta(\alpha+1), C_\delta(\alpha+2)\} \subseteq D\}$$

is stationary?

According to Shelah in [10], if there is a club-sequence as in the above question on $\kappa^+ \cap \operatorname{cf}(\kappa)$ and GCH holds, then there is a κ^+-Suslin tree. In particular, an affirmative answer to Question 1.3 would yield an affirmative answer to the following well-known open question (cf., e.g., [5, 7]).

Question 1.4. Does GCH imply that there is an ω_2-Suslin tree?

As mentioned also in [10], using the methods from [11] it is possible to provide a negative answer to the easier form of Question 1.3 where we fix a stationary $S \subseteq \kappa^+ \cap \operatorname{cf}(\kappa)$ with $(\kappa^+ \cap \operatorname{cf}(\kappa)) \setminus S$ also stationary and we ask that $\vec{C}$ be defined on S rather than on all of $\kappa^+ \cap \operatorname{cf}(\kappa)$:

Theorem 1.5. (Shelah) Suppose $\kappa \geq \omega_1$ is a regular cardinal such that $\kappa^{<\kappa} = \kappa$, $S \subseteq \kappa^+ \cap \operatorname{cf}(\kappa)$ is stationary and $S' = (\kappa^+ \cap \operatorname{cf}(\kappa)) \setminus S$ is also stationary. Then the following holds in a generic extension preserving the stationarity of both S and S' and not adding new $<\kappa$-sequences of ordinals.

1. For every club-sequence $\langle C_\delta \mid \delta \in S\rangle$ such that $\operatorname{ot}(C_\delta) = \kappa$ for all δ there is a club $D \subseteq \kappa^+$ such that

$$\{\alpha < \kappa \mid \{C_\delta(\alpha+1), C_\delta(\alpha+2)\} \subseteq D\}$$

is bounded for all $\delta \in S$.

2. For every club-sequence $\langle C_\delta \mid \delta \in S\rangle$, if it holds for all δ that
   (a) $\mathrm{ot}(C_\delta) = \kappa$ and that
   (b) $\mathrm{cf}(C_\delta(\alpha+1)) < \kappa$ for all α,

   then there is a club $D \subseteq \kappa^+$ such that $\{\alpha < \kappa \mid C_\delta(\alpha+1) \in D\}$ is bounded for all $\delta \in S$.

The main result in this paper is the following.

Theorem 1.6. Let $\omega_1 \leq \kappa < \kappa^{++} \leq \vartheta$ be regular cardinals such that $2^{<\kappa} = \kappa$, $2^\kappa = \kappa^+$ and $2^{<\vartheta} = \vartheta$. Then there is a partial order $\mathbb{P}$ with the following properties.

1. $\mathbb{P}$ is $<\kappa$-closed.

2. There is some $\Phi \in \mathbf{H}(\vartheta^+)$ such that $\mathbb{P}$ is proper with respect to all $N \preccurlyeq \mathbf{H}((2^\vartheta)^+)$ such that $\mathbb{P}, \Phi \in N$, $|N| = \kappa$ and ${}^{<\kappa}N \subseteq N$.

3. $\mathbb{P}$ is κ^{++}-Knaster.

4. $\mathbb{P}$ forces that for every club-sequence $\langle C_\delta \mid \delta \in \kappa^+ \cap \mathrm{cf}(\kappa)\rangle$ with $\mathrm{ot}(C_\delta) = \kappa$ for all δ there is a club $D \subseteq \kappa^+$ such that $\{\alpha < \kappa \mid \{C_\delta(\alpha+1), C_\delta(\alpha+2)\} \subseteq D\}$ is bounded in κ for all δ.

5. $\mathbb{P}$ forces that for every club-sequence $\langle C_\delta \mid \delta \in \kappa^+ \cap \mathrm{cf}(\kappa)\rangle$, if for all δ,
   (a) $\mathrm{ot}(C_\delta) = \kappa$, and
   (b) $\mathrm{cf}(C_\delta(\alpha+1)) < \kappa$ for all $\alpha < \kappa$,

   then there is a club $D \subseteq \kappa^+$ such that $\{\alpha < \kappa \mid C_\delta(\alpha+1) \in D\}$ is bounded in κ for all δ.

6. $\mathbb{P}$ forces $2^\mu = \vartheta$ for every $\mu \in [\kappa, \vartheta)$.

The classical notion of properness can be extended to structures which are not necessarily countable. Specifically, conclusion (2) in Theorem 1.6 says that if $N \preccurlyeq \mathbf{H}((2^\vartheta)^+)$ is such that $|N| = \kappa$, ${}^{<\kappa}N \subseteq N$, and $\mathbb{P}, \Phi \in N$, then for every $q \in \mathbb{P} \cap N$ there is an extension q' of q which is $(N, \mathbb{P})$-generic, i.e., such that q' forces $E \cap \dot{G} \cap N \neq \varnothing$ for every dense subset E of $\mathbb{P}$ belonging to N. Parts of the standard theory of properness (but not all of it)[1] extend to the general setting. In particular, if χ is a cardinal, $T \subseteq \chi$, and $\mathbb{P}$ is a partial order wich is proper for a stationary class of structures

[1] It is well-known that there can be no general preservation theorem for properness with respect to uncountable models (on the other hand, properness with respect to countable models is always preserved under countable support iterations).

N such that $N \cap \chi \in T$, then forcing with $\mathbb{P}$ preserves the stationarity of T. Note that $2^{<\kappa} = \kappa$ implies that for every stationary $T \subseteq \kappa^+ \cap \mathrm{cf}(\kappa)$ and every cardinal $\varrho \geq \kappa^+$, the set of $N \preccurlyeq \mathbf{H}(\varrho)$ such that $|N| = \kappa$, ${}^{<\kappa}N \subseteq N$ and $N \cap \kappa^+ \in T$ is stationary. Also, recall that, for a cardinal λ, a partial order $\mathbb{P}$ is λ-Knaster if for every $X \subseteq \mathbb{P}$ of size λ there is a subset of X of size λ consisting of pairwise compatible conditions.

It follows that every forcing satisfying (1)–(3) from Theorem 1.6 preserves all cofinalities and all stationary subsets of $\kappa^+ \cap \mathrm{cf}(\kappa)$. In particular, Theorem 1.6 answers Shelah's Question 1.3 negatively, but it also shows that 2^κ, where κ is the regular cardinal for which we kill the relevant club-guessing at κ^+, can be arbitrarily large.

As we briefly mentioned at the beginning, Theorem 1.6 is proved by building a certain forcing iteration with supports of size less than κ and with certain systems of submodels as side conditions.[2] These submodels are 'active', as side conditions, at an initial segment of stages of the iteration, as indicated by markers associated to them. This type of forcing construction was first used in [3] and afterwards in [2], but only for $\kappa = \omega$. Te reader can find in [3] the appropriate background and general motivation for this kind of construction, so we shall not go into that here.

The present construction is in spirit quite similar to the one in [3], replacing of course ω by κ everywhere and, as one would also expect, looking at submodels of size κ closed under $<\kappa$-sequences except of countable submodels. One the other hand, the present construction has a couple of features which were not present in the constructions from [3] or [2]:

The present construction can be viewed as an iteration, with a suitable type of symmetric systems of structures as side conditions, in which at each stage a certain book-keeping function feeds us a club-sequence $\vec{C}$ on $\kappa^+ \cap \mathrm{cf}(\kappa)$, and we shoot a certain club of κ^+ which will destroy (part of) the potential guessing character of $\vec{C}$. However, it is important for the proof to go through—and, specifically, for the properness proof (Lemma 3.12)—that we destroy these club-sequences rather slowly: More specifically, we start by fixing a partition $(S_\varrho)_{\varrho<\kappa}$ of $\kappa^+ \cap \mathrm{cf}(\kappa)$ into stationary sets and for every specific stage β of the iteration we make sure that there is exactly one $\varrho < \kappa$ such that the club D_β we add at that stage is asked to "kill" the relevant $\vec{C}$ picked at that stage only for those $\delta \in D_\beta$ which are in S_ϱ (in the end, the club witnessing that $\vec{C}$ has been "killed" everywhere, and not just on some stationary set $S \subseteq \kappa^+ \cap \mathrm{cf}(\kappa)$, is the intersection of κ—many of the clubs D_β explicitly added along the iteration). Nothing like this was needed in the constructions from [3] or [2]. It looks difficult to convey in few word why such a move is needed here; we shall simply refer the reader to the actual

[2] These are elementary submodels of $(\mathbf{H}(\vartheta), \in, T)$ for some cardinal ϑ and some suitable predicate $T \subseteq \mathbf{H}(\vartheta)$.

proof (specifically, cf. the proof of Lemma 3.12, and particularly the part of it when $t^* \in A$ is found in M^* and shown to be compatible with t).

Another sense in which the present construction differs from the one in [3] and also the one in [2] is the following: An essential feature of some of the proofs in [3] and [2] is that they are by induction. For example, in the proofs of properness, if $(\mathbb{P}_\alpha \mid \alpha < \vartheta)$ is the corresponding iteration, one proves for every $\alpha \leq \vartheta$ that if $\mathbb{P}_\beta$ has the relevant form of properness for all $\beta < \alpha$, then this is true for $\mathbb{P}_\alpha$ itself.[3] In order to run this type of proof it is crucial that the supports be finite, or otherwise the induction breaks down at stages of countable cofinality. However, given the nature of the present approach, finite support do not work as we want to have a $<\kappa$-closed forcing in the end[4] (we need $<\kappa$-supports instead), and the type of inductive approach from [3] and [2] completely breaks down. Instead, here one proves the relevant properness lemma by a direct construction.

The notation in this paper is fairly standard (cf., e.g., [4, 6]), but we shall also use pieces of notation that are not so standard and that will be introduced at the appropriate place. Given a set N, if $N \cap |N|^+$ is an ordinal, then we shall usually denote this ordinal by δ_N. Also, if $q = (F, \Delta)$, F is a function, Δ consists of pairs (N, τ), where τ is an ordinal, and β is an ordinal, *the restriction of q to β*, denoted by $q|_\beta$, is the ordered pair $(F \restriction \beta, \Delta')$, where Δ' consists of all pairs (N, τ') with $(N, \tau) \in \Delta$ and $\tau' = \min\{\tau, \sup(N \cap (\beta + 1))\}$.

The rest of the paper is structured as follows. In § 2, we adapt the notion of symmetric system from [3] to the present context and present the relevant amalgamation lemmas. In § 3.1, we introduce a forcing notion for destroying an instance of the club-guessing we are looking at here and prove the relevant density lemmas for it. Then, in § 3.2, we first construct the forcing $\mathbb{P}$ witnessing Theorem 1.6 and then prove a sequence of lemmas which together will prove the theorem.

2 Symmetric systems of submodels

Let us fix two arbitrary regular cardinals $\lambda < \chi$ for this section. In [3, § 2] we consider a certain natural notion of symmetric system of submodels and prove its relevant properties.[5] The notion of symmetric system admits a natural generalisation to higher cardinalities, which is the one we shall consider next. The theory of finite symmetric systems as developed in [3] goes through in the general setting with just notational changes.

[3] This type of inductive argument seems to be unavoidable when the goal is to build a model of a reasonably general forcing axiom (as in [3] and [2]).

[4] The reason we want a $<\kappa$-closed forcing is that we want to preserve all cardinals $\lambda \leq \kappa$.

[5] This notion was not new. The paper [3] contains older references in the literature where this same type of system appears.

Definition 2.1. Let $T \subseteq \mathbf{H}(\chi)$ and let $\mathcal{N} \subseteq \wp(\mathbf{H}(\chi))$ be such that $|\mathcal{N}| < \lambda$. $\mathcal{N}$ *is a symmetric* λ-T-*system* if and only if the following hold:

(A) For every $N \in \mathcal{N}$, $(N, \in, T) \preccurlyeq (\mathbf{H}(\chi), \in, T)$, $|N| = \lambda$, $\lambda \in N$, and ${}^{<\lambda}N \subseteq N$.

(B) Given N and N' in $\mathcal{N}$, if $\delta_N = \delta_{N'}$, then there is a (unique) isomorphism

$$\Psi_{N,N'} : (N, \in, T) \longrightarrow (N', \in, T)$$

Furthermore, $\Psi_{N,N'}$ is the identity on $N_0 \cap N_1$.

(C) For all N_0, N_1 and N_1' in $\mathcal{N}$, if $N_0 \in N_1$ and $\delta_{N_1} = \delta_{N_1'}$, then $\Psi_{N_1,N_1'}(N_0) \in \mathcal{N}$.

(D) For all N_0 and N_1 in $\mathcal{N}$, if $\delta_{N_0} < \delta_{N_1}$, then there is some $N_1' \in \mathcal{N}$ such that $\delta_{N_1'} = \delta_{N_1}$ and $N_0 \in N_1'$.

In the statement of condition (B) and throughout the paper, if N and N' are such that there is a unique isomorphism $\Psi : (N, \in) \longrightarrow (N', \in)$, we shall tend to denote this isomorphism by $\Psi_{N,N'}$. Also, we shall occasionally refer to 'symmetric λ-systems' or even 'symmetric systems', without mention of T and/or λ, in contexts where these parameters are either understood or not relevant.

The proof of the following fact is immediate. It will be used in the proof of Lemma 3.12.

Fact 2.2. For every $T \subseteq \mathbf{H}(\chi)$ as in Definition 2.1 and every λ-T-symmetric system $\mathcal{N}$, if N_0, N_1 are in $\mathcal{N}$ and $\delta_{N_0} < \delta_{N_1}$, then there is some $N_0' \in \mathcal{N} \cap N_1$ such that $\delta_{N_0'} = \delta_{N_0}$ and such that $N_0 \cap N_1 = N_0 \cap N_0'$.

Proof. Let $N_1' \in \mathcal{N}$ be as given by (D) in Definition 2.1 for the pair N_0, N_1, and let $N_0' = \Psi_{N_1',N_1}(N_0)$. Q.E.D.

Our main amalgamation lemmas for λ-T-symmetric systems will be the following. The proofs of these lemmas are identical to the proofs of the corresponding lemmas in [3], with just the obvious notational changes.

Lemma 2.3. Let $T \subseteq \mathbf{H}(\chi)$, let $\mathcal{N}$ be a λ-T-symmetric system, and let $N \in \mathcal{N}$. Then the following hold.

(i) $\mathcal{N} \cap N$ is also a λ-T-symmetric system.

(ii) If $\mathcal{W} \subseteq N$ is a λ-T-symmetric system and $\mathcal{N} \cap N \subseteq \mathcal{W}$, then

$$\mathcal{V} := \mathcal{N} \cup \{\Psi_{N,N'}(W) \,:\, W \in \mathcal{W},\, N' \in \mathcal{N},\, \delta_{N'} = \delta_N\}$$

is a λ-T-symmetric system.

Lemma 2.4. Let $T \subseteq \mathbf{H}(\chi)$, and suppose $\mathcal{N}_0 = \{N_i^0 : i < \mu\}$ and $\mathcal{N}_1 = \{N_i^1 : i < \mu\}$ are λ-T-symmetric systems for some $\mu < \lambda$. Suppose that $(\bigcup \mathcal{N}_0) \cap (\bigcup \mathcal{N}_1) = R$ and that there is an isomorphism Ψ between the structures

$$\langle \bigcup_{i<\mu} N_i^0, \in, T, R, N_i^0 \rangle_{i<\mu}$$

and

$$\langle \bigcup_{i<m} N_i^1, \in, T, R, N_i^1 \rangle_{i<\mu}$$

fixing R. Then $\mathcal{N}_0 \cup \mathcal{N}_1$ is a λ-T-symmetric system.

3 The proof

3.1 Killing a club-sequence

We shall start by introducing the following notion of rank (cf. the definition in [1] or [3]): Given a regular cardinal λ and an ordinal δ, we define the *λ-Cantor-Bendixson rank of δ*, $\mathrm{rk}^\lambda(\delta)$, by specifying that

$\mathrm{rk}^\lambda(\delta) \geq 1$ if and only if δ is a limit point of ordinals of cofinality λ, and that

if $\mu > 1$, $\mathrm{rk}^\lambda(\delta) \geq \mu$ if and only if for every $\eta < \mu$, δ is a limit of ordinals ε such that $\mathrm{cf}(\varepsilon) = \lambda$ and $\mathrm{rk}^\lambda(\varepsilon) \geq \eta$.

If $(\xi_i)_{i<\lambda}$ and $(\varrho_i)_{i<\lambda}$ are increasing sequences of ordinals and $\varrho_i \leq \mathrm{rk}^\lambda(\xi_i)$ for all i, then $\mathrm{rk}^\lambda(\sup\{\xi_i \mid i < \lambda\}) \geq \sup\{\varrho_i \mid i < \lambda\}$. In particular, if N is an elementary substructure of some $\mathbf{H}(\chi)$, $\lambda \in N$ and δ_N exists, then $\mathrm{rk}^\lambda(\delta_N) = \delta_N$. This is true because, by correctness of N inside $\mathbf{H}(\chi)$, letting $\mu = \min((N \cap \chi) \setminus \delta_N)$ if $(N \cap \chi) \setminus \delta_N \neq \varnothing$ and $\mu = \chi$ otherwise, for every $\xi \in \mu \cap N$ there is some $\xi' \geq \xi$ in N such that $\mathrm{rk}^\lambda(\xi') = \xi'$, and therefore there is an increasing sequence $(\xi_i)_{i<\mathrm{cf}(|N|)}$ of ordinals such that $\sup\{\xi_i \mid i < |N|\} = \delta_N$ and such that for all i, $\xi_i \in N \cap \delta_N$ and $\mathrm{rk}^\lambda(\xi_i) = \xi_i$.

Definition 3.1. Given a regular cardinal λ, a function $f \subseteq \lambda^+ \times \lambda^+$ is a *λ-approximation* if the following hold:

(a) $|f| < \lambda$,

(b) the function f is strictly increasing,

(c) for every $\xi \in \mathrm{dom}(f)$, $\mathrm{rk}^\lambda(f(\xi)) \geq \lambda + \xi$,

(d) for every $\xi \in \mathrm{dom}(f)$,

(d.1) if ξ is a nonzero limit ordinal such that $\mathrm{cf}(\xi) < \lambda$, then $\xi = \sup(\mathrm{dom}(f{\restriction}\xi))$ and $f(\xi) = \sup(f``\xi)$, and (d.2) if ξ is either a successor ordinal or a limit ordinal such that $\mathrm{cf}(\xi) = \lambda$, then $\mathrm{cf}(f(\xi)) = \lambda$.

It follows of course from (b) together with (d.1) and (d.2) that if f is a λ-approximation and $\xi \in \mathrm{dom}(f)$ is any nonzero limit ordinal, then $\mathrm{cf}(f(\xi)) = \mathrm{cf}(\xi)$. The following fact is an immediate consequence of the definitions.

Fact 3.2. For every regular cardinal λ, if $\mathcal{F}$ is a collection of size less than λ consisting of λ-approximations and every two members of $\mathcal{F}$ are compatible as functions, then $\bigcup \mathcal{F}$ is a λ-approximation.

Given a club-sequence $\vec{C} = \langle C_\delta \mid \delta \in \kappa^+ \cap \mathrm{cf}(\kappa)\rangle$ with $\mathrm{ot}(C_\delta) = \kappa$ for all δ, let $\mathbb{Q}$ be the following partial order: A condition in $\mathbb{Q}$ is an ordered pair (f, π) satisfying the following conditions:

(1) The function $f \subseteq \kappa^+ \times \kappa^+$ is a κ-approximation.

(2) The function π satisfies $\mathrm{dom}(\pi) \subseteq \mathrm{dom}(f) \cap \mathrm{cf}(\kappa)$ and $\pi(\xi) \in f(\xi)$ for all $\xi \in \mathrm{dom}(\pi)$.

(3) Let $\xi \in \mathrm{dom}(\pi)$ and $\zeta \in \mathrm{dom}(f) \cap \xi$ and suppose that

(i) $\sigma_0^\zeta = \max(C_{f(\xi)} \cap f(\zeta))$ in case $\max(C_{f(\xi)} \cap f(\zeta))$ exists.

(ii) $\sigma_1^\zeta = \min\{\sigma \in C_{f(\xi)} \mid \sigma \geq f(\zeta)\}$,

(iii) $\sigma_2^\zeta = \min\{\sigma \in C_{f(\xi)} \mid \sigma > \sigma_1^\zeta\}$, and

(iv) $\pi(\xi) < f(\zeta)$.

Then the following hold:

(3.0) If σ_0^ζ exists, is a successor point of $C_{f(\xi)}$, and $\pi(\xi) < \sigma_0^\zeta$, then there is some $\zeta_0 < \zeta$ such that $\{\zeta_0, \zeta_0 + 1\} \subseteq \mathrm{dom}(f)$, $f(\zeta_0) < \sigma_0^\zeta$ and $f(\zeta_0 + 1) > \sigma_0^\zeta$.

(3.1) If σ_1^ζ is a successor point of $C_{f(\xi)}$, then

if $f(\zeta) < \sigma_1^\zeta$, then $\zeta + 1 \in \mathrm{dom}(f)$ and $f(\zeta + 1) > \sigma_1^\zeta$, and
if ζ is a limit ordinal with $\mathrm{cf}(\zeta) < \kappa$, then $f(\zeta) < \sigma_1^\zeta$.

(3.2) There is some $\zeta_2 \geq \zeta$ with $\{\zeta_2, \zeta_2 + 1\} \subseteq \mathrm{dom}(f)$, $f(\zeta_2) < \sigma_2^\zeta$ and $f(\zeta_2 + 1) > \sigma_2^\zeta$.

Given $\mathbb{Q}$-conditions (f_0, π_0) and (f_1, π_1), (f_1, π_1) extends (f_0, π_0) if $f_0 \subseteq f_1$ and $\pi_0 \subseteq \pi_1$. Given a club-sequence $\vec{C}$ as above, we shall denote the corresponding forcing $\mathbb{Q}$ by $\mathbb{Q}_{\vec{C}}$. The following simple observation is an immediate consequence of (3.2) in the above definition together with (d.1) in Definition 3.1.

Fact 3.3. If $\vec{C} = \langle C_\delta \mid \delta \in \kappa^+ \cap \mathrm{cf}(\kappa)\rangle$ is a club-sequence with $\mathrm{ot}(C_\delta) = \kappa$ for all δ, $(f, \pi) \in \mathbb{Q}_{\vec{C}}$, $\xi \in \mathrm{dom}(\pi)$, and there is some $\zeta \in \mathrm{dom}(f \restriction \xi)$ such that $f(\zeta) > \pi(\xi)$, then $\sup(f``\xi)$ is a limit point of $C_{f(\xi)}$.

Lemma 3.4. Let $\vec{C} = \langle C_\delta \mid \delta \in \kappa^+ \cap \mathrm{cf}(\kappa)\rangle$ be a club-sequence such that $\mathrm{ot}(C_\delta) = \kappa$ for all δ, and let (f, π) be a condition in $\mathbb{Q}_{\vec{C}}$. Then the following hold:

(a) For every $\xi \in \mathrm{dom}(f)$ such that $\mathrm{cf}(\xi) = \kappa$ there is an extension (f, π') of (f, π) such that $\xi \in \mathrm{dom}(\pi')$.

(b) For every limit ordinal $\xi \in \mathrm{dom}(f)$, $\xi' \in \xi$ and $\alpha \in f(\xi)$ there are $\xi'' \in (\xi', \xi)$, $\beta \in (\alpha, f(\xi))$ and an extension (f', π) of (f, π) such that $\xi'' \in \mathrm{dom}(f')$ and $f'(\xi'') = \beta$.

(c) For every $\xi \in \kappa^+ \setminus \mathrm{dom}(f)$ there is an extension (f', π) of (f, π) such that $\xi \in \mathrm{dom}(f')$. Furthermore, if

1. $\mathrm{rk}^\kappa(\xi) = \xi = \kappa + \xi$,
2. $\mathrm{cf}(\xi) = \kappa$,
3. $\mathrm{ran}(f \restriction \xi) \subseteq \xi$,
4. $\mathrm{cf}(\min(\mathrm{dom}(f) \setminus \xi)) = \kappa$ in case $\min(\mathrm{dom}(f) \setminus \xi)$ exists, and
5. $[\xi, f(\xi^*)) \cap C_{f(\varrho)} = \varnothing$ whenever $\varrho \in \mathrm{dom}(\pi)$ is such that $\varrho > \xi^* = \min(\mathrm{dom}(f) \setminus \xi)$ in case $\min(\mathrm{dom}(f) \setminus \xi)$ exists,

then f' can be taken so that $f'(\xi) = \xi$.

Proof. Part (a) is obvious: Since $|f| < \kappa$ and $\mathrm{cf}(f(\xi)) = \kappa$, it suffices to let $\pi'(\xi) < f(\xi)$ be such that $f``\xi \subseteq \pi'(\xi)$.

For part (b) we may assume $\mathrm{cf}(\xi) = \kappa$ as otherwise we are done by (d.1) in the definition of κ-approximation. Assume $\mathrm{dom}(f \restriction \xi) \neq \varnothing$, and let $\bar{\xi} = \sup(\mathrm{dom}(f \restriction \xi))$ (the case $\mathrm{dom}(f \restriction \xi) = \varnothing$ is easier). We may also assume that $\max(\mathrm{dom}(f \restriction \xi))$ does not exist (otherwise the proof is again easier). Pick any successor ordinal $\xi'' \in (\xi', \xi)$ such that $\bar{\xi} + 1 < \xi''$. Since $\mathrm{cf}(f(\xi)) = \kappa$ and $\mathrm{rk}^\kappa(f(\xi)) > \kappa$, we may find $\beta \in (\alpha, f(\xi)) \setminus C_{f(\xi)}$ of cofinality κ such that

(i) β is above all members of $C_\delta \cap f(\xi)$ for all $\delta \in f``(\xi, \kappa^+)$ of cofinality $\kappa$ and above at least two $\sigma \in C_{f(\xi)}$ such that $\sigma > \sup(f``\xi)$, and

(ii) $\mathrm{rk}^\kappa(\beta) \geq \kappa + \xi''$.

Let also $\beta_0 \in (\alpha, \beta)$ of cofinality κ be such that

(iii) β_0 is above all members of $C_\delta \cap f(\xi)$ for all $\delta \in f``(\xi, \kappa^+)$ of cofinality κ,

(iv) $\mathrm{rk}^\kappa(\beta_0) \geq \kappa + \bar{\xi} + 1$, and such that

(v) β_0 is above $\max(C_{f(\xi)} \cap \beta)$.[6]

Let $(\delta_n)_{1 \leq n < \omega}$ be a strictly increasing sequence of ordinals in $(\beta, f(\xi))$ of cofinality κ such that $|(\beta, \delta_1) \cap C_{f(\xi)}| \geq 2$ and such that for all $n \geq 1$,

(vi) $\delta_n \notin C_{f(\xi)}$,

(vii) $\mathrm{rk}^\kappa(\delta_n) \geq \kappa + \xi'' + n$, and

(viii) $|(\delta_n, \delta_{n+1}) \cap C_{f(\xi)}| \geq 2$

This sequence exists since $C_{f(\xi)}$ is cofinal in $f(\xi)$ and of order type κ and since $\mathrm{rk}^\kappa(f(\xi)) > \kappa$. Now we may extend f to a function f' with domain $\mathrm{dom}(f) \cup \{\bar{\xi}, \bar{\xi}+1\} \cup \{\xi'' + n \mid n < \omega\}$ which send $\bar{\xi}$ to $\sup(f``\bar{\xi})$, $\bar{\xi}+1$ to β_0, ξ'' to β and, for every integer $n > 0$, sends $\xi'' + n$ to δ_n. Let us check that (f', π) is a condition in $\mathbb{Q}_{\vec{C}}$:

By the choice of β_0 in (iii) it is clear that for every γ in the set

$$\{\sup(f``\bar{\xi}), \beta_0, \beta\} \cup \{\delta_n \mid 0 < n < \omega\}$$

the addition of γ to the range of f will not cause any problem with condition (3) in the definition of $\mathbb{Q}_{\vec{C}}$-condition for those $\varrho \in \mathrm{dom}(\pi)$ above ξ such that $\pi(\varrho) < \gamma$. This follows from $\pi(\varrho) < f(\xi)$ together with the fact that $\max(C_{f(\varrho)} \cap f(\xi)) = \max(C_{f(\varrho)} \cap \gamma)$ if $\gamma \geq \beta_0$, that γ is a limit point of $C_{f(\varrho)}$ if $\gamma = \sup(f``\bar{\xi})$ by (3.2) applied to ϱ and to a tail of $\mathrm{dom}(f \restriction \bar{\xi})$, and that $\min\{\sigma \in C_{f(\varrho)} \mid \sigma \geq f(\xi)\} = \min\{\sigma \in C_{f(\varrho)} \mid \sigma \geq \gamma\}$ if $\gamma \geq \beta_0$.

As to condition (3) for ξ in case $\xi \in \mathrm{dom}(\pi)$, we show, for γ as above, that the corresponding instance of (3.0)–(3.2) holds. To start with, note that if $\sup(f``\bar{\xi}) > \pi(\xi)$, then $\sup(f``\bar{\xi})$ is a limit point of $C_{f(\xi)}$ by Lemma 3.3, which means that (3.0) holds for the pair ξ, $\bar{\xi}$, and that $\{\bar{\xi}, \bar{\xi}+1\}$ witnesses the relevant instances of (3.1) and (3.2) for the pair ξ, $\bar{\xi}$ by (i) and (v). For $\gamma \geq \beta_0$, the corresponding instances of (3.0)–(3.2) hold immediately by construction.

The first part of (c) can be easily established by arguing as in the proof of (b) and can be left as an exercise for the reader (in the case when ξ is a limit ordinal with $\mathrm{cf}(\xi) < \kappa$, one has to make sure of course that $\mathrm{dom}(f' \restriction \xi)$ is cofinal in ξ and $\sup(\mathrm{ran}(f' \restriction \xi)) = f'(\xi)$).

We shall just sketch the proof of the second part of (c). Suppose $\xi \notin \mathrm{dom}(f)$ satisfies the hypotheses (1)–(5). We have to show that there is an extension (f', π) of (f, π) such that $\xi \in \mathrm{dom}(f')$ and $f'(\xi) = \xi$. We may assume that $f \restriction \xi$ is nonempty and does not have a maximum and also

[6]Note that $\max(C_{f(\xi)} \cap \beta)$ indeed exists since $\mathrm{cf}(\beta) = \kappa$.

that $\xi^* = \min(\mathrm{dom}(f) \setminus \xi)$ exists (the proof in the other cases is easier). Let $\bar{\xi} = \sup(\mathrm{dom}(f \restriction \xi))$ and let $\beta_0 < \xi$ be above $\max(C_{f(\varrho)} \cap \xi)$ for every $\varrho \in \mathrm{dom}(\pi)$ such that $\varrho \geq \xi^*$. Since $\mathrm{cf}(\xi^*) = \kappa$ and $\mathrm{rk}^\kappa(f(\xi^*)) > \kappa$, we may pick a strictly increasing sequence $(\delta_n)_{1 \leq n < \omega}$ of ordinals of cofinality κ in $(\xi, f(\xi^*)) \setminus C_{f(\xi^*)}$ such that $|(\xi, \delta_1) \cap C_{f(\xi^*)}| \geq 2$ and such that for all n,

(i) $\mathrm{rk}^\kappa(\delta_n) \geq \kappa + \xi + n$, and

(ii) $|(\delta_n, \delta_{n+1}) \cap C_{f(\xi^*)}| \geq 2$.

Now it is easy to verify as in the proof of part (b) that f' is as desired, where $\mathrm{dom}(f') = \mathrm{dom}(f) \cup \{\bar{\xi}, \bar{\xi}+1\} \cup \{\xi + n \mid n < \omega\}$ extends f, sends $\bar{\xi}$ to $\sup(f``\bar{\xi})$, $\bar{\xi}+1$ to $\beta_0$, $\xi$ to itself, and send each $\xi + n$, for $n > 0$, to $\delta_n$. The verification that the act of adding either $\sup(f``\bar{\xi})$, β_0, or any δ_n, to the range of f' does not interfere with condition (3) in the definition of $\mathbb{Q}_{\vec{C}}$ relative to either ξ^*, if $\xi^* \in \mathrm{dom}(\pi)$, or relative to any $\varrho \in \mathrm{dom}(\pi)$ above ξ^* is as in the proof of part (b). Hence we just need to argue that adding ξ does not cause trouble either. Note that $\max(C_{f(\xi^*)} \cap \xi) = \max(C_{f(\xi^*)} \cap \beta_0)$ and therefore (3.0) holds for the pair ξ^*, ξ since it holds for the pair ξ^*, $\bar{\xi}+1$ as witnessed by $\{\bar{\xi}, \bar{\xi}+1\}$ (cf. the proof of part (b)). Also, if $\varrho \in \mathrm{dom}(\pi)$ is above ξ^*, $\max(C_{f(\varrho)} \cap \xi)$ exists and is a successor ordinal, and $\max(C_{f(\varrho)} \cap \xi) > \pi(\varrho)$, then (3.0) holds for ϱ, ξ because it holds for ϱ, $\bar{\xi}+1$ as witnessed again by the pair $\{\bar{\xi}, \bar{\xi}+1\}$ (again cf. the proof of part (b)). The relevant instances of (3.1) and (3.2) are satisfied automatically, as witnessed by $\{\xi, \xi+1\}$, since $\mathrm{cf}(\xi) = \kappa$ and by the choice of δ_1. Finally, let $\varrho \in \mathrm{dom}(\pi)$, $\varrho > \xi^*$, such that $\pi(\varrho) < \xi$, let $\sigma = \min(C_{f(\varrho)} \setminus \xi)$ and let $\sigma' = \min(C_{f(\varrho)} \setminus (\sigma+1))$. Since $[\xi, f(\xi^*)) \cap C_{f(\varrho)} = \emptyset$, $\sigma = \min(C_{f(\varrho)} \setminus f(\xi^*))$ and $\sigma' = \min(C_{f(\varrho)} \setminus (\sigma+1))$. But then (3.1) and (3.2) hold for ϱ, ξ because they hold for ϱ, ξ^*. Q.E.D.

The following lemma is easily proved by an easier version of the proof of Lemma 3.4.

Lemma 3.5. Let $\vec{C} = \langle C_\delta \mid \delta \in \kappa^+ \cap \mathrm{cf}(\kappa) \rangle$ be a club-sequence such that $\mathrm{ot}(C_\delta) = \kappa$ for all δ, let (f, π) and (f', π') be $\mathbb{Q}_{\vec{C}}$-conditions, and suppose there are $\eta < \xi$ such that

1. $\xi \in \mathrm{dom}(f)$, $\mathrm{cf}(\xi) = \kappa$, and $f(\xi) = \xi$,
2. $\eta > \max(C_\delta \cap \xi)$ for every $\delta \in \mathrm{ran}(f)$ above ξ of cofinality κ,
3. $\xi \notin \mathrm{dom}(\pi)$,
4. $\mathrm{dom}(f') \subseteq \xi$, and
5. $(f' \restriction \eta, \pi' \restriction \eta) = (f \restriction \xi, \pi \restriction \xi)$.

Then $(f \cup f', \pi \cup \pi')$ is a $\mathbb{Q}_{\vec{C}}$-condition.

Lemma 3.6. Let $\vec{C} = \langle C_\delta \mid \delta \in \kappa^+ \cap \mathrm{cf}(\kappa)\rangle$ be a club-sequence such that $\mathrm{ot}(C_\delta) = \kappa$ for all δ. Then the following hold:

1. $\mathbb{Q}_{\vec{C}}$ is $<\kappa$–directed closed. In fact, if $\{(f_i, \pi_i) \mid i < \mu\} \subseteq \mathbb{Q}_{\vec{C}}$ is a directed set of size less than κ, then $(\bigcup_{i<\mu} f_i, \bigcup_{i<\mu} \pi_i)$ is the greatest lower bound of $\{(f_i, \pi_i) \mid i < \mu\}$.

2. If G is generic for $\mathbb{Q}_{\vec{C}}$, then $F = \bigcup\{f \mid (\exists \pi)((f, \pi) \in G)\}$ is the enumerating function of a club $D \subseteq (\kappa^+)^{\mathbf{V}}$ such that

   (a) $\{\alpha < \kappa \mid \{C_\delta(\alpha+1), C_\delta(\alpha+2)\} \subseteq D\}$ is bounded in δ for every δ, and such that

   (b) if, in addition, $\mathrm{cf}(C_\delta(\alpha+1)) < \kappa$ for all $\alpha < \kappa$, then $\{\alpha < \kappa \mid C_\delta(\alpha+1) \in D\}$ is bounded in δ for every δ.

Proof. The first part of Lemma 3.6 follows immediately from Fact 3.2. The second part is a consequence of Lemma 3.4 together with clause (3) in the definition of $\mathbb{Q}$-condition: By parts (b) and (c) of Lemma 3.4, F is the enumerating function of a club of κ^+. By part (a), if $\xi \in \kappa^+ \cap \mathrm{cf}(\kappa)$, then for a tail of $\zeta < \xi$ it holds that if

(i) $\sigma_0^\zeta = \max(C_{F(\xi)} \cap F(\zeta))$ in case $\max(C_{F(\xi)} \cap F(\zeta))$ exists,

(ii) $\sigma_1^\zeta = \min\{\sigma \in C_{F(\xi)} \mid \sigma \geq F(\zeta)\}$, and

(iii) $\sigma_2^\zeta = \min\{\sigma \in C_{F(\xi)} \mid \sigma > \sigma_1^\zeta\}$,

then,

(A0) if σ_0^ζ exists and is a successor point of $C_{F(\xi)}$, then there is some $\zeta_0 < \zeta$ such that $F(\zeta_0) < \sigma_0^\zeta$ and $F(\zeta_0 + 1) > \sigma_0^\zeta$,

(A1) if ζ is a limit ordinal with $\mathrm{cf}(\zeta) < \kappa$ and σ_1^ζ is a successor point of $C_{F(\xi)}$, then $F(\zeta) < \sigma_1^\zeta$, and

(A2) there is some $\zeta_2 \geq \zeta$ with $F(\zeta_2) < \sigma_2^\zeta$ and $F(\zeta_2 + 1) > \sigma_2^\zeta$.

Let ζ be in this tail. It suffices to show that

1. $\min(C_{F(\xi)} \setminus (F(\zeta)+1)) \notin D$ if $F(\zeta)$ is a successor point of $C_{F(\xi)}$, that

2. $\max(C_{F(\xi)} \cap F(\zeta)) \notin D$ if $F(\zeta)$ is a double successor point of $C_{F(\xi)}$, and that

3. $F(\zeta)$ is not a successor point of $C_{F(\xi)}$ if $\mathrm{cf}(C_{F(\xi)}(\alpha)) < \kappa$ for all $\alpha < \kappa$.

For (1), note that if $F(\zeta)$ is a successor point of $C_{F(\xi)}$, then $F(\zeta) = \sigma_1^\zeta$ and $\min(C_{F(\xi)} \setminus (F(\zeta)+1)) = \sigma_2^\zeta$. But by (A2), $\sigma_2^\zeta \notin D$. For (2), we have that if $F(\zeta)$ is a double successor point of $C_{F(\xi)}$, then $\max(C_{F(\xi)} \cap F(\zeta)) = \sigma_0^\zeta$ exists and is a successor point of $C_{F(\xi)}$. But then $\sigma_0^\zeta \notin D$ by (A0). Finally, for (3), if $\mathrm{cf}(C_{F(\xi)}(\alpha)) < \kappa$ for all α and $F(\zeta)$ is a successor point of $C_{F(\xi)}$, then $\sigma_1^\zeta = F(\zeta)$. But then ζ is a limit ordinal of cofinality less than κ, and therefore $F(\zeta) < \sigma_1^\zeta$ by (A1), which is a contradiction. Q.E.D.

It is worth pointing out that if $\vec{C} = \langle C_\delta \mid \delta \in \kappa^+ \cap \mathrm{cf}(\kappa)\rangle$ is a club-sequence such that $\mathrm{cf}(C_\delta(\alpha+1)) = \kappa$ for all α, then the corresponding form of (3) in the above proof cannot be derived; in fact we cannot rule out that there are ζ which are not limit ordinals of cofinality less than κ such that $F(\zeta)$ is a successor point of $C_{F(\xi)}$. Such ζ will typically come from extending a $\mathbb{Q}_{\vec{C}}$-condition (f, π) by adding to f the pair (ζ, ζ) as in Lemma 3.4 (c) (where ζ has cofinality κ, $f``\zeta \subseteq \zeta$, $\mathrm{rk}^\kappa(\zeta) = \kappa + \zeta = \zeta$, and so on). This situation will come up in the proof of Lemma 3.12 and is the reason why the proof of Theorem 1.6 cannot be adapted to the construction of a model in which for every $\vec{C}$ as above there is a club $D \subseteq \kappa^+$ such that $\{\alpha < \kappa \mid C_\delta(\alpha+1) \in D\}$ is bounded for all δ. And of course we know by Theorem 1.1 that such a construction is impossible.

3.2 The main construction

To start with, let $(S_\varrho)_{\varrho<\kappa}$ be a partition of $\kappa^+ \cap \mathrm{cf}(\kappa)$ into stationary sets and let $\Phi : \vartheta \longrightarrow \mathbf{H}(\vartheta)$ be such that for every $a \in \mathbf{H}(\vartheta)$, $\Phi^{-1}(a) \subseteq \vartheta$ is unbounded. Φ exists by $2^{<\vartheta} = \vartheta$. For every $a \in \mathbf{H}(\vartheta)$ let also $(X_\varrho^a)_{\varrho<\kappa}$ be a partition of $\Phi^{-1}(a)$ into unbounded subsets of ϑ.

The poset $\mathbb{P}$ witnessing Theorem 1.6 will be $\mathbb{P}_\vartheta$, where $\langle \mathbb{P}_\alpha \mid \alpha \leq \vartheta\rangle$ is the sequence of posets to be defined soon.

Given an ordered pair $q = (F, \Delta)$, we shall sometimes refer to F and Δ as F_q and Δ_q, respectively. If F_q is a function, $\gamma \in \mathrm{dom}(F_q)$, and $F_q(\gamma)$ is also an ordered pair, we shall tend to write $F_q(\gamma)$ as $(f_q(\gamma), \pi_q(\gamma))$. These conventions will be typically applied to conditions or to ordered pairs in the process of becoming conditions.

Fix now $\alpha \leq \vartheta$ and suppose $\mathbb{P}_\beta$ has been defined for all $\beta < \alpha$. A condition in $\mathbb{P}_\alpha$ is an ordered pair $q = (F, \Delta)$ with the following properties.

1. $F \subseteq \alpha \times [\mathbf{H}(\kappa^+)]^{<\kappa}$ is a function of size less than κ and such that for all $\gamma \in \mathrm{dom}(F)$, $F(\gamma)$ is of the form $(f(\gamma), \pi(\gamma))$, where $f(\gamma) \subseteq \kappa^+ \times \kappa^+$ is a κ-approximation and $\pi(\gamma)$ is a function with $\mathrm{dom}(\pi(\gamma)) \subseteq \mathrm{dom}(f(\gamma)) \cap \mathrm{cf}(\kappa)$ and $\pi(\gamma)(\nu) \in f(\gamma)(\nu)$ for all $\nu \in \mathrm{dom}(\pi(\gamma))$.

2. Δ is such that

   (i) Δ is a binary relation with $\mathrm{dom}(\Delta)$ a symmetric κ-Φ-system of elementary substructures of $\mathbf{H}(\vartheta)$, and

   (ii) every member of Δ is of the form (N, τ), where $\tau \leq \alpha$ is an ordinal which is in N or is a limit point of ordinals in N.

3. $q|_\beta \in \mathbb{P}_\beta$ for all $\beta < \alpha$.

4. Suppose $\beta = \alpha + 1$. If $\Phi(\beta)$ is not a $\mathbb{P}_\beta$-name, then let $\Phi^*(\beta)$ be, say, a $\mathbb{P}_\beta$-name for the Φ–first club-sequence of the form $\langle C_\delta \mid \delta \in (\kappa^+ \cap \mathrm{cf}(\kappa))^{\mathbf{V}} \rangle$ with $\mathrm{ot}(C_\delta) = \kappa$ for all δ. If $\Phi(\beta)$ is a $\mathbb{P}_\beta$-name, then let $\Phi^*(\beta)$ be a $\mathbb{P}_\beta$-name such that $\mathbb{P}_\beta$ forces that $\Phi^*(\beta)$ is a club-sequence as above and that $\Phi^*(\beta) = \Phi(\beta)$ if $\Phi(\beta)$ is such a club-sequence. Also, for every $\delta \in \kappa^+ \cap \mathrm{cf}(\kappa)$, let $\dot{C}^\beta_\delta$ be the canonical name for the δth member of $\Phi^*(\beta)$. If $\beta \in \mathrm{dom}(F)$, then the following hold:

   (A) $q|_\beta \Vdash_{\mathbb{P}_\beta} (f(\beta), \pi(\beta)) \in \mathbb{Q}_{\Phi^*(\beta)}$.

   (B) For every N, if $(N, \alpha) \in \Delta_q$, then $\delta_N \in \mathrm{dom}(f(\beta))$ and $\delta_N = f(\beta)(\delta_N)$.

   (C) For every $\nu \in \mathrm{dom}(f(\beta))$, if $\beta \in X^{\Phi(\beta)}_\varrho$ and $f(\beta)(\nu) \notin S_\varrho$, then $\nu \notin \mathrm{dom}(\pi(\beta))$.

Given two $\mathbb{P}_\alpha$-conditions q, q', q' *extends* q if and only if $\mathrm{dom}(F_q) \subseteq \mathrm{dom}(F'_q)$ and, for each $\gamma \in \mathrm{dom}(F_q)$, we have $f_q(\gamma) \subseteq f_{q'}(\gamma)$ and $\pi_p(\gamma) \subseteq \pi_{q'}(\gamma)$, and $\Delta_q \subseteq \Delta_{q'}$.

It is clear that $\mathbb{P}_\alpha \subseteq \mathbf{H}(\vartheta)$ for all α and that $\mathbb{P}_\beta \subseteq \mathbb{P}_\alpha$ for all $\beta < \alpha$. The following lemma shows that $\langle \mathbb{P}_\alpha \mid \alpha \leq \vartheta \rangle$ is a forcing iteration, in the sense that $\mathbb{P}_\beta$ is a complete suborder of $\mathbb{P}_\alpha$ for all $\beta < \alpha$ (Corollary 3.8). The proof is essentially identical to a corresponding proof in [3]. We include this proof here for the readers' benefit, though.

Lemma 3.7. Let $\beta \leq \alpha \leq \vartheta$. Suppose $q = (F_q, \Delta_q) \in \mathbb{P}_\beta$, $r = (F_r, \Delta_r) \in \mathbb{P}_\alpha$, and $q \leq_\beta r|_\beta$. Then

$$r \wedge q := (F_q \cup (F_r \restriction [\beta, \alpha)), \Delta_q \cup \Delta_r)$$

is a condition in $\mathbb{P}_\alpha$ extending r.

Proof. The proof is by induction on $\alpha \geq \beta$. The crucial point is the use of the markers τ in the definition of the forcing. New side conditions (N, τ) appearing in Δ_q may well have the property that $N \cap [\beta, \alpha) \neq \varnothing$, but they will not impose any problematic requirements—coming from (4) (B) in the definition—on ordinals $\gamma \in \mathrm{dom}(F_r \restriction [\beta, \alpha))$. The reason is simply that $\tau \leq \beta$. The details of the proof are as follows.

The case $\alpha = \beta$ of the induction is obvious, so let us start by assuming that $\alpha = \beta^* + 1$, where $\beta^* \geq \beta$. Clearly, $r \wedge q$ satisfies (1) and (2) in the definition of $\mathbb{P}_{\beta^*+1}$. By the induction hypothesis we know that the restriction of $r \wedge q$ to β^*, that is,

$$(r \wedge q)|_{\beta^*} = (F_q \cup (F_r \restriction [\beta, \beta^*)), \Delta_q \cup \Delta_{r|_{\beta^*}})$$

is a condition in $\mathbb{P}_{\beta^*}$ extending $r|_{\beta^*}$. Therefore, $r \wedge q$ also satisfies (3). If $\beta^* \notin \mathrm{dom}(F_r)$, then $r \wedge q$ is a condition in $\mathbb{P}_{\beta^*+1}$ since (4) is automatically satisfied. If $\beta^* \in \mathrm{dom}(F_r)$, then $F_{r \wedge q}(\beta^*) = F_r(\beta^*)$, which immediately gives (4) (C) for β^*. Also, $(r \wedge q)|_{\beta^*}$ forces in $\mathbb{P}_{\beta^*}$ that $F_r(\beta^*)$ is in $\mathbb{Q}_{\Phi^*(\beta^*)}$ (since $r|_{\beta^*}$ forces this and $(r \wedge q)|_{\beta^*}$ extends $r|_{\beta^*}$). This gives (4) (A) for $q \wedge_\beta r$ and β^* in this case. Now we check that δ_N is a fixed point of $F_r(\beta^*)$ whenever $(N, \beta^* + 1) \in \Delta_q \cup \Delta_r$. For this, note that $(N, \beta^* + 1) \in \Delta_r$. Hence δ_N is a fixed point of $F_r(\beta^*)$ by (4) for r and β^*. Finally note that the induction hypothesis and the inclusion $\Delta_r \subseteq \Delta_{r \wedge q}$ together imply that $r \wedge q$ extends r.

The case when α is a nonzero limit ordinal follows directly from the induction hypothesis. Q.E.D.

Corollary 3.8. For all $\beta < \alpha \leq \kappa$, every maximal antichain in $\mathbb{P}_\beta$ is a maximal antichain in $\mathbb{P}_\alpha$, and therefore $\mathbb{P}_\beta$ is a complete suborder of $\mathbb{P}_\alpha$.

Lemma 3.9 follows immediately from the first part of Lemma 3.6.

Lemma 3.9. For every $\alpha \leq \vartheta$, $\mathbb{P}_\alpha$ is $<\kappa$-closed. In fact, if $\lambda < \kappa$ and $(q_\alpha)_{\alpha<\lambda}$ is a decreasing sequence of conditions in $\mathbb{P}_\alpha$, then the condition $q^* = (F_{q^*}, \bigcup_{\alpha<\lambda} \Delta_{q_\alpha})$, where $\mathrm{dom}(F_{q^*}) = \bigcup_{\alpha<\lambda} \mathrm{dom}(F_{q_\alpha})$ and, for all $\xi \in \mathrm{dom}(F^*)$,

$$f_{q^*}(\xi) = \bigcup \{f_{q_\alpha}(\xi) \mid \alpha < \lambda,\, \xi \in \mathrm{dom}(F_{q_\alpha})\}$$

and

$$\pi_{q^*}(\xi) = \bigcup \{\pi_{q_\alpha}(\xi) \mid \alpha < \lambda,\, \xi \in \mathrm{dom}(F_{q_\alpha})\},$$

is a greatest lower bound of $\{q_\alpha \mid \alpha < \lambda\}$.

Under the hypotheses of Lemma 3.9, we shall call the condition q^* *the canonical greatest lower bound of* $\{q_\alpha \mid \alpha < \lambda\}$.

Lemma 3.10. For all $\alpha \leq \vartheta$, $\mathbb{P}_\alpha$ is κ^{++}-Knaster.

Proof. The proof uses standard Δ-system arguments. Suppose $(q_i)_{i<\kappa^{++}}$ is a sequence of conditions in $\mathbb{P}_\alpha$. By $2^\kappa = \kappa^+$ we may assume that the collection $\{\bigcup \mathrm{dom}(\Delta_{q_i}) \mid i < \kappa^{++}\}$ forms a Δ-system with root R, i.e., for all distinct i, i', $\bigcup \mathrm{dom}(\Delta_{q_i}) \cap \bigcup \mathrm{dom}(\Delta_{q_{i'}}) = R$.

Again using $2^\kappa = \kappa^+$ we may assume as well that there are $\mu < \kappa$ and enumerations $(N^i_\varsigma)_{\varsigma<\mu}$ of Δ_{q_i} (for $i < \mu$) such that, letting $\mathfrak{N}_i = \langle \bigcup \mathrm{dom}(\Delta_{q_i}), \in, \Phi, R, N^i_\varsigma, \Phi\rangle_{\varsigma<\mu}$ for all i, we have that for all i, $i' < \kappa^{++}$, $\mathfrak{N}_i$ and $\mathfrak{N}_{i'}$ are isomorphic via a unique isomorphism that fixes R.

The first assertion follows from the fact that by $2^\kappa = \kappa^+$ there are at most κ^+-many isomorphism types for such structures. For the second assertion note that, if Ψ is the unique isomorphism between $\mathfrak{N}_i$ and $\mathfrak{N}_{i'}$, then the restriction of Ψ to $R \cap \vartheta$ has to be the identity on $R \cap \vartheta$. Since there is a bijection $\varphi : \mathbf{H}(\vartheta) \longrightarrow \vartheta$ definable in $(\mathbf{H}(\vartheta), \in, \Phi)$, we have that Ψ fixes R if and only if it fixes $R \cap \vartheta$, and therefore it fixes R.

We may assume as well that $\{\mathrm{dom}(F_{q_i}) \mid i < \kappa^{++}\}$ forms a Δ-system with root ϱ and that for all $\gamma \in \varrho$, $\delta < \kappa^+$ and all distinct i, i' in κ^{++},

(i) $\bigcup \mathrm{dom}(\Delta_{q_i}) \cap \mathrm{dom}(F_{q_{i'}}) = \bigcup \mathrm{dom}(\Delta_{q_i}) \cap \varrho$,

(ii) $(f_{q_i}(\gamma), \pi_{q_i}(\gamma)) = (f_{q_{i'}}(\gamma), \pi_{q_{i'}}(\gamma))$, and

(iii) there is some N such that $(N, \gamma+1) \in \Delta_{q_i|_{\gamma+1}}$ and $\delta_N = \delta$ if and only if there is some N such that $(N, \gamma+1) \in \Delta_{q_{i'}|_{\gamma+1}}$ and $\delta_N = \delta$.

Let $i < i' < \kappa^{++}$. Using the above paragraph, together with the fact that $\bigcup \mathrm{dom}(\Delta_{q_i}) \cap \bigcup \mathrm{dom}(\Delta_{q_{i'}})$ is a symmetric κ-Φ-system by Lemma 2.4, one can easily verify that q_i and $q_{i'}$ are compatible as witnessed by $(F_{q_i} \cup F_{q_{i'}}, \Delta_{q_i} \cup \Delta_{q_{i'}})$. In fact it is not difficult to see by induction on $\beta \leq \alpha$ that the restriction of the pair $(F_{q_i} \cup F_{q_{i'}}, \Delta_{q_i} \cup \Delta_{q_{i'}})$ to β is a condition in $\mathbb{P}_\beta$. Q.E.D.

Lemma 3.11. The forcing $\mathbb{P}$ forces $2^\lambda = \vartheta$ for every $\mathbf{V}$-cardinal $\lambda \in [\kappa, \vartheta)$.

Proof. It is easy to see that $\mathbb{P}$ adds at least ϑ-many Cohen subsets of κ. For this, let G be $\mathbb{P}$-generic. Given $\gamma < \vartheta$, let $f^G_\gamma : (\kappa^+)^{\mathbf{V}} \longrightarrow (\kappa^+)^{\mathbf{V}}$ be the function added by G at the γ-th coordinate, i.e., the union of the functions $f_q(\gamma)$ for $q \in G$ with $\gamma \in \mathrm{dom}(F_q)$. Fix $X_\gamma \subseteq f^G_\gamma(\kappa)$, $X_\gamma \in \mathbf{V}$, such that for every $\eta < f^G_\gamma(\kappa)$ and every $\nu < \kappa$ there are, both in X_γ and in $f^G_\gamma(\kappa) \setminus X_\gamma$, unboundedly many ordinals σ below $f^G_\gamma(\kappa)$ such that $\mathrm{rk}^\lambda(\sigma) \geq \kappa+\nu$. Then, $A_\gamma := \{\nu < \kappa : f^G_\gamma(\nu+1) \in X_\gamma\}$ is a Cohen subset of κ by a straightforward density argument. Furthermore, by another density argument, $A_\gamma \neq A_{\gamma'}$ for all distinct γ, γ'.

For the other inequality, note that for every cardinal $\lambda \in [\kappa^+, \vartheta)$ there are not more than ϑ^λ-many nice names for subsets of λ by the κ^{++}-c.c. of $\mathbb{P}$. But $\vartheta^\lambda = \vartheta$. Q.E.D.

We now prove a crucial properness lemma. Many of the features of our construction are there precisely to make this lemma work. Lemma 3.12 shows that $\mathbb{P}$ is proper with respect to all the relevant submodels. As mentioned shortly in the introduction, there is no general preservation theorem for properness with respect to any reasonable class of uncountable submodels. It is therefore not surprising that, unlike in most proofs of properness in the context of iterated forcing relative to countable submodels (in particular the proofs of properness in [3] and [2]), the proof of Lemma 3.12 is not by induction,[7] but instead proceeds by giving a direct[8] construction.

Lemma 3.12. The forcing $\mathbb{P}$ is proper with respect to all $N^* \preccurlyeq \mathbf{H}((2^\vartheta)^+)$ such that $\mathbb{P}, \Phi \in N^*$, ${}^{<\kappa}N^* \subseteq N^*$ and $|N^*| = \kappa$.

Proof. Let $N = N^* \cap \mathbf{H}(\vartheta)$, and note that ${}^{<\kappa}N \subseteq N$. Let $q \in N \cap \mathbb{P}$. Let F^* be the function with the same domain as F_q such that $F^*(\gamma) = (f_q(\gamma) \cup \{(\delta_N, \delta_N)\}, \pi_q(\gamma))$ for all $\gamma \in \mathrm{dom}(F_q)$ and let q^* be given by $q^* = (F^*, \Delta_q \cup \{(N, \sup(N \cap \vartheta))\})$. q^* is clearly a condition in $\mathbb{P}$ extending q. Hence it will be enough to show that q^* is $(N^*, \mathbb{P})$-generic.

For this, let $A \in N^*$ be a maximal antichain of $\mathbb{P}$, and let q^1 be a condition in $\mathbb{P}$ extending both q^* and a condition $t \in A$. Note that A is in fact in N by the κ^{++}-chain condition of $\mathbb{P}$. The goal now is of course to see that $t \in N^*$, and for this it will suffice to show that there is a condition in $A \cap N$ compatible with t.

We may extend q^1 to a condition q^2 such that for every $\gamma \in \mathrm{dom}(F_t) \cap N$,

(i) $q^2|_\gamma$ forces that $\sup(\mathrm{ran}(f_{q^2}(\gamma) \restriction \delta_N))$ is a limit point of $\dot{C}^\gamma_{\delta_N}$, and

(ii) there is an ordinal $\eta_\gamma < \delta_N$ such that $q^2|_\gamma$ forces $\dot{C}^\gamma_\delta \cap \delta_N \subseteq \eta_\gamma$ for every $\delta \in \mathrm{ran}(f_{q^2}(\gamma))$ of cofinality κ such that $\delta > \delta_N$.

The condition q^2 can be taken to be the canonical greatest lower bound of a decreasing sequence $(q^2_n)_{n<\omega}$ of conditions extending q^1 such that for all n it holds that

(iii) for all $\gamma \in \mathrm{dom}(F_t) \cap N$, $q^2_{n+1}|_\gamma$ forces that the supremum of the set $\mathrm{ran}(f_{q^2_{n+1}}(\gamma) \restriction \delta_N)$ is a limit point of $\dot{C}^\gamma_{\delta_N}$, and

(iv) for all $\gamma \in \mathrm{dom}(F_{q^2_n})$, there is an ordinal $\eta < \delta_N$ for which $q^2_{n+1}|_\gamma$ forces $\dot{C}^\gamma_\delta \cap \delta_N \subseteq \eta$ for every $\delta \in \mathrm{ran}(f_{q^2_n}(\gamma))$ of cofinality κ such that $\delta > \delta_N$.

[7] More precisely, it is not of the form "Assume, for an arbitrarily given α, that all $\mathbb{P}_\beta$ (for $\beta < \alpha$) satisfy the relevant form of properness, and then argue that $\mathbb{P}_\alpha$ satisfies it as well."

[8] The construction is certainly direct but it is, perhaps, also the hardest part of the proof of Theorem 1.6.

Given q^2_n, q^2_{n+1} can be found by first extending q^2_n to a condition r satisfying (iv)—which exists since every $\mathbb{P}_\gamma$ forces both that $\mathrm{cf}(\delta_N) = \kappa$ (since it is $<\kappa$-closed) and that, for all δ, all limit points of $\dot{C}^\gamma_\delta$ below δ have cofinality less than κ—and then extending r to a condition q^2_{n+1} satisfying (iii) as well. Let $(\gamma_i)_{i<\mu}$, for some limit ordinal $\mu < \kappa$, be an enumeration of $\mathrm{dom}(t) \cap N$ such that every member of $\mathrm{dom}(t) \cap N$ is γ_i for unboundedly many $i < \mu$ (we may assume $\mathrm{dom}(t) \cap N \neq \varnothing$ as otherwise the proof is easier). Then q^2_{n+1} can be taken to be any lower bound of a decreasing sequence $(r_i)_{i<\mu}$ of conditions extending r such that for all i, $r_i|_{\gamma_i}$ forces that $\sup(\mathrm{ran}(f_{r_i}(\gamma_i) \restriction \delta_N))$ is a limit point of $\dot{C}^{\gamma_i}_{\delta_N}$. Given i and $(r_{i'})_{i'<i}$, r_i can be found by first finding a lower bound r' of $(r_{i'})_{i'<i}$ and then extending r' in at most ω stages (as in the proof of Lemma 3.4).

Let now $\varrho < \kappa$ be such that $\gamma \notin X^{\Phi(\gamma)}_\varrho$ for every $\gamma \in \mathrm{dom}(F_t) \cap N$. Since $\mathbf{H}(\vartheta^+) \in N^*$, we can pick $M^* \in N^*$ of size κ such that ${}^{<\kappa}M^* \subseteq M^*$, $\delta_{M^*} \in S_\varrho$, $M^* \preccurlyeq \mathbf{H}(\vartheta^+)$, $\sup\{\eta_\gamma \mid \gamma \in \mathrm{dom}(F_{q^2}) \cap N\} < \delta_M$, and such that M^* contains $\mathbb{P}$, Φ, A, $f_{q^2}(\gamma) \restriction \delta_N$ for all $\gamma \in N \cap \mathrm{dom}(F_{q^2})$, and $\mathrm{dom}(\Delta_{q^2}) \cap N$. Let $M = M^* \cap \mathbf{H}(\vartheta)$ and note that ${}^{<\kappa}M \subseteq M$ and that, since $\{\mathbf{H}(\vartheta), \Phi\} \in M^*$, $(M, \in, \Phi) \preccurlyeq (\mathbf{H}(\vartheta), \in, \Phi)$.

Claim 3.13. There is a condition q^4 stronger than q^2 such that

(1) $(M, \sup(M \cap \vartheta)) \in \Delta_{q^4}$ and

(2) there is some $\eta < \delta_M$ such that for every $\gamma \in \mathrm{dom}(F_t) \cap M$, $q^4|_\gamma$ forces $\max(\dot{C}^\gamma_\delta \cap \delta_M) < \eta$ for every $\delta \in \mathrm{ran}(f_{q^4}(\gamma))$ of cofinality κ such that $\delta > \delta_M$.

Proof. By extending q^2 slightly if necessary using Lemma 2.3 (ii) we may assume that $(M, 0) \in \Delta_{q^2}$: Since $\mathrm{dom}(\Delta_{q^2}) \cap N \in M$ is a symmetric system by Lemma 2.3 (i), $(\mathrm{dom}(\Delta_{q^2}) \cap N) \cup \{M\} \in N$ is a symmetric system. By Lemma 2.3 (ii) there is a symmetric Φ-system $\mathcal{M} \supseteq \mathrm{dom}(\Delta_{q^2}) \cup \{M\}$. But now $(F_{q^2}, \Delta_{q^2} \cup \{(M', 0) \mid M' \in \mathcal{M}\})$ is a condition extending q^2 of the specified form.

We may find a condition q^3 stronger than q^2 and such that the pair $(N, \sup(M \cap \vartheta))$ is in Δ_{q^3}. This condition q^3 can be built by recursion on $\mathrm{dom}(F_{q^2}) \cap M$ using the fact that $\mathbb{P}$ is $<\kappa$-closed. The details are as follows:

q^3 is the result of taking any lower bound r^* of a certain decreasing sequence $(r_i)_{i<\bar{\mu}}$, for $\bar{\mu} < \kappa$, of conditions extending q_2, and adding $(M, \sup(M \cap \vartheta))$ to its Δ.

Let $(\bar{\gamma}_i)_{i<\bar{\mu}}$ be the strictly increasing enumeration of $\mathrm{dom}(F_{q^2}) \cap M$ (which without loss of generality we may assume nonempty). The sequence $(r_i)_{i<\bar{\mu}}$ is built using $(\bar{\gamma}_i)_{i<\bar{\mu}}$. At any given stage i of the construction, suppose that we are handed a decreasing sequence $(r_{i'})_{i'<i}$ of conditions such

that $(M, \bar{\gamma}_{i'}+1) \in \Delta_{r_{i'}|_{\bar{\gamma}_{i'}+1}}$ for all i'. Let $r = q^2$ if $i = 0$, let $r = r_{i_0}$ if $i = i_0+1$, and let r be any lower bound of $(r_{i'})_{i'<i}$ if i is a nonzero limit ordinal. Let $\bar{r} = (F_r, \Delta_r \cup \{(M, \bar{\gamma}_i)\})$. Using the fact that $(M, \bar{\gamma}_{i'}+1) \in \Delta_{r_{i'}|_{\bar{\gamma}_{i'}+1}}$ for all $i' < i$ and that $\mathrm{dom}(F_r)$ has empty intersection with the interval $[\sup\{\bar{\gamma}_{i'}+1 \mid i' < i\}, \bar{\gamma}_i)$, it is easy to check that $\bar{r}$ is indeed a condition. By Lemma 3.4 (c) we may extend $\bar{r}|_{\bar{\gamma}_i}$ to a condition $\bar{r}' \in \mathbb{P}_{\bar{\gamma}_i}$ for which there is a function $f \supseteq f_{q^2(\bar{\gamma}_i)}$ in $\mathbf{V}$ such that δ_M is a fixed point of f and $\bar{r}'$ forces $(f, \pi_{q^2(\bar{\gamma}_i)}) \in \mathbb{Q}_{\Phi^*(\bar{\gamma}_i)}$. Let now F be the function with domain $\mathrm{dom}(F_{\bar{r}'}) \cup \mathrm{dom}(F_{q_2})$ such that $F\restriction\bar{\gamma}_i = F_{\bar{r}'}$, $F\restriction(\bar{\gamma}_i, \vartheta) = F_{q^2}\restriction(\bar{\gamma}_i, \vartheta)$ and $F(\bar{\gamma}_i) = (f, \pi_{q^2(\bar{\gamma}_i)})$, and let $r_i = (F, \Delta_{\bar{r}'} \cup \{(M, \bar{\gamma}_i+1)\})$. It is easy to verify that r_i is a $\mathbb{P}$-condition extending $\bar{r}'$, and that the final move of going from a lower bound r^* of $(r_i)_{i<\bar{\mu}}$ to q^3 yields also a $\mathbb{P}$-condition.

Using again that $\mathrm{cf}(\delta_M) = \kappa$ holds in all extensions by all $\mathbb{P}_\gamma$ and that, for every relevant club-sequence $\langle C_\delta \mid \delta \in (\kappa^+ \cap \mathrm{cf}(\kappa))^{\mathbf{V}}\rangle$, every limit point of every C_δ below δ is forced to have cofinality less than κ, we may now extend q^3 to a condition q^4 for which there is an ordinal $\eta < \delta_M$ such that $\eta > \delta_Q$ for every $Q \in \mathrm{dom}(\Delta_{q^4})$ with $\delta_Q < \delta_M$, and such that for every $\gamma \in \mathrm{dom}(F_t) \cap M$ and every $\delta \in \mathrm{ran}(f_{q^4}(\gamma))$ of cofinality κ such that $\delta > \delta_M$, $q^4|_\gamma$ forces $\max(\dot{C}^\gamma_\delta \cap \delta_M) < \eta$. This condition q^4 can be obtained as in the above constructions. Q.E.D.

Let now q^4 be given by the above claim, and for any $\gamma \in \mathrm{dom}(F_t) \setminus M$ let

$$\beta_\gamma = \min(((M \cap \vartheta) \cup \{\vartheta\}) \setminus \gamma)$$

and

$$\alpha_\gamma = \sup\{\sup(Q \cap \beta_\gamma) \mid Q \in \mathrm{dom}(\Delta_{q^4}) \cap M\}$$

Note that α_γ is in M, and that β_γ is also in M if $\beta_\gamma < \vartheta$. Note also that $\mathrm{cf}(\beta_\gamma) > \kappa$, and that therefore $\alpha_\gamma < \beta_\gamma$ since every $Q \in \mathrm{dom}(\Delta_{q^4}) \cap M$ has size κ and since $\mathrm{dom}(\Delta_{q^4}) \cap M \in M$ has size less than κ, and that in fact we have that $\alpha_\gamma < \gamma$ by the choice of β_γ.

Claim 3.14. $Q \cap [\alpha_\gamma, \beta_\gamma] \cap M = \varnothing$ for every $\gamma \in \mathrm{dom}(F_t) \setminus M$ and every $Q \in \mathrm{dom}(\Delta_{q^4})$ such that $\delta_Q < \delta_M$.

Proof. Let $\gamma \in \mathrm{dom}(F_t) \setminus M$ and suppose $Q \in \mathrm{dom}(\Delta_{q^4})$ is such that $\delta_Q < \delta_M$. Then, by Fact 2.2 there is some $Q' \in \mathrm{dom}(\Delta_{q^4}) \cap M$ such that $\delta_{Q'} = \delta_Q$ and such that $Q \cap M = Q \cap Q'$. Then $\zeta \in Q \cap [\alpha_\gamma, \beta_\gamma] \cap M$ would imply $\zeta \in [\alpha_\gamma, \beta_\gamma] \cap Q'$, but this would contradict the choice of α_γ. Q.E.D.

Working now in M^*, we may find a condition $t^* \in A$ satisfying the following.

(a) There is a function $\varphi : \mathrm{dom}(F_t) \setminus M \longrightarrow M \cap \vartheta$ such that (a1) for all $\gamma \in \mathrm{dom}(F_t) \setminus M$, $\varphi(\gamma) \in [\alpha_\gamma, \beta_\gamma] \setminus (\mathrm{dom}(q^4) \cap M)$, and

(a2) $\mathrm{dom}(F_{t^*}) = (\mathrm{dom}(F_t) \cap M) \cup \mathrm{ran}(\varphi)$.

(b) For every $\gamma \in \mathrm{dom}(F_{t^*})$,

(b1) $\delta_{N'}$ is a fixed point of $f_{t^*}(\gamma)$ whenever $(N', \tau) \in \Delta_{q^4} \cap M$ is such that $\gamma \in N'$ and $\tau \geq \gamma$, and

(b2) $f_{t^*}(\gamma) \restriction \eta = f_t(\gamma) \restriction \delta_M$ and $\pi_{t^*}(\gamma) \restriction \eta = \pi_t(\gamma) \restriction \delta_M$ if $\gamma \in \mathrm{dom}(F_t)$.

(c) There is a symmetric κ-Φ-system $\mathcal{N}$ such that $\mathcal{N} \supseteq \mathrm{dom}(\Delta_{t^*}) \cup \mathrm{dom}(\Delta_{q^4} \cap M)$.

This condition $t^* \in A$ may be found by the correctness of M^* in $\mathbf{H}(\vartheta^+)$ since the existence of a condition $t^* \in A$ satisfying (a)–(c) can be expressed in $\mathbf{H}(\vartheta^+)$ by a sentence σ with a certain parameter $p \in [M^*]^{<\kappa}$ which is in M^* by the closure of M^* under $<\kappa$-sequences, and such that t witnesses the truth of σ.

By the existence of $\mathcal{N}^*$ as in (c), we get that by Lemma 2.3 (ii) there is a condition $\bar{q}^4$ extending q^4 of the form (F_{q^4}, Δ) and such that $\mathcal{N}^* \subseteq \mathrm{dom}(\Delta_{\bar{q}^4})$.[9] It remains to see that $\bar{q}^4$ and t^* are compatible. For this, we are going to build a common extension q^5 of $\bar{q}^4$ and t^* by recursion on $\mathrm{dom}(F_{t^*})$. The construction of q^5 is along the lines of the construction of q^3 from q^2 in the proof of Claim 3.13. Let $(\gamma_i^*)_{i<\mu^*}$, for some $\mu^* < \kappa$, be the strictly increasing enumeration of $\mathrm{dom}(F_{t^*})$, which we may assume is nonempty. We build a certain decreasing sequence $(q_i^5)_{i<\mu^*}$ of conditions such that for all i,

(i) q_i^5 is a condition in $\mathbb{P}$ extending $\bar{q}_4$,

(ii) q_i^5 is of the form $\bar{q}^4 \wedge r$ for some $r \in \mathbb{P}_{\gamma_i^*+1}$, and

(iii) $q_i^5|_{\gamma_i^*+1}$ $(= r)$ extends $t^*|_{\gamma_i^*+1}$.

At a given stage i of the construction we are handed a decreasing sequence $(q_{i'}^5)_{i'<i}$ of conditions such that each $q_{i'}^5$ satisfies (i)–(iii). We let $r = (F_{\bar{q}^4}, \Delta_{\bar{q}^4} \cup \Delta_{t^*|_{\gamma_0}})$ if $i = 0$, $r = q_{i'}^5$ if $i = i' + 1$, and let r be the greatest lower bound of $\{q_{i'}^5 \mid i' < i\}$ if i is a nonzero limit ordinal. It is easily checked that r is indeed a condition in $\mathbb{P}$ in each case. We consider now the following two cases.

Suppose first $\gamma_i^* \in \mathrm{dom}(F_t)$. Then $r|_{\gamma_i^*}$ forces that

$$p = (f_{q^4}(\gamma_i^*) \cup f_{t^*}(\gamma_i^*), \pi_{q^4}(\gamma_i^*) \cup \pi_{t^*}(\gamma_i^*))$$

[9] We already saw this argument in the proof of Claim 3.13.

is a condition in $\mathbb{Q}_{\Phi^*(\gamma_i^*)}$ by (b) in the choice of t^* together with Lemma 3.5 and together with the fact that $\delta_M \notin \mathrm{dom}(\pi_{q^4}(\gamma_i^*))$ by (4) (C) in the definition of $\mathbb{P}_{\gamma_i^*+1}$ for $q^4|_{\gamma_i^*+1}$ and γ_i since $f_{q^4}(\gamma_i^*)(\delta_M) = \delta_M \in S_\varrho$ and $\gamma_i^* \notin X_\varrho^{\Phi(\gamma_i^*)}$. It follows that, letting F be the function with the same domain as F_r and such that $F\restriction\gamma_i^* = F_r\restriction\gamma_i^*$, $F(\gamma_i^*) = p$, and $F\restriction(\gamma_i^*, \vartheta) = F_r\restriction(\gamma_i^*, \vartheta)$, $q_i^5 := (F, \Delta_r)$ is a condition in $\mathbb{P}$ extending r and is such that $q_i^5\restriction\gamma_i^* + 1$ extends $t^*|\gamma_i^* + 1$.

The other case is when $\gamma_i^* \notin \mathrm{dom}(F_t)$. Note that also $\gamma_i^* \notin \mathrm{dom}(F_{q^4})$ in this case by the choice of $\varphi(\gamma)$ in (a1). By Lemma 3.4 (c) together with the $<\kappa$-closure of $\mathbb{Q}_{\Phi^*(\gamma_i^*)}$ in $\mathbf{V}^{\mathbb{P}_{\gamma_i^*}}$ (Lemma 3.6 (1)), there is a function $f \supseteq f_{t^*}(\gamma_i^*)$ and an extension s of $r\restriction\gamma_i^*$ such that

(i) δ_Q is a fixed point of f for every Q such that $(Q, \gamma_i^* + 1) \in \Delta_{\bar{q}^4|_{\gamma_i^*+1}}$ and $\delta_Q \geq \delta_M$, and

(ii) r forces $(f, \pi) \in \mathbb{Q}_{\Phi^*(\gamma_i^*)}$.

Since $\gamma_i^* \notin Q$ for any Q such that $(Q, \gamma_i^* + 1) \in \Delta_{\bar{q}^4|_{\gamma_i^*+1}}$ and $\delta_Q < \delta_M$ by Claim 3.14, we have that δ_Q is a fixed point of f for every Q such that $(Q, \gamma_i^* + 1) \in \Delta_{\bar{q}^4|_{\gamma_i^*+1}} \cup \Delta_{t^*|_{\gamma_i^*+1}}$. Now we can amalgamate the relevant objects into a condition q_i^5 as in the previous case.

Finally we can take q^5 to be any lower bound of $(q_i^5)_{i<\mu^*}$. Q.E.D.

By Lemmas 3.9, 3.10 and 3.12, $\mathbb{P}$ does not collapse cofinalities. In particular, κ^+ and $\kappa^+ \cap \mathrm{cf}(\kappa)$ have the same meaning in $\mathbf{V}$ as in any generic extension by any $\mathbb{P}_\alpha$, which means that the statements of the following lemmas are not ambiguous.

The proof of the following lemma is similar to the proof of Lemma 3.4 by standard density arguments.

Lemma 3.15. Suppose $\beta < \vartheta$, G is a $\mathbb{P}_{\beta+1}$-generic filter over $\mathbf{V}$, $\langle C_\delta \mid \delta \in \kappa^+ \cap \mathrm{cf}(\kappa)\rangle$ is the interpretation of $\Phi^*(\beta)$ by the restriction G_β of G to $\mathbb{P}_\beta$, and D is the union of all sets of the form $\mathrm{ran}(f_q(\beta))$, where $q \in G$ and $\beta \in \mathrm{dom}(F_q)$. Then the following hold:

(a) D is a club of κ^+.

(b) If $\langle C_\delta \mid \delta \in \kappa^+ \cap \mathrm{cf}(\kappa)\rangle = \Phi(\beta)_{G_\beta}$ and $\varrho < \kappa$ is such that $\beta \in X_\varrho^{\Phi(\beta)}$, then for every $\delta \in D \cap S_\varrho$,

1. $\{\alpha < \delta \mid \{C_\delta(\alpha + 1), C_\delta(\alpha + 2)\} \subseteq D\}$ is bounded in δ, and
2. if, in addition, $\mathrm{cf}(C_\delta(\alpha+1)) < \kappa$ for all $\delta \in \kappa^+ \cap \mathrm{cf}(\kappa)$ and $\alpha < \kappa$, then $\{\alpha < \delta \mid C_\delta(\alpha + 1) \in D\}$ is bounded in δ.

Lemma 3.16. The forcing $\mathbb{P}$ forces that if $\langle C_\delta \mid \delta \in \kappa^+ \cap \mathrm{cf}(\kappa)\rangle$ is a club-sequence with $\mathrm{ot}(C_\delta) = \kappa$ for all δ, then there is a club $D \subseteq \kappa^+$ such that

1. $\{\alpha < \kappa \mid \{C_\delta(\alpha+1), C_\delta(\alpha+2)\} \subseteq D\}$ is bounded for every δ and such that
2. if, in addition, $\mathrm{cf}(C_\delta(\alpha+1)) < \kappa$ for all $\delta \in \kappa^+ \cap \mathrm{cf}(\kappa)$ and $\alpha < \kappa$, then $\{\alpha < \delta \mid C_\delta(\alpha+1) \in D\}$ is bounded in δ for every δ.

Proof. Let G be $\mathbb{P}$-generic and, for every $\beta < \vartheta$, let D_β be the union of all sets of the form $\mathrm{ran}(f_q(\beta))$, where $q \in G$ and $\beta \in \mathrm{dom}(F_q)$. Let $\vec{C} = \langle C_\delta \mid \delta \in \kappa^+ \cap \mathrm{cf}(\kappa)\rangle \in \mathbf{V}[G]$ be a club-sequence such that $\mathrm{ot}(C_\delta) = \kappa$ for all δ. By the κ^{++}-c.c. of $\mathbb{P}$ in $\mathbf{V}$ we may find some $\mathbb{P}$-name $\dot{x} \in \mathbf{H}(\vartheta)^{\mathbf{V}}$ such that $\dot{x}_G = \vec{C}$. Then, for every $\beta < \vartheta$ such that $\Phi(\beta) = \dot{x}$, if $\varrho < \kappa$ is such that $\beta \in X^{\dot{x}}_\varrho$, then by Lemma 3.15 D_β is a club on κ^+ such that for every $\delta \in D_\beta \cap S_\varrho$,

1. $\{\alpha < \delta \mid \{C_\delta(\alpha+1), C_\delta(\alpha+2)\} \subseteq D_\beta\}$ is bounded in δ, and
2. if, in addition, $\mathrm{cf}(C_\delta(\alpha+1)) < \kappa$ for all $\delta \in \kappa^+ \cap \mathrm{cf}(\kappa)$ and $\alpha < \kappa$, then $\{\alpha < \delta \mid C_\delta(\alpha+1) \in D_\beta\}$ is bounded in δ.

Now let $(\beta_\varrho)_{\varrho<\kappa}$ be such that $\beta_\varrho \in X^{\dot{x}}_\varrho$ for all ϱ. Then, if $D = \bigcap_{\varrho<\kappa} D_{\beta_\varrho}$, D is a club of κ^+ and, since $\{S_\varrho : \varrho < \kappa\}$ is a partition on $\kappa^+ \cap \mathrm{cf}(\kappa)$, we have that for every $\delta \in D$,

1. $\{\alpha < \delta \mid \{C_\delta(\alpha+1), C_\delta(\alpha+2)\} \subseteq D\}$ is bounded in δ, and
2. if, in addition, $\mathrm{cf}(C_\delta(\alpha+1)) < \kappa$ for all $\delta \in \kappa^+ \cap \mathrm{cf}(\kappa)$ and $\alpha < \kappa$, then $\{\alpha < \delta \mid C_\delta(\alpha+1) \in D\}$ is bounded in δ.

Q.E.D.

Lemma 3.16 concludes the proof of Theorem 1.6.

References

[1] D. Asperó and S.-D. Friedman. Large cardinals and locally defined well-orders of the universe. *Annals of Pure and Applied Logic*, 157(1):1–15, 2009.

[2] D. Asperó and M. Mota. A generalization of Martin's Axiom. Submitted.

[3] D. Asperó and M. Mota. Forcing consequences of PFA together with the continuum large. *Transactions of the American Mathematical Society*, to appear.

[4] T. J. Jech. *Set theory.* Springer Monographs in Mathematics. Springer, third millenium edition, 2003.

[5] M. Kojman and S. Shelah. μ-complete Souslin trees on μ^+. *Archive for Mathematical Logic*, 32(3):195–201, 1993.

[6] K. Kunen. *Set theory: an introduction to independence proofs*, volume 102 of *Studies in Logic and the Foundations of Mathematics*. Elsevier, 1980.

[7] A. Rinot. Jensen's diamond principle and its relatives. In L. Babinkostova, A. E. Caicedo, S. Geschke, and M. Scheepers, editors, *Set theory and its applications, Papers from the Annual Boise Extravaganzas (BEST) held in Boise, ID, 1995–2010*, volume 533 of *Contemporary Mathematics*, pages 125–156. American Mathematical Society, 2011.

[8] S. Shelah. *Cardinal arithmetic*, volume 29 of *Oxford Logic Guides*. Clarendon Press Oxford University Press, New York, 1994. Oxford Science Publications.

[9] S. Shelah. Colouring and non-productivity of $\aleph_2$-c.c. *Annals of Pure and Applied Logic*, 84(2):153–174, 1997.

[10] S. Shelah. On what I do not understand (and have something to say). I. *Fundamenta Mathematicae*, 166(1-2):1–82, 2000.

[11] S. Shelah. Not collapsing cardinals $\leq \kappa$ in $(< \kappa)$-support iterations. *Israel Journal of Mathematics*, 136:29–115, 2003.

[12] D. Soukup and L. Soukup. Club guessing for dummies, 2010. Preprint (arXiv:1003.4670).

A survey of axiom selection as a machine learning problem

Jasmin Christian Blanchette[1], Daniel Kühlwein[2*]

[1] Fakultät für Informatik, Technische Universität München, München, Germany
[2] Institute for Computing and Information Science, Radboud Universiteit Nijmegen, Nijmegen, The Netherlands

Abstract

Automatic theorem provers struggle to discharge proof obligations of interactive theorem provers. This is partly due to the large number of background facts that are passed to the automatic provers as axioms. Axiom selection algorithms predict the relevance of facts, thereby helping to reduce the search space of automatic provers. This paper presents an introduction to axiom selection as a machine learning problem and describes the challenges that distinguish it from other applications of machine learning.

1 Introduction

The foundations of modern mathematics were laid at the end of the 19th century and the beginning of the 20th century. Seminal works such as Frege's *Begriffsschrift* [8] established the notion of mathematical proofs as formal derivations in a logical calculus. In *Principia Mathematica* [51], Whitehead and Russell set out to show by example that all of mathematics can be derived from a small set of axioms using an appropriate logical calculus. Even though Gödel later showed that no consistent axiom system can capture all mathematical truth [9], *Principia* showed that normal mathematics can indeed be catered for by a formal system. Proofs could now be rigidly defined, and verifying the validity of a proof was a simple matter of checking whether the rules of the calculus were correctly applied. But formal proofs were extremely tedious to write (and read), and so they found no audience among practicing mathematicians.

1.1 Interactive Theorem Proving

With the advent of computers, formal mathematics became a more realistic proposal. *Interactive theorem provers* (ITP), or *proof assistants*, are computer programs that support the creation of formal proofs. Proofs are

*The first author is supported by the Nederlandse Organisatie voor Wetenschappelijk Onderzoek (NWO) project *Learning2Reason*. The second author is supported by the Deutsche Forschungsgemeinschaft (DFG) project *Hardening the Hammer* (grant Ni 491/14-1).

Stefan Geschke, Benedikt Löwe, Philipp Schlicht (*eds.*).
Infinity, computability, and metamathematics: Festschrift celebrating the 60th birthdays of Peter Koepke and Philip Welch. College Publications, London, 2014. Tributes, Volume 23.

Theorem. There are infinitely many primes:
for every number n there exists a prime $p > n$.

Proof (after Euclid).
Given n. Consider $k = n! + 1$, where $n! = 1 \cdot 2 \cdot 3 \cdot \ldots \cdot n$.
Let p be a prime that divides k.
For this number p we have $p > n$: otherwise $p \leq n$;
but then p divides $n!$, so p cannot divide $k = n! + 1$,
contradicting the choice of p. Q.E.D.

FIGURE 1. An informal proof that there are infinitely many prime numbers [50].

written in the input language of the ITP, which can be thought of as being at the intersection between a programming language, a logic, and a mathematical typesetting system. In an ITP proof, each statement the user makes gives rise to a proof obligation. The ITP ensures that every proof obligation is discharged by a correct proof.

ACL2 [21], Coq [3], HOL4 [41], HOL Light [14], Isabelle [37], Mizar [12], and PVS [38] are perhaps the most widely used ITPs. Figures 1 and 2 show a simple informal proof and the corresponding Isabelle proof. Virtually all ITPs provide some built-in automation in the form of *tactics* that perform arbitrarily complex reasoning. In Figure 2, the **by** command specifies which tactic should be applied to discharge the current proof obligation.

Developing proofs in ITPs usually requires a lot more work than sketching a proof with pen and paper. Nevertheless, the benefit of gaining quasi-certainty about the correctness of the proof led a number of mathematicians to adopt these systems.

The largest mechanization project is probably the ongoing formalization of the proof of Kepler's conjecture by Thomas Hales and his colleagues in HOL Light [13]. Other major undertakings are the formal proofs of the four-color theorem [10] and of the odd-order theorem [11] in Coq, both developed under Gonthier's leadership. In terms of mathematical breadth, the *Mizar Mathematical Library* [30] is perhaps the main achievement of the ITP community so far: With nearly 52,000 theorems, it covers a large portion of the mathematics taught at the undergraduate level.

1.2 Automatic Theorem Proving

In contrast to interactive theorem provers, *automatic theorem provers* (ATPs) work without human interaction. They take a problem as input, consisting of a set of *axioms* and a *conjecture*, and attempt to deduce the conjecture from the axioms. The TPTP (Thousands of Problems for Theorem Provers) library [42] has established itself as a central infrastructure

```
theorem Euclid: ∃p ∈ prime. n < p
proof–
  let ?k = n! + 1
  obtain p where prime: p ∈ prime and dvd: p dvd ?k
    using prime-factor-exists by auto
  have n < p
  proof–
    have ¬ p ≤ n
    proof
      assume p ≤ n
      with prime-g-zero have p dvd n! by (rule dvd-factorial)
      with dvd have p dvd ?k − n! by (rule dvd-diff)
      then have p dvd 1 by simp
      with prime show False using prime-nd-one by auto
    qed
    then show ?thesis by simp
  qed
  from this and prime show ?thesis ..
qed

corollary ¬ finite prime
  using Euclid by (fastsimp dest!: finite-nat-set-is-bounded simp: le-def)
```

FIGURE 2. An Isabelle proof corresponding to the informal proof of Figure 1 [50].

for exchanging ATP problems. Its main developer also organizes an annual competition, CADE's ATP Systems Competition (CASC) [43], that measures progress in this field. E [40], SPASS [49], Vampire [39], and Z3 [35] are well-known ATPs for classical first-order logic.

Some researchers apply ATPs to open mathematical problems. William McCune's proof of the Robbins conjecture using a custom ATP is the main success story [31]. More recently, ATPs have also been integrated into ITPs [5, 19, 46], where they help increase the productivity by reducing the number of manual interactions needed to carry out a proof. Instead of using a built-in tactic, the ITP translates the current proof obligation (e.g., the lemma that the user has just stated but not proved yet) into an ATP problem. If the ATP can solve it, the proof is translated to the logic of the ITP and the user can proceed. In a recent study, about 70% of the proof obligations arising in a representative Isabelle corpus could be solved by ATPs [24].

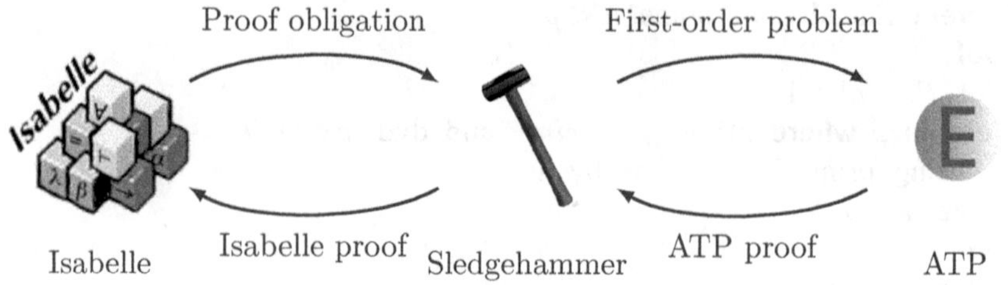

FIGURE 3. Sledgehammer [5] integrates ATPs (here E) into Isabelle.

1.3 Industrial Applications

Apart from mathematics, formal proofs are also used in industry. With the ever increasing complexity of software and hardware systems, quality assurance is a large part of the time and money budget of projects. Formal mathematics can be used to *prove* that an implementation meets a specification. Although tests might still be mandated by certification authorities, formal proofs can both drastically reduce the testing burden and increase confidence that the systems are bug-free.

AMD and Intel have been verifying floating-point procedures since the late 1990s [15, 34], partly as a consequence of the Pentium bug. Microsoft have had success applying formal verification methods to Windows device drivers [2]. One of the largest software verification projects so far is seL4, a verified operating system kernel [22].

1.4 Learning to Reason

One of the main reasons why formal mathematics and related technologies have not become mainstream yet is that developing ITP proofs is tedious. The reasoning capabilities of ATPs and ITP tactics are in many respects far behind what is considered standard for a human mathematician. Developing an interactive proof requires not only knowledge of the subject of the proof, but also of the ITP and its libraries.

One way to make users of ITPs more productive is to improve the success rate of ATPs. ATPs struggle with problems that have too many unnecessary axioms since they increase the search space. This is especially an issue when using ATPs from an ITP, where users have access to thousands of facts (axioms, definitions, lemmas, theorems, and corollaries) in the background libraries. Each fact is a potential axiom for an ATP.[1] *Axiom selection algorithms* heuristically select facts that are likely to be useful for inclusion as axioms in the problem given to the ATP.

[1] A terminological note is in order. ITP axioms are fundamental assumption in the common mathematical sense (e.g., the axiom of choice). In contrast, ATP axioms are arbitrary formulas that can be used to establish the conjecture.

Learning mathematics involves studying proofs to develop a mathematical intuition. Experienced mathematicians often know how to approach a new problem by simply looking at its statement. Assume that p is a prime number and $a, b \in \mathbb{N} - \{0\}$. Consider the following statement:

$$\text{If } p \mid ab, \text{ then } p \mid a \text{ or } p \mid b.$$

Even though mathematicians know many areas of mathematics (e.g., linear algebra, probability theory, analysis), when trying to prove the above statement they would ignore those areas and rely on their knowledge about number theory. At an abstract level, they perform axiom selection to reduce their search space.

Axiom selection algorithms typically rely on static features of the conjecture and axioms [16, 33]. For example, if the conjecture involves π and sin, they will prefer axioms that contain either of these two symbols, ideally both. The main drawback of such approaches is that they focus exclusively on formulas, ignoring the rich information contained in proofs. In particular, they do not learn from previous proofs. In this paper, we present an overview to axiom selection as a machine learning problem, an idea introduced one decade ago by Urban [44]. In a way, we are trying to teach the computer mathematical intuition.

2 Machine learning in a nutshell

This section aims to provide a high-level introduction to machine learning; for a more thorough discussion, we refer to standard textbooks [4, 29, 36].

Machine learning concerns itself with extracting information from data. Some typical examples of machine learning are listed below:

Spam classification Predict if a new email is spam.

Face detection Find human faces in a picture.

Web search Predict the websites that contain the information the user is looking for.

The results of a learning algorithm is a function that takes a new datapoint (email, picture, search query) and returns a target value (spam / not spam, location of faces, ranking of relevant websites). The learning is done by optimizing a *score function* over a *training dataset*. Typical score function are accuracy (how many emails were correctly labeled?) and the root mean square error (the Euclidean distance between the predicted values and the actual values). Elements of the training datasets are datapoints together with their expected value. For example:

Spam classification A set of emails together with their classification.

Face detection A set of pictures where all faces are marked.

Web search A set of query–websites tuples.

The performance of the learned function critically depends on the quality of the training data, as expressed by the aphorism "Garbage in, garbage out." Getting training data that is representative for the problem, and hence generalizes well, is crucial.

In addition to the training data, problem *features* are also essential. Features are the input of the prediction function and should describe the relevant attributes of the datapoint. A datapoint can have several possible *feature representations. Feature engineering* concerns itself with identifying relevant features [28]. To simplify computations, most machine learning algorithms require that the features are a (sparse) real-valued vector. Potential features are listed below.

Spam classification A list of all the words occurring in the email.

Face detection The matrix containing the color values of the pixels.

Web search The n-grams of the query.

From a mathematical point of view, most machine learning problems can be reduced to an optimization problem. Let $D \subseteq X \times T$ be a dataset consisting of datapoints and their corresponding target values. Let $\varphi : X \to \mathfrak{F}$ be a feature function that maps a datapoint to its feature representation in the *feature space* $\mathfrak{F}$ (usually a subset of $\mathbb{R}^n$ some $n \in \mathbb{N}$). Furthermore, let F be a function space and s a (convex) score function $s : D \times F \to \mathbb{R}$. Elements of F map features to the target space T—i.e., $F \subseteq (\mathfrak{F} \to T)$. One possible goal is to find the function $f \in F$ that maximizes the average score over the training set D. The main differences between various learning algorithms are the function space and the score function they use.

If the function space is too expressive, *overfitting* may occur: The learned function might perform well on the training data, but poorly on unseen data. A simple example is trying to fit a polynomial of degree $n - 1$ through n training datapoints; this will give perfect scores on the training data but is likely to yield a curve that behaves so wildly as to be useless to make predictions. The issue is well known from the world of finance, where very sophisticated models have been successfully applied to predict the past.

Regularization is used to balance function complexity with the result of the score function. To estimate how well a learning algorithm generalizes or to tune metaparameters (e.g., the regularization parameter or the maximum degree of a polynomial), *cross-validation* partitions the training data in two sets: one set used for training, the other for the evaluation.

3 Axiom selection as a machine learning problem

Using an ATP within an ITP requires a method to filter out irrelevant axioms during the creation of the ATP problem. Since most ITPs libraries contain several thousands of theorems, simply translating every library fact into an ATP axiom overwhelms the ATP; indeed, parsing huge problem files has been an issue with some ATPs. To use machine learning to create such a relevance filter, we must first answer three questions:

1. What is the goal of the learning?
2. What is the training data?
3. What are the features?

At a first glance, the goal seems clear:

> Given an ATP problem with axioms A and conjecture c, predict a subset of axioms $B \subseteq A$ that is sufficient for proving c.

But what does "sufficient" mean exactly? Clearly, the ATP's chances of success depend on which ATP is used, the time limit, and even the computer hardware. Moreover, there can be several potentially disjoint subsets from which the conjecture can be derived, reflecting the existence of alternative proofs. Which subset should then be chosen? Before we can answer these questions, we must introduce the training data.

3.1 The Training Data

The training data is extracted from the proof library of the ITP. For Isabelle, this could mean the libraries included with the prover or the *Archive of Formal Proofs* [23]; for Mizar, the *Mizar Mathematical Library* [30]. The data could also include custom libraries defined by the user or third parties.

Abstracting from its source, we assume that the training data consists of a set of facts (axioms, definitions, lemmas, theorems, corollaries) equipped with

1. a *visibility relation* that for each fact states which other facts appear before it;
2. a *dependency tree* that for each fact shows which facts were used in its proof (for lemmas, theorems, and corollaries);
3. a *formula tree representation* of each fact.

3.1.1 Example.

Figure 4 introduces a simple, constructed library. For each statement, every statement that occurs above it is visible. Axioms 1 and 2 and Definitions 1 and 2 are visible from Theorem 1, whereas Corollary 1 is not visible. Figure 5 presents the corresponding dependency tree. Finally, Figure 6 shows the formula tree of $\forall x\ x + 1 > x$.

Axiom 1. A

Axiom 2. B

Definition 1. C if and only if A

Definition 2. D if and only if C

Theorem 1. C
Proof. By Axiom 1 and Definition 1.

Corollary 1. D
Proof. By Theorem 1 and Definition 2.

FIGURE 4. A simple library.

3.2 What to Learn

Having defined the training data, we can now reconsider the initial learning goal:

> Given an ATP problem with axioms A and conjecture c, predict a subset of axioms $B \subseteq A$ that is sufficient for proving c.

In the ATP-as-tactic setting, the conjecture of the corresponding ATP problem is the current proof obligation the ITP user wants to discharge and the axioms are the visible facts. Since the score function only needs to be defined on the training dataset, the dependency tree can be used to define which axioms are sufficient. This allows us the restate the learning goal as follows:

> Given an ITP proof obligation c, predict the parents of c in the dependency tree.

For now, we ignore alternative proofs and assume that the dependencies extracted from the ITP are the dependencies that an ATP would use. Of course, predicting the exact parents is unrealistic. Treating axiom selection as a ranking rather than a subset selection problem allows more room for error and hence simplifies the problem. With this adjustment, we can state the final version of our learning goal:

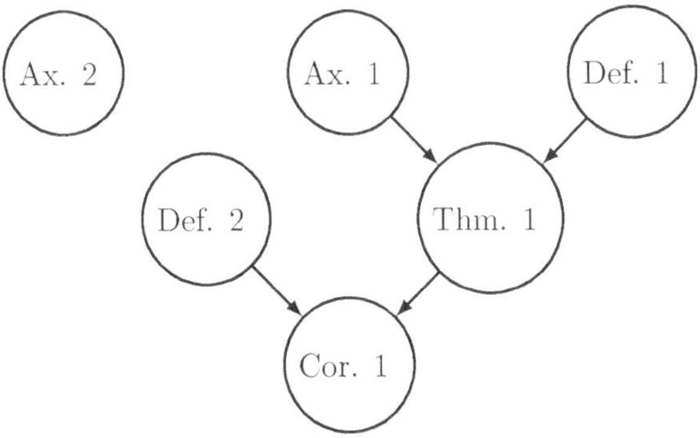

FIGURE 5. The dependency tree of the library of Figure 4, where edges denote dependency between facts.

> Given a training dataset and the formula tree of a proof obligation (Section 3.1), rank the visible facts according to their predicted usefulness.

In the training phase, the learning algorithm is allowed to learn from the proofs of all visible facts. The score function is chosen to optimize the ranks of the proof dependencies. For all facts in the training set, their corresponding dependencies should be ranked as high as possible.

It has often been observed that it is better to invoke an ATP repeatedly with different options for a short period of time (e.g., five seconds) than to let it run undisturbed until the user stops it. This optimization is called *time slicing*. Having a ranking function makes it possible to create different ATP problems for different slices, each with a different number of axioms, as illustrated in Figure 7. Slices with few axioms are more likely to find complex proofs involving a few obvious axioms, whereas those with lots of axioms might find simple proofs involving more obscure axioms.

3.3 Features

Almost all learning algorithms require the features of the input data to be a real vector. Therefore a method is needed to translate formula trees representing a proof obligation into real vectors.

3.3.1 Symbols

A simple approach is to take the set of symbols of a formula as its feature set. The symbols correspond to the node labels in the formula tree. It usually makes sense to leave out symbols corresponding to variables, since variable names are immaterial.

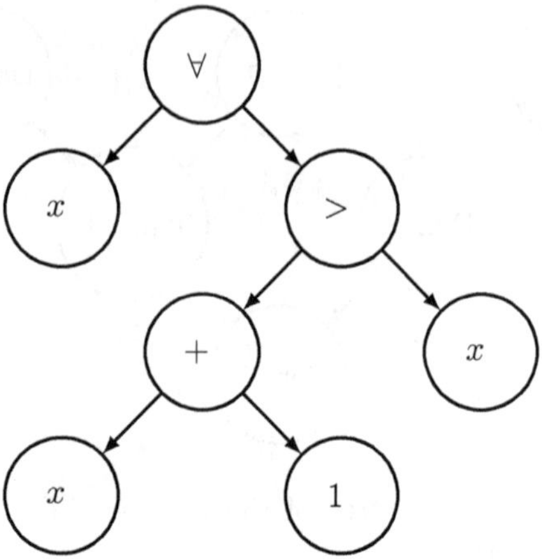

FIGURE 6. The formula tree for $\forall x\ x + 1 > x$.

Let n denote the vector size, which should be at least as large as the total number of symbols in the library. Let i be an injective index function that maps each symbol s to a positive number $i(s) \leq n$. The feature representation of a formula tree t is the binary vector $\varphi(t)$ such that $\varphi(t)(j) = 1$ if and only if the symbol with index j appears in t.

The example formula tree in Figure 6 contains the symbols $\forall$, $>$, $+$, and 1 (but not the variable x). Given $n = 10$, $i(\forall) = 1$, $i(>) = 4$, $i(+) = 6$, and $i(1) = 7$, the corresponding feature vector is $(1, 0, 0, 1, 0, 1, 1, 0, 0, 0)$.

3.3.2 Subterms and subformulas

In addition to the symbols, one can also include as features the subterms and subformulas of the formula to prove—i.e., the subtrees of the formula tree [47]. For example, the formula tree in Figure 6 has subtrees associated with x, 1, $x + 1$, $x > x + 1$, and $\forall x\ x + 1 > x$. Adding all subtrees significantly increases the size of the feature vector. Many subterms and subformulas appear only once in the library and are hence useless for making predictions. An approach to curtail this explosion is to consider only small subtrees (e.g., those with a height of at most 2 or 3).

3.3.3 Types

The formalisms supported by the vast majority of ITP systems are typed (or sorted), meaning that each term can be given a type that describes the values that can be taken by the term. Examples of types are *int*, *real*, *real* $\times$ *real*, and *real* $\to$ *real*. Adding the types that appear in the formula tree as additional features can help [18, 24]. Like terms, types can be represented as trees, and we may choose between encoding only basic types or also some or all complex subtypes.

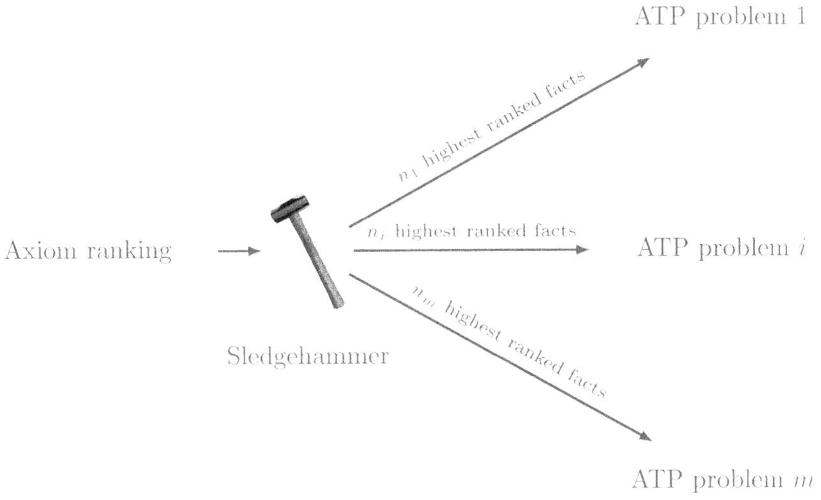

FIGURE 7. Sledgehammer generates several ATP problems (slices) from a single ranking.

3.3.4 Context

Due to the way humans develop complex proofs, the last few facts that were proved are likely to be useful in a proof of the current goal [7]. However, the machine learning algorithm might rank them poorly because they are new and hence little used, if at all. Adding the feature vectors of the nearby facts to the feature vector of the proof obligation, in a weighted fashion, is a method for ensuring that they obtain a better rank. This method is particularly useful when a formula has very few or very general features but occurs in a wider context.

4 Challenges

Axiom selection has several peculiarities that restrict which machine learning algorithms can be effectively used. In this section, we illustrate these challenges on a large fragment of Isabelle's *Archive of Formal Proofs* (AFP). The AFP benchmarks contain 165,964 facts distributed over 116 entries contributed by dozens of Isabelle users.[2] Most entries are related to computer science (e.g., data structures, algorithms, programming languages, and process algebras). The dataset was generated using Sledgehammer [24] and is available publicly at `http://www.cs.ru.nl/~kuehlwein/downloads/afp.tar.gz`.

[2] A number of AFP entries were omitted because of technical difficulties.

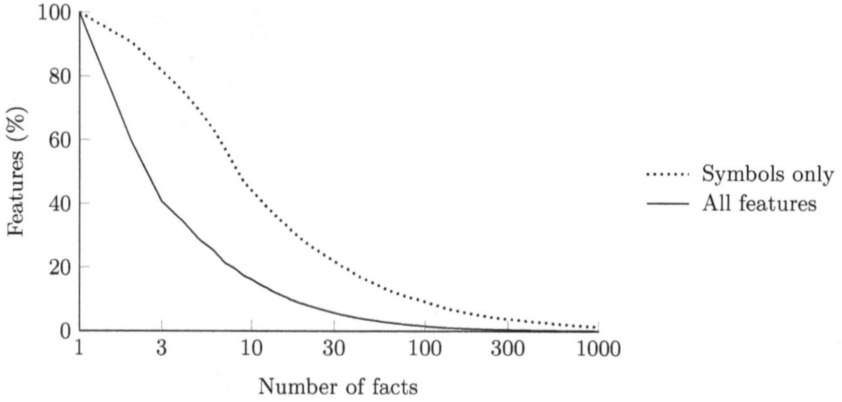

FIGURE 8. Distribution of the feature appearances in the *Archive of Formal Proofs*.

4.1 Features

The features introduced in Section 3.3 are very sparse. For example, the AFP contains 20,461 symbols. Adding small subterms and subformulas as well as basic types raises the total number of features to 328,361. Rare features can be very useful, because if two facts share a very rare feature, the likelihood that one depends on the other is very high. However, they also lead to much larger and sparser feature vectors.

Figure 8 shows the percentage of features that appear in at least x facts in the AFP, for various values of x. If we consider all features, then only 3.37% of the features appear in more than 50 facts. Taking only the symbols into account gives somewhat less sparsity, with 2.65% of the symbols appearing in more than 500 facts. Since there are 165,964 facts in total, this means that 97.35% of all symbols appear in less than 0.3% of the training data.

Another peculiarity of the axiom selection problem is that the number of features is not a priori fixed. Introducing new names for new concepts is standard mathematical practice. Hence, the learning algorithm must be able to cope with an unbounded, ever increasing feature set.

4.2 Dependencies

Like the features, the dependencies are also sparse. On average, an AFP fact depends on 5.5 other facts; 19.4% of the facts (axioms and definitions) have no dependencies at all, and 10.7% have at least 20 dependencies. Figure 9 shows the percentage of facts that are dependencies of at least x facts in

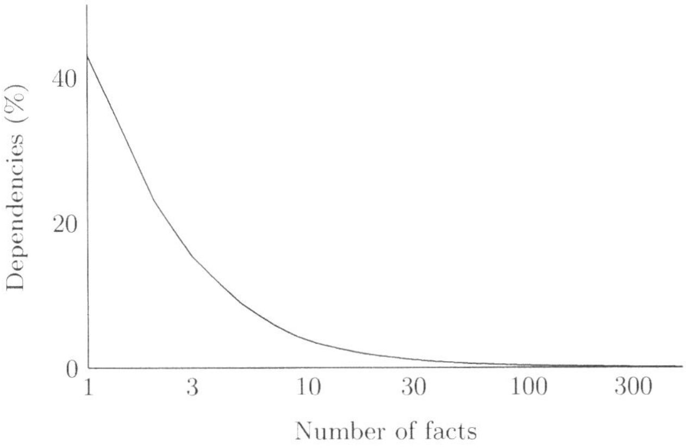

FIGURE 9. Distribution of the dependency appearances in the *Archive of Formal Proofs*.

the AFP, for various values of x. Less than half of the facts (43.0%) are a dependency in at least one other fact, and 94,593 facts are never used as dependencies. This includes 32,259 definitions as well as 17,045 facts where the dependencies could not be extracted and were hence left empty. Only 0.08% of the facts are being used as dependencies more than 500 times.

The main issue is that the dependencies in the training data might be incomplete or otherwise misleading. The dependencies extracted from the ITP are not necessarily the same as an ATP would use [1]. For example, Isabelle users can use induction in an interactive proof, and this would be reflected in the dependencies—the induction principle is itself a (higher-order) fact. But ATPs are limited to first-order logic without induction. If an alternative first-order proof is possible, this is the one that should be learned. Experiments with combinations of ATP and ITP proofs indicate that ITP dependencies are a reasonable guess, but learning from ATP dependencies yields better results [26, 47].

More generally, the training data lacks information about alternative proofs. In practice, this means that any evaluation method that relies only on the training data cannot reliably evaluate whether an axiom selection algorithm produces good predictions. There is no choice but to actually run ATPs—and even then the hardware, time limit, and version of the ATP can heavily influence the results.

4.3 Online Learning and Speed

Any algorithm for axiom selection must update its predictions model and create predictions fast. The typical use case is that of an ITP user who develops a theory fact by fact, proving each along the way. Usually these facts depend on one another, often in the familiar sequence definition–lemma–theorem–corollary. After each user input, the prediction model might need to be updated. In addition, it is not uncommon for users to alter existing definitions or lemmas, which should trigger some relearning.

Speed is essential for a axiom selection algorithm since the automated proof finding process needs to be faster than manual proof creation. The less time is spent on updating the learning model and predicting the axiom ranking, the more time can be used by ATPs. Users of ITPs tend to be impatient: If the automatic provers do not respond within half a minute or so, they usually prefer to carry out the proof themselves.

4.4 Translations between Logics

There is a wide gap between the ITPs' and the ATPs' logics. Much research has been concerned with bridging the gap between the two, by encoding ITP into ATP formulas. The main difficulties are connected with the ITPs' support for higher-order construct (e.g., quantification over functions and predicates, λ-expressions) and rich polymorphic type systems. Complete translations of both features are well known, but they lead to so much clutter that the proof search effectively grinds to a halt.

In practice, interactive problems are mostly first-order and their type information is largely irrelevant. This can be exploited to yield a lightweight translation, by encoding higher-order constructs locally (and leaving the first-order parts of the problem unchanged) [32] and by keeping a minimal amount of type information necessary to prevent the discovery of spurious proofs (i.e., proofs that are ill-typed in the ITP) [6].

5 Conclusion

This paper provided an introduction to the axiom selection problem. For further reading, we refer to [25], which reviews several algorithms on a benchmark suite derived from the *Mizar Mathematical Library*, and [19,20], which introduced a nearest-neighbor approach to axiom selection. Urban and Vyskočil give a more results-oriented introduction [48].

Machine learning has also been employed to improve other aspects of ATP reasoning. In particular, learning algorithms have been used to predict which search strategies are most likely to succeed in finding a proof [17,27, 45].

References

[1] J. Alama, D. Kühlwein, and J. Urban. Automated and human proofs in general mathematics: An initial comparison. In N. Bjørner and A. Voronkov, editors, *Logic for Programming, Artificial Intelligence, and Reasoning, 18th International Conference, LPAR-18, Mérida, Venezuela, March 11-15, 2012, Proceedings*, volume 7180 of *Lecture Notes in Computer Science*, pages 37–45. Springer-Verlag, 2012.

[2] T. Ball, E. Bounimova, V. Levin, R. Kumar, and J. Lichtenberg. The Static Driver Verifier Research Platform. In T. Touili, B. Cook, and P. Jackson, editors, *Computer Aided Verification, 22nd International Conference, CAV 2010, Edinburgh, UK, July 15-19, 2010, Proceedings*, volume 6174 of *Lecture Notes in Computer Science*, pages 119–122. Springer-Verlag, 2010.

[3] Y. Bertot and P. Castéran. *Interactive Theorem Proving and Program Development—Coq'Art: The Calculus of Inductive Constructions*. Texts in Theoretical Computer Science. Springer-Verlag, 2004.

[4] C. M. Bishop. *Pattern Recognition and Machine Learning*. Information Science and Statistics. Springer-Verlag, 2006.

[5] J. C. Blanchette, S. Böhme, and L. C. Paulson. Extending Sledgehammer with SMT solvers. *Journal of Automated Reasoning*, 51(1):109–128, 2013.

[6] J. C. Blanchette, S. Böhme, A. Popescu, and N. Smallbone. Encoding monomorphic and polymorphic types. In N. Piterman and S. Smolka, editors, *Tools and Algorithms for the Construction and Analysis of Systems, 19th International Conference, TACAS 2013, Held as Part of the European Joint Conferences on Theory and Practice of Software, ETAPS 2013, Rome, Italy, March 16-24, 2013, Proceedings*, volume 7795 of *Lecture Notes in Computer Science*, pages 493–507. Springer, 2013.

[7] M. Cramer, P. Koepke, D. Kühlwein, and B. Schröder. Premise selection in the Naproche system. In J. Giesl and R. Hähnle, editors, *Automated Reasoning, 5th International Joint Conference, IJCAR 2010, Edinburgh, UK, July 16-19, 2010, Proceedings*, volume 6173 of *Lecture Notes in Computer Science*, pages 434–440. Springer-Verlag, 2010.

[8] G. Frege. *Begriffsschrift, eine der arithmetischen nachgebildete Formelsprache des reinen Denkens*. Verlag von Louis Nebert, 1879.

[9] K. Gödel. Über formal unentscheidbare Sätze der Principia Mathematica und verwandter Systeme I. *Monatshefte für Mathematik und Physik*, 38(1):173–198, 1931.

[10] G. Gonthier. Formal proof—The four-color theorem. *Notices of the American Mathematical Society*, 55(11):1382–1393, 2008.

[11] G. Gonthier, A. Asperti, J. Avigad, Y. Bertot, C. Cohen, F. Garillot, S. L. Roux, A. Mahboubi, R. O'Connor, S. O. Biha, I. Pasca, L. Rideau, A. Solovyev, E. Tassi, and L. Théry. A machine-checked proof of the odd order theorem. In S. Blazy, C. Paulin-Mohring, and D. Pichardie, editors, *Interactive Theorem Proving, 4th International Conference, ITP 2013, Rennes, France, July 22-26, 2013, Proceedings*, volume 7998 of *Lecture Notes in Computer Science*, pages 163–179. Springer-Verlag, 2013.

[12] A. Grabowski, A. Korniłowicz, and A. Naumowicz. Mizar in a Nutshell. *Journal of Formalized Reasoning*, 3(2):153–245, 2010.

[13] T. C. Hales. Introduction to the Flyspeck project. In T. Coquand, H. Lombardi, and M.-F. Roy, editors, *Mathematics, Algorithms, Proofs*, volume 05021 of *Dagstuhl Seminar Proceedings*. Schloss Dagstuhl, 2005.

[14] J. Harrison. HOL Light: A tutorial introduction. In M. Srivas and A. Camilleri, editors, *Formal Methods in Computer-Aided Design, First International Conference, FMCAD '96, Palo Alto, California, USA, November 6-8, 1996, Proceedings*, volume 1166 of *Lecture Notes in Computer Science*, pages 265–269. Springer-Verlag, 1996.

[15] J. Harrison. Formal verification of IA-64 division algorithms. In M. Aagaard and J. Harrison, editors, *Theorem Proving in Higher Order Logics, 13th International Conference, TPHOLs 2000, Portland, Oregon, USA, August 14-18, 2000, Proceedings*, volume 1869 of *Lecture Notes in Computer Science*, pages 233–251. Springer-Verlag, 2000.

[16] K. Hoder and A. Voronkov. Sine qua non for large theory reasoning. In N. Bjørner and V. Sofronie-Stokkermans, editors, *Automated Deduction, CADE-23, 23rd International Conference on Automated Deduction, Wroclaw, Poland, July 31–August 5, 2011, Proceedings*, volume 6803 of *Lecture Notes in Computer Science*, pages 299–314. Springer-Verlag, 2011.

[17] F. Hutter, H. H. Hoos, K. Leyton-Brown, and T. Stützle. ParamILS: An automatic algorithm configuration framework. *Journal of Artificial Intelligence Research*, 36:267–306, 2009.

[18] C. Kaliszyk and J. Urban. Learning-assisted automated reasoning with Flyspeck, 2012. Preprint (arXiv:1211.7012).

[19] C. Kaliszyk and J. Urban. Automated reasoning service for HOL Light. In J. Carette, D. Aspinall, C. Lange, P. Sojka, and W. Windsteiger, editors, *Intelligent Computer Mathematics, MKM, Calculemus, DML, and Systems and Projects 2013, Held as Part of CICM 2013, Bath, UK, July 8-12, 2013, Proceedings*, volume 7961 of *Lecture Notes in Computer Science*, pages 120–135. Springer-Verlag, 2013.

[20] C. Kaliszyk and J. Urban. Stronger automation for Flyspeck by feature weighting and strategy evolution. In J. C. Blanchette and J. Urban, editors, *Third International Workshop on Proof Exchange for Theorem Proving, PxTP 2013, Lake Placid, NY, USA, June 9-10, 2013*, volume 14 of *EasyChair Proceedings in Computing*, pages 87–95. EasyChair, 2013.

[21] M. Kaufmann, P. Manolios, and J. S. Moore. *Computer-Aided Reasoning: An Approach.* Advances in Formal Methods. Kluwer Academic Publishers, 2000.

[22] G. Klein, J. Andronick, K. Elphinstone, G. Heiser, D. Cock, P. Derrin, D. Elkaduwe, K. Engelhardt, R. Kolanski, M. Norrish, T. Sewell, H. Tuch, and S. Winwood. seL4: formal verification of an operating-system kernel. *Communications of the ACM*, 53(6):107–115, 2010.

[23] G. Klein, T. Nipkow, and L. Paulson, editors. *Archive of Formal Proofs.* `http://afp.sf.net/`.

[24] D. Kühlwein, J. C. Blanchette, C. Kaliszyk, and J. Urban. MaSh: Machine learning for sledgehammer. In S. Blazy, C. Paulin-Mohring, and D. Pichardie, editors, *Interactive Theorem Proving, 4th International Conference, ITP 2013, Rennes, France, July 22-26, 2013, Proceedings*, volume 7998 of *Lecture Notes in Computer Science*, pages 35–50. Springer-Verlag, 2013.

[25] D. Kühlwein, T. Laarhoven, E. Tsivtsivadze, J. Urban, and T. Heskes. Overview and evaluation of premise selection techniques for large theory mathematics. In B. Gramlich, D. Miller, and U. Sattler, editors, *Automated Reasoning, 6th International Joint Conference, IJCAR 2012, Manchester, UK, June 26-29, 2012, Proceedings*, volume 7364 of *Lecture Notes in Computer Science*, pages 378–392. Springer-Verlag, 2012.

[26] D. Kühlwein and J. Urban. Learning from multiple proofs: First experiments. In P. Fontaine, R. A. Schmidt, and S. Schulz, editors, *Third Workshop on Practical Aspects of Automated Reasoning, PAAR-2012, Manchester, UK, June 30–July 1, 2012*, volume 21 of *EasyChair Proceedings in Computing*, pages 82–94. EasyChair, 2013.

[27] D. Kühlwein and J. Urban. MaLeS: A framework for automatic tuning of automated theorem provers, 2013. Preprint (arXiv:1308.2116).

[28] H. Liu and H. Motoda. *Feature Selection for Knowledge Discovery and Data Mining*. Kluwer Academic Publishers, 1998.

[29] D. J. MacKay. *Information Theory, Inference and Learning Algorithms.* Cambridge University Press, 2003.

[30] R. Matuszewski and P. Rudnicki. Mizar: The first 30 years. *Mechanized Mathematics and Its Applications*, 4:3–24, 2005.

[31] W. McCune. Solution of the Robbins problem. *Journal of Automated Reasoning*, 19(3):263–276, 1997.

[32] J. Meng and L. C. Paulson. Translating higher-order clauses to first-order clauses. *Journal of Automated Reasoning*, 40(1):35–60, 2008.

[33] J. Meng and L. C. Paulson. Lightweight relevance filtering for machine-generated resolution problems. *Journal of Applied Logic*, 7(1):41–57, 2009.

[34] J. S. Moore, T. W. Lynch, and M. Kaufmann. A mechanically checked proof of the AMD5_K86™ floating point division program. *IEEE Transactions on Computers*, 47(9):913–926, 1998.

[35] L. Moura and N. Bjørner. Z3: An efficient SMT solver. In C. Ramakrishnan and J. Rehof, editors, *Tools and Algorithms for the Construction and Analysis of Systems, 14th International Conference, TACAS 2008, Held as Part of the Joint European Conferences on Theory and Practice of Software, ETAPS 2008, Budapest, Hungary, March 29-April 6, 2008, Proceedings*, volume 4963 of *Lecture Notes in Computer Science*, pages 337–340. Springer-Verlag, 2008.

[36] K. P. Murphy. *Machine Learning: A Probabilistic Perspective.* MIT Press, 2012.

[37] T. Nipkow, L. C. Paulson, and M. Wenzel. *Isabelle/HOL: A Proof Assistant for Higher-Order Logic*, volume 2283 of *Lecture Notes in Computer Science*. Springer-Verlag, 2002.

[38] S. Owre and N. Shankar. A brief overview of PVS. In O. A. Mohamed, C. Muñoz, and S. Tahar, editors, *Theorem Proving in Higher Order Logics, 21st International Conference, TPHOLs 2008, Montreal, Canada, August 18-21, 2008, Proceedings*, volume 5170 of *Lecture Notes in Computer Science*, pages 22–27. Springer-Verlag, 2008.

[39] A. Riazanov and A. Voronkov. The design and implementation of VAMPIRE. *AI Communications*, 15(2-3):91–110, 2002.

[40] S. Schulz. E—A Brainiac Theorem Prover. *AI Communications*, 15(2-3):111–126, 2002.

[41] K. Slind and M. Norrish. A Brief Overview of HOL4. In O. A. Mohamed, C. Muñoz, and S. Tahar, editors, *Theorem Proving in Higher Order Logics, 21st International Conference, TPHOLs 2008, Montreal, Canada, August 18-21, 2008, Proceedings*, volume 5170 of *Lecture Notes in Computer Science*, pages 28–32. Springer-Verlag, 2008.

[42] G. Sutcliffe. The TPTP problem library and associated infrastructure. *Journal of Automated Reasoning*, 43(4):337–362, 2009.

[43] G. Sutcliffe. The 6th IJCAR automated theorem proving system competition—CASC-J6. *AI Communications*, 26(2):211–223, 2013.

[44] J. Urban. MPTP—motivation, implementation, first experiments. *Journal of Automated Reasoning*, 33(3-4):319–339, 2004.

[45] J. Urban. BliStr: The blind strategymaker, 2013. Preprint (arXiv:1301.2683).

[46] J. Urban, P. Rudnicki, and G. Sutcliffe. ATP and presentation service for Mizar formalizations. *Journal of Automated Reasoning*, 50(2):229–241, 2013.

[47] J. Urban, G. Sutcliffe, P. Pudlák, and J. Vyskočil. MaLARea SG1—Machine learner for automated reasoning with semantic guidance. In A. Armando, P. Baumgartner, and G. Dowek, editors, *Automated Reasoning, 4th International Joint Conference, IJCAR 2008, Sydney, Australia, August 12-15, 2008, Proceedings*, volume 5195 of *Lecture Notes in Computer Science*, pages 441–456. Springer-Verlag, 2008.

[48] J. Urban and J. Vyskočil. Theorem proving in large formal mathematics as an emerging AI field. In M. P. Bonacina and M. E. Stickel, editors, *Automated Reasoning and Mathematics—Essays in Memory of William W. McCune*, volume 7788 of *Lecture Notes in Computer Science*, pages 240–257. Springer-Verlag, 2013.

[49] C. Weidenbach, D. Dimova, A. Fietzke, R. Kumar, M. Suda, and P. Wischnewski. SPASS version 3.5. In R. A. Schmidt, editor, *Automated Deduction, CADE-22, 22nd International Conference on Automated Deduction, Montreal, Canada, August 2-7, 2009, Proceedings*, volume 5663 of *Lecture Notes in Computer Science*, pages 140–145. Springer-Verlag, 2009.

[50] M. Wenzel and F. Wiedijk. A comparison of Mizar and Isar. *Journal of Automated Reasoning*, 29(3-4):389–411, 2002.

[51] A. N. Whitehead and B. Russell. *Principia Mathematica.* Cambridge University Press, 2nd edition, 1927.

The lost melody phenomenon

Merlin Carl*

Fachbereich Mathematik und Statistik, Universität Konstanz, Konstanz, Germany

Abstract

A typical phenomenon for machine models of transfinite computations is the existence of so-called lost melodies, i.e., real numbers x such that the characteristic function of the set $\{x\}$ is computable while x itself is not (a real having the first property is called recognizable). This was first observed by Hamkins and Lewis for infinite time Turing machine [11], then demonstrated by Koepke and the author for ITRMs [5]. We prove that, for unresetting infinite time register machines introduced by Koepke in [18], recognizability equals computability, i.e., the lost melody phenomenon does not occur. Then, we give an overview on our results on the behaviour of recognizable reals for ITRMs as introduced in [19]. We show that there are no lost melodies for ordinal Turing machines (OTMs) or ordinal register machines (ORMs) without parameters and that this is, under the assumption that $0^{\#}$ exists, independent from ZFC. Then, we introduce the notions of resetting and unresetting α-register machines and give some information on the question for which of these machines there are lost melodies.

1 Introduction

The research on machine models of transfinite computations began with the seminal Hamkins-Lewis paper [11] on Infinite Time Turing Machines (ITTMs). These machines, which are basically classical Turing machines equipped with transfinite running time, have succesfully been applied to various areas of mathematics such as descriptive set theory [7, 27] and model theory [12] and turned out to show a variety of fascinating behaviour. A particularly interesting feature that has frequently played a role in applications is the existence of so-called lost melodies. A lost melody is a real number $x \subseteq \omega$ which is recognizable, i.e., for some ITTM-program P, the computation of P with y on the input tape (which plays the role of a real oracle for an ITTM) is defined for all y and outputs 1 iff $y = x$ and otherwise outputs 0, but not computable, i.e., no program computes the characteristic function of x. The existence of lost melodies for ITTMs was observed and proved in [11].

*We are indebted to Philipp Schlicht for sketching a proof of Lemma 2.3, a crucial hint for the proof of Theorem 2.6 and suggesting several very helpful references, in particular concerning Theorem 4.2. We also thank the anonymous referee for several corrections and suggestions that helped to considerably improve the paper.

In the meantime, a rich variety of transfinite machine types have been defined, studied and related to each other: Koepke introduced Infinite Time Register Machines [18], which were later relabeled as unresetting or weak Infinite Time Register Machines (wITRMs) when an enhanced version was considered in [5]. Further generalizations led to α-Turing machines [20] α-β-Turing machines, transfinite λ-calculus [28], the hypermachines of Friedman and Welch (basically ITTMs with a more complex limit behaviour, cf. [9]) and infinite time Blum-Shub-Smale-machines [21]. An arguably ultimate upper bound is set by Koepke's ordinal register machines (ORMs) and ordinal Turing machines (OTMs), which, using ordinal parameters, can calculate the whole of Gödel's constructible hierarchy **L**. (The paper [3] contains an argument to the effect that OTM-computability is indeed a conceptual analogue of Turing-computability in the transfinite.) For many of these machine types, the computational strength has been precisely determined.

In this paper, we are interested in the question how typical the existence of lost melodies is for models of transfinite computations. While it was shown in [5] that ITRMs, like ITTMs, have lost melodies, the question was to the best of our knowledge not considered for any other of these machine types and has in particular been open concerning wITRMs. Specifically, we focus on machine models generalizing register machines: In § 2, we prove that there are no lost melodies for unresetting ITRMs, we summarize (mostly leaving out or merely sketching proofs) in § 3 some of our earlier results on ITRM-recognizability obtained in [4] and [2] and proceed in § 4 to show that there are again no lost melodies for ordinal register and Turing machines, without ordinal parameters and that the answer for ordinal machines with parameters is undecidable under a certain set-theoretical assumption. Then, for the parameter-free case, we interpolate between these extrem cases by introducing resetting and unresetting α-register machines and show that for resetting α-register machines, lost melodies always exist. For unresetting α-register machines, the picture is quite different: It turns out that, while there are no lost melodies for $\alpha = \omega$, there exist countable values of α for which there are, but their supremum is countable, so that from some $\gamma < \omega_1$ on, lost melodies for unresetting α-register machines cease to exist.

Let us now introduce the relevant machine types, the resetting and unresetting α-register machines. (The unresetting version was originally suggested in the final paragraph of [18].) An α-register machine has finitely many registers, each of which can store a single ordinal $< \alpha$. The instructions for an α-register machine (also simply called α-machine) are the same as for the unlimited register machines of [8]: the increasing of a register content by 1, copying a register content to another register, reading out the r_ith bit of an oracle (where r_i is the content of the ith register), jumping to a certain program line provided a certain register content is 0, and stopping. Programs for α-register machines are finite sequences of instructions, as usual. The running time of an α-machine is the class of ordinals. At successor times, computations proceed as for the classical model of unlimited register machines, introduced in [8]. It remains to fix what to do at a limit time λ. We consider three possibilites, where Z_ι denotes the active program line at time ι and $R_{i\iota}$ denotes the content of the ith register at time ι:

Option 1: Z_λ and $R_{i\lambda}$ are undefined. Setting $\lambda = \omega$, this would just be a classical URM.

Option 2: $Z_\lambda := \liminf_{\iota<\lambda} Z_\iota$, $R_{i\iota} = \liminf_{\iota<\lambda} R_{i\iota}$, if the latter is $< \alpha$ and otherwise, the computation is undefined. Setting $\alpha = \omega$, these are the unresetting or weak infinite time register machines introduced in [18].[1] We call these unresetting or weak α-machines.

Option 3: $Z_\lambda := \liminf_{\iota<\lambda} Z_\iota$, $R_{i\iota} = \liminf_{\iota<\lambda} R_{i\iota}$, if the latter is $< \alpha$ and otherwise $R_{i\lambda} = 0$. Setting $\alpha = \omega$, these are the infinite time register machines (ITRMs) of [19]. We call these resetting or strong α-machines.

Most of our notation and terminology is standard. We write KP for Kripke-Platek set theory (cf., e.g., [26]), and ZF^- is Zermelo-Fraenkel set theory without the power set axioms in the version described in [10]. If P is a program and x a real, then $P^x\downarrow$ means that P, when run in the oracle x, stops, while $P^x\uparrow$ means that P in the oracle x diverges. By $P^x\downarrow = y$ we mean that $P^x(i)\downarrow$ for all $i \in \omega$ and that in the final state of $P^x(j)$, the first register contains 1 iff $j \in y$ and otherwise 0. We write $x \leq_{\mathrm{h}} y$ for hyperarithmetic reducibility, i.e., for $x \in \mathbf{L}_{\omega_1^{\mathrm{CK},y}}[y]$. By On we denote the class of ordinals, and by small greek letters we denote ordinals unless stated otherwise. The (class) function $p : \mathrm{On} \times \mathrm{On} \to \mathrm{On}$ is Cantor's pairing function.

It turns out (cf. [19]) that unresetting ω-machines are much weaker than their resetting analogue; in particular, resetting ω-machines can compute all finite iterations of the halting problem for unresetting ω-machines. We fix the following general definitions:

Definition 1.1. Let P be program of any of the machine types described above, and let x be a real. We say that P recognizes x iff $P^x\downarrow = 1$ and $P^y\downarrow = 0$ for all $y \neq x$. We say that x is recognizable by an (un)resetting α-machine iff there is a program P for such a machine that recognizes x.

[1] In the cited paper, these machines are just called infinite time register machines, without further qualification. Later on, when resetting infinite time register machines were introduced, the terminology was changed.

When the machine type is clear from the context, we merely state that x is recognizable.

2 Weak ITRMs

Proposition 2.1. Let x be wITRM-computable. Then x is wITRM-recognizable.

Proof. Let P be a wITRM-program that computes x. The idea is to compare x to the oracle bitwise. A bit of care is necessary to arrange this comparison without overflowing registers. Use a separate counting register R and two flag registers R_1^{flag} and R_2^{flag}. Initially, R and R_1^{flag} contain 0 and R_2^{flag} contains 1. In a computation step, when R contains i, compute the ith bit of x and compare it to the oracle. If these bits disagree, we stop with output 0. Otherwise, we successively set all registers but R_1^{flag} and R_2^{flag} to 0 once and then set the content of R to $i+1$ (after a register has been set to 0, it may be used to store i for this purpose) and swap the contents of R_1^{flag} and R_2^{flag}. In this way, if the number in the oracle is x, then a state will eventually occur in which R_1^{flag} and R_2^{flag} both contain 0 and R contains 0, in which case we output 1. Q.E.D.

Let us denote by wRECOG the set of reals recognizable by a weak ITRM and by wCOMP the set of reals computable by a weak ITRM. The following is the relativized version of [18, Theorem 1]:

Theorem 2.2. Let $x, y \subseteq \omega$. Then x is wITRM-computable in the oracle y iff $x \in \mathbf{L}_{\omega_1^{\text{CK},y}}[y]$. In particular, x is wITRM-computable iff $x \in \mathbf{L}_{\omega_1^{\text{CK}}}$ iff x is hyperarithmetic.

Proof. The proof given in [5] relativizes. We omit the proof to avoid what would amount to a mere repition of that proof. Q.E.D.

Lemma 2.3. Let $x \subseteq \omega$ and let $M \models \mathsf{KP}$ be such that $\omega^M = \omega$ and $x \in M$. Then $\omega_1^{\text{CK},x}$ is an initial segment of On^M.

Proof. If $x \in M$, then, as $M \models \mathsf{KP}$, we have $z \in M$ for every z which is recursive in x.

Now let $x \in M$. We have to show that every $\alpha < \omega_1^{\text{CK},x}$ belongs to the well-founded part of M. Since $M \models \mathsf{KP}$, M satisfies the recursion theorem for Σ_1-definitions. Let $z \subset \omega \times \omega$ be such that (ω, z) is a well-ordering. For all $\beta \in \text{On}$, we define, by Σ_1-recursion, a function F via $F(\beta) = \sup_z\{F(\gamma)+1 \mid \gamma < \beta\}$ if this supremum exists, and otherwise $F(\beta) = \omega$.

We show that $\text{ran}(F) = \omega$: Otherwise, we have $\text{ran}(F) \subsetneq \omega$ and $\text{ran}(F)$ is closed under z-predecessors. Since (ω, z) is a well-ordering in V, $\text{ran}(F)$

must have a z-supremum $n \in \omega$. Hence $\mathrm{ran}(F) = \{m \in \omega \mid (m, n) \in z\} \in M$: By the injectivity of F, F^{-1} is Σ_1-definable. By Σ_1-replacement, $\mathrm{ran}(F^{-1})$ is a set, hence an ordinal γ. Consequently, we have $\omega \subseteq \mathrm{ran}(F)$. We now show that $\mathrm{ran}(F \mid \gamma) = \omega$ for some $\gamma \in \mathrm{On} \cap M$. Suppose that $\omega \notin \mathrm{ran}(F)$. Then F is injective and $F^{-1} : \omega \to \mathrm{On}$ is a function, contradicting Σ_1-replacement (as $\mathrm{On} \cap M$ is not a set in M). Relativizing this argument to x, we obtain the desired result. Q.E.D.

Lemma 2.4. Let P be a wITRM-program, and let $x \subseteq \omega$. Then $P^x\uparrow$ iff there exist $\sigma < \tau < \omega_1^{\mathrm{CK},x}$ such that $Z(\tau) = Z(\sigma)$, $R_i(\tau) = R_i(\sigma)$ for all $i \in \omega$ and $R_i(\gamma) \geq R_i(\sigma)$ for all $i \in \omega$, $\sigma < \gamma < \tau$. (Here, $Z(\gamma)$ and $R_i(\gamma)$ denote the active program line and the content of register i at time γ.)

Proof. This is an easy adaption of [19, Lemma 3]. Q.E.D.

The following lemma allows us to quantify over countable ω-models of KP by quantifying over reals:

Lemma 2.5. There is a Σ^1_1-statement $\varphi(v)$ such that $\varphi(x)$ holds only if x codes an ω-model of KP and such that, for any countable ω-model M of KP, there is a code c for M such that $\varphi(c)$ holds.

Proof. Every countable ω-model M of KP can be coded by a real $c(M)$ in such a way that the $i \in \omega$ is represented by $2i$ in c and ω is represented by 1. We can then consider a set S of statements saying that a real c codes a model of KP together with $\{P_k \mid k \in \omega + 1\}$, where P_k is the statement $\forall i < k(p(2i, 2k) \in c) \wedge \forall j \exists i < k(p(j, 2k) \in c \to j = 2i)$ for $k \in \omega$ and P_ω is the statement $\forall i(p(2i, 1) \in c) \wedge \forall j \exists i(p(j, 1) \in c \to j = 2i)$. Then $\bigwedge S$ is a hyperarithmetic conjunction of arithmetic formulas in the predicate c. But such a conjunction is equivalent to a Σ^1_1-formula.

Q.E.D.

Theorem 2.6. Let x be recognizable by a wITRM. Then $\{x\}$ is a Σ^1_1-singleton.

Proof. Let P be a program that recognizes x on a wITRM. Let $\mathrm{KP}(z)$ be a Σ^1_1-formula (in the predicate z) stating that z codes an ω-model of KP with ω represented by 1 and every integer i represented by $2i$ as constructed in Lemma 2.5. Let $E(y, z)$ be a first-order formula (in the predicates y and z) stating that the structure coded by z contains y. (We can, e.g., take $E(y, z)$ to be $\exists k \forall i(z(i) \leftrightarrow z(p(2i, k))))$.) Furthermore, let $\mathrm{Acc}_P(z, y)$ be a first-order formula (in the predicates y and z) stating that $P^y\downarrow = 1$ in the structure coded by z. Finally, let $\mathrm{NC}_P(y)$ be a first-order formula (in the predicate y) stating that in the computation P^y, there are no two states s_{ι_1}, s_{ι_2} with $\iota_1 < \iota_2$ such that $s_{\iota_1} = s_{\iota_2}$ and, for every $\iota_1 < \iota < \iota_2$, the content $r_{i\iota}$ of register R_i at time ι is at least $r_{i\iota_1}$ (the content of R_i at time ι_1) and the index of the active program at time ι is not smaller than the index of the active program line at time ι_1. (This is just the cycle criterion from Lemma 2.4.)

This is possible in KP models containing x since, by Lemma 2.3 above, $\omega_1^{\mathrm{CK},x}$ is an initial segment of the well-founded part of each such model and, by Lemma 2.4, the computation either cycles before $\omega_1^{\mathrm{CK},x}$ or stops—thus the cycling or halting behaviour takes part in the well-founded part of the model and is hence absolute between such a model and $\mathbf{V}$. Now, take $\varphi(a)$ to be $\exists z(\mathrm{KP}(z) \wedge E(a,z) \wedge \mathrm{Acc}_P(z,a) \wedge \mathrm{NC}_P(a))$. This is a Σ_1^1-formula. We claim that x is the only solution to $\varphi(a)$: To see this, first note that x clearly is a solution, since $\omega_1^{\mathrm{CK},x}$ is an initial segment of every KP-model containing x by Lemma 2.3.

On the other hand, assume that $b \neq x$. In this case, as P recognizes x, we have $P^b\downarrow = 0$ in the real world, and hence, by absoluteness of wITRM-(oracle)-computations for KP-models containing the relevant oracles, also inside $\mathbf{L}_{\omega_1^{\mathrm{CK},b}}[b]$. Now $\mathbf{L}_{\omega_1^{\mathrm{CK},b}}[b]$ is certainly a countable KP-model containing b, hence a counterexample to $\varphi(b)$. Hence $\varphi(b)$ is false if $b \neq x$, as desired.

Q.E.D.

Corollary 2.7. If a real x is wITRM-recognizable, then it is wITRM-computable. Hence, there are no lost melodies for weak ITRMs.

Proof. By Kreisel's basis theorem (cf. [26, p. 75]), if a is not hyperarithmetical and $B \neq \varnothing$ is Σ_1^1, then B contains some element b such that $a \not\leq_{\mathrm{h}} b$. Now suppose that x is wITRM-recognizable. By Theorem 2.6, $\{x\}$ is Σ_1^1 and certainly non-empty. If x was not hyperarithmetical, then, by Kreisel's theorem, $\{x\}$ would contain some b such that $x \not\leq_{\mathrm{h}} b$. But the only element of $\{x\}$ is x, so $x \notin \mathrm{HYP}$ implies $x \not\leq_{\mathrm{h}} x$, which is absurd. Hence $x \in \mathrm{HYP}$. So x is wITRM-computable.

Q.E.D.

3 Resetting ITRMs

ITRM-recognizability was considered in [5, 4, 2]. We give here a summary of some of the most important results. Recall that an ITRM is different from a wITRM in that it, in case of a register overflow, resets the content of the overflowing registers to 0 and continues computing.

The following characterization of the computational strength of ITRMs with real oracles is a relativized version of the main theorem of [19]:

Theorem 3.1. A real x is ITRM-computable in the oracle y iff $x \in \mathbf{L}_{\omega_\omega^{\mathrm{CK},y}}[y]$.

We saw that computability equals recognizability for wITRMs. For ITRMs, the situation is very different. Clearly, analogous to Proposition 2.1, the computable reals are still recognizable. But, for ITRMs, the lost melody phenomenon does occur:

Theorem 3.2. There exists a real x such that x is not ITRM-computable, but ITRM-recognizable.

Proof. The real x can be taken to be a $<_{\mathbf{L}}$-minimal real coding an $\in$-minimal $\mathbf{L}_\alpha$ such that $\mathbf{L}_\alpha \models \mathsf{ZF}^-$. Cf. [5] for the details. Q.E.D.

There are more straightforward examples. In [2], it is shown that, if $(P_i \mid i \in \omega)$ is some natural enumeration of the ITRM-programs, then $h := \{i \in \omega \mid P_i\downarrow\}$, the halting number for ITRMs, is recognizable. The usual argument of course shows that it is not ITRM-computable. In the following, we denote the set of ITRM-recognizable reals by RECOG and write RECOG_n for the set of reals that are recognizable by an ITRM using at most n registers. It was shown in [4] that $\text{RECOG}_n \subsetneq \text{RECOG}$, i.e., the recognizability strength of ITRMs increases with the number of registers. This corresponds to the result established in [19] that the computational strength of ITRMs increases with the number of registers.

Using Shoenfield's absoluteness lemma, it is not hard to see that recognizable reals are always constructible (cf. [4]). We consider the distribution of recognizable reals in the canonical well-ordering $<_{\mathbf{L}}$ of the constructible universe:

Theorem 3.3. There are gaps in the ITRM-recognizable reals, i.e., there are $x, y, z \in \wp(\omega) \cap L$ such that $x <_{\mathbf{L}} y <_{\mathbf{L}} z$, $x, z \in \text{RECOG}$, but $y \notin \text{RECOG}$.

Proof. As there are only countably many ITRM-recognizable real, there must exist a countable α such that $\mathbf{L}_\alpha \models \mathsf{ZF}^-$ and $\mathbf{L}_\alpha$ contains some non-recognizable reals y. Let z be the $<_{\mathbf{L}}$-minimal code of the $\in$-minimal $\mathbf{L}_\alpha$ with these properties and let $x = 0$. Then x, y and z are as desired. The details can be found in [4]. Q.E.D.

This suggest the detailed study of the distribution of the ITRM-recognizable reals among the constructible reals, which was carried out in [4] and [2]. We summarize the main results.

An ordinal α is called Σ_1-fixed iff there exists a Σ_1-formula φ such that α is minimal with $\mathbf{L}_\alpha \models \varphi$. We also define $\sigma := \sup\{\alpha \mid \alpha \text{ is } \Sigma_1\text{-fixed}\}$ to be the supremum of the Σ_1-fixed ordinals. It is easy to see by reflection that the Σ_1-fixed ordinals are countable and that there are countably many of them (as there are only countable many formulas). Hence σ is countable as well. It can also be shown that σ is the supremum of parameter-free OTM-halting times.

Theorem 3.4. If σ is the supremum of the Σ_1-fixed ordinals, then

1. $\text{RECOG} \subseteq \mathbf{L}_\sigma$,
2. the set $\{\alpha \mid \text{RECOG} \cap (\mathbf{L}_{\alpha+1} - \mathbf{L}_\alpha) \neq \varnothing\}$ is cofinal in σ, and
3. for every $\gamma < \sigma$ there is some $\alpha < \sigma$ such that $\text{RECOG} \cap (\mathbf{L}_{\alpha+\gamma} - \mathbf{L}_\alpha) = \varnothing$.

Proof. Cf. [4]. Q.E.D.

The ITRM-computability of a real can be effectively characterized in purely set theoretical terms (namely as being an element of $\mathbf{L}_{\omega_\omega^{\text{CK}}}$). Correspondingly, we have the following necessary criterion for ITRM-recognizability:

Lemma 3.5. Let $x \in \text{RECOG}$. Then $x \in \mathbf{L}_{\omega_\omega^{\text{CK},x}}$. In particular, we have $\omega_\omega^{\text{CK},x} > \omega_\omega^{\text{CK}}$, hence $\omega_i^{\text{CK},x} > \omega_i^{\text{CK}}$ for some $i \in \omega$.

Proof. Cf. [2]. Q.E.D.

Lemma 3.5 in fact allows a machine-independent characterization of recognizability:

Theorem 3.6. Let $x \in \wp^{\mathbf{L}}(\omega)$. Then $x \in \text{RECOG}$ iff x is the unique witness for some Σ_1-formula in $\mathbf{L}_{\omega_\omega^{\text{CK},x}}$.

Proof. Cf. [2]. Q.E.D.

One might now ask where non-recognizable occur; clearly, every real in $\mathbf{L}_{\omega_\omega^{\text{CK}}}$ is recognizable, but what happens above $\omega_\omega^{\text{CK}}$? E.g., is there some $\alpha > \omega_\omega^{\text{CK}}$ such that the reals in $\mathbf{L}_\alpha$ are still all recognizable? It turns out that this is not the case and that, in fact, unrecognizables turn up wherever possible in the $\mathbf{L}$-hierarchy. As usual, we say that $\alpha \in \text{On}$ is an *index* iff $(\mathbf{L}_{\alpha+1} - \mathbf{L}_\alpha) \cap \wp(\omega) \neq \varnothing$.

Theorem 3.7. Let $\alpha \geq \omega_\omega^{\text{CK}}$ be an index. Then there exists a real $x \notin \text{RECOG}$ such that $x \in \mathbf{L}_{\alpha+1} - \mathbf{L}_\alpha$.

Proof. Cf. [2]. Q.E.D.

In light of Lemma 3.5, it is natural to concentrate the study of recognizability on reals x with $x \in \mathbf{L}_{\omega_\omega^{\text{CK},x}}$. It turns out that the distribution of recognizables becomes much tamer when we do this:

Theorem 3.8. (The 'All-Or-Nothing-Theorem') Let γ be an index. Then either all $x \in \mathbf{L}_{\gamma+1} - \mathbf{L}_\gamma$ with $x \in \mathbf{L}_{\omega_\omega^{\text{CK},x}}$ are recognizable or none of them is.

Proof. Cf. [2]. The idea is that, given a recognizable $a \in \mathbf{L}_{\gamma+1} - \mathbf{L}_\gamma$, this can be used to identify the $<_{\mathbf{L}}$-minimal code c of $\mathbf{L}_{\gamma+1}$, which can in turn be used to identify every real in $\mathbf{L}_{\gamma+1}$. Q.E.D.

4 Ordinal Machines

Ordinal Turing machines (OTMs) and ordinal register machines (ORMs) were introduced in [17] and [22], respectively, and seem to provide an upper bound on the strength of a reasonable transfinite model of computation (cf., e.g., [3] for an argument in favor of this claim). In the papers just cited, Koepke proves that, when finite sets of ordinals are allowed as parameters, these machines can compute the characteristic function of a set x of ordinals iff $x \in \mathbf{L}$. In particular a real x is computable by such a machine iff $x \in \mathbf{L}$. We formulate our results from now on for OTMs only, as they carry over verbatim to ORMs. To clarify the role of the parameters, we give a separate definition for recognizability by OTMs with parameters: we say that a real $x \subseteq \omega$ is *parameter-OTM-recognizable* iff for some OTM-program P with a finite sequence $\vec{\gamma}$ of ordinal parameters and every $y \subseteq \omega$, $P^y\downarrow = 1$ iff $x = y$ and otherwise $P^y\downarrow = 0$. Trivially, there are no lost melodies for parameter-OTMs in $\mathbf{L}$. If $\omega_1^{\mathbf{L}}$ is not collapsed, this continues to hold:

Theorem 4.1. Assume that $\omega_1^{\mathbf{L}} = \omega_1$. Let $y \subseteq \omega$ be parameter-OTM-recognizable. Then $y \in \mathbf{L}$. Consequently, each such y is OTM-computable (with parameters).

Proof. Let P and $\vec{\gamma} \in \mathrm{On}^{<\omega}$ be such that $P^x(\vec{\gamma})\downarrow = 1$ iff $x = y$ and $P^x(\vec{\gamma})\downarrow = 0$, otherwise. Coding the finite sequence $\vec{\gamma}$ as a single ordinal, we may assume without loss of generality that $\vec{\gamma}$ consists of a single ordinal γ. Observe that the computation c of $P^y(\gamma)$ is an element of $\mathbf{L}[y]$. Let $\alpha > \gamma$ be both a limit of admissible ordinals and of index ordinals large enough such that $c \in \mathbf{L}_\alpha[y]$. Then c is Σ_1-definable over $\mathbf{L}_\alpha[x]$ in the parameters x and γ and $\mathbf{L}_\alpha[x]$ believes that c is the computation of $P^x(\gamma)$. Let H be the Σ_1-Skolem hull of $\{\{c\} \cup \{y\} \cup \{\gamma\}\}$ in $\mathbf{L}_\alpha[y]$, and let $\pi : H \to_{coll} M$ be the transitive collapse of H. Then H is countable in $\mathbf{L}[y]$, and hence so is M. As $y \subseteq \omega$, condensation holds in the $\mathbf{L}[y]$-hierarchy, so $M = \mathbf{L}_{\bar{\alpha}}[y]$ for some $\bar{\alpha} \leq \alpha$, where $\bar{\alpha}$ is a limit of admissibles and a limit of indices. Furthermore, $\mathbf{L}_{\bar{\alpha}}[y]$ believes that $\pi(c)$ is the computation of $\pi(P)^{\pi(y)}(\pi(\gamma))$ and that $\pi(P)^{\pi(y)}(\pi(\gamma))\downarrow = 1$. But as $y \subseteq \omega$ and P is just (coded by) a natural number, we have $\pi(P) = P$ and $\pi(y) = y$.

Thus $\mathbf{L}_{\bar{\alpha}}[y] \models P^y(\pi(\gamma))\downarrow = 1$. By absoluteness of computations, it follows that indeed $P^y(\pi(\gamma))\downarrow = 1$. Moreover, by Σ_1-absoluteness, y is the unique element $z \in \mathbf{L}_{\bar{\alpha}}[y]$ such that $\mathbf{L}_{\bar{\alpha}}[y] \models P^z(\pi(\gamma))\downarrow = 1$.

But as $\gamma' := \pi(\gamma) \in \mathbf{L}_{\bar{\alpha}}[y]$ and the latter is countable in $\mathbf{L}[y]$ and transitive, γ' is countable $\mathbf{L}[y]$. As $\omega_1^{\mathbf{L}} = \omega_1$, γ' is also countable in $\mathbf{L}$.

Consequently, there is a $<_{\mathbf{L}}$-minimal constructible real z that codes γ'. The statement $\psi := \exists x P^x(\gamma')\downarrow = 1$ is now a Σ_1-statement in the parameter z. By a theorem of Jensen and Karp (cf. [14, § 5]), this statement is absolute between $\mathbf{L}_\zeta$ and $\mathbf{V}_\zeta$ whenever ζ is a limit of admissibles and $\mathbf{L}_\zeta$ contains the relevant parameters. As $\bar{\alpha}$ is a limit of indices and $\gamma' < \bar{\alpha}$, we have $z \in \mathbf{L}_{\bar{\alpha}}$.

So ψ is absolute between $\mathbf{L}_{\bar{\alpha}}[y]$ and $\mathbf{L}_{\bar{\alpha}}$, hence $\mathbf{L}_{\bar{\alpha}} \models \exists x P^x(\gamma')\downarrow = 1$, so there is $x \in \mathbf{L}_{\bar{\alpha}}$ such that $\mathbf{L}_{\bar{\alpha}} \models P^x(\gamma')\downarrow = 1$.

By absoluteness of computations between transitive structures again, it follows that $P^x(\gamma')\downarrow = 1$ also holds in $\mathbf{L}_{\bar{\alpha}}[y]$. But as we saw above, y is the unique element of $\mathbf{L}_{\bar{\alpha}}[y]$ with this property. Thus we must have $x = y$. Hence $y \in \mathbf{L}_{\bar{\alpha}} \subseteq L$.

Q.E.D.

By a very similar argument, we can also see that all reals recognizable with parameters that are countable in $\mathbf{L}$ are parameter-OTM-computable. On the other hand, if the universe is much unlike $\mathbf{L}$, lost melodies for parameter-OTMs can occur:

Theorem 4.2. Assume that $0^\#$ exists. Then there is a lost melody for parameter-OTMs. In fact, $0^\#$ is parameter-OTM-recognizable in the parameter ω_1.

Proof. By [15, Theorem 14.11], the relation $x = 0^\#$ is Π^1_2, so $x \neq 0^\#$ is Σ^1_2. Furthermore, Σ^1_2-relations are absolute between transitive models of KP containing ω_1 (cf., e.g., [27, Corollary 1]). Now, let $\alpha > \omega_1$ be minimal such that $M := \mathbf{L}_\alpha[0^\#] \models \mathsf{KP}$. Then $\mathbf{L}[0^\#]$ contains a bijection $f : \omega_1 \leftrightarrow M$. Hence, M is coded by $r := \{p(\iota_1, \iota_2) \mid \iota_1, \iota_2 < \omega_1 \wedge f(\iota_1) \in f(\iota_2)\} \in \mathbf{L}[\mathbf{0}^\#]$. To recognize $0^\#$ with an OTM when ω_1 is given as a parameter, we proceed as follows: Given a real x in the oracle, search through the subsets of ω_1 in $\mathbf{L}[x]$ (by a similar procedure used in the proof of Theorem 4.4) for a set c coding a KP-model M' of the form $\mathbf{L}_\beta[x]$ that contains ω_1. As such sets exist in $\mathbf{L}[x]$, such a c will eventually be found. Once this has happened, check, using c, whether $M' \models x = 0^\#$. If not, then, by absoluteness, $x \neq 0^\#$, otherwise $x = 0^\#$. Q.E.D.

Taken together, the last two theorems readily yield:

Corollary 4.3. If $0^\#$ exists, then it is undecidable in ZFC whether there are lost melodies for parameter-OTMs.

We do not know if the large cardinal assumption can be weakened or dropped, i.e., if we can prove the relative consistency of the existence of lost melodies for parameter-OTM in ZFC. From now on, when we talk about OTMs, we always mean the parameter-free case. What happens if we consider OTMs without ordinal parameters? It turns out that then, there are no lost melodies:

Theorem 4.4. Let $x \subseteq \omega$ and P be an OTM-program such that, for each $y \subseteq \omega$, we have $P^y\downarrow = 1$ iff $y = x$ and $P^y\downarrow = 0$, otherwise. Then x is OTM-computable (without parameters).

Proof. In [16], it is shown that every constructible set of ordinals is uniformly computable from an appropriate finite set of ordinal parameters. Hence, there is a program Q which, for every input $\vec{\alpha}$, a finite sequence of ordinals, computes the characteristic function of a set x of ordinals in a such a way that for every constructible $x \subseteq \mathrm{On}$, there exists $\vec{\gamma}_x$ such that Q computes the characteristic function of x on input $\vec{\gamma}_x$. We shall use Q to search through the constructible reals, looking for some $x \subseteq \mathrm{On}$ such that $P^{x\cap\omega}\downarrow = 1$. To do this, we use some natural enumeration $(\vec{\gamma}_\iota \mid \iota \in \mathrm{On})$ of finite sequences of ordinals and carry out the following procedure for each $\iota \in \mathrm{On}$. First, find $\vec{\gamma}_\iota$, and let x_ι be the set of ordinals whose characteristic function is computed by Q on input $\vec{\gamma}_\iota$. Then check, using P, whether $P^{x_\iota\cap\omega}\downarrow = 1$. As $P^y\downarrow$ for all $y \subseteq \omega$, this will eventually be determined. If $P^{x_\iota\cap\omega}\downarrow = 1$, then x is found and we can write it on the tape. Otherwise, continue with $\iota + 1$. In this way, every constructible real will eventually be checked. By Lemma 4.1, x must be constructible, hence x will at some point be considered, identified and written on the tape. Thus x is computable. Q.E.D.

In fact, by almost the same reasoning, a much weaker assumption on x is sufficient:

Corollary 4.5. Let $x \subseteq \omega$ and P be an OTM-program such that, for each $y \subseteq \omega$, we have $P^y\downarrow$ iff $y = x$ and $P^y\uparrow$, otherwise. Then x is OTM-computable (without parameters).

Proof. First, observe that, by Shoenfield absoluteness, such an x must be an element of $\mathbf{L}$. Now, we use a slight modification of the proof of Theorem 4.4: Again, we use a program Q to successively write all constructible sets of naturals to the tape. But now, we let P run simultaneously on all the written reals. At some point, x will be written to the tape and at some later point, P will halt on it. When that happens, just copy the real on which P halted to the beginning of the tape, thus writing x. This can then be used to decide every bit of x. Q.E.D.

An easy reflection argument shows that a halting OTM- (and ORM-)computation with a real oracle always has a countable running time. Our results above hence in fact hold for unresetting ω_1-machines as well.

In the parameter-free case, this shows that, for extremely strong models of computation, the lost melody phenomenon is no longer present. This motivates a further inspection what exactly is necessary for the existence of lost melodies.

5 α-register machines

Recall that, for $\alpha \in \mathrm{On}$, let a resetting/unresetting α-register machine works like an ITRM or wITRM with the difference that a register may now contain an arbitrary ordinal $< \alpha$. Hence, an ITRM is a resetting ω-register machine and a wITRM is an unresetting ω-register machine. This generalization was suggested at the end of [18].

We denote by wCOMP_α, COMP_α, wRECOG_α and RECOG_α the set of reals computable by an unresetting α-register machine, computable by a resetting α-register machine, recognizable by an unresetting α-register machine and recognizable by a resetting α-register machine, respectively.

We have seen that $\mathrm{wCOMP}_\omega = \mathrm{wRECOG}_\omega$, $\mathrm{COMP}_\omega \subsetneq \mathrm{RECOG}_\omega$, and that lost melodies for unresetting machines vanish when the register contents are unbounded. Hence, we ask:

> For which α are there lost melodies for resetting/unresetting α-register machines?

We start with the following easy observation:

Lemma 5.1. Let α be an ordinal.

(1) For infinite α, $\mathrm{wCOMP}_\alpha \subseteq \mathrm{wRECOG}_\alpha$ and $\mathrm{COMP}_\alpha \subseteq \mathrm{RECOG}_\alpha$.

(2) For arbitrary α, $\mathrm{wCOMP}_\alpha \subseteq \mathrm{COMP}_\alpha$ and $\mathrm{wRECOG}_\alpha \subseteq \mathrm{RECOG}_\alpha$.

Proof. (1) As $\alpha \geq \omega$, we can again compute a real x and compare it to the oracle bitwise.

(2) A terminating computation by an unresetting α-machine will run exactly the same on a resetting α-machine. Q.E.D.

Lemma 5.2. Let $\alpha > \beta$ be ordinals. Assume that there is an unresetting α-program P such that $P(b)\downarrow = 1$ iff $b = \beta$ and $P(b)\downarrow = 0$, otherwise. Then $\mathrm{COMP}_\beta \subseteq \mathrm{wCOMP}_\alpha$.

Proof. Given α, β and P as in the assumptions, let $y \in \mathrm{COMP}_\beta$, and let Q be a resetting β-program computing y. To compute y on an unresetting α-machine, we describe how to simulate Q on such a machine. Assume that Q uses k registers. Reserve k registers $R_1^Q, \ldots, R_k^Q$ of the unresetting α-machine. Then, we proceed as follows: At successor steps, simply carry out Q on $R_1^Q, \ldots, R_k^Q$. At limit steps of the Q-computation, check, using

P, whether any of these registers contains β. If so, reset these register contents to 0 and proceed, otherwise proceed without any modifications. This simulates Q on an unresetting α-machine.

To recognize limit steps in the computation of Q, reserve two extra registers, R_1 and R_2; initially, let R_1 contain 1 and R_2 contain 0. Whenever a step of Q is carried out, swap their contents. Whenever their contents are equal, set R_1 again to 1 and R_2 to 0. In this way, the contents of R_1 and R_2 will be equal iff the Q-computation has just reached a limit stage. Q.E.D.

5.1 The unresetting case

Lemma 5.3. Let $\alpha < \beta$ be ordinals. Then $\text{wCOMP}_\alpha \subseteq \text{wCOMP}_\beta \subseteq \text{wCOMP}_{\omega_1}$ and $\text{wRECOG}_\alpha \subseteq \text{wRECOG}_\beta \subseteq \text{wRECOG}_{\omega_1}$.

Proof. If $\alpha < \beta$, then terminating unresetting α-computations work exactly the same on unresetting β-machines. Q.E.D.

We have seen above that $\text{wCOMP}_\omega = \text{wRECOG}_\omega$. We shall see now that that this happen again for ω_1 and in fact for all but countably many countable ordinals α.

Lemma 5.4. $\text{wCOMP}_{\omega_1} = \text{wRECOG}_{\omega_1}$.

Proof. This follows from Theorem 4.4, as wCOMP_{ω_1} and wRECOG_{ω_1} are just the set of ORM-computable and ORM-recognizable reals (without ordinal parameters), respectively. Q.E.D.

Theorem 5.5. There is $\beta < \omega_1$ such that there are no lost melodies for unresetting γ-machines whenever $\gamma \geq \beta$.

Proof. Let β be large enough such that $\text{wCOMP}_\beta = \text{wCOMP}_{\omega_1}$ and $\text{wRECOG}_\beta = \text{wRECOG}_{\omega_1}$. (This is possible by monotonicity and the fact that there are only countably many programs.) Then, for all $\gamma \geq \beta$, we have $\text{wCOMP}_\gamma = \text{wCOMP}_{\omega_1} = \text{wRECOG}_{\omega_1} = \text{wRECOG}_\gamma$ by Lemma 5.4. Q.E.D.

Our next goal is to show that there are ordinals α for which $\text{wCOMP}_\alpha \subsetneq \text{wRECOG}_\alpha$, i.e., for which the lost melody phenomenon does occur:

Lemma 5.6. There is an unresetting $\omega+1$-program P such that $P(x)\downarrow = 1$ iff $x = \omega$ and $P(x)\downarrow = 0$, otherwise.

Proof. Let R_1 be the register containing x. Use a register R_2 to successively count upwards from 0. Use a flag to check whether the machine is in a limit state. Eventually, the content of R_1 is reached. If this happens in a limit step, then R_1 contains ω, otherwise, it does not. Q.E.D.

Lemma 5.7. $\text{wCOMP}_{\omega+1} = \text{COMP}_\omega$ and $\text{wRECOG}_{\omega+1} = \text{RECOG}_\omega$, i.e., unresetting $\omega + 1$-machines are equivalent in computational and recognizability strength to ITRMs.

Proof Sketch. One direction follows from Lemma 5.6 and Lemma 5.2.

For the other direction, we show that a resetting ω-machine (i.e., an ITRM) can simulate an unresetting $(\omega+1)$-machine. To see this, proceed as follows: Let P be a program for an unresetting $\omega+1$-machine. Assume that P uses k registers $R'_1, ..., R'_k$. We set up an ITRM-program in the following way: Reserve $R_1, ..., R_k$ for the simulation of P. In the simulation, let 0 represent ω and let $i+1$ represent i for all $i \in \omega \setminus \{0\}$. Whenever P requires that the content of R'_i is set to the value 0, set R_i to 1. When P requires that the content of R'_i is increased by 1 and this content is currently 0, stop. Otherwise, run P on $R_1, ..., R_k$ in the usual way. Q.E.D.

Theorem 5.8. $\text{wCOMP}_{\omega+1} \neq \text{wRECOG}_{\omega+1}$, i.e., there are lost melodies for unresetting $\omega + 1$-machines.

Proof. This follows immediately from Lemma 5.7, since, by the lost melody theorem for ITRMs, we have $\text{COMP}_\omega \neq \text{RECOG}_\omega$. Q.E.D.

Arguments similar to the proof of Lemma 5.7 show that a resetting ω-machine can in fact simulate an unresetting $(\omega + i)$-machine for all $i \in \omega$ (and more). On the other hand, it can be shown that this is no longer the case for unresetting α-machines when $\alpha > \omega_\omega^{\text{CK}}$ is exponentially closed: Coding $x \in \mathbf{L}_\alpha$, $x = \{y \in \mathbf{L}_\beta \mid \mathbf{L}_\beta \models \varphi_n(y, \vec{\gamma})\}$ (where $\vec{\beta}$ is a finite sequence of ordinals and $\beta < \alpha$) by $(\alpha, n, \vec{\gamma})$ and using techniques similar to those developed in [23], we can evaluate arbitrary statements about the coded elements inside $\mathbf{L}_\alpha$ with an unresetting α-machine. This allows us to search through $\wp(\omega) \cap \mathbf{L}_\alpha$ with such a machine. As in the proof of Theorem 4.4, one can now conclude that all reals in $\mathbf{L}_\alpha$ recognizable by an unresetting α-machine are already computable by such a machine. We also saw that $\text{RECOG}_\omega \subseteq \text{wRECOG}_\alpha$ for $\alpha > \omega$. Now, the minimal real code $c := \text{cc}(\mathbf{L}_{\omega_\omega^{\text{CK}}})$ of $\mathbf{L}_{\omega_\omega^{\text{CK}}}$ is an element of $\mathbf{L}_{\omega_\omega^{\text{CK}}+2}$, and hence of $\mathbf{L}_\alpha$. The real c is easily seen to be ITRM-recognizable, but as $c \notin \mathbf{L}_{\omega_\omega^{\text{CK}}}$, it is not ITRM-computable. But $c \in \text{RECOG}_\omega \cap \mathbf{L}_\alpha \subseteq \text{wRECOG}_\alpha \cap \mathbf{L}_\alpha \subseteq \text{wCOMP}_\alpha$. So $c \in \text{wCOMP}_\alpha - \text{COMP}_\omega$.

We saw that $\text{wCOMP}_{\omega+1} \subsetneq \text{wRECOG}_{\omega+1}$ and there is a countable β such that $\text{wCOMP}_\gamma = \text{wRECOG}_\gamma$ when $\gamma > \beta$. We do not know if there are gaps in the ordinals for which lost melodies exist, i.e., if there are $\omega + 1 < \gamma < \delta$ such that $\text{wCOMP}_\gamma = \text{wRECOG}_\gamma$, but $\text{wCOMP}_\delta \subsetneq \text{wRECOG}_\delta$.

5.2 The resetting case

Note first that the computational strength for various values of α much higher than in the unresetting case:

Theorem 5.9. Let P_i be some natural enumeration of the ORM-programs. There is $\alpha < \omega_1$ such that some resetting α-machine can solve the halting problem for parameter-free ORMs (i.e., unresetting ω_1-machines), i.e., there is an unresetting α-program Q such that $Q(i)\downarrow = 1$ iff $P_i(0)$ stops and $Q(i)\downarrow = 0$ iff $P_i(0)$ diverges.

Proof. Let α_1 be the supremum of the register contents occurring in any halting parameter-free ORM-computation, let α_2 be the supremum of the parameter-free ORM-halting times and let $\alpha := \max\{\alpha_1, \alpha_2\}$ (of course, as all registers are initially 0 and a register content can be increased at most by 1 in one step, we shall have $\alpha_1 \leq \alpha_2$; it is not hard to see that in fact $\alpha_1 = \alpha_2$).

Now consider ORM-programs with a fixed number n of registers. Then a resetting α-machine can solve the halting problem for such programs by simply simulating the given program P in the registers $R_1, ..., R_n$, while using a further register R_{n+1} as a clock by increasing its content by 1 whenever a step in the simulation is carried out. If any of the registers $R_1, ..., R_n, R_{n+1}$ overflows, then P does not halt and we output 0; otherwise, the simulation reaches the halting state and we output 1.

A register overflow can be detected as follows: If a register R has overflown, then the machine must be in a limit state (which can be detected by flags in the usual way) and R must contain 0. In this situation, either there has been an overflow or the prior content of R has been 0 cofinally often in the current running time. This can be distinguished by an extra register R' whose content is set to 0 whenever R contains 0 and to 1, otherwise. Hence, if R' contains 0 in a limit state, then the content of R must have been 0 cofinally often.

Now, by [23], there is a universal ORM, so we have an effective method how to find, for every ORM-program P, an ORM-program with the same halting behaviour, but using only 12 registers. This, in combination with the halting problem solver for programs with any fixed number of registers, solves the halting problem for ORMs. Q.E.D.

The same holds when one considers the recognizability strength. To show this, we need some preliminaries.

Lemma 5.10. There is an ITRM-program R such that, for each real x coding an ordinal $\alpha < \omega_1$ according to $f : \alpha \rightarrow \omega$ injective, R^x changes the content of the register R_1 exactly $\alpha + 1$ many times.

Proof. By [19, Lemma 2], the program P defined there to test the oracle for well-foundedness of the coded relation runs for at least β many steps when the oracle codes a well-ordering of length β. Roughly, P uses a stack to represent a finite descending sequence and attempts to continue it. We reserve a separate register R_1 and flip its content between 0 and 1 whenever a new element is put on the stack in the computation of P^x. The argument for [19, Lemma 2] shows that the content of R_1 will be changed at least α many times. If this happens exactly α many times, we simply set up our program to flip the content of R_1 once more after P has stopped. If it happens more than α many times, then some finite sequence $\vec{s}$ of natural numbers is the $\alpha + 1$th sequence that is put on the stack and we set up our program to stop once $\vec{s}$ has appeared on the stack. Q.E.D.

Corollary 5.11. Let $\alpha < \omega_1$. There is a resetting α-program I which, given a real x coding an ordinal γ, halts with output 1 iff $\gamma = \alpha$ and otherwise halts with output 0.

Proof. As α-register machines can simulate ITRMs, we can use Lemma 5.10 to obtain a program R that (run in the oracle x) changes the content of register R_1 exactly $\alpha+1$ many times. We use a separate register R_2 that starts with content 0 and whose content is incremented by 1 whenever the content of R_1 is changed. Now, if R_2 overflows and the content of R_1 is changed afterwards without R halting, then $\alpha < \gamma$. If, on the other hand, R stops without R_2 having overflown, we have $\gamma < \alpha$. If neither happens, i.e., if R_2 overflows and the next change of the content of R_1 is followed by R halting, then $\alpha = \gamma$. These scenarios are easy to detect. Q.E.D.

Theorem 5.12. There exists $\alpha < \omega_1$ and $x \subseteq \omega$ such that $x \in \text{RECOG}_\alpha$, but $x \notin \text{wRECOG}_{\omega_1}$.

Proof. Let τ be the supremum of stages containing new ORM-recognizables. Let $\alpha + 1 > \tau$ be an index such that $\mathbf{L}_\alpha \models \mathsf{ZF}^-$ and let $r := \text{cc}(\mathbf{L}_\alpha)$ be the $<_\mathbf{L}$-minimal real coding $\mathbf{L}_\alpha$. It is well known that this implies $\text{cc}(\mathbf{L}_\alpha) \in \mathbf{L}_{\alpha+2}$ (cf., e.g., [1]). Then r is recognizable by a resetting α-machine. To see this, first note that the property of being the minimal code of an index $\mathbf{L}$-stage can be checked by an ITRM using the strategy described in the proof of the lost melody theorem for ITRMs in [5]. We saw above that resetting α-machines can simulate ITRMs for all $\alpha \geq \omega$, hence this can be carried out by a resetting $\alpha + 1$-machine. It only remains to test whether the coded stage $\mathbf{L}_\zeta$ is indeed $\mathbf{L}_\alpha$. This can be done by using Corollary 5.11 to test whether the order type of $\text{On} \cap \mathbf{L}_\zeta$ is α. Q.E.D.

Theorem 5.13. Let $\alpha < \omega_1$, and let $\delta > \alpha$ be such that δ is a limit of indices, but not itself an index. Then any α-machine computation (with empty input and oracle) either halts in less than δ many steps or does not halt at all.

Proof. This is an adaption of the argument given in [19] for ITRMs. As δ is a limit of indices, but not an index, it follows (cf., e.g., [6, 25, 24]) that $\mathbf{L}_\delta \models \mathsf{ZF}^- +$"for all x there is a surjective $f : \omega \to x$" and hence that (cf. [13]) $\varrho_\omega^\delta = \delta$, where ϱ_ω^α denotes the ultimate projectum of $\mathbf{L}_\alpha$. We claim that there is no $f : \xi \to \delta$ with unbounded range and $\xi < \delta$ definable over $\mathbf{L}_\delta$. To see this, assume that there is such an f. By assumption, there is, for every $\beta < \delta$ an index between β and δ and hence $\mathbf{L}_\delta$ contains a $<_\mathbf{L}$-minimal bijection g_β between ω and β. Define a surjective map $\bar{f} : \xi \times \omega \to \delta$ via $\bar{f}(\iota, j) = g_{f(\iota)}(j)$. Let h be a bijection between $\xi \times \omega$ and $\xi\omega$ and define a surjective $\tilde{f} : \xi\omega \to \delta$ by $\tilde{f} := \hat{f} \circ h^{-1}$. As $\xi < \delta$ and $\mathbf{L}_\delta \models \mathsf{ZF}^-$, we also have $\xi\omega < \delta$, and $\tilde{f}$ is certainly definable over $\mathbf{L}_\delta$. Hence a surjection from some $\zeta < \delta$ onto δ (and hence onto $\mathbf{L}_\delta$ is definable over $\mathbf{L}_\delta$, so that $\varrho_\omega^\delta < \delta = \varrho_\omega^\delta$, a contradiction.

Now, there is a natural injection from the states of an α-machine into α^ω, as the state can be given by a finite tuple of ordinals $< \alpha$ representing the register contents and a single natural number representing the active program line. Such a map j is definable over $\mathbf{L}_{\alpha^\omega}$ and hence certainly an element of $\mathbf{L}_\delta$.

Now let P be an α-program, and let C be the computation of P, restricted to the first δ many steps. For a machine state s, let γ_s denote $\sup\{\beta < \delta \mid C(\beta) = s\}$.

Assume first that $\{\beta < \delta \mid \gamma_{C(\beta)} < \delta\}$ is cofinal in δ, i.e., there are cofinally many states that appear only on boundedly many times. Then we can define, over $\mathbf{L}_\delta$, a partial map $f : \alpha\omega \to \delta$ by letting $f(\xi) = \gamma_{j^{-1}(\xi)}$ if $j^{-1}(\xi)$ is defined and $\gamma_{f^{-1}(\xi)} < \delta$ and otherwise $f(\xi) = 0$. By assumption, f has unbounded range in δ, which contradicts our observation above.

Hence, we may assume that there is some $\gamma < \delta$ such that every machine state assumed after time γ appears at cofinally in δ many times. Suppose that P uses $n \in \omega$ many registers. The possible machine states are hence elements of $\prod_{i=1}^n \alpha \times \omega$. Let us partially order the set S of machine states occurring in the computation after time γ by letting $(\beta_1, ..., \beta_n, k) \leq_\mathrm{s} (\gamma_1, ..., \gamma_n, l)$ iff $k \leq l$ and $\beta_i \leq \gamma_i$ for all $i \in \{1, ..., n\}$. It is easy to see that $\leq_\mathrm{s}$ is well-founded.

For each two states $Z_1, Z_2 \in S$, there is $Z_3 \in S$ such that $Z_1 \leq_\mathrm{s} Z_3$ and $Z_1 \leq_\mathrm{s} Z_3$: To see this, observe that we can define over $\mathbf{L}_\delta$ a strictly increasing map $\sigma : \omega \to \delta$ such that $C(\sigma(2i)) = Z_1$ and $C(\sigma(2i+1)) = Z_2$ for all $i \in \omega$. By our observation above, $\mathrm{ran}(\sigma)$ must be bounded in δ, so let $\bar{\delta} := \sup(\mathrm{ran}(\sigma))$. Then $C(\bar{\delta})$ is as desired.

Now, by well-foundedness of $\leq_\mathrm{s}$, S must contain a minimal element Z. It is easy to see that Z is in fact unique: For if Z_1 and Z_2 were two distinct minimal elements of S, then by our last observation, we would have $Z_3 \in S$ with $Z_3 \leq Z_1$ and $Z_3 \leq Z_2$. As $Z_1 \neq Z_2$, one of the inequalities would have to be strict, contradicting the minimality of Z_1 and Z_2.

Hence Z is assumed cofinally in δ many times, while all other states occurring after time γ are $\geq_s Z$. Consequently, the machine state at time δ is again Z and it is easy to see that the computation cycles. Hence, a resetting α-machine computation either halts before time δ or does not halt at all.

Q.E.D.

Corollary 5.14. $\mathrm{COMP}_\alpha \subseteq \mathbf{L}_\delta$, where δ is the minimal limit of indices above α which is not itself an index.

Proof. Since δ is not an index, every subset of ω definable over $\mathbf{L}_\delta$ is an element of $\mathbf{L}_\delta$. Now let $x \in \mathrm{COMP}_\alpha$, and let P be a resetting α-program that computes x, i.e., $P(i)\downarrow = 1$ if $i \in x$ and $P(i)\downarrow = 0$ if $i \notin x$ for all $i \in \omega$. By Theorem 5.13 and as $P(i)\downarrow$ for all $i \in \omega$, the halting time of $P(i)$ must be smaller than δ for all $i \in \omega$. Hence $i \in x$ is expressed over $\mathbf{L}_\delta$ by an $\in$-formula stating the existence of a halting P-computation with input i and output 1. Consequently, we must have $x \in \mathbf{L}_\delta$. Q.E.D.

This allows us to show that there are lost melodies for resetting α-machines for all infinite $\alpha < \omega_1$:

Theorem 5.15. Let $\alpha < \omega_1$ be infinite. Then there $\mathrm{COMP}_\alpha \neq \mathrm{RECOG}_\alpha$, i.e., there is a lost melody for resetting α-machines.

Proof. Given $\alpha < \omega_1$, let r_α be the $<_{\mathbf{L}}$-minimal real coding an $\mathbf{L}$-level $\mathbf{L}_\gamma$ such that γ is a limit of indices but not itself an index, $\gamma + 1$ is an index and $\mathbf{L}_\gamma$ contains a real coding α. Then we must also have $r_\alpha \notin \mathrm{COMP}_\alpha$ by Corollary 5.14. We show that $r_\alpha \in \mathrm{RECOG}_\alpha$ by an argument similar to the proof of the lost melody theorem for ITRMs. Let x be given in the oracle. First, we can—even with an ITRM—check whether x codes an $\mathbf{L}$-level $\mathbf{L}_\zeta$ with cofinally many indices. If not, $x \neq r_\alpha$. If so, the methods developed in the proof of the lost melody theorem for ITRMs allow us to compute from x the truth predicate for $\mathbf{L}_{\zeta+2}$, which allows us to check whether ζ and $\zeta+1$ are indices. If ζ is an index or $\zeta+1$ is not, then $x \neq r_\alpha$. Otherwise, we need to check whether $\zeta > \alpha$ (this suffices to guarantee the existence of a real coding α, since at this point we already know that ζ is a limit of indices). This can be done as follows: Inside r_α, α must be coded by some natural number i that can be given to our program in advance. So we test whether i codes an ordinal ϑ in x. If not, then $x \neq r_\alpha$. Now, we can easily compute from i and x a real y coding the order type ϑ (just delete every $p(k, j) \in x$

with $\{p(k,i), p(j,i)\} \not\subseteq x$) and then use Corollary 5.11 to check whether y codes α. If not, then $x \neq r_\alpha$. Otherwise, we know that i codes α inside r_α.

Next, we check whether there is any $\alpha < \zeta' < \zeta$ with the same properties. If yes, then $x \neq r_\alpha$. Otherwise, we know that x codes $\mathbf{L}_\gamma$ and it remains to check the $<_\mathbf{L}$-minimality of x. As $\mathbf{L}_{\zeta+1}$ is an index, we know that the minimal real coding $\mathbf{L}_\zeta$ must be an element of $\mathbf{L}_{\zeta+2}$. As we just mentioned, we can, given x, evaluate the truth predicate for $\mathbf{L}_{\zeta+2}$. Hence, we can search through (the code of) $\mathbf{L}_{\zeta+2}$ until we find the $<_\mathbf{L}$-minimal real coding $\mathbf{L}_\zeta$ and compare it with x. If these reals disagree, then $x \neq r_\alpha$, otherwise $x = r_\alpha$. So r_α is recognizable.

This proves that r_α is a lost melody for resetting α-machines.

It remains to see that such an $\mathbf{L}$-level $\mathbf{L}_\gamma$ exists. To see this, let $\gamma > \alpha$ be a a minimal limit of indices, and let $\alpha < \delta < \gamma$ be an index. Let x be a real such that $x \in \mathbf{L}_{\delta+1} - \mathbf{L}_\delta$. Then the elementary hull H of $\{x\}$ in $\mathbf{L}_\gamma$ is (isomorphic to) an $\mathbf{L}$-level $\mathbf{L}_\beta$ with cofinally many indices which contains x, where $\beta \leq \gamma$. It follows that $\beta = \gamma$ and that in fact $H = \mathbf{L}_\gamma$. This hull is definable over $\mathbf{L}_{\gamma+1}$, so that we get a bijection between ω and $\mathbf{L}_\gamma$ in $\mathbf{L}_{\gamma+2}$ by the standard finestructural arguments. Hence $\gamma + 1$ is indeed an index, so γ is as desired. Q.E.D.

Note that, by Shoenfield absoluteness, we have $\wp^\mathbf{L}(\omega) = \mathrm{wCOMP}_{\omega_1} \subseteq \mathrm{COMP}_{\omega_1} \subseteq \mathrm{RECOG}_{\omega_1} \subseteq \wp^\mathbf{L}(\omega)$, so that there can be no lost melodies for $\alpha \geq \omega_1$.

6 Conclusion and further work

We have seen that lost melodies exist for a resetting α-machines iff $\alpha < \omega_1$ is infinite and that for unresetting α-machines, lost melodies do not exist for $\alpha = \omega$, do exist for $\alpha = \omega + 1$ and cease to exist from some countable ordinal on. In the special case of resetting ω-machines or ITRMs, the recognizables allow for a detailed analysis among the constructible reals and show several surprising regularities. In the parameter-OTM-case, we reach the limits of ZFC. In general, the relation between the computability and recognizability strength of transfinite models of computation seems to be far from trivial.

In this paper, we have restricted our attention to reals, as these can be dealt with by all models in question and can hence be used as a basis for comparison. One could equally well consider subsets of other ordinals, which might be more appropriate for some models.

Once we do this, interesting questions arise, even for classical Turing machines: Consider, for example, Turing programs using at most n states and symbols for some $n \in \mathbb{N}$. Let us say that a natural number k is n-computable iff there is such a Turing program that outputs k when run on the empty input, and let us say that k is n-recognizable iff there is such a

Turing program that stops with output 1 on the input k and with output 0 on all other integers. Are there infinitely many $n \in \mathbb{N}$ for which there exists $l \in \mathbb{N}$ which is n-recognizable, but not n-computable? This provides a kind of a miniaturization of the question for the existence of lost melodies.

Another topic one might pursue is to consider the various generalizations of Turing machines (ITTMs, α-Turing machines, α-β-Turing machines).

Bibliography

[1] G. Boolos and H. Putnam. Degrees of unsolvability of constructible sets of integers. *Journal of Symbolic Logic*, 33:497–513, 1968.

[2] M. Carl. Optimal results on ITRM-recognizability. Preprint (arXiv:1306.5128v1).

[3] M. Carl. Towards a Church-Turing-thesis for infinitary computations. In P. Bonizzoni, V. Brattka, G. Della Vedova, and B. Löwe, editors, *The Nature of Computation: Logic, Algorithms, Applications, 1st July–5th July, Università degli Studi di Milano-Bicocca, Milan, Italy, Informal Proceedings*, pages 30–39. Università degli Studi di Milano-Bicocca, 2013. Unpublished.

[4] M. Carl. The distribution of ITRM-recognizable reals. *Annals of Pure and Applied Logic*, to appear.

[5] M. Carl, T. Fischbach, P. Koepke, R. Miller, M. Nasfi, and G. Weckbecker. The basic theory of infinite time register machines. *Archive for Mathematical Logic*, 49(2):249–273, 2010.

[6] C. T. Chong. A recursion-theoretic characterization of constructible reals. *Bulletin of the London Mathematical Society*, 9(3):241–244, 1977.

[7] S. Coskey. Infinite-time Turing machines and Borel reducibility. In K. Ambos-Spies, B. Löwe, and W. Merkle, editors, *Mathematical Theory and Computational Practice, 5th Conference on Computability in Europe, CiE 2009, Heidelberg, Germany, July 19-24, 2009, Proceedings*, volume 5635 of *Lecture Notes in Computer Science*, pages 129–133. Springer-Verlag, 2009.

[8] N. Cutland. *Computability, An introduction to recursive function theory.* Cambridge University Press, 1980.

[9] S.-D. Friedman and P. D. Welch. Hypermachines. *Journal of Symbolic Logic*, 76(2):620–636, 2011.

[10] V. Gitman, J. D. Hamkins, and T. A. Johnstone. What is the theory ZFC without power set? Submitted.

[11] J. D. Hamkins and A. Lewis. Infinite time Turing machines. *Journal of Symbolic Logic*, 65(2):567–604, 2000.

[12] J. D. Hamkins, R. Miller, D. Seabold, and S. Warner. Infinite time computable model theory. In S. B. Cooper, B. Löwe, and A. Sorbi, editors, *New computational paradigms. Changing conceptions of what is computable*, pages 521–557. Springer-Verlag, 2008.

[13] R. B. Jensen. The fine structure of the constructible hierarchy. *Annals of Mathematical Logic*, 4:229–308, 1972.

[14] R. B. Jensen and C. Karp. Primitive recursive set functions. In D. S. Scott, editor, *Axiomatic Set Theory*, volume XIII (Part 1) of *Proceedings of Symposia in Pure Mathematics*, pages 143–176. American Mathematical Society, 1971.

[15] A. Kanamori. *The higher infinite. Large cardinals in set theory from their beginnings*. Springer Monographs in Mathematics. Springer, 2nd edition, 2003.

[16] P. Koepke. Computing a model of set theory. In S. B. Cooper, B. Löwe, and L. Torenvliet, editors, *New Computational Paradigms. First Conference on Computability in Europe, CiE 2005, Amsterdam, The Netherlands, June 8-12, 2005, Proceedings*, volume 3526 of *Lecture Notes in Computer Science*, pages 223–232. Springer-Verlag, 2005.

[17] P. Koepke. Turing computations on ordinals. *Bulletin of Symbolic Logic*, 11:377–397, 2005.

[18] P. Koepke. Infinite time register machines. In A. Beckmann, U. Berger, B. Löwe, and J. V. Tucker, editors, *Logical Approaches to Computational Barriers. Second Conference on Computability in Europe, CiE 2006, Swansea, UK, June 30-July 5, 2006, Proceedings*, volume 3988 of *Lecture Notes in Computer Science*, pages 257–266. Springer-Verlag, 2006.

[19] P. Koepke and R. Miller. An enhanced theory of infinite time register machines. In A. Beckmann, C. Dimitracopoulos, and B. Löwe, editors, *Logic and Theory of Algorithms. 4th Conference on Computability in Europe, CiE 2008, Athens, Greece, June 15-20, 2008, Proceedings*, volume 5028 of *Lecture Notes in Computer Science*, pages 306–315. Springer-Verlag, 2008.

[20] P. Koepke and B. Seyfferth. Ordinal machines and admissible recursion theory. *Annals of Pure and Applied Logic*, 160(3):310–318, 2009.

[21] P. Koepke and B. Seyfferth. Towards a theory of infinite time Blum-Shub-Smale machines. In S. B. Cooper, A. Dawar, and B. Löwe, editors, *How the World Computes, Turing Centenary Conference and 8th Conference on Computability in Europe, CiE 2012, Cambridge, UK, June 18-23, 2012, Proceedings*, volume 7318 of *Lecture Notes in Computer Science*, pages 310–318. Springer-Verlag, 2012.

[22] P. Koepke and R. Siders. Computing the recursive truth predicate on ordinal register machines. In A. Beckmann, U. Berger, B. Löwe, and J. V. Tucker, editors, *Logical Approaches to Computational Barriers*, volume 7-2006 of *Computer Science Research Reports*, pages 160–169. Swansea University, 2006. Unpublished.

[23] P. Koepke and R. Siders. Register computations on ordinals. *Archive for Mathematical Logic*, 47(6):529–548, 2008.

[24] S. Leeds and H. Putnam. An intrinsic characterization of the hierarchy of constructible sets of integers. In R. O. Gandy and C. E. M. Yates, editors, *Logic Colloquium '69, Proceedings of the Summer School and Colloquium in Mathematical Logic, Manchester, August 1969*, volume 61 of *Studies in Logic and the Foundations of Mathematics*, pages 311–350. North-Holland, Amsterdam, 1971.

[25] W. Marek and M. Srebrny. Gaps in the constructible universe. *Annals of Pure and Applied Logic*, 6:359–394, 1973/74.

[26] G. E. Sacks. *Higher recursion theory*. Perspectives in Mathematical Logic. Springer-Verlag, 1990.

[27] P. Schlicht and B. Seyfferth. Tree representations and ordinal machines. *Computability*, 1(1):45–57, 2012.

[28] B. Seyfferth. *Three models of ordinal computability*. PhD thesis, Rheinische Friedrich-Wilhelms-Universität Bonn, 2012.

Modelling the usage of partial functions and undefined terms using presupposition theory

Marcos Cramer

Interdisciplinary Centre for Security, Reliability and Trust, University of Luxembourg, Luxembourg, Luxembourg

Abstract

We describe how the linguistic theory of presuppositions can be used to analyse and model the usage of partial functions and undefined terms in mathematical texts. We compare our account to other accounts of partial functions and undefined terms, showing how our account models the actual usage of partial functions and undefined terms more faithfully than existing accounts. The model described in this paper has been developed for the *Naproche* system, a computer system for proof-checking mathematical texts written in controlled natural language, and has largely been implemented in this system.

1 Introduction

Partial functions are ubiquitous in mathematical practice. For example, the division function / over the complex numbers (or any other field) is partial, since z_1/z_2 is only defined if $z_2 \neq 0$. Another prominent example is the square root function over the real numbers, which is only defined for non-negative arguments. Partial functions appear in very basic mathematics, for example the subtraction function over the natural numbers, as well as in more advanced mathematics, for example the Lebesgue integral function $\int$ over real functions, which maps Lebesgue integrable functions to their Lebesgue integrals but is undefined on other real functions. As can be seen from these examples, partial functions can be either unary or of higher arity; for simplifying the exposition, we shall concentrate on unary partial functions for the rest of this introduction.

The usual notion of the *domain* of a function separates into two distinct notions in the case of partial functions: The *domain of application* of a partial function consists of the values to which it may be applied, and the *domain of definition* of a partial function consists of the values at which it is defined. For example, the square root function over the reals mentioned above has $\mathbb{R}$ as its domain of application; its domain of definition is the set $\mathbb{R}_0^+$ of non-negative reals. The domain of definition is always a subset of the domain of application. A partial function is called a *total function* if its domain of definition is identical to its domain of application.

When a mathematical expression denoting a partial function is applied to a term denoting an element outside its domain of definition, the resulting

Stefan Geschke, Benedikt Löwe, Philipp Schlicht (*eds.*).
Infinity, computability, and metamathematics: Festschrift celebrating the 60th birthdays of Peter Koepke and Philip Welch. College Publications, London, 2014. Tributes, Volume 23.

term is an *undefined term*. Thus, if $\surd$ denotes the square root function over the reals, $\sqrt{-1}$ is an undefined term. An undefined term does not refer to any mathematical object.

The common formal systems like first-order logic and simple type theory do not have any means to formalize partial functions and undefined terms in them. Standard one-sorted first-order logic allows for function symbols, but these necessarily denote total functions. Both the domain of application and the domain of definition of a function denoted by a function symbol must coincide with the *domain of discourse*, i.e., with the domain specified by the structure used for interpreting all terms and formulae in the formalism, which is also the *domain of quantification* over which the quantifiers range. In multi-sorted first-order logic as well as in simple type theory, one can have functions with different domains of application, but the domain of definition of any given function always coincides with its domain of application.

A number of formal systems have been proposed to account for the common usage of partial functions and undefined terms. Prominent examples include Michael Beeson's *Logic of Partial Terms* [1] and William Farmer's closely related *Partial First-Order Logic* (PFOL) [5], which we sketch below. A more recent approach by Freek Wiedijk and Jan Zwanenburg [13], the usage of *domain conditions* on top of standard first-order logic, comes very close to our approach; but the authors did not notice, or at least did not explicitly mention, the close relation of their approach to the linguistic theory of presuppositions, which helps clarifying some further points, as we shall show in this paper.

This paper is partially based on [4], where we discussed the phenomenon of presuppositions in mathematical texts to a linguistic readership, and thereby already explained our approach to partial functions and undefined terms. In contrast to [4], this paper aims at a readership from the Formal Mathematics community, and for the first time compares our approach to approaches originating from this community.

Before explaining our own approach, we sketch William Farmer's Partial First-Order Logic in § 2.[1] In § 3 we explain the context in which we developed our approach, the proof checking algorithm of the Naproche system. In § 4 we sketch the linguistic theory of presuppositions and introduce the terminology from this theory that we shall need in this paper. § 5 contains those part of our account of how to use presupposition theory to model the usage partial functions and undefined terms that have been implemented in the Naproche system. In § 6 we sketch Freek Wiedijk's and Jan Zwanenburg's approach to use domain conditions to account for partial functions

[1]The main reason for introducing Farmer's approach before our own is that we shall make use of the syntax of Partial First-Order Logic for presenting our own approach in the following sections.

and undefined terms and compare it to our approach. In § 7, we show how our account from § 5 can be extended by making use of a detail of presupposition theory that we ignored in § 5, the possibility to accommodate presuppositions. We then compare this extended account with Farmer's Partial First-Order Logic.

2 Partial First-Order Logic

In [5], William Farmer defined Partial First-Order Logic (PFOL). It allows for partial functions and undefined terms, and is based on the following three tenets:

1. Variables and constants are defined terms.
2. The application of a function to an undefined argument is undefined.
3. Formulas are always true or false. The application of a predicate is false if any of its arguments is undefined.

We now turn to the formal definition of PFOL. We shall use the nowadays more usual symbols $\rightarrow$, $\leftrightarrow$ and ι where Farmer used $\supset$, $\equiv$ and I. The first two are the well known connectives from standard first-order logic; ι is used for formalizing definite description: $\iota x\ \varphi(x)$ corresponds to the natural language expression "the x such that $\varphi(x)$". Additionally to the already mentioned connectives $\rightarrow$ and $\leftrightarrow$, PFOL has the standard connectives $\neg$, $\wedge$ and $\vee$, the quantifiers $\forall$ and $\exists$ and the identity relation symbol $=$. Further symbols are used for abbreviating PFOL formulae: $t\downarrow$ (read "t is defined") abbreviates $t = t$; $t\uparrow$ (read "t is undefined") abbreviates $\neg t = t$; and $t_1 \simeq t_2$ (read "t_1 and t_2 are quasi-equal") abbreviates $t_1\downarrow \vee\, t_2\downarrow \rightarrow t_1 = t_2$.

Farmer defines the semantics of PFOL in a way very analogous to the standard definition of the semantics of first-order logic. Terms and formulae are interpreted in a *model*, which consists of a domain D and interpretations for the constants, function symbols and relation symbols. While the interpretations of constants and n-ary relation symbols are as in first-order logic, namely elements of D and total functions from D^n to $\{T, F\}$, respectively, the interpretations of n-ary function symbols now do no longer have to be total function from D^n to D, but may be partial functions from D^n to D (i.e., with D^n as domain of application, but any subset of D^n as domain of definition). It is straightforward to define the semantics of formulae in such a model according to the three tenets listed above.

3 Proof checking mathematical texts in the Naproche system

Before we can go on to explain our approach to partial functions and undefined terms, we first need to clarify the context in which it was developed, namely the proof checking of mathematical texts in the Naproche system.

The Naproche system (cf. [3, 2]) is a computer program that can check the correctness of mathematical texts written in a *controlled natural language*, the *Naproche CNL*. A controlled natural language is a subset of a natural language defined through a formal grammar. By "checking the correctness" we mean that it tries to establish all proof steps found in the text based on the information gathered from previous parts of the text, in a similar way as a mathematician reads a (foundational) mathematical text if asked not to use his mathematical knowledge originating from other sources. For checking single proof steps, the Naproche system makes use of state-of-the-art *automated theorem provers* (ATPs). Given a set of *premises*[2] and a *conjecture*, an ATP tries to find either a proof that the premises logically imply the conjecture, or build a model for the premises and the negation of the conjecture, which shows that that they do not imply it. A conjecture together with a set of axioms handed to an ATP is called a *proof obligation*.

The Naproche system first translates an input text into a semantic representation format called *Proof Representation Structure* (PRS), an adaptation of *Discourse Representation Structures* [9], a common representation format in formal linguistics. The actual proof checking is performed on PRSs. For the sake of simplicity, we shall in this paper assume that the Naproche CNL input is translated into a formal language that is syntactically identical with the language of PFOL, i.e., standard first-order syntax with an additional term construction principle for terms of the form $\iota x\ \varphi(x)$ for representing definite descriptions.

Since first-order and PFOL formulae are usually used to formally express single statements and not complete texts, we need to say some words about how complete texts are translated into this formal language. In the simplified exposition for this paper, we shall leave out a number of constructs used for structuring mathematical texts, e.g., theorem-proof blocks, and concentrate on simple texts consisting of axioms, local assumptions and assertions. The concatenation of sentences is usually rendered by conjoining their respective translations with $\wedge$. A special case are axioms and local assumptions: When an axiom appears in a text, the part of the text starting at the axiom is translated by a formula of the form $\varphi \rightarrow \vartheta$, where φ is the translation of the axiom and ϑ is the translation of the text following the axiom. The translation of local assumptions, which are marked by one of the keywords "assume", "suppose" and "let" in the Naproche CNL, is similar, only that one has to take into account the *scope* of the assumption:

[2]In the ATP community, the term "axiom" is usually used for what we call "premise" here; the reason for our deviation from the standard terminology is that in the context of our work the word "axiom" means a completely different thing, namely an axiom stated inside a mathematical text that is to be checked by the Naproche system. The premises that we are considering here can originate either from axioms, from definitions or from previously proved results.

In mathematical texts, an assumption always has a scope, which starts at the assumption and contains all assertions made under that assumptions. The end of the scope of an assumption is usually not marked in a special way in the natural language of mathematics, but in the Naproche CNL it is usually marked with a sentence starting with the keyword "thus".[3] The scope of an assumption is translated as $\varphi \to \vartheta$, where φ is the translation of the assumption and ϑ is the translation of the rest of the scope of the assumption. The translation of the scope of an assumption is embedded into the translation of a complete text as if the whole scope of the assumption were a single sentence.

Before explaining the treatment of partial functions and undefined terms in the proof checking, we shall now first explain the basic functioning of the proof checking algorithm with total functions and without undefined terms. The proof checking algorithm keeps track of a list of first-order formulae considered to be true, called *premises*, which gets continuously updated during the checking process. Each assertion is checked by an ATP based on the currently active premises.

Below we list how the algorithm proceeds on an input formula φ depending on the form of φ. We use Γ to denote the list of premises considered true before encountering the formula in question, and Γ' to denote the list of premises considered true after encountering the formula in question. A proof obligation checking that φ follows from Γ will be denoted by $\Gamma \vdash^? \varphi$. For any given formula φ, we denote by $\mathrm{FI}(\varphi)$ the *formula image* of φ, which is a list of first-order formulae representing the content of φ; the definitions of $\mathrm{FI}(\varphi)$ and of the checking algorithm are mutually recursive, as specified below. (In the case of this proof checking algorithm with total functions and without undefined terms, the conjunction of the formulae in $\mathrm{FI}(\varphi)$ is always logically equivalent to φ; this will, however, no longer be the case once we consider the extension of the proof checking algorithm to partial functions and undefined terms.)

If φ is atomic, check $\Gamma \vdash^? \varphi$ and set Γ' to be Γ, φ.

If φ is of the form $\varphi_1 \wedge \varphi_2$, check φ_1 with premise list Γ and φ_2 with the premise list that is active after checking φ_1; set Γ' to be the premise list that is active after checking φ_2.

If φ is of the form $\varphi_1 \to \varphi_2$, check φ_2 with premise list $\Gamma \cup \mathrm{FI}(\varphi_1)$ and set Γ' to be $\Gamma \cup \{\bigwedge \mathrm{FI}(\varphi_1) \to \psi \mid \psi \in \mathrm{FI}(\varphi_2)\}$.

[3]The scope of an assumption also ends when the proof inside which the assumption was introduced is ended with a "Qed". But since the simplified fragment of the Naproche CNL that we are currently considering does not contain theorem-proof blocks, we can ignore this special case as well as similar cases relating to other structural constructs of mathematical texts that we are now ignoring (e.g., case distinctions).

If φ is of the form $\neg\psi$, check $\Gamma \vdash^? \neg \bigwedge \mathrm{FI}(\psi)$ and set Γ' to be $\Gamma, \neg \bigwedge \mathrm{FI}(\psi)$.

If φ is of the form $\varphi_1 \vee \varphi_2$ or $\varphi_1 \leftrightarrow \varphi_2$, check $\Gamma \vdash^? \bigwedge \mathrm{FI}(\varphi_1) \vee \bigwedge \mathrm{FI}(\varphi_2)$ or $\Gamma \vdash^? \bigwedge \mathrm{FI}(\varphi_1) \leftrightarrow \bigwedge \mathrm{FI}(\varphi_2)$, respectively; set Γ' to be $\Gamma, \bigwedge \mathrm{FI}(\varphi_1) \vee \bigwedge \mathrm{FI}(\varphi_2)$ or $\Gamma, \bigwedge \mathrm{FI}(\varphi_1) \leftrightarrow \bigwedge \mathrm{FI}(\varphi_2)$, respectively.

If φ is of the form $\exists x\ \psi$ or $\forall x\ \psi$, check $\Gamma \vdash^? \exists x\ \bigwedge \mathrm{FI}(\psi)$ or $\Gamma \vdash^? \forall x\ \bigwedge \mathrm{FI}(\psi)$, respectively; set Γ' to be $\Gamma, \exists x\ \bigwedge \mathrm{FI}(\psi)$ or $\Gamma, \forall x\ \mathrm{FI}(\psi)$, respectively.

For computing $\mathrm{FI}(\varphi)$, the algorithm proceeds analogously to the checking of φ, only that no proof obligations are sent to the ATP: The updated premise lists are still computed, and $\mathrm{FI}(\varphi)$ is defined to be $\Gamma' - \Gamma$, where Γ is the active premise list before processing φ and Γ' is the active premise list after processing φ. This is implemented by allowing the algorithm to process a formula φ in two different modes: The Check-Mode described above for checking the content of φ, and the No-Check-Mode, which refrains from sending proof obligations to the ATP, but still expands the premise list in order to compute $\mathrm{FI}(\varphi)$.

4 Presuppositions

Loosely speaking, a *presupposition* of some utterance is an implicit assumption that is taken for granted when making the utterance. In the linguistic literature on presupposition theory, presuppositions are generally accepted to be triggered by certain lexical items called *presupposition triggers*. Here are some common examples of presupposition triggers:

Definite descriptions: In English, definite descriptions are marked by the definite article "the", possessive pronouns or genitives. The presupposition of a definite description of the form "the F" is that there is a unique object with property F.

Factive verbs, e.g., "regret", "realize" and "know". For example, the presupposition of "A knows φ" is that φ holds true.

Change of state verbs, e.g., "stop" and "begin". For example, the presupposition of "A stops doing x" is that A did x before.

In mathematical texts, most of the presupposition triggers discussed in the linguistic literature, e.g., factive verbs and change of state verbs, are not very common or even completely absent. Definite descriptions, however, do appear in mathematical texts as presupposition triggers (e.g., "the smallest natural number n such that $n^2 - 1$ is prime"). The presupposition of a definite description "the F" can be divided into two separate presuppositions:

One existential presupposition, claiming that there is at least one F, and one uniqueness presupposition, claiming that there is at most one F.

Furthermore, there is a kind of presupposition trigger which does not exist outside mathematical texts: Expressions denoting partial functions. For example, the division symbol "/" triggers the presupposition that its second argument is non-zero; and the square root function over the reals triggers the presupposition that its argument is non-negative.[4]

Presupposition projection is the way in which presuppositions triggered by expressions within the scope of some operator have to be evaluated outside this scope. Consider for example the following three sentences:

$$\tfrac{1}{x+1} \in A \text{ and } x \neq 0. \tag{1}$$

$$\frac{1}{x+1} < \frac{1}{x}. \tag{2}$$

$$\text{If } \tfrac{1}{x+1} \in A \text{ and } x \neq 0, \text{ then } \tfrac{1}{x} \in A. \tag{3}$$

We see that (1) and (3) presuppose that $x+1 \neq 0$ and (2) presupposes that $x+1 \neq 0$ and $x \neq 0$. So (3) inherits the presupposition of (1), but does not inherit the additional presupposition of (2). The precise way in which presuppositions project under various operators has been discussed at great length in the literature (cf., e.g., [11, 8] for overviews of this dispute). Our formal treatment of presuppositions in mathematical texts turns out to have equivalent predictions (cf. [4]) about presupposition projection to Irene Heim's approach to presuppositions (cf. [7]).

Presupposition accommodation is what we do if we find ourselves faced with a presupposition the truth of which we cannot establish in the given context: We add the presupposition to the context, in order to be able to process the sentence that presupposes it. For example, if I say "John's wife is a philosopher" to someone who does not know that John has a wife, they will accommodate the fact that John has a wife, i.e., add this presupposition to the context in which they interpret the sentence. Note that presupposition accommodation is not always possible: If someone knows that John does not have a wife, they will not be able to accommodate the presuppositions of the example sentence. This is called *presupposition failure* and results in an inability to make sense of the sentence.

[4]In Naproche, the division function / with its usual presupposition triggering features can be introduced in a proof text by a sentence of the following form: "For all real numbers x, y such that $y \neq 0$, there exists a real number $\frac{x}{y}$ such that $y \cdot \frac{x}{y} = x$." This paper is not concerned with the issue of how such partial functions are introduced, but with how they are used once they have been introduced. For more details on how partial functions are introduced, the interested reader should consult [2].

5 Proof checking with presuppositions

In this section, we shall describe how presuppositions can be handled in the proof checking algorithm of the Naproche system if we do not take care of the possibility of presupposition accommodation. The proof checking algorithm described in this section is called PCC (*Presuppositional Proof-Checking*). In § 7, we shall describe how this proof checking algorithm can be adapted to allow for presupposition accommodation.

Most accounts of presupposition make reference to the *context* in which an utterance is uttered, and claim that presuppositions have to be satisfied in the context in which they are made. There are different formalizations of how a context should be conceptualized. For enabling the Naproche proof checking algorithm described in the previous section to handle presuppositions, it is an obvious approach to use the list of active premises as the context in which our presuppositions have to be satisfied.

In ordinary non-mathematical discourse, assertion usually are expected to provide new information, i.e., not to be logically implied by the available knowledge. In mathematical texts, on the other hand, assertions are expected to be logically implied by the available knowledge rather than adding something logically new to it. Because of this peculiarity of mathematical texts, both presuppositions and assertions in proof texts have to follow logically from the context. For a sentence like "$\frac{1}{x+1}$ is negative" to be legitimately used in a mathematical text, both the fact that $x + 1 \neq 0$ and the negativity of $\frac{1}{x+1}$ must be inferable from the context.

This parallel treatment of presuppositions and assertions, however, does not necessarily hold for presupposition triggers that are subordinated by a logical operation like negation or implication. For example, in the sentence "A does not contain $\frac{1}{x}$", the presupposition that $x \neq 0$ does not get negated, whereas the containment assertion does. This is explained in the following way: In order to make sense of the negated sentence, we first need to *make sense of* what is inside the scope of the negation. In order to make sense of some expression, all presuppositions of that expression have to follow from the current context. The presuppositions triggered by $\frac{1}{x}$ are inside the scope of the negation, so they have to follow from the current context. The containment assertion, however, does not have to follow from the current context, since it is not a presupposition, and since it is negated rather than being asserted affirmatively.

In our implementation, *making sense of a something* corresponds to processing it with the proof checking algorithm, whether in the Check-Mode or in the No-Check-Mode. So according to the above explanation, presuppositions, unlike assertions, also have to be checked when encountered in the No-Check-Mode.

For example, the formula representing the sentence (4) is (5).

$$A \text{ does not contain } \tfrac{1}{x}. \tag{4}$$

$$\neg\text{contains}(A, \frac{1}{x}) \tag{5}$$

When the checking algorithm encounters the negated formula, it needs to find the formula image of the formula in the scope of the negation, for which it will process this formula in No-Check-Mode. Now $\frac{1}{x}$ triggers the presupposition that $x \neq 0$, which has to be checked despite being in No-Check-Mode. So we send the proof obligation (6) to the ATP.

$$\Gamma \vdash^? x \neq 0 \tag{6}$$

Finally, the proof obligation that we want for the assertion of sentence (4) is (7):

$$\Gamma, x \neq 0 \vdash^? \neg\text{contain}(A, \frac{1}{x}) \tag{7}$$

In order to get this, we need to use the non presuppositional formula image $\{\text{contain}(A, \frac{1}{x})\}$ of the formula in the scope of the negation: The non-presuppositional formula image is defined to be the subset of formulae of the formula image that do not originate from presuppositions. When extending the above proof checking algorithm to an algorithm capable of handling presuppositions, we have to use this non-presuppositional formula image wherever we used the formula image in the original proof checking algorithm. The presupposition premises which get pulled out of the formula image have to be added to the list of premises that were active before starting to calculate the formula image (cf. the treatment of the presupposition premise $x \neq 0$ in (7)).

When proof checking ι terms, a new constant symbol is used in the premises that make reference to the unique object presupposed by the ι term. Consider for example sentence (8), whose translation is (9). When proof checking (9), the presuppositions triggered by the ι term trigger the proof obligations (10) and (11), and the assertion of the sentence is checked by proof obligation (12); here c is the newly introduced constant symbol that refers to the unique object presupposed by the ι term.

$$A \text{ does not contain the empty set.} \tag{8}$$

$$\neg\text{contain}(A, \iota x\ \text{empty}(x) \wedge \text{set}(x)) \tag{9}$$

$$\Gamma \vdash \exists x(\text{empty}(x) \wedge \text{set}(x)) \tag{10}$$

$$\Gamma \cup \{\text{empty}(c) \wedge \text{set}(c)\} \vdash \forall y(\text{empty}(y) \wedge \text{set}(y) \rightarrow y = c) \quad (11)$$

$$\Gamma \cup \{\text{empty}(c) \wedge \text{set}(c), \forall y(\text{empty}(y) \wedge \text{set}(y) \rightarrow y = c)\} \vdash \neg\text{contain}(A, c) \quad (12)$$

When presuppositions are triggered inside the scope of a quantifier, a somewhat more sophisticated approach is needed; cf. [4] or [2].

6 First-order logic with domain conditions

In [13], Wiedijk and Zwanenburg introduced an approach to partial functions and undefined terms which they termed *first-order logic with domain conditions*. Their approach turns out to be equivalent to PCC, only that their approach does not admit terms representing definite descriptions.[5] However, in the next section, we shall discuss how the possibility to accommodate presuppositions affects our approach: It will turn out that granted this possibility, our approach becomes practically much more similar to Farmer's approach than to Wiedijk's and Zwanenburg's.

In Wiedijk's and Zwanenburg's approach, one can use standard first-order syntax for formalizing talk about partial functions. In order to avoid potential problems caused by undefined terms, they define a set $\mathcal{DC}(\varphi)$ of *domain conditions* for every first-order formula φ. Domain conditions are judgements of the form $\Gamma \vdash \psi$. The idea is that if one wants to prove a statement about partial functions formalized by a first-order formula φ, one should not only prove φ but also establish the judgements in $\mathcal{DC}(\varphi)$.

There is an almost perfect correspondence between the domain conditions of a formula φ and the proof obligations triggered by presuppositions in our account. In simple examples, they coincide completely. For example, (13) has domain condition (14), and its presupposition triggers the proof obligation (15) in our account:

$$x > 0 \rightarrow \frac{1}{x} > 0 \quad (13)$$

$$x > 0 \vdash x \neq 0 \quad (14)$$

$$x > 0 \vdash^{?} x \neq 0 \quad (15)$$

This is no surprise if one looks at the formal definition of domain conditions: Domain conditions are defined relative to a context, which is a list of first-order formulae.[6] So Wiedijk and Zwanenburg actually define $\mathcal{DC}_\Gamma(\varphi)$

[5] We ignore the if-then-else construct that Wiedijk and Zwanenburg added to first-order logic for their approach, as they at any rate consider this addition not essential.

[6] Actually, Wiedijk and Zwanenburg define a context to be a list of variables and first-order formulae which satisfies the condition that all free variables in a formula in the context are previously listed as variables in the context. However, dropping the variables from their contexts does not alter their account in relevant way.

for a context Γ and a formula φ, and the set $\mathcal{DC}(\varphi)$ of absolute domain conditions can be identified with $\mathcal{DC}_{\varnothing}(\varphi)$. The contexts in their account are updated in a similar way as the premise lists in our account. For example, $\mathcal{DC}_{\Gamma}(\varphi \rightarrow \psi)$ is defined to be $\mathcal{DC}_{\Gamma}(\varphi) \cup \mathcal{DC}_{\Gamma,\varphi}(\psi)$. Here, the context Γ gets updated to Γ, φ for calculating the domain conditions triggered inside ψ, in a similar way as in our account the premise list Γ gets updated to Γ, φ when proof-checking the ψ inside $\varphi \rightarrow \psi$.

However, there is one difference between the ways their contexts and our premise lists get updated, which we illustrate through an example. The set of domain conditions of (16) is (17), and the set of proof obligations triggered by presuppositions in (16) is (18):[7]

$$\frac{1}{x} > 0 \rightarrow \frac{1}{x+1} > 0 \tag{16}$$

$$\{\vdash x \neq 0; \frac{1}{x} > 0 \vdash x + 1 \neq 0\} \tag{17}$$

$$\{\vdash x \neq 0; x \neq 0, \frac{1}{x} > 0 \vdash^{?} x + 1 \neq 0\} \tag{18}$$

As the proof checking algorithm processes $\frac{1}{x} > 0$, the premise list gets updated not only by $\frac{1}{x} > 0$, but also by its presupposition $x \neq 0$. Hence this presupposition additionally appears among the premises of the second proof obligation in (18), whereas it does not appear on the left hand side of the second domain condition in (17) according to the above cited definition of the domain conditions of an implication. But $x \neq 0$ must at any rate be provable because of the first domain condition and first proof obligation in (17) and (18), respectively, so that this syntactic difference is semantically irrelevant.

Since the domain conditions of a formula are thus semantically equivalent to the proof obligations triggered by presuppositions in our account, Wiedijk's and Zwanenburg's account is essentially equivalent to our account as described so far.

7 Accommodation of presuppositions[8]

Recall that accommodating a presupposition means adding it to the context of the utterance in case we cannot establish it in this context. One commonly distinguishes between *global* and *local* accommodation of presuppositions. Global accommodation is the process of altering the global

[7] We use semicolons to separate the elements in these sets, since commas are already used for separating the formulae in contexts or premise lists.

[8] Unlike our approach as discussed in § 5, the adaptations to our approach presented in this section have not yet been implemented in the Naproche system.

context in such a way that the presupposition in question can be justified; local accommodation on the other hand involves only altering some local context, leaving the global context untouched. It is a generally accepted principle of presupposition theory that in usual discourse, global accommodation is *ceteris paribus* preferred over local accommodation.

In the introduction, we mentioned the peculiarity of mathematical texts that new assertions do not add new information (in the sense of logically not inferable information) to the context. Here "context" does not refer to our formal definition of context as a list of formulae. Instead, the *context* of a sentence in a mathematical texts should now be understood to be the set of models in which the axioms, definitions and assumptions in whose scope the sentence is made hold. This definition of *context* is analogous to the definition of *context* in various linguistic theories (e.g., Heim's theory of presupposition [7]) as a set of possible worlds that are under consideration when an utterance is made. When mathematicians state axioms, they limit the context, i.e., the set of models they consider, to the set where the axioms hold. Similarly, when they make local assumptions, they temporarily limit the context. But when making assertions, these assertions are thought be logically implied by what has been assumed and proved so far, so they do not further limit the context.

The modification of the context in the case of local assumptions is certainly a modification of the local context. For the sake of giving a unified treatment, it is useful to view the modification of the context in the case of axioms also as a modification of the local context, only that the mathemtician is planning to stay in this locally modified context for the rest of the text. With this understading of local as opposed to global contexts, one may succinctly state the pragmatic principle mentioned above in terms of contexts as follows: In a mathematical text, the global context may not be altered.

This pragmatic principle implies that global accommodation is not possible in mathematical texts, since global accommodation implies adding something new to the global context. Local accommodation, on the other hand, is allowed, and does occur in real mathematical texts:

> Suppose that f has n derivatives at x_0 and n is the smallest positive integer such that $f^{(n)}(x_0) \neq 0$. [12]

This is a local assumption. The projected existential presupposition of the definite description "the smallest positive integer such that $f^{(n)}(x_0) \neq 0$" is that for any function f with some derivatives at some point x_0, there is a smallest positive integer n such that $f^{(n)}(x_0) \neq 0$. Now this is not valid in real analysis, and we cannot just assume that it holds using global accommodation. Instead, we make use of local accommodation, thus adding the accommodated fact that there is a smallest such integer for f to the assumptions that we make about f with this sentence.

When we accommodate a presupposition locally in this way, it no longer triggers a proof obligation. Instead, the content of the presupposition is added to the formula image of the atomic formula that triggered the presupposition.

7.1 Two ways to handle accommodation in presuppositional proof checking

We now consider two possible ways of handling accommodation of presuppositions in a proof checking algorithm:

According to the linguistic theory of presuppositions, accommodation should only be performed if necessary. In the case of our proof checking algorithm, this means that we should always first try to establish the proof obligation triggered by the presupposition. If that fails, we accommodate the presupposition on the most local level possible (i.e., in the scope of the atomic formula inside which the presupposition was triggered). We call the proof checking algorithm that handles accommodation in this way PCC+FlexAcc (*Presuppositional Proof-Checking with Flexible Accommodation*).

In order to compare presuppositional proof checking with Farmer's PFOL, we shall additionally consider a proof checking algorithm called PCC+ImmAcc (*Presuppositional Proof-Checking with Immediate Accommodation*), which works as follows: PCC+ImmAcc does not produce any proof obligations from presuppositions, but always directly accommodates presuppositions on the most local level possible.

First we want to point out that PCC+ImmAcc has the same implications as Farmer's account:[9] In the case of an atomic formula $R(t_1, \ldots, t_n)$, the presuppositions triggered by the arguments $t_1, \ldots, t_n$ would become part of the the formula image of $R(t_1, \ldots, t_n)$, so that this formula image has precisely the semantics that Farmer gives to atomic formulae: It can only be true if all the presuppositions triggered by the arguments $t_1, \ldots, t_n$ hold, i.e., only if $t_1, \ldots, t_n$ are all defined terms.

Unlike PCC+ImmAcc, PCC+FlexAcc treats accommodation in the same way as does the linguistic theory of presuppositions. As we shall illustrate through an example from a real mathematical text, we believe that the linguistically motivated proof checking algorithm PCC+FlexAcc gives rise to a better account of how mathematicians actually use potentially un-

[9]Since Farmer does not actually define a proof checking algorithm, we need to make precise what we mean by this: We mean that if one were to define a proof checking algorithm on PFOL in a canonical way, i.e., in a way completely analogous to the way we defined the non-presuppositional proof checking algorithm over standard first-order logic in § 3.

defined terms arising from partial functions and definite descriptions than does Farmer's PFOL or PCC+ImmAcc.

7.2 PPC+FlexAcc and PPC+ImmAcc compared on an example

As an example to compare PCC+FlexAcc and PCC+ImmAcc, we use the proof of Theorem 2 of Edmund Landau's *Foundations of Analysis* [10]:

> **Theorem 2:** $x' \neq x$.
> **Proof:** Let $\mathfrak{M}$ be the set of all x for which this holds true.
> I) By Axiom 1 and Axiom 3, $1' \neq 1$; therefore 1 belongs to $\mathfrak{M}$.
> II) If x belongs to $\mathfrak{M}$, then $x' \neq x$, and hence by Theorem 1, $(x')' \neq x'$, so that x' belongs to $\mathfrak{M}$.
> By Axiom 5, $\mathfrak{M}$ therefore contains all the natural numbers, i.e., we have for each x that $x' \neq x$.

The first sentence of the proof is an assumption, which is marked by the keyword "Let". It introduces a new symbol $\mathfrak{M}$ to the discourse and at the same time assumes that this symbol refers to the same object as the definite description "the set of all x for which this $[x' \neq x]$ holds true".[10] The newly introduced symbol has a scope, which is the part of the text in which it may be used. The scope of a symbol introduced in an assumption generally coincides with the scope of the assumption. Since $\mathfrak{M}$ is still used in the last sentence of the proof, this sentence still belongs to the scope of the assumption. Since the proof inside which the assumption was made ends at that sentence, this must be the last sentence in the scope of the assumption.

According to the explanations we gave for translating texts with assumptions into the formal representation language on which the proof checking algorithm is defined, the proof of this theorem gets translated into a formula φ of the form

$$\mathfrak{M} = \iota s\, (\text{set}(s) \wedge \forall x (x \in s \leftrightarrow x' \neq x)) \\ \rightarrow \vartheta \wedge (\forall n\, (n \in \mathbb{N} \rightarrow n \in \mathfrak{M}) \wedge \forall x\, x' \neq x),$$

where ϑ is the translation of the text between the assumption and the last sentence of the proof. The ι term in this formula triggers the presupposition that there is a unique s satisfying $\text{set}(s) \wedge \forall x (x \in s \leftrightarrow x' \neq x)$.

[10] The introduction of new symbols in assumptions can be accounted for using linguistic theories of dynamic quantifiers, e.g., Discourse Representation Theory [9] or Dynamic Predicate Logic [6]. Cf. [2] for details about how to use Dynamic Predicate Logic to account for this phenomenon prevalent in the language of mathematics. As pointed out in footnote 11 below, we shall have to make use of one detail of this account; we nevertheless refrain from explaining this account in detail, as this would go beyond the scope of this paper.

We want to compare how PCC+FlexAcc and PCC+ImmAcc check the cited theorem-proof block. So far, we have not said anything about proof-checking theorem-proof blocks. For the sake of the current exposition, we can translate the whole theorem-proof block by $\varphi \wedge \forall x\ x' \neq x$ (here φ is the formula representing the proof as spelled out above, and $\forall x\ x' \neq x$ is the translation of the theorem statement, which is implicitly quantified universally). This means that the proof checking algorithm will first check the translation of the proof, and then check the translation of the theorem statement using the premise list that is active at the end of checking the translation of the proof. Furthermore, we shall assume that the premise list that is active before checking the theorem-proof block contains premises that encode the basic set theory that is needed for understanding the proof. This basic set theory has to be strong enough for proving the presupposition triggered by the ι term.

In both PCC+FlexAcc and PCC+ImmAcc, $\mathfrak{M} = \iota s\ (\text{set}(s) \wedge \forall x(x \in s \leftrightarrow x' \neq x))$ will be processed in No-Check-Mode. In PCC+FlexAcc, the presupposition triggered by the ι term will be checked immediately after the ι term has been parsed. By our assumption about basic set theory being encoded in the premise list, the presupposition will be successfully checked. Let us assume that the rest of the proof is checked successfully. The premise list that is active right after checking the last sentence in the proof contains the formula $\forall x\ x' \neq x$. One might be tempted to think that this premise is therefore contained in the premise list that is active when checking the theorem assertion $\forall x\ x' \neq x$, in which case checking the theorem assertion would become trivial. However, note that after checking the (translation of the) last sentence of the proof, the scope of the assumption closes. After closing this assumption, the active premise list no longer contains the premise $\forall x\ x' \neq x$, but instead contains $\forall \mathfrak{M}\ (\mathfrak{M} = c \rightarrow \forall x\ x' \neq x)$, where c is the constant symbol newly introduced due to the ι term.[11] This premise trivially implies $\forall x\ x' \neq x$, so the theorem assertion still follows trivially from the premise list that is active after checking the proof.

When reading "for every x, $x' \neq x$" at the end of the proof, a mathematical reader would have the feeling that the proof is finished, since this is what had to be established. The reason why we intuitively feel that the proof is already finished when Landau writes "for every x, $x' \neq x$" is that we do not really feel the assumption at the beginning of the proof as an assumption, but just as an introduction of the temporary constant $\mathfrak{M}$ with a defined meaning. We can account for this intuition in the framework of the theory developed in this paper as follows:

[11] According to the rules for proof checking an implication, the premise list that is active after closing the assumption would have to contain a formula of the form $\mathfrak{M} = c \rightarrow \forall x\ x' \neq x$. The additional quantifier $\forall \mathfrak{M}$ comes from the account of introducing new symbols that was mentioned in footnote 10 above.

The non-presuppositional content of the assumption, represented by the premise $\mathfrak{M} = c$, is trivial. After retracting the assumption, we have the premise $\forall \mathfrak{M}\ (\mathfrak{M} = c \to \forall x\ x' \neq x)$ in the active premise list. Since the non-presuppositional content of the assumption is trivial and leads to this premise trivially equivalent to the desired result, we do not feel the assumption to have any content at all, and hence do not feel it to be an assumption in the first place.

Let us now consider how PCC+ImmAcc processes the theorem-proof block under consideration. In PCC+ImmAcc, the presupposition ψ of the ι term will not be checked when processing the assumption, but will be added to the premise list together with the premise $\mathfrak{M} = c$ that expresses the non-presuppositional content of the assumption. At the end of the proof, we then have $\forall \mathfrak{M}\ (\psi \wedge \mathfrak{M} = c \to \forall x\ x' \neq x)$ instead of $\forall \mathfrak{M}\ (\mathfrak{M} = c \to \forall x\ x' \neq x)$ in the active premise list. Now using the fact that ψ follows from the basic set theory being encoded in the initial premise list, we shall still be able to deduce $\forall x\ x' \neq x$ from the premise list that is active when processing the theorem assertion. But now the usage of the assumed basic set theory to establish ψ is at a point in the proof where a mathematical reader would not feel that anything needs to be proved. So on this account, we get a wrong prediction as to where in a proof certain facts should be established.

Besides being theoretically unappealing, such wrong predictions can also make the proof checking less feasible: It may happen that the position of the proof where the second account wrongly requires a presupposition to be established is not a trivial deduction step as in the above example, but requires some reasoning. In PCC+ImmAcc, this would mean that the presupposition has to be established together with this additional reasoning needed at that step, in a single proof obligation. This might make the proof obligation too hard to be established by the ATP used by the system. At the assumptions where presuppositions have to be established in PCC+FlexAcc, on the other hand, one never needs to establish something else at the same time.

8 Conclusion

We have shown how the linguistic theory of presuppositions can be used to analyse and model the usage of partial functions and undefined terms in mathematical texts. We have considered three possible proof checking algorithms taking care of presuppositions, PCC, PCC+ImmAcc and PCC+FlexAcc, which differ in the way they treat the phenomenon of presupposition accommodation. Wiedijk's and Zwanenburg's account involving *domain conditions* is essentially equivalent to PCC, i.e., to not allowing presuppositions to be accommodated. Farmer's Partial First-Order Logic is equivalent to PCC+ImmAcc, i.e., to locally accommodating all presuppositions, without checking first whether they could be discharged. PCC+FlexAcc treats accommodation in the same way as linguistic presupposition theory, i.e., accommodates presuppositions only if they cannot be discharged. We have argued that PCC+FlexAcc provides an account of the usage of partial functions and undefined terms that models the intuitions and actual usage by mathematicians more faithfully than Farmer's or Wiedijk's and Zwanenburg's account.

References

[1] M. J. Beeson. *Foundations of constructive mathematics, Metamathematical studies*, volume 6 of *Ergebnisse der Mathematik und ihrer Grenzgebiete*. Springer-Verlag, Berlin, 1985.

[2] M. Cramer. *Proof-checking mathematical texts in controlled natural language*. PhD thesis, Rheinische Friedrich-Wilhelms-Universität Bonn, 2013.

[3] M. Cramer, B. Fisseni, P. Koepke, D. Kühlwein, B. Schröder, and J. Veldman. The Naproche project, controlled natural language proof checking of mathematical texts. In N. Fuchs, editor, *Controlled Natural Language, Workshop on Controlled Natural Language, CNL 2009, Marettimo Island, Italy, June 8-10, 2009, Revised Papers*, volume 5972 of *Lecture Notes in Computer Science*, pages 170–186. Springer-Verlag, 2010.

[4] M. Cramer, D. Kühlwein, and B. Schröder. Presupposition projection and accommodation in mathematical texts. In M. Pinkal, I. Rehbein, S. Schulte im Walde, and A. Storrer, editors, *Semantic Approaches in Natural Language Processing, Proceedings of the Conference on Natural Language Processing 2010*, pages 29–36. universaar, 2010.

[5] W. M. Farmer. Reasoning about partial functions with the aid of a computer. *Erkenntnis*, 43(3):279–294, 1995.

[6] J. Groenendijk and M. Stokhof. Dynamic predicate logic. *Linguistics and Philosophy*, 14(1):39–100, 1991.

[7] I. Heim. On the projection problem for presuppositions. In M. Barlow, F. D., and M. Westcoat, editors, *Proceedings of the Second West Coast Conference on Formal Linguistics*, pages 114–125. Stanford Linguistics Department, 1983.

[8] N. Kadmon. *Formal Pragmatics, Semantics, Pragmatics, Presupposition, and Focus*. Wiley-Blackwell, 2001.

[9] H. Kamp and U. Reyle. *From Discourse to Logic: Introduction to Model-theoretic Semantics of Natural Language*, volume 42 of *Studies in Linguistics and Philosophy*. Kluwer Academic Publishers, 1993.

[10] E. Landau. *Foundations of Analysis, The Arithmetic of Whole, Rational, Irrational and Complex Numbers*. Chelsea Publishing Company, 1951. Translated by F. Steinhardt.

[11] S. C. Levinson. *Pragmatics*. Cambridge Textbooks in Linguistics. Cambridge University Press, 1983.

[12] W. F. Trench. *Introduction to Real Analysis*. Pearson Education, 2003.

[13] F. Wiedijk and J. Zwanenburg. First order logic with domain conditions. In D. A. Basin and B. Wolff, editors, *Theorem Proving in Higher Order Logics, 16th International Conference, TPHOLs 2003, Rom, Italy, September 8-12, 2003, Proceedings*, volume 2758 of *Lecture Notes in Computer Science*, pages 221–237. Springer-Verlag, 2003.

The foundation axiom and elementary self-embeddings of the universe

Ali Sadegh Daghighi[1], Mohammad Golshani[2], Joel David Hamkins[3,4], Emil Jeřábek[5,*]

[1] School of Mathematics, Amirkabir University of Technology, Tehran, Iran
[2] School of Mathematics, Institute for Research in Fundamental Sciences (IPM), Tehran, Iran
[3] Departments of Mathematics, Philosophy, & Computer Science, The Graduate Center of The City University of New York, New York NY, United States of America
[4] Department of Mathematics, College of Staten Island, Staten Island NY, United States of America
[5] Institute of Mathematics, Akademie věd České republiky, Praha, Czech Republic

Abstract

We consider the role of the foundation axiom and various anti-foundation axioms in connection with the nature and existence of elementary self-embeddings of the set-theoretic universe.

We shall investigate the role of the foundation axiom and the various anti-foundation axioms in connection with the nature and existence of elementary self-embeddings of the set-theoretic universe. All the standard proofs of the well-known Kunen inconsistency [12], for example, the theorem asserting that there is no nontrivial elementary embedding of the set-theoretic universe to itself, make use of the axiom of foundation (cf. [11, 9]), and this use is essential, assuming that ZFC is consistent, because there are models of $\mathsf{ZFC}^{-\mathsf{f}}$ that admit nontrivial elementary self-embeddings and even nontrivial definable automorphisms. Meanwhile, a fragment of the Kunen inconsistency survives without foundation as the claim in $\mathsf{ZFC}^{-\mathsf{f}}$ that there is no nontrivial elementary self-embedding of the class of well-founded sets. Nevertheless, some of the commonly considered anti-foundational theories, such as the Boffa theory BAFA, prove outright the existence of nontrivial automorphisms of the set-theoretic universe, thereby refuting the Kunen assertion in these theories. On the other hand, several other common anti-foundational theories, such as Aczel's anti-foundational theory $\mathsf{ZFC}^{-\mathsf{f}}+\mathsf{AFA}$ and Scott's theory $\mathsf{ZFC}^{-\mathsf{f}}+\mathsf{SAFA}$, reach the opposite conclusion by proving

*This inquiry grew out of a question posed by the first author on MathOverflow [5] and the subsequent exchange posted by the third and fourth authors there. The second author's research has been supported by a grant from IPM (No. 91030417); the third author's research has been supported in part by Simons Foundation grant 209252, by PSC-CUNY grant 66563-00 44 and by grant 80209-06 20 from the CUNY Collaborative Incentive Award program; the fourth author has been supported by grant IAA100190902 of GA AV ČR, Center of Excellence CE-ITI under the grant P202/12/G061 of GA ČR, and RVO: 67985840.

Stefan Geschke, Benedikt Löwe, Philipp Schlicht (*eds.*).
Infinity, computability, and metamathematics: Festschrift celebrating the 60th birthdays of Peter Koepke and Philip Welch. College Publications, London, 2014. Tributes, Volume 23.

that there are no nontrivial elementary embeddings from the set-theoretic universe to itself. Thus, the resolution of the Kunen inconsistency in set theory without foundation depends on the specific nature of one's anti-foundational stance.

In this article, we should like to extend those results and examine the full range of possibility of these various anti-foundational axioms with the existence of such self-embeddings, showing for example that there are models of set theory having automorphisms, but no elementary self-embeddings (Theorem 4.3), or having elementary self-embeddings, but no automorphisms (Theorem 5.1), or having a prescribed group of automorphisms (Theorem 5.5), among other possibilities.

1 Foundation and anti-foundation

The *axiom of foundation* is one of the standard axioms of the Zermelo–Fraenkel (ZFC) axiomatization of set theory, asserting that the set membership relation $\in$ is well-founded, so that every nonempty set has an $\in$-minimal member. Thus, the foundation axiom allows for proofs by $\in$-induction, and indeed the axiom is equivalent to the $\in$-induction scheme. To give one immediate consequence, the axiom of foundation refutes $x \in x$ for any set x, for in this case $\{x\}$ would have no $\in$-minimal element; in particular, the axiom of foundation rules out the existence of *Quine atoms*, sets x for which $x = \{x\}$. Beyond this, the axiom of foundation girds much of the large-scale conceptual framework by which many set-theorists understand the cumulative universe of all sets. For example, one uses it to prove in ZF that every set appears at some level $\mathbf{V}_\alpha$ of the von Neumann hierarchy, for if every element of a set x appears in some $\mathbf{V}_\alpha$, then x itself appears in $\mathbf{V}_{\beta+1}$, where β is larger than the ranks of the elements of x, and so by $\in$-induction every set appears. Similarly, the axiom of foundation implies that every transitive set A is *rigid*, meaning that there is no nontrivial isomorphism of $\langle A, \in\rangle$ with itself, for if $\pi\colon A \to A$ is an isomorphism and π fixes every element $b \in a$ for some $a \in A$, then it follows easily that π must also fix a, and so π is the identity function by $\in$-induction.

Meanwhile, there are several commonly considered anti-foundational set theories, which we shall now briefly review. Most of these theories include the base theory we denote by $\mathsf{ZFC}^{-\mathsf{f}}$, which consists of all axioms of ZFC (including the collection and separation axiom schemes) except for the axiom of foundation.[1] Similarly, $\mathsf{GBC}^{-\mathsf{f}}$ denotes GBC without the axiom of

[1] Just as in the case of set theory $\mathsf{ZFC}^{-\mathsf{p}}$ without the power set axiom (cf. [8]), one should take care to use the collection axiom plus separation in $\mathsf{ZFC}^{-\mathsf{f}}$, rather than merely the replacement axiom, because these are no longer equivalent without foundation, although they are again equivalent without foundation in the presence of any of IE, BAFA, $\mathbf{V} = \mathrm{WF}(t)$ for a set t, or global choice. To see the inequivalence, consider a model with

foundation. In particular, please note that the theory $\mathsf{GBC}^{-\mathsf{f}}$ includes the global axiom of choice, which asserts that there is a class relation that is a set-like well-ordering of all sets; this will be important in a few of our arguments. We shall sometimes also desire the global choice axiom in a context closer to $\mathsf{ZFC}^{-\mathsf{f}}$; in order to achieve this, we expand the language of set theory with a new function symbol C, and work in the theory we denote $\mathsf{ZFGC}^{-\mathsf{f}}$, which has all the $\mathsf{ZFC}^{-\mathsf{f}}$ axioms including instances of the collection and separation schemes for the expanded language, plus the assertion that $C\colon \mathrm{Ord} \to \mathbf{V}$ is a bijection of the class of ordinals with the class of all sets.

Perhaps the most commonly used axiom in non-well-founded set theory is the *anti-foundation axiom*, denoted AFA, first investigated by Forti and Honsell [7] and then popularized by Aczel [1] and Barwise and Moss [3]. The axiom has found numerous applications in computer science and formal semantics. Viale [18] investigated the constructible universe under AFA. In one formulation, AFA asserts that every directed graph $\langle A, e\rangle$ has a unique *decoration*, which is a mapping $a \mapsto f(a)$ of the elements of A to sets, such that

$$f(a) = \{f(b) \mid b\ e\ a\}$$

for every $a \in A$. This axiom therefore extends Mostowski's observation on well-founded relations to apply universally to all directed graphs. For example, by considering a graph with exactly one point and an edge from that point to itself, it follows that under AFA, there is a unique Quine atom $a = \{\,a\,\}$.

Since there are many equivalent formulations of the anti-foundation axiom, let us mention a few more of them. Define that a *partial (inverse) bisimulation* between directed graphs $\langle A, e\rangle$ and $\langle A', e'\rangle$ is a relation $\sim\, \subseteq A \times A'$ such that:[2]

1. For every $x \in A$ and $x', y' \in A'$ such that $x \sim x'$ and $y'\ e'\ x'$, there exists a $y \in A$ such that $y\ e\ x$ and $y \sim y'$.

2. For every $x' \in A'$ and $x, y \in A$ such that $x \sim x'$ and $y\ e\ x$, there exists a $y' \in A'$ such that $y'\ e'\ x'$ and $y \sim y'$.

Such a bisimulation is *total* if $\mathrm{dom}(\sim) = A$ and $\mathrm{ran}(\sim) = A'$, in which case the two structures are called *bisimilar*. The anti-foundation axiom AFA is

ω many Quine atoms, but take only those sets in some $\mathrm{WF}(t)$ for any finite set t of such atoms, so that the full set of atoms never appears; the replacement axiom holds in this model, but not collection.

[2]These are technically *inverse* bisimulations, which have the appropriate directionality for expressing AFA using the $\in$ relation, as we have done here, but we shall drop this 'inverse' qualifier. Other accounts of AFA use the usual bisimulation directionality, but instead invert the set membership relation to $\ni$, in effect inverting the direction of all the graphs here.

equivalent to the statement that every binary relation $\langle A, e\rangle$ is bisimilar with $\langle t, \in\rangle$ for a unique transitive set t. Thus, the axiom generalizes the situation in ZF, where Mostowski's argument shows that every well-founded directed graph is bisimilar to a unique transitive set; under AFA we get this for all directed graphs.

For another variant of AFA, define that an *accessible pointed graph* is a triple $\langle A, e, a\rangle$ where $e \subseteq A \times A$ and a is an element of A to which every element of A is related by the reflexive transitive closure of e. For example, the *canonical picture* of a set a is $\langle \mathrm{TC}(\{a\}), \in, a\rangle$, and this is an accessible pointed graph. Again generalizing the situation in ZF for well-founded relations, the anti-foundation axiom AFA is equivalent to the assertion that an accessible pointed graph $\langle A, e, a\rangle$ is isomorphic to a canonical picture of a set if and only if it is strongly extensional, meaning that every partial bisimulation from $\langle A, e\rangle$ to itself agrees with the identity relation on A.

Scott [16] considered an anti-foundation axiom—following Aczel we shall denote it by SAFA—based on trees instead of arbitrary accessible pointed graphs. An accessible pointed graph $\langle A, e, a\rangle$ is a *tree* if every vertex has a unique e-path to a. For example, the *canonical tree picture* of a set a is the tree whose vertices are finite sequences $\langle x_0, \dots, x_n\rangle$ where $x_0 = a$, and $x_{i+1} \in x_i$ for every $i < n$; we may think of the tree as growing downwards, so that child nodes correspond to elements. In ZF, one may easily prove that the canonical tree picture of a set has no nontrivial automorphisms, since every such automorphism would give rise to an automorphism of the transitive closure of the set. The Scott anti-foundation axiom SAFA generalizes this to the non-wellfounded realm by asserting that a tree is isomorphic to a canonical tree picture of a set if and only if it has no nontrivial automorphism. To illustrate, observe that under SAFA, there is precisely one Quine atom: there cannot be two, because if a and b were distinct Quine atoms, then the canonical tree picture of the doubleton set $\{\, a, b \,\}$ would have a nontrivial automorphism, swapping a and b; and there must be at least one, because the tree consisting of a single infinite descending chain is rigid, and so there must be sets a_n with $a_n = \{\, a_{n+1} \,\}$; but these must all be equal, because if $a_n \neq a_m$, then the tree picture of the doubleton $\{\, a_n, a_m \,\}$ consists of two descending chains joined at the top, and this tree is not rigid.

More generally, we say that two transitive sets s and t are isomorphic if there is an isomorphism of the structure $\langle s, \in\rangle$ with $\langle t, \in\rangle$. In the case of non-transitive sets, we say that two sets x and y are isomorphic, if they can be placed into transitive sets having an isomorphism mapping x to y, or equivalently, if the transitive closures of $\{\, x \,\}$ and $\{\, y \,\}$ are isomorphic by an isomorphism mapping x to y. Thus, two sets are isomorphic if and only if their canonical pictures are isomorphic as accessible pointed graphs.

Finsler [6] developed non-well-founded set theory based on the informal principle that the universe of sets is maximal, subject to maintaining the axioms of extensionality and of *isomorphism extensionality*:

$$\text{Isomorphic sets are equal.} \qquad (\mathsf{IE})$$

This axiom implies immediately that there is at most one Quine atom, since any two would be isomorphic. Aczel [1] formalized Finsler's idea in an axiom denoted FAFA, which asserts that an accessible pointed graph $G = \langle A, e, a \rangle$ is isomorphic to a canonical picture of a set if and only if it is extensional (that is, satisfies the axiom of extensionality) and the accessible pointed graph Gu of points accessing u is not isomorphic to Gv for distinct $u, v \in A$.

Actually, Aczel introduced an entire family of axioms $\mathsf{AFA}^{\sim}$, one for each so-called regular bisimulation concept $\sim$. Each axiom characterizes canonical pictures as accessible pointed graphs satisfying a version of extensionality appropriate for $\sim$, and particular choices of $\sim$ yield the axioms AFA, SAFA, and FAFA. Each $\mathsf{AFA}^{\sim}$ can be thought of as consisting of two parts: existence (stating that certain accessible pointed graph correspond to canonical pictures) and uniqueness (asserting a strengthening of the extensionality axiom). Larger $\sim$ lead to weaker existence assertions and stronger uniqueness assertions. The extremes are FAFA and AFA, which correspond to the smallest and largest regular bisimulation, respectively. In particular, the uniqueness part of every $\mathsf{AFA}^{\sim}$ implies IE, which is the uniqueness part of FAFA.

Boffa [4] introduced a theory which maximizes the universe of sets with respect to the plain axiom of extensionality, thus badly violating the isomorphism extensionality axiom IE. A weak version of Boffa's axiom postulates that every extensional binary relation $\langle a, e \rangle$ is isomorphic to $\langle t, \in \rangle$ for a transitive set t. The drawback of this statement is that it provides no easy way of extending existing sets: even if $\langle a, e \rangle$ already includes a set $\langle t_0, \in \rangle$ as its transitive part, the isomorphism from a to t does not have to preserve it, as there may be many sets isomorphic to t_0. For this reason, the full Boffa's axiom (the axiom of *superuniversality*), denoted here BAFA, asserts that for every transitive set t_0 and every extensional binary relation $\langle a, e \rangle$ that end-extends $\langle t_0, \in \rangle$, meaning $t_0 \subseteq a$ and $\in \restriction t_0 = e \cap (a \times t_0)$, there exists a transitive set t, and an isomorphism from $\langle a, e \rangle$ to $\langle t, \in \rangle$ that is the identity on t_0. Thus, the axiom asserts a kind of saturation property for the transitive sets, namely, that they realize the types expressed by extensional binary relations end-extending a given transitive set. For example, under BAFA, there must be a proper class of Quine atoms, since we can extend the canonical picture of a given set of Quine atoms by a relation that describes what it would be like to have one more, or κ many more for any cardinality κ, and this new relation must be realized in a transitive set, which will have corresponding additional actual Quine atoms.

This level of saturation causes a high degree of homogeneity in any set-theoretic universe satisfying BAFA, where we have many distinct but isomorphic copies of whatever structure is produced. Such homogeneity, in turn, can cause a difficulty in class-length constructions by transfinite recursion, since the constructed objects are rarely unique, and so one cannot usually pick out the precise continuation of a given transfinite construction without a class choice principle. For this reason, it is convenient to include in Boffa's theory the global axiom of choice, which allows one to make choices in such a context, relative to a fixed class well-ordering of the sets. Thus, we work with BAFA over the theory $\mathsf{ZFGC}^{-\mathsf{f}}$ or $\mathsf{GBC}^{-\mathsf{f}}$, which includes the global axiom of choice. Boffa's theory has been used as a basis of formalization of nonstandard analysis by Ballard and Hrbáček [2].

In this article, we shall also make a few arguments in the theory asserting that the universe of sets is generated by a set or class of Quine atoms. Such theories are closely connected with the permutation models of set theory. Permutation models were originally constructed for the set theory ZFA with atoms (objects with no elements, but distinct from the empty set), which requires weakening of the axiom of extensionality. Alternatively, one can replace atoms with Quine atoms, simply by redefining the $\in$-relation on the atoms to make them all into self-singletons, that is, into Quine atoms; in this way extensionality is preserved at the expense of dropping foundation, which may be considered a less drastic deviation from the axioms of ZFC. Let us explain the precise meaning of being "generated". For any transitive class A, we define the *cumulative hierarchy over* A, the class denoted $\mathrm{WF}(A)$, as the elements appearing in the following recursive hierarchy:

$$\begin{aligned} \mathrm{WF}_0(A) &= A, \\ \mathrm{WF}_{\alpha+1}(A) &= \wp(\mathrm{WF}_\alpha(A)), \\ \mathrm{WF}_\lambda(A) &= \bigcup_{\alpha<\lambda} \mathrm{WF}_\alpha(A) \quad \text{for limit } \lambda \end{aligned}$$

For example, the class of well-founded sets is simply $\mathrm{WF}(\varnothing)$. When A is a proper class, note that we take only the sub*sets* of the previous stage. In weaker set theories that may not be able to formalize this recursion on classes directly, we may equivalently define that $\mathrm{WF}(A)$ is the union of $\mathrm{WF}(t)$ for all sets $t \subseteq A$, and similarly $\mathrm{WF}_\alpha(A) = \bigcup_{t\subseteq A} \mathrm{WF}_\alpha(t)$. Every set $x \in \mathrm{WF}(A)$ has a corresponding rank, the least ordinal stage α for which $x \in \mathrm{WF}_{\alpha+1}(A)$. Such a rank function leads to a weak form of the axiom of foundation in $\mathrm{WF}(A)$, namely, every set x having an element y in $\mathrm{WF}(A)$ has such an element y of least rank, and such an element y will either be in A or have no elements in common with x, since any such element would have a lower rank than y in x. In particular, every infinite $\in$-descending sequence containing a set in $\mathrm{WF}(A)$ must eventually reach an element of A.

The class $\mathrm{WF}(A)$ is transitive, as well as *full*, meaning that every subset of $\mathrm{WF}(A)$ is an element of $\mathrm{WF}(A)$, and indeed, $\mathrm{WF}(A)$ is the smallest full class containing A. When $\mathbf{V} = \mathrm{WF}(A)$, then we say that the universe is *generated* by A, and in some of our arguments below, we shall consider models generated by a class of Quine atoms. Let us introduce the notation At to refer always to the class of Quine atoms, so that $\mathbf{V} = \mathrm{WF}(\mathrm{At})$ just in case the universe is generated by its Quine atoms.

All of the anti-foundational theories we consider in this article are equiconsistent with ZFC and therefore also with GBC. A convenient tool for showing the consistency of non-well-founded set theories is *Rieger's theorem* [15] (cf. [1]), which shows in $\mathsf{ZFC}^{-\mathsf{f}}$ that if M is a class endowed with a relation $E \subseteq M \times M$ that is extensional, set-like (meaning that the E-predecessors of any $a \in M$ form a set) and full (in the sense that every subset of M is the E-extension of some $a \in M$), then $\langle M, E\rangle$ satisfies all the axioms of $\mathsf{ZFC}^{-\mathsf{f}}$. If global choice is available, then we may also expand $\langle M, E\rangle$ to a model of global choice. In particular, any full transitive class, such as $\mathrm{WF}(A)$ for any transitive class A, is a model of $\mathsf{ZFC}^{-\mathsf{f}}$.

With suitable choices for such relations E (as explained in [1]), we may arrange that $\langle M, E\rangle$ is extensional, set-like and full, while also satisfying AFA, SAFA, FAFA, or BAFA, whichever we prefer. In this way, inside any model of GBC we may construct models of $\mathsf{GBC}^{-\mathsf{f}}$ plus any of these anti-foundational theories. Furthermore, the WF of any full model $\langle M, E\rangle$ is isomorphic to the WF of the universe in which it is constructed, and one may also incorporate classes into this phenomenon. It follows that all these anti-foundational theories T are conservative over ZFC (and hence also GBC), in the sense that $T \vdash \varphi^{\mathrm{WF}}$ if and only if $\mathsf{ZFC} \vdash \varphi$ for any first-order formula φ, since every model of ZFC will arise as the WF of a model of the anti-foundational theory T, and similarly every model of GBC arises as the well-founded part (including both sets and classes) of a model of T.

We have mentioned that in any model of $\mathsf{ZFGC}^{-\mathsf{f}} + \mathsf{BAFA}$, there are a proper class of Quine atoms, and for any class A of Quine atoms, we may form the transitive class $\mathrm{WF}(A)$ inside this model. In particular, the theories asserting $\mathsf{GBC}^{-\mathsf{f}}$ plus "the universe is generated by a proper class of Quine atoms," or "the universe is generated by a set of five Quine atoms," or "... by a set of precisely $\aleph_6$ many Quine atoms," and so on, respectively, for any definable cardinality whose definition is absolute to $\mathrm{WF}(A)$, are each equiconsistent with ZFC and conservative over ZFC for assertions about the well-founded sets. One may also establish this directly from Rieger's theorem rather than via BAFA.

2 Some meta-mathematical issues

A number of meta-mathematical issues arise in any formalization of the Kunen inconsistency (we refer the readers to the discussion in the preliminary section of [9]). The most obvious issue, of course, is that the quantifier involved in the assertion "there is no nontrivial elementary embedding j" is a second-order quantifier, not directly formalizable in the usual first-order theories such as ZFC. Many set theorists prefer to interpret all talk of classes in ZFC as referring to the first-order definable classes, and with such a formalization, the Kunen inconsistency becomes a scheme, asserting of each possible definition of j that it isn't an elementary embedding of the universe. (For example, Kanamori [11] adopts this approach.) Nevertheless, this interpretation seems to miss much of the substance of the theorem, because there is an elementary proof of this formulation of the Kunen inconsistency, a simple diagonal argument not relying on the axiom of choice or any of the other combinatorial methods usually associated with the Kunen inconsistency (cf. [17] and further discussion in [9]).

So it seems natural to use a second-order set theory, such as Gödel–Bernays or Kelly–Morse set theory. In GBC, we have class quantifiers for expressing "$\exists j$," but then a second problem arises, namely, that it is not directly possible to express the assertion "j is elementary" in GBC. Kunen himself originally formulated his theorem in Kelley–Morse set theory KM precisely for this reason, since KM proves the existence of a full satisfaction class for first-order truth, making the assertion "j is elementary" expressible in KM. The Kelley-Morse theory, however, is strictly stronger theory than ZFC, even for arithmetical statements, and has a strictly higher consistency strength. Meanwhile, one can actually carry out Kunen's argument in the weaker theory GBC and even in ZFC(j), which is equiconsistent with and conservative over ZFC, by weakening the full elementarity of j to the assertion merely that "j is Σ_1-elementary," which is formalizable in the first order language of set theory with a predicate for the class j. The point is that Kunen's argument actually proves this stronger statement, that there is no nontrivial Σ_1-elementary embedding j from $\mathbf{V}$ to $\mathbf{V}$, and indeed, no Δ_0-elementary cofinal embedding from $\mathbf{V}$ to $\mathbf{V}$ (cf. the discussion in [9]). Part of the reason for this is a lemma of Gaifman's, which asserts that if $j\colon \mathbf{V} \to \mathbf{V}$ is Σ_1-elementary, then it is Σ_n-elementary for every meta-theoretic natural number n, by an induction carried out in the meta-theory. One issue here, however, is that Gaifman's lemma makes use of the fact that in ZF, every Σ_1-elementary embedding $j\colon \mathbf{V} \to \mathbf{V}$ is cofinal, in the sense that $\bigcup j[\mathbf{V}] = \mathbf{V}$, and this is no longer necessarily true in the absence of foundation, even with full elementarity, as shown in Theorem 4.3 statement 3, although we may still assert that $\bigcup j[\mathbf{V}]$ is full. It is relatively consistent with $\mathsf{ZFC}^{-\mathsf{f}}$ that there exists a Σ_1-elementary embedding $j\colon \mathbf{V} \to \mathbf{V}$ which is not elementary (although we omit the proof), and so this approach to formalizing the Kunen inconsistency statement does not fully succeed.

Some set theorists note that the proofs of the Kunen inconsistency show in fact that there can be no nontrivial elementary embedding of the form $j\colon \mathbf{V}_{\lambda+2} \to \mathbf{V}_{\lambda+2}$, a stronger statement that is expressible in the first-order language of set theory and provable in ZFC with no talk of classes. This formulation of the Kunen assertion, however, is not suitable in a context without foundation, since there could be an embedding $j\colon \mathbf{V} \to \mathbf{V}$ which is nontrivial, but only on ill-founded sets (as in Theorem 4.1), and so ruling out $j\colon \mathbf{V}_{\lambda+2} \to \mathbf{V}_{\lambda+2}$ does not settle the question for $j\colon \mathbf{V} \to \mathbf{V}$ in the anti-foundational context.

In this article, we shall take a pragmatic approach. Rather than attempting to give a universally applicable definition of the Kunen assertion, we shall instead simply present the strongest results we can prove for each of the particular anti-foundational theories on a case-by-case basis. Our results on the nonexistence of embeddings usually apply already to Σ_1-elementary embeddings, and we formulate them generally over $\mathsf{GBC}^{-\mathsf{f}}$. Results on the existence of nontrivial embeddings provide definable embeddings in the $\mathsf{ZFC}^{-\mathsf{f}}$ or $\mathsf{ZFGC}^{-\mathsf{f}}$ versions of the theories, in which case the elementarity of the embedding can be formalized as an infinite schema. Sometimes the theory proves the existence of definable classes with a stronger property (given by a single formula) which implies the infinite elementarity schema; in particular, this is the case for *automorphisms* of the universe, meaning bijections $j\colon \mathbf{V} \to \mathbf{V}$ satisfying $x \in y \leftrightarrow j(x) \in j(y)$. A similar situation will arise with the elementary embeddings that we construct by threading a back-and-forth system of embeddings from one model to another.

Let us stress that throughout the paper, elementary embeddings and isomorphisms are only supposed to be elementary with respect to the first-order language of set theory; they need not be elementary with respect to the second-order language of classes, when working in $\mathsf{GBC}^{-\mathsf{f}}$, or with respect to the language including the global choice bijection of Ord with $\mathbf{V}$, when working in $\mathsf{ZFGC}^{-\mathsf{f}}$.

3 The Kunen inconsistency on the well-founded sets

Let us now finally begin in earnest with the observation that a fragment of the Kunen inconsistency survives in set theory without the foundation axiom as the claim that there are no nontrivial elementary self-embeddings on the well-founded part of the universe.

Theorem 3.1. Work either in $\mathsf{GBC}^{-\mathsf{f}}$ or in $\mathsf{ZFC}^{-\mathsf{f}}$. Then there is no nontrivial Σ_1-elementary embedding $j\colon \mathrm{WF} \to \mathrm{WF}$. In particular, every Σ_1-elementary embedding $j\colon \mathbf{V} \to \mathbf{V}$ fixes every well-founded set: $j(x) = x$ for all $x \in \mathrm{WF}$. Furthermore, the range $j[\mathbf{V}]$ of any such embedding is a transitive full class.

Proof. Consider the structure $\mathcal{W}$ obtained by restricting the universe to have only the objects in WF and to have only the classes A with $A \subseteq \mathrm{WF}$. Since WF is a definable class in the full universe, it follows that $A \cap \mathrm{WF}$ is a class whenever A is, and it is an elementary exercise to see that $\mathcal{W}$ is a model of GBC. In particular, if $j\colon \mathrm{WF} \to \mathrm{WF}$ were a nontrivial Σ_1-elementary embedding of WF, then j would be a class in $\mathcal{W}$ and furthermore would be a nontrivial Σ_1-elementary embedding of the entire set-theoretic universe from the perspective of $\mathcal{W}$, contrary to the original Kunen inconsistency, which shows that there can be no such embedding in any model of GBC.

Alternatively, if $j\colon \mathrm{WF} \to \mathrm{WF}$ were nontrivial and Σ_1-elementary, then since $\mathrm{WF} \models \mathsf{ZFC}$ it follows that j must have a critical point κ, and then one can define the critical sequence $\kappa_{n+1} = j(\kappa_n)$ and $\lambda = \sup_n \kappa_n$. It follows easily that $j(\lambda) = \lambda$ and consequently $j \restriction \mathbf{V}_{\lambda+2}\colon \mathbf{V}_{\lambda+2} \to \mathbf{V}_{\lambda+2}$ is an elementary embedding from $\mathbf{V}_{\lambda+2}$ to itself, which violates one of the ZFC formulations of the Kunen inconsistency inside WF.

For the further claims of the theorem, we claim that if $j\colon \mathbf{V} \to \mathbf{V}$ is Σ_1-elementary, then $j \restriction \mathrm{WF}\colon \mathrm{WF} \to \mathrm{WF}$ is also Σ_1-elementary and hence trivial by the arguments of the previous paragraphs. To see this, note first that $x \in \mathrm{WF}$ is Π_1 definable, since $x \in \mathrm{WF}$ just in case there is no infinite $\in$-descending sequence starting from x, and so $j[\mathrm{WF}] \subseteq \mathrm{WF}$. (And the reader may find it helpful to note that the Σ_1-elementarity of j implies that j preserves true Σ_2 assertions.) Similarly, $j[\mathrm{Ord}]$ is cofinal in Ord, since being an ordinal is also Π_1 definable (note that the usual Δ_0 definitions of being an ordinal in ZFC do not work in the anti-foundational context). Since $x \in \mathrm{WF}_\alpha$ just in case there is a ranking function, a function $f: t \to \alpha$ for some transitive set t for which $x \in t$ and $a \in b \implies f(a) < f(b)$. Since membership in WF_α is therefore a Σ_1 property in parameter α, it follows that $y = \mathrm{WF}_\alpha$ has complexity $\Sigma_1 \wedge \Pi_1$ in y and α, namely, every set in y has an α-ranking function and every set with an α-ranking function is in y, and so $j(\mathrm{WF}_\alpha) = \mathrm{WF}_{j(\alpha)}$. To see that $j \restriction \mathrm{WF}\colon \mathrm{WF} \to \mathrm{WF}$ is Σ_1-elementary, notice that if there is an existential witness on the right side, $\mathrm{WF} \models \exists x\, \varphi(x, j(u))$, where φ is Δ_0, then the witness x is found in some WF_β, which is contained in some $\mathrm{WF}_{j(\alpha)}$, since $j[\mathrm{Ord}]$ is cofinal in Ord, and so $\mathbf{V} \models \exists x \in \mathrm{WF}_{j(\alpha)}\, \varphi(x, j(u))$, which implies $\exists x \in \mathrm{WF}_\alpha\, \varphi(x, u)$ by the Σ_1-elementarity of j, and so $\mathrm{WF} \models \exists x\, \varphi(x, u)$ as desired.

In order to show that the range of j is transitive, assume $x \in j(u)$, and by the axiom of choice fix a bijection $f\colon \kappa \to u$ with some ordinal κ. It follows that $j(f)\colon j(\kappa) \to j(u)$ is also a bijection, and so $x = j(f)(\alpha)$ for some ordinal α. But since j is the identity on ordinals, it follows that $\alpha = j(\alpha)$ and so $x = j(f)(\alpha) = j(f)(j(\alpha)) = j(f(\alpha))$, which places x into the range of j, as desired.

Finally, to verify fullness, suppose $x \subseteq j[\mathbf{V}]$. Let $u = j^{-1}[x]$, so that $x = j[u]$. Clearly $j[u] \subseteq j(u)$, but since $j(u) \subseteq j[\mathbf{V}]$ by transitivity, we achieve also the converse, and so $j(u) = j[u] = x$. Q.E.D.

The fact that the range of such an embedding j must be transitive makes them totally different in character than the kinds of embeddings that are usually considered in the ZFC context, which never have transitive range.

4 Theories with nontrivial self-embeddings of the universe

We shall now prove that ZFC without the foundation axiom is relatively consistent with the existence of nontrivial automorphisms of the set-theoretic universe. In particular, if ZFC is consistent, then the Kunen inconsistency assertion is not provable in ZFC or GBC without making use of the axiom of foundation. We begin with the theories where the universe is generated by its Quine atoms, a situation where the automorphisms and elementary embeddings have a very transparent structure, admitting a complete description.

Theorem 4.1. Work in $\mathsf{GBC}^{-\mathsf{f}}$ or $\mathsf{ZFC}^{-\mathsf{f}}$ and assume that the universe is generated by its Quine atoms, so that $\mathbf{V} = \mathrm{WF}(\mathrm{At})$.

1. If $j\colon \mathbf{V} \to \mathbf{V}$ is an automorphism of the universe $\mathbf{V}$, then $j \restriction \mathrm{At}$ is a permutation of the class of atoms At.
2. Every permutation $\sigma\colon \mathrm{At} \to \mathrm{At}$ of the atoms has a unique extension to an automorphism $\overline{\sigma}\colon \mathbf{V} \to \mathbf{V}$ of the entire universe.

Proof. The first assertion is clear, as the class of Quine atoms is definable. For the second assertion, suppose that σ is a permutation of the class of Quine atoms At. We may extend this to a permutation $\overline{\sigma}$ of all of WF(At) by the following recursion:

$$\overline{\sigma}(x) = \begin{cases} \sigma(x) & \text{if } x \in \mathrm{At}, \\ \overline{\sigma}[x] & \text{otherwise} \end{cases} \qquad (*)$$

We may view this directly as a recursion on the set-like well-founded relation $\in \setminus \mathrm{id}_{\mathrm{At}}$, or alternatively, as defining $\overline{\sigma} \restriction \mathrm{WF}(\mathrm{At})_\alpha$ by recursion on the rank α. It follows easily by induction that $\overline{\sigma} \restriction \mathrm{WF}(\mathrm{At})_\alpha$ is an $\in$-preserving automorphism of $\mathrm{WF}(\mathrm{At})_\alpha$, for every ordinal α, and so $\overline{\sigma}\colon \mathbf{V} \to \mathbf{V}$ is an automorphism of the entire universe $\mathbf{V}$. Furthermore, any automorphism extending σ must obey $(*)$, and so $\overline{\sigma}$ is the unique such extension of σ to all of $\mathbf{V}$. Q.E.D.

In particular, if At consists of at least two Quine atoms, then we have a permutation swapping two of them (leaving the rest in place), and so we may extend this to an automorphism of the entire universe WF(At), which will be definable from the Quine atoms to be swapped. In Theorem 4.3, we shall use the recursion $(*)$ even when $\sigma\colon \mathrm{At} \to \mathrm{At}$ is merely injective, yielding an embedding $\overline{\sigma}\colon \mathrm{WF}(\mathrm{At}) \to \mathrm{WF}(\mathrm{At})$.

Corollary 4.2. Work in $\mathsf{ZFC}^{-\mathsf{f}}$. If the universe is generated by its Quine atoms, $\mathbf{V} = \mathrm{WF}(\mathrm{At})$, and there are at least two Quine atoms, then there is a nontrivial automorphism of $\mathbf{V}$, definable from parameters.

If the universe is generated from exactly two atoms, then we don't need any parameters to define the automorphism, since in this case σ and hence $\overline{\sigma}$ are both definable without parameters, since there is only one nontrivial permutation of a two-element set. The theorem generalizes from automorphisms to elementary embeddings as follows.

Theorem 4.3. Work in $\mathsf{GBC}^{-\mathsf{f}}$ or $\mathsf{ZFC}^{-\mathsf{f}}$, and assume the universe is generated by its Quine atoms, so that $\mathbf{V} = \mathrm{WF}(\mathrm{At})$.

(1) If $j\colon \mathbf{V} \to \mathbf{V}$ is a Σ_1-elementary embedding, then $j{\restriction}\mathrm{At}$ is an injection of At to At, and $j = \overline{j{\restriction}\mathrm{At}}$ using the notation from $(*)$.

(2) If At is a set, every Σ_1-elementary embedding $j\colon \mathbf{V} \to \mathbf{V}$ is an automorphism.

(3) (Assuming global choice.) If the class of Quine atoms At is a proper class and $\sigma\colon \mathrm{At} \to \mathrm{At}$ is injective, then $j = \overline{\sigma}\colon \mathbf{V} \to \mathbf{V}$ is an elementary embedding, that is, Σ_n-elementary for every particular natural number n.

Proof. (1) The first assertion is clear as $j(a) \in j(a)$ for any $a \in \mathrm{At}$. Also, $j(x) \supseteq j[x]$. On the other hand, since $j[\mathbf{V}]$ is transitive by theorem 3.1, every element of $j(x)$ is of the form $j(y)$, where necessarily $y \in x$, and so $j(x) = j[x]$, which implies that j obeys $(*)$.

(2) The property of being the set of all atoms is Π_1, and so $j(\mathrm{At}) = \mathrm{At}$. But since the range of j is transitive, it follows that $j[\mathrm{At}] = \mathrm{At}$ and hence $j{\restriction}\mathrm{At}$ is a permutation, which implies that $j = \overline{j{\restriction}\mathrm{At}}$ is an automorphism.

(3) This statement is a theorem scheme, a separate statement for each meta-theoretic natural number n. Assume that At is a proper class, with an injection $\sigma\colon \mathrm{At} \to \mathrm{At}$. Let $j = \overline{\sigma}\colon \mathbf{V} \to \mathbf{V}$ be the corresponding embedding arising from σ via $(*)$, and let $A = \sigma[\mathrm{At}]$, which is a proper class subclass of At. Since inductively j is an isomorphism of $\mathrm{WF}_\alpha(\mathrm{At})$ with $\mathrm{WF}_\alpha(A)$, for every ordinal α, it follows that j is an isomorphism of WF(At) with WF(A). In order to see that j is Σ_n-elementary from $\mathbf{V}$ to $\mathbf{V}$, therefore, it therefore

suffices to show that $\mathrm{WF}(A) \prec_{\Sigma_n} \mathrm{WF}(\mathrm{At})$. For this, we verify the Tarski–Vaught criterion: assume that a Σ_n statement $\varphi(u, x)$ holds in $\mathbf{V}$ with $u \in \mathrm{WF}(A)$ and $x \in \mathbf{V}$, we have to find $v \in \mathrm{WF}(A)$ such that $\varphi(u, v)$. Fix sets $a \subseteq A$ and $a \subseteq b \subseteq \mathrm{At}$ such that $u \in \mathrm{WF}(a)$ and $x \in \mathrm{WF}(b)$. Using global choice, we can find an injection of b into A identical on a, which we can extend to a permutation $\tau\colon \mathrm{At} \to \mathrm{At}$. Since $\overline{\tau}$ is an automorphism identical on $\mathrm{WF}(a)$, we have $\varphi(u, \overline{\tau}(x))$, where $\overline{\tau}(x) \in \mathrm{WF}(\tau[b]) \subseteq \mathrm{WF}(A)$. Q.E.D.

We remark that statement (2) applies more generally: if $\mathbf{V} = \mathrm{WF}(t)$ for some set t, then every Σ_1-elementary embedding $j\colon \mathbf{V} \to \mathbf{V}$ is an automorphism: since $\mathbf{V} = \mathrm{WF}(t)$ is a Π_1 property of t, asserting that every infinite $\in$-descending sequence contains an element of t, it follows that $\mathbf{V} = \mathrm{WF}(j(t))$, which implies $\mathbf{V} = j[\mathbf{V}]$ as $j[\mathbf{V}]$ is full.

Corollary 4.4. Work in $\mathsf{ZFC}^{-\mathsf{f}}$ and assume that the universe is generated from a proper class of Quine atoms. Then there is an embedding $j\colon \mathbf{V} \to \mathbf{V}$, definable from parameters and Σ_n-elementary for every particular n, which is not an automorphism.

Proof. The result follows from Theorem 4.3 statement 3, if we have global choice. But let us prove it without global choice, using only the usual first-order axiom of choice. Since At is a proper class, we can prove the existence of at least n atoms for every $n \in \omega$ by induction on n. Using collection and choice, we can thus find an infinite countable set $\{a_n : n \in \omega\} \subseteq \mathrm{At}$. Define $\sigma\colon \mathrm{At} \to \mathrm{At}$ by $\sigma(a_n) = a_{n+1}$ and $\sigma(a) = a$ otherwise. This injection extends as in Theorem 4.3 to an isomorphism $\overline{\sigma}$ of $\mathrm{WF}(\mathrm{At})$ with $\mathrm{WF}(A)$, where $A = \mathrm{At} \setminus \{a_0\}$. In the proof of the elementarity of $\mathrm{WF}(A)$ in $\mathrm{WF}(\mathrm{At})$ in Theorem 4.3, we can assume $b = a \cup \{a_0\}$, hence we can define τ as a transposition of a_0 with any atom in $\mathrm{At} \setminus b$ without using any choice.

Thus, the embedding j we produce arises via $(*)$ from an injection on At, and our argument shows as a theorem scheme that any embedding arising this way is Σ_n-elementary for any meta-theoretic natural number n. Q.E.D.

As we mentioned after Corollary 4.2, in $\mathsf{ZFC}^{-\mathsf{f}} + \mathbf{V} = \mathrm{WF}(\mathrm{At})$, where At has precisely two Quine atoms, the unique nontrivial automorphism from Theorem 4.1 is definable without parameters, as it is induced by swapping the two atoms. On the other hand, if $|\mathrm{At}| \neq 2$, there is no nontrivial parameter-free definable Σ_1-elementary embedding, because any parameter-free definable class is preserved by all automorphisms, but the only function $\mathrm{At} \to \mathrm{At}$ that commutes with every permutation is the identity.

Let us now move on to Boffa's theory, which turns out to be rich in elementary embeddings and automorphisms. We shall start with some general remarks on basic consequences of BAFA. First, the axiom can be equivalently formulated in the following more convenient form: if $\langle a, e\rangle$ is an extensional binary relation which is an end-extension of $\langle a_0, e_0\rangle$, and f_0 is an isomorphism from $\langle a_0, e_0\rangle$ to $\langle t_0, \in\rangle$ with transitive t_0, there exists an isomorphism $f\colon \langle a, e\rangle \to \langle t, \in\rangle$ extending f_0 with transitive t.

Using global choice, BAFA implies a generalized Mostowski collapse lemma, asserting that every set-like extensional relation E on a class A is isomorphic to the $\in$ relation on some transitive class T. (The usual Mostowski collapse applies only to well-founded relations.) This can be shown by writing A as the union of an increasing chain $\{a_\alpha : \alpha \in \mathrm{Ord}\}$ of E-transitive subsets $a_\alpha \subseteq A$, and constructing an increasing chain of isomorphisms $f_\alpha\colon \langle a_\alpha, E\restriction a_\alpha\rangle \to \langle t_\alpha, \in\rangle$ with transitive t_α. Its union is then an isomorphism of $\langle A, E\rangle$ to $T = \bigcup_\alpha t_\alpha$. This has immediate consequences for the construction of elementary embeddings of $\mathbf{V}$ into transitive classes M. For example, every ultrapower $\langle \mathbf{V}^I/U, \in_U\rangle$ of the universe by an ultrafilter U on a set I gives rise to the corresponding set-like extensional relation $\in_U$, which therefore is realized as a transitive class $\langle M, \in\rangle \cong \langle \mathbf{V}^I/U, \in_U\rangle$ via the generalized Mostowski collapse, and so the ultrapower map provides an elementary embedding $j\colon \mathbf{V} \to M$ into a transitive class M, even when U is not countably complete, which of course does not happen in ZFC. This feature is essential for the development of nonstandard analysis in the framework of [2]. However, one cannot construct a nontrivial elementary embedding into $\mathbf{V}$ itself in this way.

An important special case of BAFA is that every isomorphism of transitive sets can be extended to an isomorphism whose domain and range contains any given set. This means that the class of all isomorphisms of transitive sets forms a *back-and-forth system* from $\mathbf{V}$ to $\mathbf{V}$ (cf. [13]), and consequently BAFA proves (as a scheme) that every isomorphism of transitive sets preserves the truth of any particular formula. With global choice, one can carry out the full back-and-forth construction:

Theorem 4.5. $\mathsf{ZFGC}^{-\mathsf{f}}+\mathsf{BAFA}$ proves that every isomorphism of transitive sets can be extended to an automorphism of the universe. In particular, there exist nontrivial automorphisms.

Proof. Let f_0 be any isomorphism of transitive sets. Using global choice and the back-and-forth property we just mentioned above, we may construct an increasing chain $\{f_\alpha : \alpha \in \mathrm{Ord}\}$ of isomorphisms of transitive sets such that every set eventually belongs to $\mathrm{dom}(f_\alpha) \cap \mathrm{ran}(f_\alpha)$ as α becomes large. The union $j = \bigcup_{\alpha\in\mathrm{Ord}} f_\alpha$ is therefore an automorphism of $\mathbf{V}$ extending f_0. If f_0 itself is nontrivial, such as an isomorphism of one Quine atom with another, we thereby ensure that j is nontrivial. Q.E.D.

Remark 4.6. It is not difficult to prove in BAFA that if a is any set and $b \notin \mathrm{WF}(\mathrm{TC}(a))$, then there is a proper class of sets isomorphic to b by an isomorphism fixing a. In view of Theorem 4.5, this means that the orbit of b under automorphisms of $\mathbf{V}$ fixing a is a proper class, suggesting a rich Galois theory here.

Turning from automorphisms to general elementary embeddings, we have the following criterion.

Theorem 4.7. Work in $\mathsf{ZFGC}^{-\mathsf{f}} + \mathsf{BAFA}$ or $\mathsf{GBC}^{-\mathsf{f}} + \mathsf{BAFA}$. The following are equivalent for any class M:

(1) $M = j[\mathbf{V}]$ for some elementary embedding $j\colon \mathbf{V} \to \mathbf{V}$.

(2) $M = j[\mathbf{V}]$ for some Σ_1-elementary embedding $j\colon \mathbf{V} \to \mathbf{V}$.

(3) M is transitive and isomorphic to $\mathbf{V}$.

(4) M is a full transitive model of BAFA.

Proof. The formalization of statement (1) is problematic, since we cannot express it directly even as a scheme. What we mean by including it here is that, first, it contains statement (2) as an immediate special case; and second, if statement (4) holds, then using M and the global choice function we may define a particular class embedding $j\colon \mathbf{V} \to \mathbf{V}$ for which $M = j[\mathbf{V}]$ and prove that any class satisfying this definition is Σ_n-elementary for any particular natural number n. The implication (1) $\to$ (2) is thereby clear, and (3) $\to$ (4) also, since isomorphisms are truth-preserving for any given assertion. The implication (2) $\to$ (3) follows from Theorem 3.1.

For the remaining implication (4) $\to$ (1), let I be the class of all isomorphisms $f\colon t \to s$ of transitive sets, with $s \subseteq M$. We claim that I is a back-and-forth system from $\mathbf{V}$ to M. The back direction follows directly from BAFA; for the forth direction, let $f \in I$ be as above, and $t' \supseteq t$ be a transitive set. Using the axiom of choice, we can construct an isomorphism $g\colon \langle t', \in\rangle \to \langle a, e\rangle$ extending f, where $\langle a, e\rangle$ is an extensional structure end-extending $\langle s, \in\rangle$ such that $a \setminus s \subseteq \mathrm{Ord}$. Since M is full, $\langle a, e\rangle \in M$, hence there is an isomorphism $h\colon \langle a, e\rangle \to \langle s', \in\rangle$ extending id_s for some transitive set $s' \in M$ using $M \models \mathsf{BAFA}$. Then $f' = h \circ g \in I$ extends f to t'.

Using transfinite recursion as in the proof of Theorem 4.5, we can construct an isomorphism $j\colon \mathbf{V} \simeq M$, which will be definable from M and the global choice function. Moreover, if $a \in M$, then $\mathrm{id}_{\mathrm{TC}(\{a\})} \in I$ can be extended in the same way to an isomorphism of $\mathbf{V}$ to M fixing a. This implies (as a scheme) that M and $\mathbf{V}$ must agree on the truth of any particular formula having parameters in M, and so $j\colon \mathbf{V} \to \mathbf{V}$ is an elementary embedding, in the sense that it is Σ_n-elementary for any particular n. Q.E.D.

Theorem 4.8. In the theory $\mathsf{ZFGC}^{-\mathsf{f}} + \mathsf{BAFA}$, there is a definable class elementary embedding $j\colon \mathbf{V} \to \mathbf{V}$, which is not an automorphism.

Proof. This is technically a theorem scheme, since we assert that the particular class embedding $j\colon \mathbf{V} \to \mathbf{V}$ that we define is Σ_n-elementary for every particular natural number n. The class j will be defined relative to the global choice function that is available in $\mathsf{ZFGC}^{-\mathsf{f}}$. Let $A = (\mathbf{V} \times \{0\}) \cup \{a\}$, $a = \langle 1, 1\rangle$ and $E = \{\langle\langle x, 0\rangle, \langle y, 0\rangle\rangle : x \in y\} \cup \{\langle a, a\rangle\}$; thus, $\langle A, E\rangle$ is a disjoint union of an isomorphic copy of $\mathbf{V}$ and an additional Quine atom. By the generalized Mostowski collapse, there is an isomorphism $F\colon \langle A, E\rangle \to \langle T, \in\rangle$ to a transitive class T. Let j be the composition of F with the natural inclusion of $\mathbf{V}$ in A. Then j is an isomorphism of $\mathbf{V}$ to a transitive class $M = T \setminus \{F(a)\} \subsetneq \mathbf{V}$, and M is an elementary substructure of $\mathbf{V}$ by Theorem 4.7, hence $j\colon \mathbf{V} \to \mathbf{V}$ is an elementary embedding which is not an automorphism. Q.E.D.

Although the nontrivial automorphisms and elementary embeddings arising in Theorems 4.5 and 4.8 were definable, these definitions made essential use of the global choice function, in order to carry out the transfinite back-and-forth construction. We now show that one cannot define such embeddings in the first-order language of set theory alone:

Theorem 4.9. $\mathsf{ZFC}^{-\mathsf{f}} + \mathsf{BAFA}$ proves that there is no nontrivial Δ_0-elementary embedding $j\colon \mathbf{V} \to \mathbf{V}$ of the universe with itself that is definable (with set parameters) in the first-order language of set theory.

Proof. This is a theorem scheme, a separate statement for each possible definition, asserting that it does not define a nontrivial Δ_0-elementary embedding of the universe $\mathbf{V}$ to itself. Assume that such an embedding j is definable using parameters from a transitive set u. Since set isomorphisms preserve any particular formula, Remark 4.6 implies that $j(x) \in \mathrm{WF}(u \cup \mathrm{TC}(x))$ for every x. Given any set x, let y be a set such that $y = \{y, x\}$ using Boffa's axiom. Then $j(y) = \{j(y), j(x)\}$, and $j(y) \in \mathrm{WF}(u \cup \mathrm{TC}(y)) = \mathrm{WF}(u \cup \mathrm{TC}(\{x\}) \cup \{y\})$. Since $j(y) \in j(y)$, the least α such that $j(y) \in \mathrm{WF}_\alpha(u \cup \mathrm{TC}(\{x\}) \cup \{y\})$ must be 0, so $j(y) \in u \cup \mathrm{TC}(\{x\}) \cup \{y\}$. Since $j^{-1}[u \cup \mathrm{TC}(\{x\})]$ is a set, but there is a proper class of solutions to $y = \{y, x\}$, we may assume that we choose y so that $j(y) \notin u \cup \mathrm{TC}(\{x\})$. But in this case $j(y) = y$ and consequently $j(x) = x$. Q.E.D.

Much of the material we have presented in this section on BAFA has been already known; for more information on automorphisms and elementary embeddings in $\mathsf{ZFGC}^{-\mathsf{f}} + \mathsf{BAFA}$, we refer the reader to [14, 10].

5 Automorphisms and elementary embeddings of the universe

In the previous section, we produced models exhibiting several patterns of possibility for the existence of nontrivial automorphisms and nontrivial elementary embeddings of the universe. Specifically, we have a model with nontrivial automorphisms, but no other nontrivial elementary embeddings (Corollary 4.2 and Theorem 4.3 statement 2); we have a model with nontrivial automorphisms and other nontrivial elementary embeddings (Theorem 4.3 statement 3); and we have the models of ZFC, which have no nontrivial automorphisms and no nontrivial elementary embeddings (by the Kunen inconsistency itself). We should like now to round out these possibilities with the missing case, namely, a model of set theory having nontrivial elementary embeddings, but no nontrivial automorphisms.

Theorem 5.1. If ZFC is consistent, there is a model of $\mathsf{GB}^{-\mathsf{f}} + \mathsf{AC}$ having a class $j\colon \mathbf{V} \to \mathbf{V}$ that is an elementary embedding of the universe to itself, but in which no class is a nontrivial automorphism of the universe.

Proof. Let us explain more precisely what we mean. Assuming ZFC is consistent, we shall construct a model of $\mathsf{GB}^{-\mathsf{f}} + \mathsf{AC}$ that has a class j that is a Σ_n-elementary embedding $j\colon \mathbf{V} \to \mathbf{V}$ of the universe to itself, for every meta-theoretic natural number n, but no class in the model is a nontrivial automorphism of the universe. Actually, we may make a more uniform claim, producing a model of $\mathsf{GBC}^{-\mathsf{f}} + \mathsf{BAFA}$, for which there is a formula $\Phi(X)$ such that the model satisfies the single statement "the collection of classes satisfying Φ is closed under the class-formation axioms of $\mathsf{GB}^{-\mathsf{f}}$, and there is a non-identical function $j\colon \mathbf{V} \to \mathbf{V}$ preserving all classes satisfying Φ".

If ZFC is consistent, then there is a model of $\mathsf{GBC}^{-\mathsf{f}} + \mathsf{BAFA}$, and we shall work inside that model. Therefore, assume $\mathsf{GBC}^{-\mathsf{f}} + \mathsf{BAFA}$ and let $\overline{\mathrm{Ord}} = \mathrm{Ord} \cup \{\infty\}$, where $\alpha < \infty$ for all $\alpha \in \mathrm{Ord}$. Fix a bijection between a proper class A of Quine atoms, and the class of all sequences $a\colon \omega \to \overline{\mathrm{Ord}}$ such that $a(m) = \infty$ for all but finitely many m. We will identify $a \in A$ with the corresponding sequence in notation. Let $<$ denote the lexicographic order on A, namely,

$$a < b \leftrightarrow \exists n \in \omega\, (a{\restriction}n = b{\restriction}n \wedge a(n) < b(n)),$$

and let $A_n = \{a \in A : \forall m \geq n\ a(m) = \infty\}$. Notice that $\langle A, < \rangle$ is a dense linear order with largest element ∞^* (the constant ∞ sequence) and no least element, whereas $\langle A_n, < \rangle$ is a well-order isomorphic to the lexicographic order on $\overline{\mathrm{Ord}}^n$.

Let us also fix a set $r_{a,b}$ such that $r_{a,b} = \langle r_{a,b}, a, b \rangle$ for every $a, b \in A$ such that $a < b$. Note that the sets $r_{a,b}$ are pairwise distinct, and they are not in $\mathrm{WF}(A)$. For any subclass $B \subseteq A$, denote

$$\mathrm{WF}^{<}(B) = \mathrm{WF}(B \cup \{r_{a,b} : a, b \in B, a < b\}).$$

Let $M_n = \mathrm{WF}^<(A_n)$ and $M = \bigcup_{n\in\omega} M_n$. The first-order part of our desired model will be $\langle M, \in\rangle$. The purpose of adding the sets $r_{a,b}$ is to make $<$ definable in M, since $a < b \leftrightarrow M \models a \neq b \wedge \exists x\, x = \langle x, a, b\rangle$.

The classes of our model will be the subclasses of M that are invariant under certain partial isomorphisms, which we now describe. Let $\mathcal{P}^M(A)$ be the class of all subsets of A that belong to M, that is, subsets of A_n for some $n \in \omega$. Let I denote the class of all order-preserving isomorphisms $f\colon u \to v$ where $u, v \in \mathcal{P}^M(A)$ and $\infty^* \in u, v$.

Lemma 5.2. Let $f \in I$, $f\colon u \to v$, $v \subseteq A_n$. For every $u \subseteq u' \in \mathcal{P}^M(A)$, there exists $f' \in I$ such that $f \subseteq f'$ and $f'\colon u' \to v'$ where $v' \subseteq A_{n+1}$.

Proof. Since u' is well-ordered, we can find an increasing enumeration $u' \setminus u = \{a'_\alpha : \alpha < \gamma\}$. For each $\alpha < \gamma$, let a_α be the smallest element of u larger than a'_α (note that $a'_\alpha < \infty^* \in u$, so a_α exists). By assumption, $(f(a_\alpha))(m) = \infty$ for all $m \geq n$; let $f'(a'_\alpha)$ be the sequence in A_{n+1} which differs from $f(a_\alpha)$ only in the nth coordinate, where we put $(f'(a'_\alpha))(n) = \alpha$. Then f' has the required properties. Q.E.D.

Every order-isomorphism $f\colon u \to v$ from I uniquely extends to an $\in$-isomorphism $\overline{f}\colon \mathrm{WF}^<(u) \to \mathrm{WF}^<(v)$, since $\overline{f}(r_{a,b}) = r_{f(a),f(b)}$ is the unique set in M satisfying $x = \langle x, f(a), f(b)\rangle$, and then we use well-founded recursion on $\in$ to make $\overline{f}(x) = \overline{f}[x]$. More generally, for any $k \in \omega$, let $s_k\colon A \to A$ denote the shift operator $(s_k(a))(n) = a(n+k)$, and let $\overline{f}^k$ be the unique isomorphism $\mathrm{WF}^<(s_k^{-1}[u]) \to \mathrm{WF}^<(s_k^{-1}[v])$ such that for any $a \in s_k^{-1}[u]$, $\overline{f}^k(a)\restriction k = a\restriction k$ and $s_k(\overline{f}^k(a)) = f(s_k(a))$. That is, we leave the first k elements of a unchanged and apply f to the tail of the sequence. If $k \in \omega$ and $u \in \mathcal{P}^M(A)$, we say that a relation $R \subseteq M^r$ is *k-invariant*, if $\overline{f}^k$ preserves R for every $f \in I$, meaning that

$$R(x_1, \ldots, x_r) \leftrightarrow R(\overline{f}^k(x_1), \ldots, \overline{f}^k(x_r))$$

for all $x_1, \ldots, x_r \in \mathrm{dom}(\overline{f}^k)$.

Observe that R is k-invariant as an r-ary relation if and only if it is as a (unary) class of r-tuples, because $\overline{f}^k$ is an $\in$-isomorphism. If R is k-invariant, it is also k'-invariant for any $k' > k$.

Our model will be $\mathcal{M} = \langle M, \mathcal{X}, \in\rangle$, where $\mathcal{X}$ is the collection of all subclasses of M that are k-invariant for some $k \in \omega$. Formally, $\mathcal{X}$ is not an object of any kind in our model; rather, we are defining a (parametric) interpretation of $\mathsf{GB}^{-\mathsf{f}}$ in $\mathsf{GBC}^{-\mathsf{f}}+\mathsf{BAFA}$, delimiting classes of the interpreted theory by a formula of the background theory.

Lemma 5.3. (1) For every $k \in \omega$, the collection of k-invariant relations is closed under first-order definability and contains $\in$.

(2) $\mathcal{M} \models \mathsf{GB}^{-\mathsf{f}} + \mathsf{AC}$.

(3) $\mathcal{M} \models$ "every automorphism of $\langle \mathbf{V}, \in \rangle$ is the identity".

Proof. (1) Closure under Boolean operations is trivial. Define $S(x)$ as $\exists y\, R(x,y)$, so that $S = \mathrm{dom}(R)$, where R is k-invariant, and fix $f \in I$, $f\colon u \to v$, and $x \in \mathrm{WF}^{<}(s_k^{-1}[u])$. If $S(x)$, fix a $y \in M$ such that $R(x,y)$, and $u' \in \mathcal{P}^M(A)$ such that $y \in \mathrm{WF}^{<}(u')$. By Lemma 5.2, there is $g \supseteq f$ in I such that $\mathrm{dom}(g) \supseteq s_k[u']$. Then $y \in \mathrm{dom}(\overline{g}^k)$, hence $R(\overline{g}^k(x), \overline{g}^k(y))$, which implies $S(\overline{g}^k(x))$, where $\overline{g}^k(x) = \overline{f}^k(x)$. The other direction is symmetric.

(2) Being a full transitive class, each $\mathrm{WF}^{<}(u)$ is a model of $\mathsf{ZFC}^{-\mathsf{f}}$, hence $\mathcal{M}$ is a model of $\mathsf{ZFC}^{-\mathsf{f}}$ without collection. In $\mathcal{M}$, any set is a class as every $x \in \mathrm{WF}^{<}(u)$ with $u \subseteq A_n$ is n-invariant, and $\mathcal{M}$ satisfies the class formation axioms of $\mathsf{GB}^{-\mathsf{f}}$ by (1), it remains to show that it satisfies collection. Assume that $\forall x \in z\, \exists y\, R(x,y)$, where $R \subseteq M^2$ is k-invariant, and $z \in M_k$. If $x \in z$ and $y \in M$ satisfy $R(x,y)$, fix $u \in \mathcal{P}^M(A)$ such that $y \in \mathrm{WF}^{<}(u)$. By Lemma 5.2, there exists $f \in I$ such that $\mathrm{dom}(f) \supseteq s_k[u]$ and $\mathrm{ran}(f) \subseteq A_{k+1}$. Then $\overline{f}^k(x) = x$ as $f(\infty^*) = \infty^*$, $\overline{f}^k(y) \in M_{k+1}$, and $R(x, \overline{f}^k(y))$. Thus $\forall x \in z\, \exists y \in M_{k+1}\, R(x,y)$. Using collection in the background theory, there is a subset $w \subseteq M_{k+1}$ such that $\forall x \in z\, \exists y \in w\, R(x,y)$; we have $w \in M_{k+1} \subseteq M$ as M_{k+1} is full.

(3) Let $j\colon M \to M$ be a k-invariant automorphism. Since A and $<$ are definable in M, it follows that j restricts to an automorphism of $\langle A, < \rangle$. Fix $n \geq k$; we claim that $j[A_n] = A_n$. Assume for contradiction that $j(a) = b$ where $a \in A_n \not\ni b$ (the other case is symmetric). Then $s_n(a) = \infty^* \neq s_n(b)$. Let f be an order-preserving function with domain $\{\infty^*, s_n(b)\}$ such that $f(s_n(b)) \neq s_n(b)$. Then $\overline{f}^n(a) = a$ and $\overline{f}^n(b) \neq b$, contradicting the n-invariance of j. Thus, j restricts to an automorphism of $\langle A_n, < \rangle$. Since this is a well order, it follows that $j \restriction A_n = \mathrm{id}$, and this implies $j \restriction M_n = \mathrm{id}$. Since $n \geq k$ was arbitrary, the entire automorphism is trivial $j = \mathrm{id}$, as desired. Q.E.D.

Let $\sigma\colon A \to A$ be the function such that $(\sigma(a))(0) = 1 + a(0)$ and $s_1(\sigma(a)) = s_1(a)$, so that $\sigma(\langle a_0, a_1, a_2, \dots \rangle) = \langle 1 + a_0, a_1, a_2, \dots \rangle$. Since σ is an order-preserving bijection from A to

$$A' = \{a \in A : a(0) \neq 0\},$$

it has a unique extension to an isomorphism $j\colon M \to \mathrm{WF}^{<}(A')$, namely j is the directed union of the isomorphisms $\overline{f}\colon \mathrm{WF}^{<}(u) \to \mathrm{WF}^{<}(\sigma[u])$, where $\infty^* \in u \in \mathcal{P}^M(A)$ and $f = \sigma \restriction u$.

Lemma 5.4. Both σ and j are 1-invariant and consequently belong to $\mathcal{X}$. The embedding j preserves all 0-invariant classes, and in particular, $j\colon M \to M$ is a nontrivial elementary embedding.

Proof. The 1-invariance of σ is clear from the definition, as

$$\begin{aligned}\sigma(\overline{f}^1(\langle a_0, a_1, a_2, \dots\rangle)) &= \sigma(\langle a_0, f(\langle a_1, a_2, \dots\rangle)\rangle) \\ &= \langle 1 + a_0, f(\langle a_1, a_2, \dots\rangle)\rangle \\ &= \overline{f}^1(\langle 1 + a_0, a_1, a_2, \dots\rangle) \\ &= \overline{f}^1(\sigma(\langle a_0, a_1, a_2, \dots\rangle)),\end{aligned}$$

and j is definable in $\langle M, \in, \sigma\rangle$, hence it is also 1-invariant by Lemma 5.3.

The preservation of 0-invariant relations follows from the fact that if $x \in \mathrm{WF}^{<}(u)$ with $\infty^* \in u$, then $f := \sigma{\restriction}u \in I$, and $\overline{f}^0(x) = j(x)$. This implies any instance of elementarity, since the relation defined by any formula without parameters is 0-invariant by Lemma 5.3 and hence preserved by j. Q.E.D.

Theorem 5.1 now follows from Lemmas 5.3 and 5.4, and the proof is complete. Q.E.D.

Extending the idea of the previous argument, notice that we may extend $\sigma_k(\langle a_0, a_1, \dots\rangle) = \langle a_0, \dots, a_{k-1}, 1 + a_k, a_{k+1}, \dots\rangle$ to a nontrivial $(k+1)$-invariant embedding $j_k\colon M \to M$ which preserves all k-invariant classes. In particular, we may form the structure $\langle M, \in, j_0, j_1, j_2, \dots\rangle$, which is a model of $\mathsf{ZFC}^{-\mathsf{f}}(j_0, j_1, \dots)$ + "there is no nontrivial automorphism of $\langle \mathbf{V}, \in\rangle$" + "$j_k$ is a nontrivial elementary self-embedding of $\langle \mathbf{V}, \in, j_0, \dots, j_{k-1}\rangle$" for every k.

Let us introduce some notation to help summarize what we've done. Let $\mathrm{Aut}(\mathbf{V})$ denote the collection of automorphisms of $\mathbf{V}$ and, informally, let $\mathrm{Eem}(\mathbf{V})$ the collection of elementary embeddings of $\mathbf{V}$. This latter notation is informal, because in light of the issues mentioned in § 2, we are not actually able to express the property "$j\colon \mathbf{V} \to \mathbf{V}$ is an elementary embedding" as a single assertion about the class j, even in the full second-order language of Gödel–Bernays set theory. We are able to express that a class $j\colon \mathbf{V} \to \mathbf{V}$ is Σ_n-elementary for any particular natural number n in the meta-theory, and in this way we can say of any specific class j that "$j\colon \mathbf{V} \to \mathbf{V}$ is elementary" as an infinite scheme of statements, and this scheme-theoretic treatment of elementarity suffices for many applications. Meanwhile, in contrast, the property "$j\colon \mathbf{V} \to \mathbf{V}$ is an automorphism" *is* a first-order expressible property of j, and one can prove that any such automorphism is Σ_n-elementary for any particular natural number n in the meta-theory.

We have produced models of $\mathsf{ZFC}^{-\mathsf{f}}$ realizing all four separating refinements of the fact that $\{\mathrm{id}_{\mathbf{V}}\} \subseteq \mathrm{Aut}(\mathbf{V}) \subseteq \mathrm{Eem}(\mathbf{V})$.

1. $\{\mathrm{id}_{\mathbf{V}}\} = \mathrm{Aut}(\mathbf{V}) = \mathrm{Eem}(\mathbf{V})$. Models of ZFC have no nontrivial automorphisms or elementary self-embeddings of universe, and indeed no nontrivial Σ_1-elementary self-embeddings.

2. $\{\mathrm{id}_{\mathbf{V}}\} \subsetneq \mathrm{Aut}(\mathbf{V}) = \mathrm{Eem}(\mathbf{V})$. If $\mathbf{V} = \mathrm{WF}(A)$ for a set A of at least two Quine atoms, then there are nontrivial automorphisms, but no other nontrivial elementary embeddings.

3. $\{\mathrm{id}_{\mathbf{V}}\} = \mathrm{Aut}(\mathbf{V}) \subsetneq \mathrm{Eem}(\mathbf{V})$. The model of Theorem 5.1 has no nontrivial automorphisms, but does have a nontrivial elementary embedding.

4. $\{\mathrm{id}_{\mathbf{V}}\} \subsetneq \mathrm{Aut}(\mathbf{V}) \subsetneq \mathrm{Eem}(\mathbf{V})$. If $\mathbf{V} = \mathrm{WF}(A)$ for a proper class of Quine atoms, then there are nontrivial automorphisms, as well as non-automorphic nontrivial elementary embeddings.

In the case of statement (2), the model $\mathbf{V} = \mathrm{WF}(A)$ for a set A of Quine atoms, we have $\mathrm{Aut}(\mathbf{V}) = \mathrm{Aut}(A)$, the permutation group of the set A. In fact, let us now show that every group can arise this way.

Theorem 5.5. Assume $\mathsf{GBC}^{-\mathsf{f}} + \mathsf{BAFA}$. Then for every group G, there is a transitive set A_G whose automorphism group is isomorphic to G, and the automorphism group of the corresponding cumulative universe $\mathrm{WF}(A_G)$ generated over this set is also isomorphic to G, in the sense that every automorphism of A_G extends to a unique automorphism of $\mathrm{WF}(A_G)$ and every automorphism of $\mathrm{WF}(A_G)$ arises this way.

Proof. Work in $\mathsf{GBC}^{-\mathsf{f}} + \mathsf{BAFA}$, and fix any group G, which we may assume is in WF, since it has an isomorphic copy there. Let A_G consist of the transitive closure of the following objects:

1. Quine atoms a_g for every $g \in G$, and

2. Sets $r_{g,h}$ satisfying $r_{g,h} = \langle r_{g,h}, a_g, h, a_{gh}\rangle$ for every $g, h \in G$.

If $j\colon A_G \to A_G$ is an automorphism of A_G, then j must fix every element of G, as these are in WF, and it must permute the Quine atoms of A_G. It follows that $j(a_g) = a_{\pi(g)}$ for some permutation π of G and furthermore that $j(r_{g,h}) = \langle j(r_{g,h}), a_{\pi(g)}, h, a_{\pi(gh)}\rangle = r_{\pi(g),h}$. Thus, the embedding j is determined uniquely by π, and we may see also that $\pi(gh) = \pi(g)h$, which implies that $\pi(h) = gh$ for every h, where $g = j(1)$. Conversely, for every $g \in G$, the permutation $\pi(h) = gh$ extends to an automorphism j_g of A_G. Furthermore, $j_{gh} = j_g \circ j_h$, and so $g \mapsto j_g$ is an isomorphism of G

with $\mathrm{Aut}(A_G)$, meaning the automorphisms of the structure $\langle A_G, \in\rangle$. So the automorphism group of the transitive set A_G is isomorphic to G. Every automorphism of A_G extends canonically to an automorphism of $\mathrm{WF}(A_G)$ via $(*)$, and conversely, every automorphism of $\mathrm{WF}(A_G)$ arises from an automorphism of A_G, since it must permute the Quine atoms and the sets of the form r_{gh}. So in $\mathrm{WF}(A_G)$, the full automorphism group of the universe is definably isomorphic with G, in the sense that every automorphism of A_G extends to an automorphism of $\mathrm{WF}(A_G)$ and every automorphism $\mathrm{WF}(A_G)$ arises as such an extension. Q.E.D.

6 Theories without nontrivial self-embeddings

Let us show now in contrast that there are no nontrivial elementary embeddings or automorphisms under Aczel's anti-foundation axiom and similar anti-foundational theories where equality of sets is determined by the isomorphism type of the underlying $\in$-relation on the hereditary members of the set.

Theorem 6.1. Under $\mathsf{GBC}^{-\mathsf{f}} + \mathsf{IE}$, there is no nontrivial Σ_1-elementary embedding $j\colon \mathbf{V} \to \mathbf{V}$ of the universe to itself.

Proof. Suppose that $j\colon \mathbf{V} \to \mathbf{V}$ is a Σ_1-elementary embedding of the universe to itself. Take any $x \in \mathbf{V}$, and let t be the transitive closure of $\{x\}$. Since the range $j[\mathbf{V}]$ is transitive by Theorem 3.1, it follows that $j[t]$ is also transitive, and $j\restriction t$ is an isomorphism of $\langle t, \in\rangle$ with $\langle j(t), \in\rangle$. By IE, therefore, $j\restriction t$ is the identity, and so $j(x) = x$. Q.E.D.

What is going on is this: under the IEf principle, every transitive set t is determined by the isomorphism type of the underlying directed graph $\langle t, \in\rangle$; but by the axiom of choice, this graph has an isomorphic copy in the well-founded universe WF, which is consequently fixed by j. Thus, $\langle t, \in\rangle$ and $\langle j(t), \in\rangle$ are both isomorphic to the same graph and hence to each other, and so $j(t) = t$. Applying this to the transitive closure t of $\{\, x \,\}$, it follows that $j(x) = x$ for every set x.

Corollary 6.2. Under $\mathsf{GBC}^{-\mathsf{f}} + \mathsf{AFA}$, $\mathsf{GBC}^{-\mathsf{f}} + \mathsf{SAFA}$, $\mathsf{GBC}^{-\mathsf{f}} + \mathsf{FAFA}$, and more generally, $\mathsf{GBC}^{-\mathsf{f}} + \mathsf{AFA}^{\sim}$ for any regular bisimulation concept $\sim$, there is no nontrivial Σ_1-elementary embedding of the universe.

The proof of Theorem 6.1 makes a fundamental use of the axiom of choice, both in order to know that $j\restriction\mathrm{WF}$ is trivial and to find a surjection from a well-founded set to a given set. The proof of Theorem 6.1 applies as is to $\mathsf{GB}^{-\mathsf{f}} + \mathsf{IE}$ if we know a priori that $j[\mathbf{V}]$ is transitive, in particular $\mathsf{GB}^{-\mathsf{f}} + \mathsf{IE}$ proves that there are no nontrivial *automorphisms* of the universe. However, this does not resolve the case of general elementary embeddings, so it is natural to inquire whether one may prove Theorem 6.1 without the axiom of choice.

Question 6.3. Is it consistent with $\mathsf{GB}^{-\mathsf{f}} + \mathsf{AFA}$, $\mathsf{ZF}^{-\mathsf{f}} + \mathsf{AFA}$, or a similar extension of $\mathsf{ZF}^{-\mathsf{f}} + \mathsf{IE}$ that there is a nontrivial elementary embedding of the universe to itself?

In light of the fact that it remains a prominent open question whether one can prove the Kunen inconsistency in GB, that is, with the axiom of foundation but without the axiom of choice, we shouldn't expect an easy negative answer to this question. But perhaps one might hope for a positive answer by building a suitable model of AFA without the axiom of choice, where some ill-founded sets have no well-founded copies of their hereditary $\in$-graphs and can be subject to nontrivial embeddings.

References

[1] P. Aczel. *Non-well-founded sets*, volume 14 of *CSLI Lecture Notes*. Stanford University, 1988.

[2] D. Ballard and K. Hrbáček. Standard foundations for nonstandard analysis. *Journal of Symbolic Logic*, 57(2):741–748, 1992.

[3] J. Barwise and L. Moss. *Vicious circles. On the mathematics of non-wellfounded phenomena*, volume 60 of *CSLI Lecture Notes*. CSLI Publications, 1996.

[4] M. Boffa. *Forcing et négation de l'axiome de fondement*, volume 2/40,7 of *Mémoires de la Classe des science, Académie royale de Belgique, Collection in-8°*. Palais des Académies, 1972.

[5] A. S. Daghighi. Is there any large cardinal beyond Kunen inconsistency? Posting on `mathoverflow.net`, 8 July 2013.

[6] P. Finsler. Über die Grundlagen der Mengenlehre, I. *Mathematische Zeitschrift*, 25:683–713, 1926.

[7] M. Forti and F. Honsell. Set theory with free construction principles. *Annali della Scuola Normale Superiore di Pisa. Classe di Scienze. Serie IV*, 10(3):493–522, 1983.

[8] V. Gitman, J. D. Hamkins, and T. A. Johnstone. What is the theory ZFC without power set? Submitted.

[9] J. D. Hamkins, G. Kirmayer, and N. L. Perlmutter. Generalizations of the Kunen inconsistency. *Annals of Pure and Applied Logic*, 163(12):1872–1890, 2012.

[10] E. Jeřábek. Reflexe v neregulárních univerzech. Master's thesis, Charles University, Prague, 2001.

[11] A. Kanamori. *The higher infinite, Large cardinals in set theory from their beginnings.* Springer Monographs in Mathematics. Springer, 2nd edition, 2003.

[12] K. Kunen. Saturated ideals. *Journal of Symbolic Logic*, 43(1):65–76, 1978.

[13] D. Marker. *Model Theory: An Introduction*, volume 217 of *Graduate Texts in Mathematics.* Springer-Verlag, 2002.

[14] P. Pajas. Endomorfismy, invariantní třídy a nestandardní principy v neregulárním universu množin. Master's thesis, Charles University, Prague, 1999.

[15] L. Rieger. A contribution to Gödel's axiomatic set theory. I. *Czechoslovak Mathematical Journal*, 7(82):323–357, 1957.

[16] D. S. Scott. A different kind of model for set theory, 1960. Unpublished paper given at the 1960 Stanford Congress of Logic, Methodology and Philosophy of Science.

[17] A. Suzuki. No elementary embedding from **V** into **V** is definable from parameters. *Journal of Symbolic Logic*, 64(4):1591–1594, 1999.

[18] M. Viale. The cumulative hierarchy and the constructible universe of ZFA. *Mathematical Logic Quarterly*, 50(1):99–103, 2004.

Sets of good indiscernibles and Chang conjectures without choice*

Ioanna M. Dimitriou

Institut für Medizinische Biometrie, Informatik und Epidemiologie, Universitätsklinikum Bonn, Bonn, Germany

Abstract

With the help of sets of good indiscernibles above a certain height, we show that Chang conjectures involving four, finitely many, or an ω-sequence of cardinals have a much lower consistency strength with ZF than they do with ZFC. We shall prove equiconsistency results for *any* finitely long Chang conjecture that starts with the successor of a regular cardinal. In particular, any Chang conjecture of the form $(\kappa_n, \ldots, \kappa_0) \twoheadrightarrow (\lambda_n, \ldots, \lambda_0)$, where κ_n is the successor of a regular cardinal, in a model of ZF, is equiconsistent to the existence of $(n-1)$-many Erdős cardinals in a model of ZFC.

For Chang conjectures of length ω we shall see that ZF+"$\bigcup \kappa_n = \bigcup \lambda_n$"+ "$\kappa_n > \lambda_n$ for at least one $n \in \omega$" + $(\ldots, \kappa_n, \ldots, \kappa_0) \twoheadrightarrow (\ldots, \lambda_n, \ldots, \lambda_0)$, is equiconsistent to the theory ZFC+"a measurable cardinal exists".

We shall use symmetric forcing to create models of ZF from models of ZFC and the Dodd-Jensen core model for the other way around. All theorems in this paper are theorems of ZFC, unless otherwise stated.

1 Introduction

Since Rowbottom's doctoral thesis in 1964, in which he connects large cardinal properties with model theoretic transfer properties, there has been extensive research on the connection between these two fields of mathematical logic. The property that lies in the centre of these investigations is that of good indiscernibility.

Definition 1.1. For a structure $\mathcal{A} = \langle A, \ldots \rangle$, where A is a set of ordinals, a set $I \subseteq A$ is called a set of indiscernibles if for every $n \in \omega$, every n-ary formula φ in the language for $\mathcal{A}$, and every $\alpha_1, \ldots, \alpha_n, \alpha'_1, \ldots, \alpha'_n$ in I, if $\alpha_1 < \cdots < \alpha_n$ and $\alpha'_1 < \cdots < \alpha'_n$, then

$$\mathcal{A} \models \varphi(\alpha_1, \ldots, \alpha_n) \text{ iff } \mathcal{A} \models \varphi(\alpha'_1, \ldots, \alpha'_n).$$

*This paper is a revised part of the third chapter in my PhD thesis [4]. My PhD project with Peter Koepke was partially supported by DFG-NWO collaboration grant KO 1353-5/1, DN62-630 between the University of Amsterdam and the University of Bonn. I am grateful to Peter Koepke for the long discussions and his guidance, especially on the core model arguments.

Stefan Geschke, Benedikt Löwe, Philipp Schlicht (*eds.*).
Infinity, computability, and metamathematics: Festschrift celebrating the 60th birthdays of Peter Koepke and Philip Welch. College Publications, London, 2014. Tributes, Volume 23.

The set I is called a set of *good indiscernibles* iff it is as above and moreover we allow parameters that lie below $\min\{\alpha_1, \ldots, \alpha_n, \alpha'_1, \ldots, \alpha'_n\}$, i.e., if moreover for every $x_1, \ldots, x_m \in A$ such that

$$x_1, \ldots, x_m \leq \min\{\alpha_1, \ldots, \alpha_n, \alpha'_1, \ldots, \alpha'_n\}$$

and every $(n+m)$-ary formula φ,

$$\mathcal{A} \models \varphi(x_1, \ldots, x_m, \alpha_1, \ldots, \alpha_n) \text{ iff } \mathcal{A} \models \varphi(x_1, \ldots, x_m, \alpha'_1, \ldots, \alpha'_n).$$

The existence of sets of good indiscernibles for first order structures ensures Chang conjectures.

Definition 1.2. For infinite cardinals $\kappa_0 < \kappa_1 < \cdots < \kappa_n$ and $\lambda_0 < \lambda_1 < \cdots < \lambda_n$, a Chang conjecture is the statement

$$(\kappa_n, \ldots, \kappa_0) \twoheadrightarrow (\lambda_n, \ldots, \lambda_0),$$

which we define to mean that for every first order structure $\mathcal{A} = \langle \kappa_n, \ldots \rangle$ with a countable language there is an elementary substructure $\mathcal{B} \prec \mathcal{A}$ of cardinality λ_n such that for every $i \leq n$, $|\mathcal{B} \cap \kappa_i| = \lambda_i$.

Since the structures we shall consider will always be wellorderable, we shall implicitly assume that they have complete sets of Skolem functions. Thus we shall always be able to take Skolem hulls, even when the axiom of choice (AC) is not available. According to Vaught [19], the model theoretic relation

$$(\omega_2, \omega_1) \twoheadrightarrow (\omega_1, \omega)$$

was first considered by Chang, and it is referred to as "the original Chang conjecture".

Under AC there has been extensive research in the connection between Chang conjectures and Erdős cardinals.

Definition 1.3. For ordinals α, β, γ, the partition relation

$$\beta \to (\alpha)^{<\omega}_{\gamma}$$

means that for any partition $f : [\beta]^{<\omega} \to \gamma$ of the set of finite subsets of β into γ many sets, there exists an $X \in [\beta]^{\alpha}$, i.e., a subset of β with ordertype α, that is homogeneous for $f \restriction [\beta]^n$ for each $n \in \omega$, i.e., for every $n \in \omega$, $|f``[X]^n| = 1$.

For an infinite ordinal α, the α-Erdős cardinal $\kappa(\alpha)$ is the least κ such that $\kappa \to (\alpha)^{<\omega}_2$. When there is an infinite ordinal α such that $\kappa = \kappa(\alpha)$ we may call κ an Erdős cardinal.

As we see in [17, 1.8(1)], Silver proved in unpublished work that if the ω_1-Erdős cardinal exists, then we can force the original Chang conjecture to be true. Kunen in [14] showed that for every $n \in \omega$, $n \geq 1$, the consistency of the Chang conjecture

$$(\omega_{n+2}, \omega_{n+1}) \twoheadrightarrow (\omega_{n+1}, \omega_n)$$

follows from the consistency of the existence of a huge[1] cardinal. Donder, Jensen, and Koppelberg in [6] showed that if the original Chang conjecture is true, then the ω_1-Erdős cardinal exists in an inner model. According to [17], the same proof shows that for any infinite cardinals κ, λ, the Chang conjecture

$$(\kappa^+, \kappa) \twoheadrightarrow (\lambda^+, \lambda)$$

implies that there is an inner model in which the μ-Erdős cardinal exists, where $\mu = (\lambda^+)^V$. According to the same source, for many other regular cardinals κ, a Chang conjecture of the form

$$(\kappa^+, \kappa) \twoheadrightarrow (\omega_1, \omega)$$

is equiconsistent with the existence of the ω_1-Erdős cardinal [17, 1.10].

Chang conjectures involving higher successors on the right hand side have consistency strength stronger than an Erdős cardinal. Donder and Koepke showed in [7] that for $\kappa \geq \omega_1$, if

$$(\kappa^{++}, \kappa^+) \twoheadrightarrow (\kappa^+, \kappa),$$

then $0^\dagger$ exists, which implies that there is an inner model with a measurable cardinal. A year later, Levinski published [16] in which the existence of $0^\dagger$ is derived from each of the following Chang conjectures:

1. for any infinite κ and any $\lambda \geq \omega_1$, the Chang conjecture $(\kappa^+, \kappa) \twoheadrightarrow (\lambda^+, \lambda)$
2. for any natural number $m > 1$ and any infinite κ, λ the Chang conjecture $(\kappa^{+m}, \kappa) \twoheadrightarrow (\lambda^{+m}, \lambda)$, and
3. for any singular cardinal κ, the Chang conjecture $(\kappa^+, \kappa) \twoheadrightarrow (\omega_1, \omega)$.

In 1988, Koepke improved on some of these results by deriving the existence of inner models with sequences of measurable cardinals [13] from Chang conjectures of the form $(\kappa^{++}, \kappa^+) \twoheadrightarrow (\kappa^+, \kappa)$ for $\kappa \geq \omega_1$. Since then, much stronger large cardinal lower bounds have been found for Chang conjectures of this form under AC. Schindler showed in [18] that an inner model with a

[1]The definition of a huge cardinal can be found in [12, p. 331].

strong cardinal[2] exists, if one assumes the Chang conjecture $(\omega_{n+2}, \omega_{n+1}) \twoheadrightarrow (\omega_{n+1}, \omega_n)$ plus $2^{\omega_{n-1}} = \omega_n$, for any $1 < n < \omega$, and Cox in [3] got an inner model with a weak repeat measure[3] from the Chang conjecture $(\omega_3, \omega_2) \twoheadrightarrow (\omega_2, \omega_1)$.

Finally, under the axiom of choice we may also get inconsistency from certain Chang conjectures. As we see in [17, 1.6], finite gaps cannot be increased, e.g.,

$$(\omega_5, \omega_4) \twoheadrightarrow (\omega_3, \omega_1)$$

is inconsistent. If we remove AC from our assumptions this picture changes drastically. In this paper, we shall get successors of regular cardinals with Erdős-like properties using symmetric forcing, thereby all sorts of Chang conjectures will become 'accessible'.

The connection between Erdős cardinals and Chang conjectures lies in the existence of sets of good indiscernibles. Koepke, strengthening a result of Silver [12, Theorem 9.3], proved in [1, Proposition 8] that for limit ordinals α, $\kappa \to (\alpha)^{<\omega}_2$ is equivalent to the existence of a set $X \in [\kappa]^\alpha$ of good indiscernibles for any first order structure $\mathcal{M} = \langle M, \ldots \rangle$ with a countable language and $M \supseteq \kappa$.

This result has been used in connection with core model arguments, and there the ordertype of the set of good indiscernibles is important. For our proofs here it is helpful to also specify at which height does the set of good indiscernibles lie.

Definition 1.4. For a cardinal κ an ordinal $\alpha \leq \kappa$ and an ordinal $\vartheta < \kappa$ we define the partition property

$$\kappa \to^{\vartheta} (\alpha)^{<\omega}_2$$

to mean that for every first order structure $\mathcal{A} = \langle \kappa, \ldots \rangle$ with a countable language, there is a set $I \in [\kappa \setminus \vartheta]^\alpha$ of good indiscernibles for $\mathcal{A}$. We call such a κ an Erdős-like cardinal with respect to ϑ, α, or just an Erdős-like cardinal, if there are such ϑ, α.

We used this notation and called these cardinals Erdős-like, due to the strong connection between Erdős and Erdős-like cardinals.

Lemma 1.5 (ZF)**.** Let κ, ϑ be infinite cardinals such that $\kappa > \vartheta$ and let $\alpha \leq \kappa$ be a limit ordinal. The following are equivalent:

(a) $\kappa \to^{\vartheta} (\alpha)^{<\omega}_2$

(b) For any partition $f : [\kappa]^{<\omega} \to 2$ there is a homogeneous set $I \in [\kappa \setminus \vartheta]^\alpha$.

---

[2] For a definition of a strong cardinal, cf., e.g., [12, p. 358].

[3] A definition of a weak repeat measure can be found in [3, Definition 11] and in [9].

Consequently, $\kappa \to^{\vartheta} (\alpha)_2^{<\omega}$ implies that the Erdős cardinal $\kappa(\alpha)$ exists. Moreover, the existence of $\kappa(\alpha)$ implies that for every $\vartheta' < \kappa(\alpha)$, $\kappa(\alpha) \to^{\vartheta'} (\alpha)_2^{<\omega}$.

Proof. Assume (a) and let $f : [\kappa]^{<\omega} \to 2$ be arbitrary. Consider the structure $\mathcal{A} = \langle \kappa, f{\restriction}[\kappa]^n \rangle_{n\in\omega}$, where each $f{\restriction}[\kappa]^n$ is considered as a relation. Clearly, any set $I \in [\kappa \setminus \vartheta]^\alpha$ of good indiscernibles for $\mathcal{A}$ is also a homogeneous set for f, so (b) holds.

Now assume that (b) holds and let $\mathcal{A} = \langle \kappa, \ldots \rangle$ be an arbitrary first order structure with a countable language. Using the same proof as [1, Proposition 8] we get a set of good indiscernibles $X \in [\kappa \setminus \vartheta]^\alpha$ for $\mathcal{A}$. Therefore (a) holds.

Now assume that the α-Erdős cardinal $\kappa(\alpha) =: \mu$ exists. Let $\vartheta' < \mu$ be arbitrary and let $g : [\vartheta']^{<\omega} \to 2$ be a partition without any homogeneous sets. Let $\mathcal{A} = \langle \kappa, \ldots \rangle$ be an arbitrary first order structure with a countable language and consider the structure $\bar{\mathcal{A}} := \mathcal{A}^\frown \langle \vartheta', g \restriction [\vartheta']^n \rangle_{n\in\omega}$, where ϑ' and each $g{\restriction}[\vartheta']^n$ are considered as relations. By [1, Proposition 8] there is a set $I \in [\mu]^\alpha$ of good indiscernibles for $\bar{\mathcal{A}}$. There must be at least one $x \in I \setminus \vartheta$ otherwise I would be a homogeneous set for g. By indiscernibility, every element of I is above ϑ. Therefore $\mu \to^{\vartheta'} (\alpha)_2^{<\omega}$. Q.E.D.

The existence of an Erdős-like cardinal implies all sorts of four cardinal Chang conjectures.

Lemma 1.6 (ZF). If $\kappa \geq \vartheta, \lambda$ are cardinals such that $\kappa \neq \vartheta$ and $\kappa \to^{\vartheta} (\lambda)_2^{<\omega}$, then for all infinite $\varrho \leq \lambda \cap \vartheta$, the Chang conjecture $(\kappa, \vartheta) \twoheadrightarrow (\lambda, \varrho)$ holds.

Proof. Let $\mathcal{A} = \langle \kappa, \ldots \rangle$ be an arbitrary first order structure with a countable language. Since κ is Erdős-like with respect to ϑ, λ, there is a set $I \in [\kappa \setminus \vartheta]^\lambda$ of good indiscernibles for $\mathcal{A}$. Let $\varrho \leq \lambda \cap \vartheta$ be arbitrary and let $\mathrm{Hull}(I \cup \varrho)$ be the $\mathcal{A}$-Skolem hull of $I \cup \varrho$. By [10, 1.2.3] we have that

$$|\mathrm{Hull}(I \cup \varrho)| \leq |I \cup \varrho| + |L| = \lambda.$$

But $\lambda = |I \cup \varrho| \leq |\mathrm{Hull}(I \cup \varrho)|$, thus $\mathrm{Hull}(I \cup \varrho)$ has cardinality λ. Because all the indiscernibles lie above ϑ and because they are good indiscernibles, they are indiscernibles with respect to parameters below ϑ. So

$$\varrho \leq |\mathrm{Hull}(I \cup \varrho) \cap \vartheta| \leq \omega \cdot \varrho = \varrho.$$

So the elementary substructure $\mathrm{Hull}(I \cup \varrho) \prec \mathcal{A}$ is as we wanted and $(\kappa, \vartheta) \twoheadrightarrow (\lambda, \varrho)$ holds. Q.E.D.

At this point we should note that Chang conjectures do not imply that some cardinal is Erdős (or Erdős-like), and therefore justify our consistency strength investigation. For this we'll need that Chang conjectures are preserved under c.c.c.-forcing. This is a well known fact but we could not find a reference for it, so we attach a proof here. We assume basic knowledge of forcing as presented, e.g., in [15] or [11, Chapter 14].

Proposition 1.7. Let V be a model of ZFC in which for the cardinals $\kappa, \vartheta, \lambda, \varrho$, the Chang conjecture $(\kappa, \lambda) \twoheadrightarrow (\lambda, \varrho)$ holds. Assume also that P is a c.c.c.-forcing. If G is a P-generic filter, then $(\kappa, \lambda) \twoheadrightarrow (\lambda, \varrho)$ holds in $V[G]$ as well.

Proof. Let $\mathcal{A} = \langle \kappa, f_i, R_j, c_k \rangle_{i,j,k \in \omega} \in V[G]$ be arbitrary. Since the language of $\mathcal{A}$ is countable, let $\{\exists x \varphi_n(x) \; ; \; n \in \omega\}$ enumerate the existential formulas of $\mathcal{A}$'s language in a way such that for every $n \in \omega$, the arity $\mathrm{ar}(\varphi_n) = k_n$ of φ_n is less than n. For every $n \in \omega$ let g_n be the Skolem function that corresponds to φ_n, and let $\dot{g}_n$ be a nice name for g_n as a subset of κ^{k_n}. Since $\dot{g}_n$ is a nice name, it is of the form

$$\dot{g}_n := \bigcup \{\{\check{x}\} \times A_x \; ; \; x \in \kappa^{k_n}\}.$$

Where each A_x is an antichain of P and since P has the c.c.c., each A_x is countable. For each $x \in \kappa^{k_n}$, let $A_x := \{p_{x,0}, p_{x,1}, p_{x,2}, \dots\}$. In V define for each $n \in \omega$ a function $g_n : \kappa^{k_n - 1} \times \omega \to \kappa$ as follows:

$$g_n(\alpha_1, \dots, \alpha_{k_n - 1}, \ell) := \begin{cases} \beta & \text{if } p_{\{\alpha_1, \dots, \alpha_{k_n-1}, \beta\}, \ell} \Vdash \dot{g}_n(\check{\alpha}_1, \dots, \check{\alpha}_{k_n - 1}) = \check{\beta} \\ 0 & \text{otherwise.} \end{cases}$$

In V consider the structure $\mathcal{C} := \langle \kappa, g_n \rangle_{n \in \omega}$. Using the Chang conjecture in V take a Chang substructure $\langle B, g_n \rangle_{n \in \omega} \prec \mathcal{C}$. But then in $V[G]$ we have that $\mathcal{B} := \langle B, f_i, R_j, c_k \rangle_{i,j,k \in \omega} \prec \mathcal{A}$ is the elementary substructure we were looking for. Q.E.D.

Lemma 1.8. Let $\kappa, \vartheta, \lambda, \varrho$ be infinite cardinals in a model V of ZFC, such that $\kappa \geq \lambda$, $\kappa > \vartheta$, and $\lambda \cap \vartheta \geq \varrho$, and assume that $(\kappa, \vartheta) \twoheadrightarrow (\lambda, \varrho)$. Then there is a generic extension where $(\kappa, \vartheta) \twoheadrightarrow (\lambda, \varrho)$ holds and κ is not the λ-Erdős.

Proof. If κ is not the λ-Erdős in V, then we are done. So assume that $\kappa = \kappa(\lambda)$ in V. Let $\mu \geq \kappa$ and consider the partial order $P := \{p : \mu \times \omega \rightharpoonup 2 \; ; \; |p| < \omega\}$, where $\rightharpoonup$ denotes a partial function. This partial order adds μ many Cohen reals and has the c.c.c. so all cardinals are preserved by this forcing. By Proposition 1.7, the Chang conjecture is preserved as well. Now let G be a P-generic filter. We have that $(2^\omega)^{V[G]} \geq \mu > \kappa$. We shall show that in $V[G]$, $\kappa \not\rightarrow (\omega_1)^2_2$ so κ is not ξ-Erdős for any $\xi \geq \omega_1$.

Let $\mathbb{R}$ denote the set of reals and let $g : \kappa \to \mathbb{R}$ be injective. Define $F : [\kappa]^2 \to 2$ by

$$F(\{\alpha, \beta\}) := \begin{cases} 1 & \text{if } g(\alpha) <_{\mathbb{R}} g(\beta) \\ 0 & \text{otherwise} \end{cases}$$

If there were an ω_1-sized homogeneous set for F, then $\mathbb{R}$ would have an ω_1-long strictly monotonous $<_{\mathbb{R}}$-chain which is a contradiction. Q.E.D.

This connection between an Erdős-like cardinal and four cardinal Chang conjectures extends also to longer Chang conjectures.

Lemma 1.9 (ZF). Assume that $\lambda_0 < \lambda_1 < \cdots < \lambda_n$ and $\kappa_0 < \kappa_1 < \cdots < \kappa_n$ are cardinals such that $\kappa_i \to^{\kappa_{i-1}} (\lambda_i)_2^{<\omega}$. Then the Chang conjecture

$$(\kappa_n, \ldots, \kappa_0) \twoheadrightarrow (\lambda_n, \ldots, \lambda_0) \text{ holds.}$$

Proof. Let $\mathcal{A} = \langle \kappa_n, \ldots \rangle$ be an arbitrary first order structure in a countable language, and let $\{f_j \ ; \ j \in \omega\}$ be a complete set of Skolem functions for $\mathcal{A}$. Since $\kappa_n \to^{\kappa_{n-1}} (\lambda_n)_2^{<\omega}$ holds, let $I_n \in [\kappa_n \setminus \kappa_{n-1}]^{\lambda_n}$ be a set of good indiscernibles for $\mathcal{A}$. To take the next set of indiscernibles I_{n-1} we must make sure that it is, in a sense, compatible with I_n. That is, the Skolem hull of $I_n \cup I_{n-1}$ must not contain more than λ_{n-2} many elements below κ_{n-2}.

To do this, we shall enrich the structure $\mathcal{A}$ with functions, e.g., $f_j(e_1, e_2, x_1, x_2)$, for some f_j with arity $\mathrm{ar}(f_j) = 4$ and some $e_1, e_2 \in I_n$. Since f_j takes ordered tuples as arguments we must consider separately the cases $f_j(e_1, x_1, e_2, x_2)$, $f_j(e_1, x_1, x_2, e_2)$, etc..

Formally, let $\bar{I}_n := \{e_1, e_2, \ldots\}$ be the first ω-many elements of I_n. For every $s < \omega$ let $\{g_{s,t} \ ; \ t < s!\}$ be an enumeration of all the permutations of s, and for every $t \in s!$ let

$$h_{s,t}(x_1, \ldots, x_s) := (x_{g_{s,t}(1)}, \ldots, x_{g_{s,t}}(s)).$$

For every $j < \omega$, every $k < \mathrm{ar}(f_j)$, and every $\ell \in \mathrm{ar}(f_j)!$ define a function $f_{j;k;\ell} : {}^{\mathrm{ar}(f_j)}\kappa_n \to \kappa_n$ by

$$f_{j;k;\ell}(x_1, \ldots, x_{\mathrm{ar}(f_j)-k}) := f_j(h_{\mathrm{ar}(f_j),\ell}(x_1, \ldots, x_{\mathrm{ar}(f_j)-k}, e_1, \ldots, e_k)).$$

Consider the structure

$$\mathcal{A}_{n-1} := \mathcal{A}^\frown \langle f_{i;c;t} \rangle_{j<\omega, k<\mathrm{ar}(f_j), \ell<\mathrm{ar}(f_j)!}.$$

Since $\kappa_{n-1} \to^{\kappa_{n-2}} (\lambda_{n-1})_2^{<\omega}$, let $I_{n-1} \in [\kappa_{n-1} \setminus \kappa_{n-2}]^{\lambda_{n-1}}$ be a set of good indiscernibles for $\mathcal{A}_{n-1}$.

Claim 1.10. For any infinite set $Z \subseteq \kappa_{n-2}$ of size λ_{n-2},

$$|\mathrm{Hull}_{\mathcal{A}}(I_n \cup I_{n-1} \cup Z) \cap \kappa_{n-2}| = \lambda_{n-2}.$$

Proof. Let $\bar{I}_{n-1} := \{e'_1, e'_2, \ldots\}$ be the first ω-many elements of I_{n-1}. The domain of the $\mathcal{A}$-Skolem Hull $\mathrm{Hull}_{\mathcal{A}}(I_n \cup I_{n-1} \cup Z)$ is the set

$$X := \{f_j(\alpha_1, \ldots, \alpha_{\mathrm{ar}(f_j)}) \; ; \; j < \omega \text{ and } \alpha_1, \ldots, \alpha_{\mathrm{ar}(f_j)} \in I_n \cup I_{n-1} \cup Z\}.$$

If for some $x = f_j(\alpha_1, \ldots, \alpha_{\mathrm{ar}(f_j)}) \in X \cap \kappa_{n-2}$ there are elements of I_n among $\alpha_1, \ldots, \alpha_{\mathrm{ar}(f_j)}$, then since I_n is a set of indiscernibles for $\mathcal{A}$ and $\alpha_1, \ldots, \alpha_{\mathrm{ar}(f_j)}$ are finitely many, we can find $\alpha'_1, \ldots, \alpha'_{\mathrm{ar}(f_j)} \in \bar{I}_n \cup I_{n-1} \cup Z$ such that

$$x = f_j(\alpha_1, \ldots, \alpha_{\mathrm{ar}(f_j)}) = f_j(\alpha'_1, \ldots, \alpha'_{\mathrm{ar}(f_j)}).$$

We rewrite the tuple $(\alpha'_1, \ldots, \alpha'_{\mathrm{ar}(f_j)})$ so that the elements of $\bar{I}_n$ (if any) appear in ascending order at the end:

$$\{\alpha'_1, \ldots, \alpha'_n\} = \{\beta_1, \ldots, \beta_{\mathrm{ar}(f_j)-k}, e_1, \ldots, e_k\}.$$

Let $(\beta_1, \ldots, \beta_{\mathrm{ar}(f_j)-k}, e_1, \ldots, e_k)$ be a permutation of $(\alpha'_1, \ldots, \alpha'_{\mathrm{ar}(f_j)})$, so for some $\ell < \mathrm{ar}(f_j)!$,

$$(\alpha'_1, \ldots, \alpha'_{\mathrm{ar}(f_j)}) = h_{\mathrm{ar}(f_j),\ell}(\beta_1, \ldots, \beta_{\mathrm{ar}(f_j)-k}, e_1, \ldots, e_k).$$

But then

$$\begin{aligned} x &= f_j(\alpha'_1, \ldots, \alpha'_{\mathrm{ar}(f_j)}) \\ &= f_j(h_{\mathrm{ar}(f_j),\ell}(\beta_1, \ldots, \beta_{\mathrm{ar}(f_j)-k}, e_1, \ldots, e_k) \\ &= f_{j,k,\ell}(\beta_1, \ldots, \beta_{\mathrm{ar}(f_j)-k}). \end{aligned}$$

Therefore,

$$\begin{aligned} X \cap \kappa_{n-2} = \{&f_{j,k,\ell}(\beta_1, \ldots, \beta_{\mathrm{ar}(f_j)-k}) < \kappa_{n-2} \; ; \\ &j < \omega, k < \mathrm{ar}(f_j), \ell \in \mathrm{ar}(f_j)!, \text{ and } \beta_1, \ldots, \beta_{\mathrm{ar}(f_j)-k} \in I_{n-1} \cup Z\}. \end{aligned}$$

But I_{n-1} is a set of good indiscernibles for $\mathcal{A}_{n-1}$, i.e., it is a set of indiscernibles for formulas with parameters below $\min I_{n-1} > \kappa_{n-2}$, therefore a set of indiscernibles for formulas with parameters from Z as well. Thus in the equation above we may replace I_{n-1} with $\bar{I}_{n-1}$. It is easy to see then that the set $X \cap \kappa_{n-2}$ has size λ_{n-2}. Q.E.D. (Claim 1.10)

Continuing like this we get for each $i = 1, \ldots, n$ a set $I_i \in [\kappa_i \setminus \kappa_{i-1}]^{\lambda_i}$ of good indiscernibles for $\mathcal{A}$ with the property that for every infinite $Z \subseteq \kappa_{i-1}$, of size λ_{i-1},

$$|\mathrm{Hull}_{\mathcal{A}}(I_n \cup \cdots \cup I_i \cup Z) \cap \kappa_{i-1}| = \lambda_{i-1}.$$

So let $I := \bigcup_{i=1,\ldots,n} I_i$ and take $\mathcal{B} := \mathrm{Hull}_{\mathcal{A}}(I \cup \lambda_0)$. By [10, 1.2.3], we have that

$$\lambda_n = |I \cup \lambda_0| \leq |\mathrm{Hull}_{\mathcal{A}}(I \cup \lambda_0)| \leq |I \cup \varrho| + \omega = \lambda_n.$$

Because for each $i = 1, \ldots, n$ we have that $I_i \in [\kappa_i \setminus \kappa_{i-1}]^{\lambda_i}$ and by the way we defined the I_i, we have that

$$|\mathrm{Hull}(I \cup \lambda_0) \cap \kappa_i| = \lambda_i.$$

So the substructure $\mathrm{Hull}(I \cup \lambda_0) \prec \mathcal{A}$ is such as we wanted for our Chang conjecture to hold. Q.E.D. (Lemma 1.9)

1.1 Countable coherent sequences of sets of good indiscernibles

In the proof of Lemma 1.9 we had to construct our sets of good indiscernibles in a way that they are compatible. When we have countably many such sets and the axiom of choice is not available, we need that these sets of good indiscernibles are, in a way, coherent.

Definition 1.11. Let $\langle \kappa_i \ ; \ i < \omega \rangle$ and $\langle \lambda_i \ ; \ 0 < i < \omega \rangle$ be strictly increasing sequences of cardinals, let $\kappa := \bigcup_{i<\omega} \kappa_i$, and let $\mathcal{A} = \langle \kappa, \ldots \rangle$ be a first order structure with a countable language. A $\langle \lambda_i \ ; \ 0 < i < \omega \rangle$-coherent sequence of good indiscernibles for $\mathcal{A}$ with respect to $\langle \kappa_i \ ; \ i < \omega \rangle$ is a sequence $\langle A_i \ ; \ 0 < i < \omega \rangle$ such that

1. for every $0 < i < \omega$, $A_i \in [\kappa_i \setminus \kappa_{i-1}]^{\lambda_i}$, and
2. if $x, y \in [\kappa]^{<\omega}$ are such that $x = \{x_1, \ldots, x_n\}$, $y = \{y_1, \ldots, y_n\}$, $x, y \subseteq \bigcup_{0<i<\omega} A_i$, and for every $0 < i < \omega$ $|x \cap A_i| = |y \cap A_i|$, then for every $(n + \ell)$-ary formula φ in the language of $\mathcal{A}$ and every $z_1, \ldots, z_\ell < \min\{x_1, \ldots, x_n, y_1, \ldots, y_n\}$,

$$\mathcal{A} \models \varphi(z_1, \ldots, z_\ell, x_1, \ldots, x_n) \iff \mathcal{A} \models \varphi(z_1, \ldots, z_\ell, y_1, \ldots, y_n).$$

We say that the sequence $\langle \kappa_i \ ; \ i < \omega \rangle$ is a coherent sequence of the Erdős-like cardinals $\kappa_{i+1} \to^{\kappa_i} (\lambda_{i+1})_2^{<\omega}$ iff for every structure $\mathcal{A} = \langle \kappa, \ldots \rangle$ with a countable language, there is a $\langle \lambda_i \ ; \ 0 < i < \omega \rangle$-coherent sequence of good indiscernibles for $\mathcal{A}$ with respect to $\langle \kappa_i \ ; \ i < \omega \rangle$.

Similarly to coherent sequences of Erdős-like cardinals, we have coherent sequences of Erdős cardinals.

Definition 1.12. Let $\lambda_1 < \cdots < \lambda_i < \ldots$ and $\kappa_0 < \cdots < \kappa_i < \ldots$ be cardinals and let $\kappa := \bigcup_{i<\omega} \kappa_i$. We say that the sequence $\langle \kappa_i \ ; \ i < \omega \rangle$ is a coherent sequence of Erdős cardinals with respect to $\langle \lambda_i \ ; \ 0 < i < \omega \rangle$ if for every $\gamma < \kappa_1$ and every $f : [\kappa]^{<\omega} \to \gamma$ there is a sequence $\langle A_i \ ; \ 0 < i < \omega \rangle$ such that

1. for every $0 < i < \omega$, $A_i \in [\kappa_i \setminus \kappa_{i-1}]^{\lambda_i}$, and

2. if $x, y \in [\kappa]^{<\omega}$ are such that $x, y \subseteq \bigcup_{i<\omega} A_i$ and for every $0 < i < \omega$ $|x \cap A_i| = |y \cap A_i|$, then $f(x) = f(y)$.

Such a sequence $\langle A_i \ ; \ 0 < i < \omega \rangle$ is called a $\langle \lambda_i \ ; \ 0 < i < \omega \rangle$-coherent sequence of homogeneous sets for f with respect to $\langle \kappa_i \ ; \ i < \omega \rangle$. Note that the 0th element of a coherent sequence of Erdős cardinals need not be an Erdős cardinal, and indeed, none of the κ_n need satisfy the minimality requirement of the usual Erdős cardinals.

Coherent sequences of Ramsey cardinals are such sequences. In [2, Theorem 3] a coherent sequence of Ramsey cardinals in ZF is forced from a model of ZFC with one measurable cardinal. Similar as in Lemma 1.5, we get that coherent sequences of Erdős and Erdős-like cardinals are equivalent.

Lemma 1.13 (ZF). Let $\lambda_1 < \cdots < \lambda_i < \ldots$ and $\kappa_0 < \cdots < \kappa_i < \ldots$ be infinite cardinals. The following are equivalent:

(a) The sequence $\langle \kappa_i \ ; \ i < \omega \rangle$ is a coherent sequence of the Erdős-like cardinals $\kappa_{i+1} \to^{\kappa_i} (\lambda_{i+1})^{<\omega}_2$.

(b) The sequence $\langle \kappa_i \ ; \ i < \omega \rangle$ is a coherent sequence of Erdős cardinals with respect to $\langle \lambda_i \ ; \ 0 < i < \omega \rangle$.

To show that when $\bigcup_{i\in\omega} \kappa_i = \bigcup_{i\in\omega} \lambda_i$, then such a coherent sequence of Erdős or Erdős-like cardinals in a model of ZF is equiconsistent with a measurable cardinal in a model of ZFC, we shall use results that involve the infinitary Chang conjecture.

Definition 1.14. For cardinals $\kappa_0 < \cdots < \kappa_n < \ldots$ and $\lambda_0 < \cdots < \lambda_n < \ldots$, with $\kappa_n \geq \lambda_n$ for all n, define the infinitary Chang conjecture

$$(\kappa_n)_{n\in\omega} \twoheadrightarrow (\lambda_n)_{n\in\omega}$$

to mean that for every first order structure $\mathcal{A} = \langle \bigcup_{n\in\omega} \kappa_n, f_i, R_j, c_k \rangle_{i,j,k\in\omega}$ there is an elementary substructure $\mathcal{B} \prec \mathcal{A}$ with domain B such that for all $n \in \omega$, $|B \cap \kappa_n| = \lambda_n$.

Sometimes, when this uniform notation is not convenient, we shall write the infinitary Chang conjecture as

$$(\ldots, \kappa_n, \ldots, \kappa_0) \twoheadrightarrow (\ldots, \lambda_n, \ldots, \lambda_0).$$

The infinitary Chang conjecture is connected to Jónsson cardinals.

Definition 1.15. A cardinal κ is called Jónsson if for every first order structure with domain κ and a countable language, there is a proper elementary substructure of cardinality κ.

In [8, §12 (4)] we read:

> Assuming that $2^{\aleph_0} < \aleph_\omega$, Silver showed that the cardinal $\aleph_\omega$ is Jónsson iff there is an infinite subsequence $\langle \kappa_n \ ; \ n \in \omega \rangle$ of the $\aleph_n$'s such that the infinitary Chang conjecture of the form
>
> $$(\ldots, \kappa_n, \kappa_{n-1}, \ldots, \kappa_1) \twoheadrightarrow (\ldots, \kappa_{n-1}, \kappa_{n-2}, \ldots, \kappa_0)$$
>
> holds. It is not known how to get such a sequence of length 4.

Clearly, if $\kappa = \bigcup_{n\in\omega} \kappa_n = \bigcup_{n\in\omega} \lambda_n$ and for some $n \in \omega$ $\kappa_n \neq \lambda_n$, then $(\kappa_n)_{n\in\omega} \twoheadrightarrow (\lambda_n)_{n\in\omega}$ implies that κ is a singular Jónsson cardinal of cofinality ω. We can get an infinitary Chang conjecture from a coherent sequence of Erdős-like cardinals as in Lemma 1.9, but without having to take care of the "compatibility" of the sets of indiscernibles, since here they are coherent.

Lemma 1.16. (ZF) Let $\langle \kappa_n \ ; \ n < \omega \rangle$ and $\langle \lambda_n \ ; \ 0 < n < \omega \rangle$ be increasing sequences of cardinals, and let $\kappa = \bigcup_{n<\omega} \kappa_n$. If $\langle \kappa_i \ ; \ i < \omega \rangle$ is a coherent sequence of cardinals with the property $\kappa_{n+1} \to^{\kappa_n} (\lambda_{n+1})_2^{<\omega}$, then the Chang conjecture

$$(\kappa_n)_{n\in\omega} \twoheadrightarrow (\lambda_n)_{n\in\omega}$$

holds.

2 Forcing good sets of indiscernibles to lie between regular cardinals and their successors.

Here we assume knowledge of symmetric forcing as presented in [4, §§ 2 & 3 of Chapter 2]. This technique is used to produce models of ZF+¬AC, called symmetric models, starting from a model of ZFC. In particular, we shall use the generalised Jech model $V(G)$ and its property of satisfying the approximation lemma, i.e., that all sets of ordinals in $V(G)$ are included in some "initial" ZFC model. One may think of this symmetric model as being the closed under set theoretic operations union of generic extensions (the aforementioned initial ZFC models) that are made with the Lévy collapses

$$E_\alpha := \{p : \eta \rightharpoonup \alpha \ ; \ |p| < \eta\}$$

for $\alpha < \kappa$ and η a fixed regular cardinal. This "union" does not contain a generic object for the entire Lévy collapse $\mathbb{P} := \{p : \eta \rightharpoonup \kappa \ ; \ |p| < \eta\}$, so that κ becomes the successor of η in the symmetric model $V(G)$.

2.1 Finitely many sets of good indiscernibles

First, let us take a look at the case of just one set of good indiscernibles being forced to lie between a regular cardinal ϑ and its successor ϑ^+.

One approach would be to take a cardinal κ with the property $\kappa \to^{\vartheta} (\alpha)_2^{<\omega}$ and, using the generalised Jech model, symmetrically collapse κ to become ϑ^+. But to use the theorems for this model and preserve the property $\kappa \to^{\vartheta} (\alpha)_2^{<\omega}$, κ would have to satisfy certain large cardinal properties, e.g., inaccessibility. Erdős-like cardinals such as κ are far from inaccessible. In fact, for every $\kappa' \geq \kappa$, $\kappa' \to^{\vartheta} (\alpha)_2^{<\omega}$ holds. So we construct a model of $\mathsf{ZF} + \neg\mathsf{AC}$ by starting with an Erdős cardinal. This is not too bad since, by Lemma 1.5, Erdős cardinals and Erdős-like cardinals are mutually existent.

First, we shall show that an Erdős cardinal is Erdős-like after small forcing.

Lemma 2.1. If V is a model of ZFC+"$\kappa = \kappa(\alpha)$ exists" for some limit ordinal α, if $\mathbb{P}$ is a partial order such that $|\mathbb{P}| < \kappa$, and G is a generic filter, then in $V[G]$, for any $\vartheta < \kappa$ the property $\kappa \to^{\vartheta} (\alpha)_2^{<\omega}$ holds.

Proof. Let $\mathcal{A} = \langle \kappa, \ldots \rangle \in V[G]$ be an arbitrary structure in a countable language and $\vartheta < \kappa$ be arbitrary. Let $g : [\vartheta]^{<\omega} \to 2$ be a function in the ground model that has no homogeneous sets (in the ground model) of ordertype λ, and consider the structure

$$\bar{\mathcal{A}} = \mathcal{A}^\frown \langle \vartheta, g \restriction [\vartheta]^n \rangle_{n \in \omega},$$

where ϑ, and each $g \restriction [\vartheta]^n$ is considered as a relation. Let $\{\varphi_n \ ; \ n < \omega\}$ enumerate the formulas of the language of $\bar{\mathcal{A}}$ so that each φ_n has $k(n) < n$ many free variables. Define $f : [\kappa]^{<\omega} \to 2$ by $f(\xi_1, \ldots, \xi_n) = 1$ iff $\mathcal{A} \models \varphi_n(\xi_1, \ldots, \xi_{k(n)})$ and $f(\xi_1, \ldots, \xi_n) = 0$ otherwise. We call this f the function that describes truth in $\mathcal{A}$. Let $\dot{f}$ be a $\mathbb{P}$-name for f. Since κ is inaccessible in V, $|\wp(\mathbb{P})| < \kappa$ in V. In V define the function $h : [\kappa]^{<\omega} \to \wp(\mathbb{P})$ by

$$h(x) := \{p \in \mathbb{P} \ ; \ p \Vdash \dot{f}(\check{x}) = 0\}.$$

By [12, Proposition 7.15] let $A \in [\kappa]^\lambda$ be homogeneous for h. Note that since we have attached g to $\mathcal{A}$, $A \subseteq \kappa \setminus \vartheta$. We shall show that A is homogeneous for f in $V[G]$, and therefore a set of good indiscernibles for $\mathcal{A}$.

Let $n \in \omega$ and $x \in [A]^n$ be arbitrary.

If $h(x) = \varnothing$, then for all $p \in \mathbb{P}$, $p \nVdash \dot{f}(\check{x}) = \check{0}$. So for some $p \in G \cap E_\gamma$, $p \Vdash \dot{f}(\check{x}) = \check{1}$ and so the colour of $[A]^n$ is 1.

If $h(x) \neq \varnothing$ and $h(x) \cap G \neq \varnothing$, then the colour of $[A]^n$ is 0.

If $h(x) \neq \varnothing$ and $h(x) \cap G = \varnothing$, then assume for a contradiction that for some $y \in [A]^n$, $f(x) \neq f(y)$. Without loss of generality say $f(y) = 0$. But then there is $p \in G$ such that $p \Vdash \dot{f}(\check{y}) = \check{0}$ so $\varnothing \neq h(y) \cap G = h(x) \cap G$, contradiction.

So in $V[G]$, $\kappa \rightarrow^{\vartheta} (\lambda)_2^{<\omega}$ holds. Q.E.D.

We can use this to get the following.

Lemma 2.2. If V is a model of $\mathsf{ZFC}+$ "$\kappa = \kappa(\lambda)$ exists", then for any regular cardinal $\eta < \kappa$, there is a symmetric model $V(G)$ of ZF in which for every $\vartheta < \kappa$, $\eta^+ \rightarrow^{\vartheta} (\lambda)_2^{<\omega}$ holds.

Proof. Let $\eta < \kappa$ be a regular cardinal, and construct the generalised Jech model $V(G)$ (cf. [4, § 3 of Chapter 1] and the beginning of this section) that makes $\kappa = \eta^+$. The approximation lemma holds in this model. Let $\vartheta < \kappa$ be arbitrary. Let $\mathcal{A} = \langle \kappa, \ldots \rangle$ be an arbitrary first order structure with a countable language and let $\dot{\mathcal{A}} \in \mathsf{HS}$ be a name for $\mathcal{A}$ with support E_γ for some $\eta < \gamma < \kappa$. By the approximation lemma

$$\mathcal{A} \in V[G \cap E_\gamma].$$

Note that $|E_\gamma| < \kappa$ therefore by Lemma 2.1, for every $\vartheta < \kappa$ the property $\kappa \rightarrow^{\vartheta} (\lambda)_2^{<\omega}$ holds in $V(G)$. Therefore the structure $\mathcal{A}$ has a set of indiscernibles $A \in [\kappa \setminus \vartheta]^{\lambda}$ and $A \in V[G \cap E_\gamma] \subseteq V(G)$. Q.E.D.

By Corollary 1.6 we get the following.

Corollary 2.3. If V is a model of ZFC with a cardinal κ that is the λ-Erdős cardinal, then for any $\eta < \kappa$ regular cardinal there is a symmetric model $V(G)$ in which for every $\vartheta < \kappa$ with $\vartheta \leq \eta$, and $\varrho \leq \lambda \cap \vartheta$

$$(\eta^+, \vartheta) \twoheadrightarrow (\lambda, \varrho) \text{ holds.}$$

Note that as with many of our forcing constructions here, this η could be *any* predefined regular ordinal of V. So we get an infinity of consistency strength results, some of them looking very strange for someone accustomed to the theory ZFC, such as the following.

Corollary 2.4. If $V \models \mathsf{ZFC}+$"$\kappa(\omega_{12})$ exists", then there is a symmetric model $V(G) \models \mathsf{ZF} + (\omega_{13}, \omega_{12}) \twoheadrightarrow (\omega_{12}, \omega_5)$.

Or even stranger:

Corollary 2.5. If $V \models \mathsf{ZFC}+$"$\kappa(\omega_\omega)$ exists", then there is a symmetric model $V(G) \models \mathsf{ZF} + (\omega_{\omega+3}, \omega_\omega) \twoheadrightarrow (\omega_\omega, \omega_2)$.

To get Chang conjectures that involve more than four cardinals we shall have to collapse the Erdős cardinals simultaneously. We shall give an example in which the Chang conjecture

$$(\omega_4, \omega_2, \omega_1) \twoheadrightarrow (\omega_3, \omega_1, \omega)$$

is forced from a model of ZFC with two Erdős cardinals. Before we do that let us see a very useful proposition.

Proposition 2.6. Assume that $V \models \mathsf{ZFC}+$"$\kappa = \kappa(\lambda)$ exists", $\mathbb{P}$ is a partial order such that $|\mathbb{P}| < \kappa$, and $\mathbb{Q}$ is a partial order that doesn't add subsets to κ. If G is $\mathbb{P} \times \mathbb{Q}$-generic, then for every $\vartheta < \kappa$,

$$V[G] \models \kappa \rightarrow^{\vartheta} (\lambda)_2^{<\omega}.$$

Proof. Let $\mathcal{A} = \langle \kappa, \ldots \rangle \in V[G]$ be an arbitrary structure with a countable language. By [15, Chapter VII, Lemma 1.3], $G = G_1 \times G_2$ for some G_1 $\mathbb{P}$-generic and some G_2 $\mathbb{Q}$-generic. Since $\mathbb{Q}$ does not add subsets to κ, we have that $\mathcal{A} \in V[G_1]$. By Lemma 2.1 we get that $\kappa \rightarrow^{\vartheta} (\lambda)_2^{<\omega}$ in $V[G_1] \subset V[G]$ and from that we get a set $H \in [\kappa \setminus \vartheta]^{\lambda}$ of indiscernibles for $\mathcal{A}$ with respect to parameters below ϑ, and $H \in V[G]$,. Therefore $V[G] \models \kappa \rightarrow^{\vartheta} (\lambda)_2^{<\omega}$. Q.E.D.

To get the desired Chang conjecture, we shall construct a symmetric model that can also be used to create a model of ZF with successive alternating measurable and non-measurable cardinals (cf. [4, § 4 of Chapter 1] for $\varrho = 2$).

Lemma 2.7. (ZFC) Assume that $\kappa_1 = \kappa(\omega_1)$, and $\kappa_2 = \kappa(\kappa_1^+)$ exist. Then there is a symmetric extension of V in which $\mathsf{ZF} + \omega_4 \rightarrow^{\omega_2} (\omega_3)_2^{<\omega} + \omega_2 \rightarrow^{\omega_1} (\omega_1)_2^{<\omega}$.

Consequently,

$$(\omega_4, \omega_2, \omega_1) \twoheadrightarrow (\omega_3, \omega_1, \omega)$$

holds in V as well.

Proof. Let $\kappa_1' = (\kappa_1^+)^V$ and define

$$\mathbb{P} := \{p : \omega_1 \rightharpoonup \kappa_1 \ ; \ |p| < \omega_1\} \times \{p : \kappa_1' \rightharpoonup \kappa_2 \ ; \ |p| < \kappa_1'\}.$$

Let $\mathcal{G}_1$ be the full permutation group of κ_1 and $\mathcal{G}_2$ the full permutation group of κ_2. We define an automorphism group $\mathcal{G}$ of $\mathbb{P}$ by letting $a \in \mathcal{G}$ iff for some $a_1 \in \mathcal{G}_1$ and $a_2 \in \mathcal{G}_2$,

$$a((p_1, p_2)) := (\{(\xi_1, a_1(\beta_1)) \ ; \ (\xi_1, \beta_1) \in p_1\}, \{(\xi_2, a_2(\beta_2)) \ ; \ (\xi_2, \beta_2) \in p_2\}).$$

Let I be the symmetry generator that is induced by the ordinals in the product of intervals $(\omega_1, \kappa_1) \times (\kappa_1', \kappa_2)$, i.e.,

$$I := \{E_{\alpha,\beta} \ ; \ \alpha \in (\omega_1, \kappa_1) \text{ and } \beta \in (\kappa_1', \kappa_2)\},$$

where

$$E_{\alpha,\beta} := \{(p_1 \cap (\omega_1 \times \alpha), p_2 \cap (\kappa_1' \times \beta)) \ ; \ (p_1, p_2) \in \mathbb{P}\}.$$

This I is a projectable symmetry generator with projections

$$(p_1, p_2) \restriction^* E_{\alpha,\beta} = (p_1 \cap (\omega_1 \times \alpha), p_2 \cap (\kappa_1' \times \beta)).$$

Take the symmetric model $V(G) = V(G)^{\mathcal{F}_I}$. It is easy to see that the approximation lemma holds for this model. This construction can be illustrated as in Figure 1.

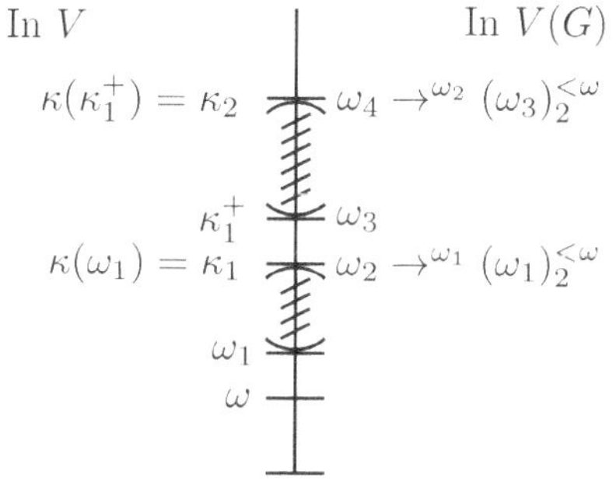

FIGURE 1.

With the standard arguments we can show that in $V(G)$ we have that $\kappa_1 = \omega_2$ and $\kappa_2 = \omega_4$. We want to show that moreover $\kappa_2 \to^{\kappa_1} (\kappa_1')_2^{<\omega}$ and $\kappa_1 \to^{\omega_1} (\omega_1)_2^{<\omega}$.

For the first partition property let $\mathcal{A} = \langle \kappa_2, \ldots \rangle$ be an arbitrary structure in a countable language and let $\dot{\mathcal{A}} \in \mathsf{HS}$ be a name for $\mathcal{A}$ with support $E_{\alpha,\beta}$. By the approximation lemma we have that $\mathcal{A} \in V[G \cap E_{\alpha,\beta}]$. Since $|E_{\alpha,\beta}| < \kappa_2$, by Lemma 2.1 we have that $V[G \cap E_{\alpha,\beta}] \models \kappa_2 \to^{\kappa_1} (\kappa_1')_2^{<\omega}$ therefore there is a set $A \in [\kappa_2 \setminus \kappa_1]^{\kappa_1'}$ of indiscernibles for $\mathcal{A}$ with respect to parameters below κ_1, and $A \in V[G \cap E_{\alpha,\beta}] \subseteq V(G)$.

For the second partition property let $\mathcal{B} = \langle \kappa_1, \ldots \rangle$ be and arbitrary structure in a countable language and let $\dot{\mathcal{B}} \in \mathsf{HS}$ be a name for $\mathcal{B}$ with support $E_{\gamma,\delta}$. We have that $E_{\gamma,\delta} = \{p : \omega_1 \rightharpoonup \gamma \ ; \ |p| < \omega_1\} \times \{p : \kappa_1' \rightharpoonup \delta \ ; \ |p| < \kappa_1'\}$, $|\{p : \omega_1 \rightharpoonup \gamma \ ; \ |p| < \omega_1\}| < \kappa_1$, and $\{p : \kappa_1' \rightharpoonup \delta \ ; \ |p| < \kappa_1'\}$ does not add subsets to κ_1. Therefore by Proposition 2.6 we get that $V[G \cap E_{\gamma,\delta}] \models \kappa_1 \to^{\omega_1} (\omega_1)_2^{<\omega}$ so there is a set $B \in [\kappa_1 \setminus \omega_1]^{\omega_1}$ of indiscernibles for $\mathcal{B}$ with respect to parameters below ω_1, and $B \in V[G \cap E_{\gamma,\delta}] \subseteq V(G)$.

So in $V(G)$ we have that $\omega_4 \to^{\omega_2} (\omega_3)_2^{<\omega}$ and $\omega_2 \to^{\omega_1} (\omega_1)_2^{<\omega}$ thus by Lemma 1.9 we have that in $V(G)$ the Chang conjecture $(\omega_4, \omega_2, \omega_1) \twoheadrightarrow (\omega_3, \omega_1, \omega)$ holds. Q.E.D.

Note that, the gap in these cardinals is necessary for this method to work. Collapsing further would destroy their partition properties. Keeping this in mind it is easy to see how to modify this proof to get any desired Chang conjecture

$$(\kappa_n, \ldots, \kappa_0) \twoheadrightarrow (\lambda_n, \ldots, \lambda_0)$$

with the κ_i and the λ_i being any predefined successor cardinals, as long as we mind the gaps.

2.2 Coherent sequences of sets of good indiscernibles

We can do the above for the infinitary version as well, using a finite support product forcing of such collapses, for a coherent sequence of Erdős cardinals $\langle \kappa_n \ ; \ n \in \omega \rangle$ with respect to $\langle \kappa_n^+ \ ; \ n < \omega \rangle$, and with $\kappa_0 = \omega_1$. In that case, we shall end up with a model of

$$\mathsf{ZF} + \neg\mathsf{AC}_\omega + (\omega_{2n+1})_{n<\omega} \twoheadrightarrow (\omega_{2n})_{n<\omega}.$$

Lemma 2.8. Let $\langle \kappa_n \ ; \ n < \omega \rangle$ and $\langle \lambda_n \ ; \ 0 < n < \omega \rangle$ be increasing sequences of cardinals such that $\langle \kappa_n \ ; \ n < \omega \rangle$ is a coherent sequence of Erdős cardinals with respect to $\langle \lambda_n \ ; \ n < \omega \rangle$. If $\mathbb{P}$ is a partial order of cardinality $< \kappa_1$ and G is $\mathbb{P}$-generic, then in $V[G]$, $\langle \kappa_n \ ; \ n < \omega \rangle$ is a coherent sequence of cardinals with the property $\kappa_{n+1} \to^{\kappa_n} (\lambda_{n+1})_2^{<\omega}$.

Proof. Let $\kappa = \bigcup_{n \in \omega}$ and let $\mathcal{A} = \langle \kappa, \ldots \rangle \in V[G]$ be an arbitrary structure in a countable language. Let $\{\varphi_n \ ; \ n < \omega\}$ enumerate the formulas of the language of $\mathcal{A}$ so that each φ_n has $k(n) < n$ many free variables. Define $f : [\kappa]^{<\omega} \to 2$ by $f(\xi_1, \ldots, \xi_n) = 1$ iff $\mathcal{A} \models \varphi_n(\xi_1, \ldots, \xi_{k(n)})$ and $f(\xi_1, \ldots, \xi_n) = 0$ otherwise. Let $\dot{f}$ be a $\mathbb{P}$-name for f. In V define a function $g : [\kappa]^{<\omega} \to \wp(\mathbb{P})$ by

$$g(x) = \{p \in \mathbb{P} \ ; \ p \Vdash \dot{f}(\check{x}) = \check{0}\}.$$

Since $|\mathbb{P}| < \kappa_1$ and κ_1 is inaccessible in V, $|\wp(\mathbb{P})| < \kappa_1$. So there is a $\langle \lambda_n \ ; \ 0 < n < \omega \rangle$-coherent sequence of homogeneous sets for g with respect to $\langle \kappa_n \ ; \ n \in \omega \rangle$. The standard arguments show that this is a $\langle \lambda_n \ ; \ 0 < n < \omega \rangle$-coherent sequence of homogeneous sets for f with respect to $\langle \kappa_n \ ; \ n \in \omega \rangle$, therefore a $\langle \lambda_n \ ; \ 0 < n < \omega \rangle$-coherent sequence of indiscernibles for $\mathcal{A}$ with respect to $\langle \kappa_n \ ; \ n \in \omega \rangle$. Q.E.D.

The model used for this following proof is again the model of [4, § 4 of Chapter 1], this time for $\varrho = \omega$.

Lemma 2.9. (ZFC) Let $\langle \kappa_n ; n \in \omega \rangle$ be a coherent sequence of Erdős cardinals with respect to $\langle \lambda_n ; 0 < n \in \omega \rangle$, where $\kappa_0 = \omega_1$. Then there is a symmetric model $V(G)$ in which $\langle \omega_{2n} ; n \in \omega \rangle$ is a coherent sequence of cardinals with the property $\omega_{2n+2} \to^{\omega_{2n}} (\omega_{2n+1})_2^{<\omega}$.

Consequently, in $V(G)$

$$(\ldots, \omega_{2n}, \ldots, \omega_4, \omega_2, \omega_1) \twoheadrightarrow (\ldots, \omega_{2n-1}, \ldots, \omega_3, \omega_1, \omega)$$

holds as well, and $\aleph_\omega$ is a Jónsson cardinal.

Proof. Let $\kappa = \bigcup_{0<n<\omega} \kappa_n$, for every $0 < n < \omega$ let $\kappa_n' = \kappa_n^+$, and let $\kappa_0' = \omega_1$. For every $0 < n < \omega$ let

$$\mathbb{P}_n := \{p : \kappa_{n-1}' \rightharpoonup \kappa_n ; |p| < \kappa_{n-1}'\},$$

and take the finite support product of these forcings

$$\mathbb{P} := \prod_{0<n<\omega}^{\text{fin}} \mathbb{P}_n.$$

For each $0 < n < \omega$ let G_n be the full permutation group of κ_n and define an automorphism group $\mathcal{G}$ of $\mathbb{P}$ by $a \in \mathcal{G}$ iff for every $n \in \omega$ there exists $a_n \in \mathcal{G}_n$ such that

$$a(\langle p_n ; n \in \omega \rangle) := \langle \{(\xi, a_n(\beta)) ; (\xi, \beta) \in p_n\} ; n \in \omega \rangle.$$

For every finite sequence of ordinals $e = \langle \alpha_1, \ldots, \alpha_m \rangle$ such that for every $i = 1, \ldots, m$ there is a distinct $0 < n_i < \omega$ such that $\alpha_i \in (\kappa_{n_i-1}', \kappa_{n_i})$, define

$$E_e := \{\langle p_{n_i} \cap (\kappa_{n_i-1}', \alpha_i) ; \alpha_i \in e \rangle ; \langle p_{n_i} ; i = 1, \ldots, m \rangle \in \mathbb{P}\},$$

and take the symmetry generator

$$I := \{E_e ; e \in \prod_{0<n<\omega}^{\text{fin}} (\kappa_{n-1}', \kappa_n)\}.$$

This is a projectable symmetry generator with projections

$$\langle p_j ; 0 < j < \omega \rangle \restriction^* E_e = \langle p_{n_i} \cap (\kappa_{n_i-1}', \alpha_i) ; \alpha_i \in e \rangle.$$

Take the symmetric model $V(G) = V(G)^{\mathcal{F}_I}$. The approximation lemma holds for $V(G)$. As usual we can show that in $V(G)$, for each $0 < n < \omega$ we have that $\kappa_n = \kappa_{n-1}'^{+}$, i.e., for every $0 < n < \omega$, $\kappa_n = \omega_{2n}$ and $\kappa_n' = \omega_{2n+1}$.

It remains to show that $\langle \kappa_i \; ; \; i \in \omega \rangle$ is a coherent sequence of cardinals with the property $\kappa_{n+1} \to^{\kappa_n} (\lambda_{n+1})_2^{<\omega}$. Let $\mathcal{A} = \langle \kappa, \ldots \rangle$ be an arbitrary structure in a countable language and let the function $f : [\kappa]^{<\omega} \to 2$ describe the truth in $\mathcal{A}$, as in the proofs of Lemma 2.1 and Lemma 2.8. Let $\dot{f} \in \mathsf{HS}$ be a name for f with support E_e. Let $e = \{\alpha_1, \ldots, \alpha_m\}$ and for each $i = 1, \ldots, m$ let n_i be such that $\alpha \in (\kappa'_{n_i-1}, \kappa_{n_i})$. By the approximation lemma,

$$f \in V[G \cap E_e],$$

i.e., f is forced via $\bar{\mathbb{P}} = \prod_{i=1}^{m} \{p : \kappa'_{n_i-1} \rightharpoonup \kappa_{n_i} \; ; \; |p| < \kappa'_{n_i-1}\}$.

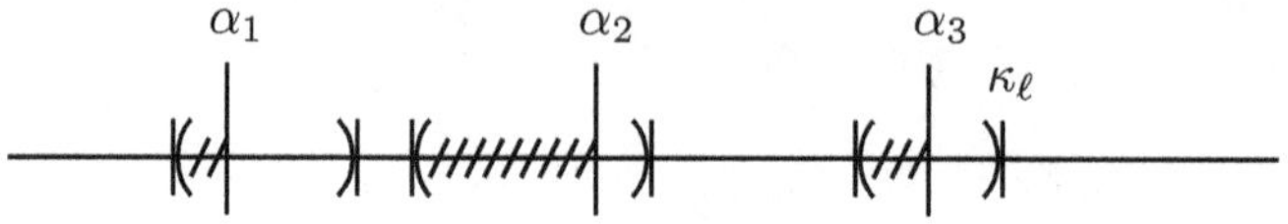

FIGURE 2.

Let $\ell := \max\{n_i \; ; \; \alpha_i \in e\}$. We're in a situation as in the image above, which is an example for $m = 3$. Since $|\mathbb{P}| < \kappa_\ell$, by Lemma 2.8 there is a $\langle \lambda_n \; ; \; \ell \leq n < \omega \rangle$-coherent sequence of indiscernibles for $\mathcal{A}$ with respect to $\langle \kappa_n \; ; \; \ell - 1 \leq n \in \omega \rangle$, i.e., a sequence $\langle A_n \; ; \; \ell \leq n < \omega \rangle$ such that for every $\ell \leq n < \omega$, $A_n \subseteq \kappa_n \setminus \kappa_{n-1}$ is of ordertype λ_n, and if $x, y \in [\kappa]^{<\omega}$ are such that $x = \{x_1, \ldots, x_m\}$, $y = \{y_1, \ldots, y_m\}$, $x, y \in \bigcup_{\ell \leq n < \omega} A_n$, and for every $\ell \leq n < \omega$, $|x \cap A_n| = |y \cap A_n|$, then for every $m + k$-ary formula φ in the language of $\mathcal{A}$, and every $z_1, \ldots, z_k$ less than $\min \bigcup_{\ell \leq n < \omega} A_n$,

$$\mathcal{A} \models \varphi(z_1, \ldots, z_k, x_1, \ldots, x_m) \iff \mathcal{A} \models \varphi(z_1, \ldots, z_k, y_1, \ldots, y_m).$$

Now we shall get sets of indiscernibles from the remaining cardinals $\kappa_1, \ldots, \kappa_{\ell-1}$ step by step, making them coherent as we go along. Before we get the rest of the A_n, note that by Proposition 2.6 we have that for every $0 < n < \ell$,

$$V[G \cap E_e] \models \kappa_n \to^{\kappa_{n-1}} (\lambda_n)_2^{<\omega}.$$

Let us see how to get $A_{\ell-1}$. For every $\ell \leq n < \omega$, let $\bar{A}_n$ be the first ω-many elements of A_n. There are only countably many $x \in [\kappa]^{<\omega}$ such that $x \subseteq \bigcup_{\ell \leq n < \omega} \bar{A}_n$. For every $i, j \in \omega$, and every $x \in [\kappa]^{<\omega}$ such that $x = \{x_1, \ldots, x_m\} \subseteq \bigcup_{\ell \leq n < \omega} \bar{A}_n$ and $m < i, j$, let

$$f_{i,x}(v_1, \ldots, v_{i-m}) := f_i(v_1, \ldots, v_{i-m}, x_1, \ldots, x_m), \text{ and}$$
$$R_{j,x}(v_1, \ldots, v_{j-m}) := R_j(v_1, \ldots, v_{j-m}, x_1, \ldots, x_m).$$

Consider the structure

$$\mathcal{A}' := \mathcal{A}^\frown \langle f_{i,x}, R_{j,x} \rangle_{i,j<\omega, x \in [\kappa]^{<\omega}, x = \{x_1, \ldots, x_m\} \subseteq \bigcup_{\ell \leq n < \omega} \bar{A}_n, m < i,j}.$$

Since $\kappa_{\ell-1} \rightarrow^{\kappa_{\ell-2}} (\lambda_{\ell-1})_2^{<\omega}$, there is a set $A_{\ell-1} \in [\kappa_{\ell-1} \setminus \kappa_{\ell-2}]^{\lambda_{\ell-1}}$ of indiscernibles for $\mathcal{A}'$ with respect to parameters below $\kappa_{\ell-2}$. By the way we defined $\mathcal{A}'$, the sequence $\langle A_n \; ; \; \ell-1 \leq n < \omega \rangle$ is a $\langle \lambda_n \; ; \; \ell-1 \leq n < \omega \rangle$-coherent sequence of indiscernibles for $\mathcal{A}$ with respect to $\langle \kappa_n \; ; \; \ell-2 \leq n < \omega \rangle$.

Continuing in this manner we get a sequence $\langle A_n \; ; \; 0 < n < \omega \rangle$ that is a $\langle \lambda_n \; ; \; 0 < n < \omega \rangle$-coherent sequence of indiscernibles for $\mathcal{A}$ with respect to $\langle \kappa_n \; ; \; n < \omega \rangle$, and such that $\langle A_n \; ; \; 0 < n < \omega \rangle \in V[G \cap E_e] \subseteq V(G)$.

Therefore we have that in $V(G)$, $\langle \omega_{2n} \; ; \; n \in \omega \rangle$ is a coherent sequence of cardinals with the property $\omega_{2n+2} \rightarrow^{\omega_{2n}} (\omega_{2n+1})_2^{<\omega}$. By Corollary 1.16 we have that in $V(G)$

$$(\ldots, \omega_{2n}, \ldots, \omega_4, \omega_2, \omega_1) \twoheadrightarrow (\ldots, \omega_{2n-1}, \ldots, \omega_3, \omega_1, \omega)$$

holds. Consequently, $\aleph_\omega$ is a Jónsson cardinal. Q.E.D.

Note that in this model the axiom of choice fails badly. In particular, by [4, Lemma 1.37] $\mathsf{AC}_{\omega_2}(\wp(\omega_1))$ is false. As mentioned after the proof of that lemma, one can get the infinitary Chang conjecture plus the axiom of dependent choice with the construction in the proof of [2, Theorem 5]. With that we get the following.

Lemma 2.10. Let $V_0 \models \mathsf{ZFC}+$"there exists a measurable cardinal κ". Let $n < \omega$ be fixed but arbitrary. There is a generic extension V of V_0, a forcing notion $\mathbb{P}$, and a symmetric model N such that

$$N \models \mathsf{ZF} + \mathsf{DC} + (\omega_n)_{0<n<\omega} \twoheadrightarrow (\omega_n)_{n<\omega} + \text{"}\aleph_\omega \text{ is Jónsson"}$$

3 Getting good sets of indiscernibles from Chang conjectures for successors of regular cardinals in ZF

In this section, we shall work with the Dodd-Jensen core model $\mathbf{K}^{\mathrm{DJ}}$ to get strength from the principles we are looking at. We shall start from a model of ZF with a Chang conjecture, and we shall get an Erdős cardinal in $\mathbf{K}^{\mathrm{DJ}}$. To be able to use the known theorems about $\mathbf{K}^{\mathrm{DJ}}$, which involve AC, we shall build $\mathbf{K}^{\mathrm{DJ}}$ inside $\mathbf{HOD}$. We shall then use the Chang conjecture to get a certain elementary substructure $K' \subseteq \mathbf{HOD}$. With a clever manoeuvre, found in [2, Proposition 8], which says $(\mathbf{K}^{\mathrm{DJ}})^{\mathbf{HOD}} = (\mathbf{K}^{\mathrm{DJ}})^{\mathbf{HOD}[K']}$ we can get this structure into $(\mathbf{K}^{\mathrm{DJ}})^{\mathbf{HOD}}$ and use the core model's structured nature to get a set of good indiscernibles.

We assume basic knowledge of this core model, as presented in [6, 5, 7], or in the short collection of the relevant results from these papers found in [4, § 2 of Chapter 3]. We shall start by looking at four cardinal Chang conjectures.

Theorem 3.1. Assume ZF and let η be a regular cardinal. If for some infinite cardinals ϑ, λ, and ϱ such that $\eta^+ > \vartheta, \lambda > \varrho$ and $\mathrm{cf}(\lambda) > \omega$ the Chang conjecture $(\eta^+, \vartheta) \twoheadrightarrow (\lambda, \varrho)$ holds, then $\kappa(\lambda)$ exists in the Dodd-Jensen core model $(\mathbf{K}^{\mathrm{DJ}})^{\mathbf{HOD}}$, and $(\mathbf{K}^{\mathrm{DJ}})^{\mathbf{HOD}} \models (\eta^+)^V \to (\lambda)^{<\omega}_2$.

Proof. Let $\kappa = (\eta^+)^V$ and in $\mathbf{K}^{\mathrm{DJ}}$ let $g : [\kappa]^{<\omega} \to 2$ be arbitrary and consider the structure $\langle \mathbf{K}^{\mathrm{DJ}}_\kappa, \in, D \cap K^{\mathrm{DJ}}_\kappa, g\rangle$, where D is a class[4] such that $\mathbf{K}^{\mathrm{DJ}} = \mathbf{L}[D]$. This is included in our structure in order to use the results in [5] and [6]. We want to find a set of indiscernibles for this structure in $\mathbf{K}^{\mathrm{DJ}}$. Using our Chang conjecture in V we get an elementary substructure

$$\mathcal{K}' = \langle K', \in, D \cap K', g'\rangle \prec \langle \mathbf{K}^{\mathrm{DJ}}_\kappa, \ldots\rangle$$

such that $|K'| = \lambda$ and $|K' \cap \vartheta| = \varrho$. Since K' is wellorderable it can be seen as a set of ordinals. We attach K' to $\mathbf{HOD}$, getting $\mathbf{HOD}[K']$. By [2, Proposition 8], $(\mathbf{K}^{\mathrm{DJ}})^{\mathbf{HOD}} = (\mathbf{K}^{\mathrm{DJ}})^{\mathbf{HOD}[K']}$.

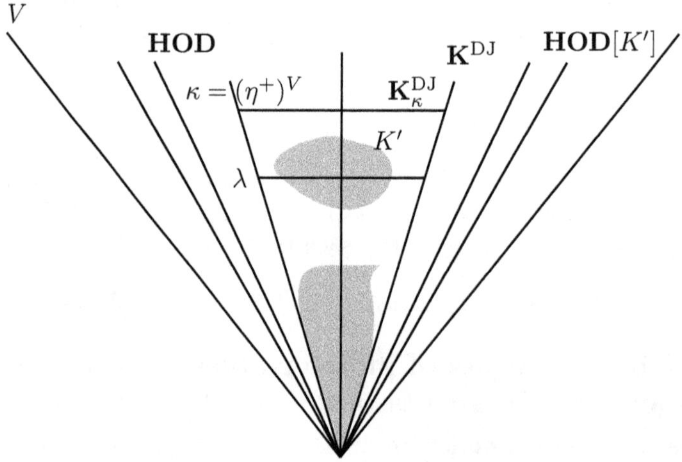

FIGURE 3.

We are now, and for the rest of this proof, working in $\mathbf{HOD}[K']$. Let $\langle \bar{K}, \in, A'\rangle$ be the Mostowski collapse of $\mathcal{K}'$, with $\pi : \bar{K} \to K'$ being an elementary embedding. We distinguish two cases:

Case 1. If $\bar{K} = \mathbf{K}^{\mathrm{DJ}}_{\lambda'}$ for some λ'. Then the map $\pi : \mathbf{K}^{\mathrm{DJ}}_{\lambda'} \to \mathbf{K}^{\mathrm{DJ}}_\kappa$ is elementary.

Since $\lambda \geq \omega_1$, by [7, Lemma 2.9], there is a non trivial elementary embedding of $\mathbf{K}^{\mathrm{DJ}}$ to $\mathbf{K}^{\mathrm{DJ}}$ with critical point α. By [7, 1.5] this means that there is an inner model with a measurable cardinal β, such that if $\alpha < \omega_1$,

[4] For details, cf. [5, Definition 6.3].

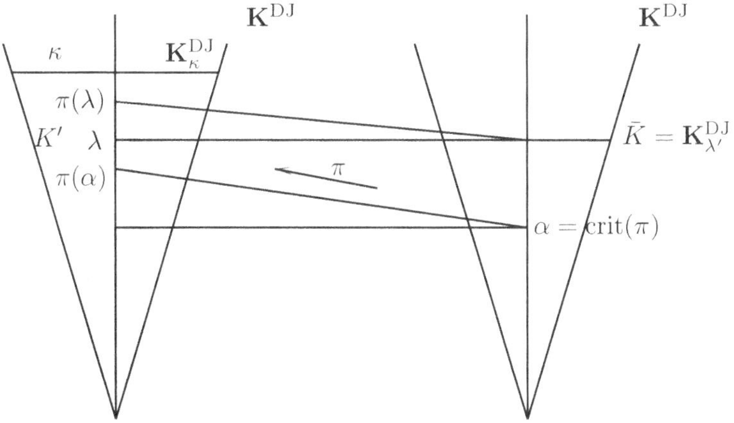

FIGURE 4.

then $\beta \leq \omega_1$ and if $\alpha \geq \omega_1$, then $\beta < \alpha^+$. Because $\alpha = \mathrm{crit}(\pi)$ and $|\bar{K}| = \lambda$, $\alpha < \lambda^+$.

Let us take a closer look at this inner model. Let U be a normal measure for β in the inner model M, define $\bar{U} := U \cap \mathbf{L}[U]$ and build $\mathbf{L}[\bar{U}]$. It is known that then $\mathbf{L}[\bar{U}] \models$"$\bar{U}$ is a normal ultrafilter over β" (cf. [12, Exercise 20.1]). We also have that $\mathbf{L}[\bar{U}] \models \mathsf{ZFC}$ which by [12, Lemma 20.5] means that $\langle \mathbf{L}[\bar{U}], \in, \bar{U}\rangle$ is iterable. Recall that such a structure $\langle \mathbf{L}[\bar{U}], \in, \bar{U}\rangle$ is called a β-model and that for a regular cardinal ν, C_ν is the club filter over ν. According to [12, Corollary 20.7], if there is a β-model, if ν is a regular cardinal above β^+, and if $\bar{C}_\nu = C_\nu \cap \mathbf{L}[C_\nu]$, then $\mathbf{L}[\bar{C}_\nu]$ is a ν-model.

Now, we have that if $\alpha < \omega$, then $\beta \leq \omega_1$, so $\beta \leq \lambda$. If $\alpha \geq \omega_1$, then $\beta < \alpha^+ \leq \lambda^+$, so again $\beta \leq \lambda$. If $\beta = \lambda$, then $\mathbf{L}[\bar{U}] \models$"$\lambda$ is Ramsey". By [7, 1.6], $\wp(\lambda) \cap \mathbf{L}[\bar{U}] = \wp(\lambda) \cap \mathbf{K}^{\mathrm{DJ}}$. So λ is Ramsey in K and we're done.

So assume that $\beta < \lambda$. Then $\beta^+ < \kappa$. We need a regular cardinal $\nu > \beta^+$ such that $\lambda \leq \nu \leq \kappa$. If κ is regular, let $\nu = \kappa$. If κ is singular, then κ is a limit cardinal so there is such a regular cardinal ν (e.g., $\nu = \lambda^{++}$).

Then by [7, 1.6] we have that $\mathbf{L}[\bar{C}_\nu] \models$"$\nu$ is Ramsey". Because [7, 1.6] says that $\wp(\nu) \cap \mathbf{L}[\bar{C}_\nu] = \wp(\nu) \cap \mathbf{K}^{\mathrm{DJ}}$, this ν is Ramsey in $\mathbf{K}^{\mathrm{DJ}}$. But this implies that in $\mathbf{K}^{\mathrm{DJ}}$, $\kappa \to (\lambda)^{<\omega}_2$.

Case 2. If $\bar{K} \neq \mathbf{K}^{\mathrm{DJ}}_\lambda$ for any λ'. By [7, Lemma 2.1] $\mathbf{K}^{\mathrm{DJ}}_\kappa \models \mathbf{V}{=}\mathbf{K}^{\mathrm{DJ}}$. Since $\bar{K}$ is elementary with $\mathbf{K}^{\mathrm{DJ}}_\kappa$, $\bar{K} \models \mathbf{V}{=}\mathbf{K}^{\mathrm{DJ}}$. This is because being the lower part of a premouse is a property describable by a formula. Let $x \in \bar{K}$. Since $\bar{K} \models \mathbf{V}{=}\mathbf{K}^{\mathrm{DJ}}$, there must be some M such that

$$\bar{K} \models \text{“}M \text{ is an iterable premouse and } x \in \mathrm{lp}(M)\text{”}.$$

So $\mathbf{K}^{\mathrm{DJ}}_{\kappa} \models$ "M is an iterable premouse" and by [7, Lemma 1.16], $\pi(M)$ is an iterable premouse in $\mathbf{HOD}[K']$. Since $\pi \upharpoonright M \to \pi(M)$ is elementary and $\pi(M)$ is an iterable premouse, by [7, Lemma 1.17], M is an iterable premouse in $\mathbf{HOD}[K']$. Thus $x \in \mathbf{K}^{\mathrm{DJ}}$, so $\bar{K} \subseteq \mathbf{K}^{\mathrm{DJ}}$. But then, since $\mathbf{K}^{\mathrm{DJ}}_{\lambda} \neq \bar{K}$ for any λ', and $\bar{K}$ has cardinality λ, there must be an iterable premouse $M \notin \bar{K}$ and a $z \in \mathbf{K}^{\mathrm{DJ}}_{\lambda} \setminus \bar{K}$ such that $\mathrm{lp}(M) \cap (K_\lambda \setminus \bar{K}) \neq \varnothing$, $z \in \mathrm{lp}(M)$, and $M \in \mathbf{K}^{\mathrm{DJ}}_{\lambda}$. Fix M.

Claim 3.2. If $\delta > \lambda$ is a regular cardinal, then for every iterable premouse $N \in \bar{K}$, $N_\delta \in M_\delta$.

Proof. Since $M \in K_\lambda$ and $|\bar{K}| = \lambda$ by [7, Lemma 1.13] we have that for every regular cardinal $\delta > \lambda$ and every iterable premouse $N \in \bar{K}$, N_δ and M_δ are comparable. Assume for a contradiction that for some $N \in \bar{K}$, $M_\delta \subseteq N_\delta$. Then $z \in \mathrm{lp}(N_\delta)$ and since $z \in \mathbf{K}^{\mathrm{DJ}}_{\lambda}$, for some $\xi < \lambda$, $z \in \mathrm{lp}(N_\xi)$. But since $N_\xi \in \bar{K}$, $z \in N_\xi \in \bar{K}$ which is transitive so $z \in \bar{K}$, contradiction. Q.E.D. (Claim 3.2)

We want such a $\delta \leq \kappa$. As before, if κ is regular, then take $\delta = \kappa$, and if κ is singular, then take $\delta = \lambda^+$. Look at M_δ. By Claim 1 we have that $\bar{K} \subseteq M_\delta$. So $g' \in M_\delta$.

Let $\langle M_i, \pi_{ij}, \gamma_i, U_i \rangle_{i \leq j < \delta}$ be the δ-iteration of M. By [7, Lemma 1.14] there is some $x \in M$ and $\vec{\varrho} \in {}^{<\omega}\{\gamma_i \,;\, i < \delta\}$ such that $g' = \pi_{0,\delta}(x)(\vec{\varrho})$. Let $\mathcal{C} = \{\gamma_i \,;\, i < \lambda\}$. By the same lemma there is a sequence $\langle k_n \,;\, n < \omega \rangle \in {}^{\omega}2$ such that for every $n < \omega$, $g'``[\mathcal{C}]^n = \{k_n\}$.

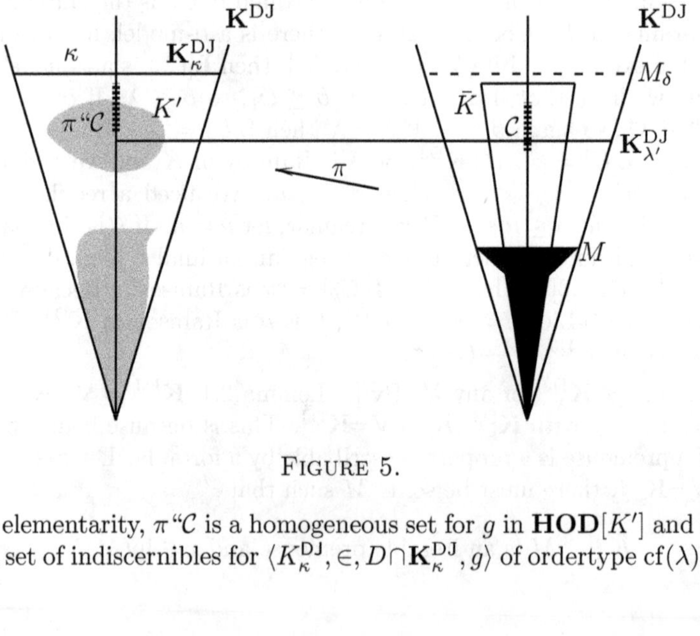

FIGURE 5.

By elementarity, $\pi``\mathcal{C}$ is a homogeneous set for $g$ in $\mathbf{HOD}[K']$ and $\pi``\mathcal{C}$ is a good set of indiscernibles for $\langle K^{\mathrm{DJ}}_{\kappa}, \in, D \cap \mathbf{K}^{\mathrm{DJ}}_{\kappa}, g \rangle$ of ordertype $\mathrm{cf}(\lambda) \geq \omega_1$.

By Jensen's indiscernibility lemma [6, Lemma 1.3] there is a homogeneous set for g of ordertype λ in $\mathbf{K}^{\mathrm{DJ}}$. Q.E.D. (Theorem 3.1)

Therefore we have the following:

Theorem 3.3. The theory $\mathsf{ZF}+(\kappa,\vartheta) \twoheadrightarrow (\lambda,\varrho)+\mathrm{cf}(\lambda) > \omega$ is equiconsistent with the theory $\mathsf{ZFC}+$"$\kappa(\lambda)$ exists".

In the proof of Lemma 1.9 we see how to combine finitely many sets of indiscernibles to make them coherent. Using this we get the following.

Lemma 3.4. Assume ZF and let $\kappa_n > \cdots > \kappa_0$, $\lambda_n > \cdots > \lambda_0$ be regular cardinals, such that the Chang conjecture $(\kappa_n,\ldots,\kappa_0) \twoheadrightarrow (\lambda_n,\ldots,\lambda_0)$ holds, then for each $i = 1,\ldots,n$, $\kappa(\lambda_i)$ exists in the Dodd-Jensen core model $(\mathbf{K}^{\mathrm{DJ}})^{\mathbf{HOD}}$ and $(\mathbf{K}^{\mathrm{DJ}})^{\mathbf{HOD}} \models \forall i \in \{1,\ldots,n\}(\kappa_i \to (\lambda_i)^{<\omega}_2)$.

Theorem 3.5. For every finite n, the theory

$$\mathsf{ZF} + \text{"}(\kappa_n,\ldots,\kappa_0) \twoheadrightarrow (\lambda_n,\ldots,\lambda_0)\text{"}$$

is equiconsistent with the theory $\mathsf{ZFC}+$"$\kappa(\lambda_n^{+(n-1)})$ exists", where λ_0 is the last cardinal appearing on the Chang conjecture.

3.1 The infinitary case

For the infinitary version, recall that if $\bigcup_{n\in\omega}\kappa_n = \bigcup_{n\in\omega}\lambda_n$ and $\kappa_n > \lambda_n$ for at least one $n \in \omega$, then

$$(\kappa_n)_{n<\omega} \twoheadrightarrow (\lambda_n)_{n<\omega}$$

implies that $\kappa := \bigcup_{n\in\omega}\kappa_n$ is a singular Jónsson cardinal. In [2, Theorem 6] it is proved that if κ is a singular Jónsson cardinal in a model of ZF, then κ is measurable in some inner model. As a corollary to that we get the following.

Corollary 3.6. If $\langle\kappa_n\ ;\ n\in\omega\rangle$ and $\langle\lambda_n\ ;\ n\in\omega\rangle$ are increasing sequences of cardinals such that $\bigcup_{n\in\omega}\kappa_n = \bigcup_{n\in\omega}\lambda_n$ and $\kappa_n > \lambda_n$ for at least one $n\in\omega$, then the infinitary Chang conjecture $(\kappa_n)_{n<\omega} \twoheadrightarrow (\lambda_n)_{n<\omega}$ implies that there is an inner model in which $\kappa = \bigcup_{n\in\omega}\kappa_n$ is measurable.

Since we can force such a coherent sequence of Erdős cardinals by starting with a measurable cardinal (cf. [2, Theorem 3]) we have the following.

Theorem 3.7. The theory $\mathsf{ZF}+$"an infinitary Chang conjecture holds with the supremum of the left hand side cardinals being the same as the supremum of the right hand side cardinals" is equiconsistent with the theory $\mathsf{ZFC}+$"a measurable cardinal exists".

We conjecture that if the supremum of the κ_n is strictly bigger than the supremum of the λ_n, then the consistency strength of such an infinitary Chang conjecture in ZF is weaker. To prove lower bounds for the consistency strength of the existence of a set of good indiscernibles between a singular cardinal and its successor, more complex core models, for stronger large cardinal axioms, must be employed. Some results on this direction can be found in the author's PhD thesis [4, § 3.4], and a paper on the subject is in preparation.

References

[1] A. Apter and P. Koepke. Making all cardinals almost Ramsey. *Archive for Mathematical Logic*, 47:769–783, 2008.

[2] A. W. Apter and P. Koepke. The consistency strength of $\aleph_\omega$ and $\aleph_{\omega_1}$ being Rowbottom cardinals without the axiom of choice. *Archive for Mathematical Logic*, 45:721–737, 2006.

[3] S. D. Cox. Consistency strength of higher Chang's conjecture, without CH. *Archive for Mathematical Logic*, 50:759–775, 2011.

[4] I. M. Dimitriou. *Symmetric Models, Singular Cardinal Patterns, and Indiscernibles.* PhD thesis, Rheinische Friedrich-Wilhelms-Universität Bonn, 2011.

[5] A. Dodd and R. B. Jensen. The core model. *Annals of Mathematical Logic*, 20:43–75, 1981.

[6] H. D. Donder, R. B. Jensen, and B. Koppelberg. Some applications of the core model. In R. B. Jensen and A. Prestel, editors, *Set theory and model theory, Proceedings of a Symposium held in Bonn, June 13, 1979*, volume 872 of *Lecture Notes in Mathematics*, pages 55–97. Springer, 1979.

[7] H. D. Donder and P. Koepke. On the consistency strength of "accessible" Jónsson cardinals and of the weak Chang conjecture. *Annals of Pure and Applied Logic*, 25:233–261, 1983.

[8] M. Foreman. Ideals and generic elementary embeddings. In M. Foreman and A. Kanamori, editors, *Handbook of set theory, Volume 2*, pages 885–1147. Springer, 2010.

[9] M. Gitik. Some results on the nonstationary ideal. *Israel Journal of Mathematics*, 92:61–112, 1995.

[10] W. Hodges. *A shorter model theory*. Cambridge University Press, 1997.

[11] T. J. Jech. *Set theory*. Springer Monographs in Mathematics. Springer, third millenium edition, 2003.

[12] A. Kanamori. *The higher infinite. Large cardinals in set theory from their beginnings*. Springer Monographs in Mathematics. Springer, 2nd edition, 2003.

[13] P. Koepke. Some applications of short core models. *Annals of Pure and Applied Logic*, 37(2):179–204, 1988.

[14] K. Kunen. Saturated ideals. *Journal of Symbolic Logic*, 43:65–76, 1978.

[15] K. Kunen. *Set theory: an introduction to independence proofs*, volume 102 of *Studies in Logic and the Foundations of Mathematics*. Elsevier, 1980.

[16] J. P. Levinski. Instances of the conjecture of Chang. *Israel Journal of Mathematics*, 48(2–3):225–243, 1984.

[17] J. P. Levinski, M. Magidor, and S. Shelah. Chang's conjecture for $\aleph_\omega$. *Israel Journal of Mathematics*, 69(2):161–172, 1990.

[18] R.-D. Schindler. On a Chang conjecture. *Israel Journal of Mathematics*, 99:221–230, 1997.

[19] R. L. Vaught. Models of complete theories. *Bulletin of the American Mathematical Society*, 69:299–313, 1963.

An ordinal λ-calculus

Tim Fischbach[1], Benjamin Seyfferth[2]

[1] Mathematisches Institut, Rheinische Friedrich-Wilhelms-Universität Bonn, Bonn, Germany
[2] Fachbereich Mathematik, Technische Universität Darmstadt, Darmstadt, Germany

Abstract

We are going to generalize classical λ-calculus to the ordinal domain. Our reasoning is centered around a generalization of Church numerals, i.e., terms that define the n-fold application of their first argument to the second, to numerals for transfinite ordinals. Once the new class of ordinal λ-terms is established, we define a transfinite procedure to assign to a given ordinal λ-term a normal form if one exists. This normal form procedure is compatible with the classical case, i.e., will find normal forms for classical terms whenever they exist. We go on to prove a confluence property for our procedure. The calculus thus defined is tied into the existing framework of ordinal computability: Using our terms to define a class of functions on the ordinals, we show that this class is identical with the class of $\boldsymbol{\Sigma}_1(L)$-definable functions on Ord.

1 Introduction

Ordinal computability is the study of models of classical computability lifted to the ordinal domain. Particular attention is paid to the elementary computational steps of thus defined transfinite computations and the subtle differences of the different models and their liftings. In his Diploma thesis [4], the first author compared ordinal Turing machines (OTMs) [8] and ordinal register machines (ORMs) [10] to existing liftings of classical recursion schemes, namely Kripke's equation calculus and ordinal min-recursive functions [7]. He showed the equivalence of all approaches on admissible ordinals. This lifting of the most common approaches to classical, finite computability theory is clearly missing a λ-calculus variant, which we shall define here. We prove its equivalence to the aforementioned models, further strengthening the idea of an ordinal Church-Turing thesis. The results of this paper have been presented as an extended abstract (without proofs) in [5] and are contained in the second author's PhD thesis [12] which was supervised by Peter Koepke.

2 Notation from classical λ-calculus

As a reference on classical λ-calculus the authors used the monograph by Barendregt [1]. Terms in classical λ-calculus are formed over the alphabet

Stefan Geschke, Benedikt Löwe, Philipp Schlicht (*eds.*).
Infinity, computability, and metamathematics: Festschrift celebrating the 60th birthdays of Peter Koepke and Philip Welch. College Publications, London, 2014. Tributes, Volume 23.

$\{\lambda, .,), (\} \cup \{v_k \mid k \in \omega\}$ by the following rules (we allow lower case letters to stand in for variables such as v_k):

(1) Every variable x is a term.

(2) If M is a term and x is a variable, then $\lambda x.M$ is a term.

(3) If M and N are terms, then so is (M, N).

We abbreviate terms of the form $\lambda x.\lambda y.M$ as $\lambda xy.M$. The *subterm* relation $S \subseteq T$ is the transitive closure of the relation $M \subseteq M$, $M \subseteq \lambda x.M$, and $M, N \subseteq (M, N)$. With respect to the expression $\lambda x.$, the notion of x as a *bound* or *free* variable has its intended meaning.

We want to identify terms that arise from each other by renaming of bound variables. By $M\frac{N}{x}$ we denote the syntactic substitution of every occurrence of the free variable x in M by the term N. If we want to replace a single instance S of some subterm of T by some term R, we write the result as $T\left[\frac{R}{S}\right]$. Whenever we write such a substitution or replacement, adequate renaming of bound variables is implied to avoid variable conflicts.

The implied interpretation of λ-terms is the following: A term $(\lambda x.M, N)$ is to be interpreted as 'the application of the function $M(x)$ to N'. Accordingly, a *rule of conversion* is defined. The above term may be transformed into $M\frac{N}{x}$. More generally, for any term T, any subterm of T of the form $(\lambda x.M, N)$ may be replaced by $M\frac{N}{x}$, i.e., we may transform T into $T\left[\frac{M\frac{N}{x}}{(\lambda x.M,N)}\right]$. We call such a transformation an application of ***β*-*conversion*** or of the ***β*-*rule*** on T. A subterm of T of the form $(\lambda x.M, N)$ is called a *redex (reducible expression)* of T.

A term S is in *normal form*, if **β**-conversion cannot be applied to it. S is *a normal form of* some term T, if S is in normal form and can be obtained from T by possibly repeated applications of the **β**-rule. There are terms without normal forms, e.g., $(\lambda x.(x, x), \lambda x.(x, x))$. The classical theory proves that the normal form of a term is uniquely determined if it exists and can be found by a certain pattern of applications of **β**-conversion. This was first proved in [2]. We denote the normal form of a term T as $\overline{T}$.

The calculus can be fitted with various semantics. From the perspective of computability theory, maybe one of the most important ones is the λ-definability of functions on the natural numbers. There are several ways of modeling natural numbers as λ-terms; consider the following:

$$\underline{\underline{0}} = \lambda fx.x$$
$$\underline{\underline{1}} = \lambda fx.(f, x)$$
$$\vdots$$
$$\underline{\underline{n}} = \lambda fx.\underbrace{(f, (f, (\ldots (f,}_{n\text{-times}} x) \ldots)$$
$$\vdots$$

The terms thus defined are referred to as *Church numerals*. A partial function $f : \mathbb{N} \supset \operatorname{dom} f \to \mathbb{N}$ is called λ-*definable* if there is a λ-term F such that for all $n \in \operatorname{dom} f$:

$$\overline{(F, \underline{\underline{n}})} = \underline{\underline{f(n)}}$$

The class of λ-definable functions is identical to the class of Turing computable functions. By virtue of the Church-Turing thesis we also speak of the class of computable functions.

3 λI-calculus

The basis for our generalization of λ-calculus shall be given by the λ**I**-calculus as described in [1, Chapter 9]. The λ**I**-terms form a subset of the λ-terms and are formed by replacing formation rule (2) by the following:

(2) If M is a term and x is a variable that appears free in M, then $\lambda x.M$ is a term.

With λ**I**, trivial applications ('forget the argument') are impossible, as terms of the form $\mathbf{K} = \lambda xy.x$ are illegal. So, in general, case distinctions (returning one of several arguments depending on the situation) or constant functions cannot be defined in λ**I**. For functions on numerals, however, this can be circumvented, exploiting the syntactic structure of Church numerals.

Definition 3.1. Set $\mathbf{I} = \lambda y.y$. This is a λ**I**-term defining the unary identity function. [1]

E.g., the numeral $\underline{\underline{0}}$ could be replaced by $\underline{\underline{0'}} = \lambda fx.(((f, \mathbf{I}), \mathbf{I}), x)$. Let $\underline{\underline{n'}} = \underline{\underline{n}}$ for all $n > 0$. Note that, since $((\underline{\underline{n'}}, \mathbf{I}), \mathbf{I})$ reduces to $\mathbf{I}$ for every n', the normal form of $((\underline{\underline{0'}}, \underline{\underline{n'}}), \underline{\underline{m'}})$ is $\underline{\underline{m'}}$. So, on numerals, the meaning '0-fold application of the first argument to the second' is retained.

In contrast to λ**I**-terms, the full set of λ-terms is sometimes referred to as λ**K**-terms. Omitting further details which can be found in [1], we state the following:

Fact 3.2. The $\lambda\mathbf{I}$-definable functions on natural numbers coincide with the $\lambda\mathbf{K}$-definable ones. [1]

4 Ordinal λ-terms

Our approach revolves around the idea of generalizing Church numerals from 'the n-fold application of f to x' to 'the α-fold application of f to x'. The intent behind that idea is that terms defining the successor function or arithmetic should generalize to the successor function on ordinals or ordinal arithmetic.

In developing our theory, we briefly considered introducing terms of transfinite length, but the asymmetry of ordinals—a limit ordinal has a right neighbor (a least larger ordinal) but no left neighbor (largest smaller ordinal)—limits the intuitive use of syntactical operations on such strings. Instead, we introduce symbols for ordinals on term level and we propose the following generalization of λ-terms to ordinal λ-terms.

Definition 4.1. Over the alphabet $\Sigma_{\mathrm{Ord}} = \{\lambda, .,), (\} \cup \{v_k \mid k \in \omega\} \cup \{{}^\alpha \mid \alpha \in \mathrm{Ord}\}$ define the set $\mathrm{Term}_{\mathrm{Ord}}$ of *ordinal λ-terms* by

(1) Every variable x is an ordinal λ-term.

(2) If M is an ordinal λ-term and x appears free in M, then $\lambda x.M$ is an ordinal λ-term.

(3) If α is an ordinal and M and N are ordinal λ-terms, then so is ${}^\alpha(M, N)$.

We often write (M, N) instead of ${}^1(M, N)$.

Informally, we refer to ordinal λ-terms just as *terms*.

We introduce an equivalence relation $\simeq_v$ on terms, identifying all terms that can be obtained from each other by renaming of bound variables. If V is a finite set of variables with largest element v_i, define for every equivalence its *v-minimal term over V* as the one where all bound variables are named v_{i+1}, v_{i+2}, v_{i+3}, etc. from left to right. For a term T, we denote by T_v^V its v-minimal term over V. We simply write v-minimal and T_v if $V = \varnothing$.

Definition 4.2. The *(ordinal) Church numerals* are of the form $\underline{\alpha} = \lambda fx.{}^\alpha(f, x)$ for $\alpha \in \mathrm{Ord}$. More generally, we refer to all terms $\simeq_v$-equivalent to some $\underline{\alpha}$ as Church numerals.

The intended meaning of terms like ${}^\beta(M, {}^\alpha(M, N))$ and ${}^{\alpha+\beta}(M, N)$ is the same. So, we want to identify all the terms of the form

$$ {}^{\alpha_{k-1}}(M, {}^{\alpha_{k-2}}(M, \ldots {}^{\alpha_0}(M, N) \ldots)) $$

with

$$ {}^{\alpha_0+\alpha_1+\ldots+\alpha_{k-1}}(M, N). $$

Let T be a term. We call a replacement of all subterms of T that are of the former form by their equivalent terms of the latter form a *contraction of applications of* T. We define an equivalence relation $\simeq_a$ by identifying every term T with the terms resulting from contractions of its applications and closing transitively. For a given term T we define its *a-minimal term* T_a as the shortest term a-equivalent to T.

In order to define an equivalent to the $\boldsymbol{\beta}$-normal form, we define a transfinite procedure that for every term either finds a term we shall call its normal form or diverges, the latter which we shall interpret as the term not having a normal form.

5 Normal form derivation

The normal form derivation will be a transfinite procedure. We declare a kind of limit convergence for our ordinal λ-terms.

Definition 5.1. Consider a term M as a finite sequence of symbols in Σ_{Ord}, i.e., $M : n \to \Sigma_{\text{Ord}}$ for some natural number $n \in \omega$. Let $(j_i \mid 0 \leq i < k)$ be the increasing sequence of those $j < n$ with $M(j) \in \{\ ^{\alpha} \mid \alpha \in \text{Ord}\}$. Let $\vec{\alpha} = (\alpha_0, \ldots, \alpha_{k-1})$ be some sequence of ordinals. Define

(i) the *flesh* of M as $\text{fl}(M) = (M(j_i) \mid i < k)$

(ii) the *skeleton* of M as $\text{sk}(M) : n \to \text{ran}(M)$ with

$$\text{sk}(M)(j) := \begin{cases} M(j), & \text{if } j \notin \{j_i \mid 0 \leq i < k\} \\ ^{1}, & \text{if } j \in \{j_i \mid 0 \leq i < k\}. \end{cases}$$

(iii) the *insertion of* $\vec{\alpha}$ *in* M as

$$M[\vec{\alpha}] = \begin{cases} M(j), & \text{if } j \notin \{j_i \mid 0 \leq i < k\} \\ ^{\alpha_i}, & \text{if } j = j_i \text{ for some } i < k. \end{cases}$$

Note that $\text{sk}(M)[\text{fl}(M)] = M$.

Definition 5.2. Let α be a limit ordinal.

(i) Let $s : \alpha \to \text{Ord}^n$ be a sequence of n-tuples of ordinals for some $n < \omega$. Define the *pointwise limes inferior* by

$$\liminf_{\beta \to \alpha} s(\beta) := (\liminf_{\beta \to \alpha} s(\beta)_0, \ldots, \liminf_{\beta \to \alpha} s(\beta)_{n-1}).$$

(ii) Let $s : \alpha \to \text{Term}_{\text{Ord}}$ be a sequence of terms. We say that *the skeletons of* s *converge*, if there is an a-minimal and v-minimal skeleton S such that there is a $\gamma < \alpha$ and for all $\gamma < \beta < \alpha$ we have $\mathrm{sk}(s(\beta))_a \simeq_v S$. We shall call S the *limit skeleton for s*. If V is a finite set of variables, S may be chosen v-minimal over V; we then speak of the limit skeleton *over V*.

(iii) Let $s : \alpha \to \Sigma^*_{\mathrm{Ord}}$ be a sequence of terms whose skeletons converge to its limit skeleton S. Let γ be minimal such that $\mathrm{sk}(s(\beta)_a) = S$ for $\gamma < \beta < \alpha$. Then the *syntactical limes inferior* of s exists and is defined by

$$\liminf_{\beta\to\alpha} s(\beta) := \mathrm{sk}(S)[\liminf_{\beta\to\alpha,\beta>\gamma} \mathrm{fl}(s(\beta)_a)]$$

If S is the limit skeleton over some finite set of variables V, we speak of the syntactical limes inferior over V, written $\liminf^V_{\beta\to\alpha} s(\beta)$

In the following we give a deterministic procedure to arrive at a given ordinal λ-term's normal form. Every step can be seen to correspond to one application of the classical $\boldsymbol{\beta}$-rule. In the classical $\lambda\mathbf{I}$-calculus, any pattern of iterated application of the $\boldsymbol{\beta}$-rule eventually yields a normal form. The idea of α-fold application of one term to another implies a transfinite length of applications of a $\boldsymbol{\beta}$-rule, and we want to make use of the limit notions for terms we just defined. We chose to give up the nondeterministic freedom of the finite case to produce stabilizing and natural behavior at limits. In turn, we get some of that freedom back by proving a weak confluence property in § 6. There, we also conjecture a stronger property that would enable us to perform finitely many arbitrary deviations from the algorithm.

Before we give a rigorous definition, it might be helpful to look at the process for arriving at a normal form we have in mind a bit more graphically. The algorithm will maintain a stack. Each stack element is a term whose normal form is to be determined. The bottom element of the stack is the original term T we wish to reduce to its normal form by some generalizations of the $\boldsymbol{\beta}$-rule. The next element shall be the *leftmost redex* of T, i.e., a leftmost subterm S of the form $S = {}^\alpha(\lambda x.M, N)$ where $\alpha > 0$. A redex is called leftmost if its operating λ, i.e., the λ that is spelled out in the representation of S above, appears to the left of all other operating λ's of redexes of T. So, S is put on the second stack level. Recursively, the algorithm will determine the normal form of S. In the mean time, the stack will get built up and torn down again and as soon as S's normal form $\overline{S}$ is found, our stack will contain exactly two elements: T as the bottom one with $\overline{S}$ on top. The next step will be to remove $\overline{S}$ from the stack, replace S in T with $\overline{S}$ and start the procedure over for the resulting term. Eventually, the bottom element will contain no more redexes and a normal form is found.

So how does the algorithm proceed to determine the normal form of some redex $S = {}^\alpha(\lambda x.M, N)$? We retain the intuition of 'the α-fold application of M to N' by the following procedure: Determine, by putting on the stack consecutively, the normal forms of the approximations ${}^1(\lambda x.M, N)$, ${}^2(\lambda x.M, N)$, etc. At limit times, syntactical inferior limits are taken (if they exist, otherwise the normal form procedure breaks down). More precisely, instead of ${}^{\gamma+1}(\lambda x.M, N)$, we evaluate the a-equivalent term

$$(\lambda x.M, {}^{\gamma}(\lambda x.M, N)),$$

substituting the term N' we determined in the previous steps as the normal form of ${}^{\gamma}(\lambda x.M, N)$, which in the end gives us $(\lambda x.M, N')$ to evaluate. We can now rely on an application of what is known as the β-rule in the finite case to end up with $M\frac{N'}{x}$ which is what is being put on the stack instead of ${}^{\gamma+1}(\lambda x.M, N)$.

Finally, we have to deal with our stack length becoming infinite. This might happen via the use of terms that work like $(\lambda x.(x, x), \lambda x.(x, x))$ and may be used as so-called fixed-point combinators in recursive definitions. For instance, one usually implements unbounded search by a term describing the following function $Q(\alpha)$: 'if condition P holds on α then return α, else evaluate and return $Q(\alpha+1)$'. A normal form procedure for $Q(0)$ will build up a stack of height β if β is the least ordinal such that P holds. If the stack length approaches a limit ξ, we shall define the stack content at level ξ in the following way: First, identify the first (from the bottom) term on the stack whose skeleton appears cofinally often below ξ as skeleton of terms on the stack. Now, set as the term on level ξ the syntactical lim inf over all terms on the stack with this skeleton. In the above example, we shall 'try α' (i.e., put $Q(\alpha)$ on the stack), then do some steps to determine whether P holds for α and if not 'try $\alpha + 1$'. In the next limit we want to 'try the first limit after α', i.e., put $Q(\alpha+\omega)$ on the stack. The term $Q(0)$ representing 'try 0' obviously is the first whose skeleton appears cofinally often. Theorem 7.8 will confirm this behavior. To avoid problems with backtracking downwards in the ordinals, we shall keep track of the stack height on which the first term that lends its skeleton to the limit appears. That way, as soon as a normal form is found (in our example the least β such that P holds) and has been handed down step by step until the stack is torn down to some limit height, we can propagate this normal form directly downwards to whenever $Q(0)$ was put on the stack.

Remark 5.3. Terms of the form ${}^0(\lambda x.M, N)$ are not treated as reducible; they are unchanged by the algorithm save for internal modifications of M and N and may vanish only through contractions of applications. One could argue that the intended meaning behind such a term is simply N (the 0-fold application of M to N) but such transformations would reintroduce terms of the form $\lambda xy.x$, violating the boundaries of $\lambda\mathbf{I}$-calculus.

For our purposes, there is an added benefit of not resolving 0-fold applications: The numeral $\underline{0} = \lambda fx.^{0}(f,x)$ is of the same syntactical form as the other numerals, enabling $\underline{0}$ to be a possible value of a syntactical lim inf of numerals. However, not setting $\underline{0}$ apart from the other numerals introduces a difficulty with arithmetic: Several classical algorithms for arithmetic (predecessor of natural numbers, subtraction, etc.) rely heavily on a term that recognizes the numeral for 0 among all the other numerals. We resolve this issue by expanding our calculus by a term that defines equality on ordinals in Definition 5.6.

For the following rigorous definition, some additional information is coded into the stack elements, but the above structure remains valid. In general, we consider a stack to be a sequence indexed by a successor ordinal. Operations changing the stack are either restricting the sequence to another successor ordinal or adding a new element on top. The algorithm below will define the behavior when infinitely many end extensions are carried out. The stack elements will be tuples $(\cdot)$, so we write a stack of tuples as $\langle(\cdot),(\cdot),\ldots,(\cdot)\rangle$. We use the symbol $\sqcup$ to denote the composition of two stacks.

We need to elaborate how renaming of bound variables will be handled in the normal form procedure. Whenever a substitution of the form $M\frac{N}{x}$ or a replacement of the form $M\left[\begin{smallmatrix}N\\P\end{smallmatrix}\right]$ is carried out, we rename bound variables adequately: In every renaming, new variables are chosen as to not accidentally bind a free variable of some term farther down on the stack. At limits, we chose the limit terms υ-minimally, while avoiding the variables of certain lower stack levels: If all stack levels are finite, we can certainly avoid all of the only finitely many variables used below. If an infinite stack level is reached, we only require the variables up to the level from which the skeleton is lent to the limit to be avoided. Inductively, at any given point in time we only need to avoid finitely many variables.

Definition 5.4. Let N be a term. A function $s : \vartheta \to (\Sigma^*_{\mathrm{Ord}} \times \Sigma^*_{\mathrm{Ord}} \times \mathrm{Ord} \times \mathrm{Ord} \times \mathrm{Ord})^{<\mathrm{Ord}}$ is called a *normal form derivation of* N if it satisfies the following conditions:

(a) $s(0) = \langle(N, \varnothing, 0, 0, 0)\rangle$.

(b) Let $\nu < \vartheta$.

(i) If the topmost element of $s(\nu)$ is of the form $(N', \varnothing, 0, 0, \delta)$, and N' is not a-minimal, and $R = {}^{\beta}(P, {}^{\alpha}(P, Q))$ be the leftmost subterm on which a contraction of applications may be carried out, then $\nu + 1 < \vartheta$ and $s(\nu+1) = s(\nu)\restriction\nu \sqcup \langle(N\left[\begin{smallmatrix}R\\ \alpha+\beta(P,Q)\end{smallmatrix}\right], \varnothing, 0, 0, \delta)\rangle$.

(ii) If the topmost element of $s(\nu)$ is of the form $(N', \varnothing, 0, 0, \delta)$, and N' is a-minimal and contains a subterm of the form ${}^{\gamma}(\lambda x.P, Q)$ where $\gamma > 0$, and ${}^{\alpha}(\lambda x.M, N'')$ is the leftmost such subterm, then $\nu + 1 < \vartheta$ and $s(\nu + 1) = s(\nu) \sqcup \langle (N'', \lambda x.M, \alpha, 0, \delta) \rangle$.

(iii) If the topmost element of $s(\nu)$ is of the form $(N'', \lambda x.M, \alpha, \beta, \delta)$ with $\beta < \alpha$, then $\nu+1 < \vartheta$ and $s(\nu+1) = s(\nu) \sqcup \langle (M\frac{N''}{x}, \varnothing, 0, 0, \delta) \rangle$.

(iv) If the two topmost elements of $s(\nu)$ are of the form

$$\langle (N'', \lambda x.M, \alpha, \beta, \delta), (N', \varnothing, 0, 0, \delta) \rangle$$

and N' is a-minimal and does not have a subterm of the form ${}^{\gamma}(\lambda x.P, Q)$ with $\gamma > 0$, then $\nu + 1 < \vartheta$ and $s(\nu + 1) = s(\nu) \upharpoonright (|s(\nu)| - 2) \sqcup \langle (N', \lambda x.M, \alpha, \beta + 1, \delta) \rangle$.

(v) If $|s(\nu)|$ is successor of a limit and the topmost element of $s(\nu)$ is of the form $(N', \varnothing, 0, 0, \delta)$ and N' is a-minimal and does not have a subterm of the form ${}^{\gamma}(\lambda x.P, Q)$ with $\gamma > 0$ and $s(\nu)(\delta) = (N'', \varnothing, 0, 0, \gamma)$, then $s(\nu + 1) = s(\nu) \upharpoonright \delta \sqcup \langle (N', \varnothing, 0, 0, \gamma) \rangle$.

(vi) If the two topmost elements of $s(\nu)$ are of the form $\langle (N', \varnothing, 0, 0, \delta), (N'', \lambda x.M, \alpha, \alpha, \delta) \rangle$, then $\nu + 1 < \vartheta$ and

$$s(\nu) \upharpoonright (n - 2) \sqcup \langle (\tilde{N}, \varnothing, 0, 0, \delta) \rangle,$$

where $\tilde{N}$ arises from N' by replacing the leftmost subterm of the form ${}^{\gamma}(\lambda x.P, Q)$ with N''.

(vii) If none of the above conditions hold, apparently we have $s(\nu) = (N', \varnothing, 0, 0, 0)$, N' is a-minimal and does not have any redexes, and $|s(\nu)| = 1$. Then $\vartheta = \nu + 1$ and N' is called *the result of the normal form derivation of* N, written $\overline{N} = N'$.

(c) Let ξ be a limit ordinal such that $s \upharpoonright \xi$ is defined.

(i) If $\liminf_{\nu<\xi} |s(\nu)| = \gamma$ and the sequence $(\nu)_{\nu<\xi \wedge |s(\nu)|=\gamma}$ is unbounded in ξ, note that the stacks $(s(\nu)\upharpoonright\gamma)_{\nu<\xi \wedge |s(\nu)|=\gamma}$ are eventually constant some stack s of length γ. Let V be the set of variables of the terms on the stack levels up to γ. If the skeletons of $(s(\nu)(\gamma))_{\nu<\xi \wedge |s(\nu)|=\gamma}$ converge, then

$$s(\xi) = s \sqcup \Big\langle \Big(\liminf{}^{V}_{\nu<\xi \wedge |s(\nu)|=\gamma} \operatorname{pr}_0(s(\nu)), \\ \varnothing, 0, 0, \liminf_{\nu<\xi \wedge |s(\nu)|=\gamma} \operatorname{pr}_4(s(\nu)) \Big) \Big\rangle .$$

If they do not converge, then $\xi = \vartheta$ and the normal form derivation of N is said to *diverge*, written $\overline{N} \uparrow$.

(ii) If, on the other hand, $\liminf_{\nu<\xi} |s(\nu)| = \gamma$, η is a limit, and $(|s(\nu)|)_{\nu<\xi}$ is unbounded in η, we require that for every $\gamma < \eta$ there is a time $\nu < \xi$ such that for every $\mu > \nu$ we have $s(\mu)\restriction\gamma = s(\nu)\restriction\gamma$. This way we obtain a *limit stack* t of length η. It remains to be determined what element is to be put on top of this limit stack at time ξ. Choose the first (i.e., with minimal γ) skeleton $u = \operatorname{sk}\operatorname{pr}_0(t(\gamma))$ of $(\operatorname{pr}_0(t(\nu)))_{\nu<\eta\wedge\text{“}\nu\text{ is even”}}$ such that either u or an v- or a-equivalent term appears unboundedly often as skeleton in the first coordinate of t. Let V be the set of variables of the terms on the stack levels up to γ. Set

$$s(\xi) = t \sqcup \Big\langle \Big(\liminf{}^{V}_{\nu<\eta\wedge\text{“}\nu\text{ is even”}\wedge\operatorname{sk}\operatorname{pr}_0(t(\nu))=u}\operatorname{pr}_0(t(\nu)),\\ \varnothing, 0, 0, \gamma\Big)\Big\rangle\,.$$

If no limit stack t exists or no skeleton appears unboundedly often in t, then $\xi = \vartheta$ and the normal form derivation of N is said to *diverge*, written $\overline{N}\uparrow$.

If no normal form derivation exists for some term N, we also say that the normal form derivation of N *diverges*, written $\overline{N}\uparrow$.

Definition 5.5. Let $\vec{\alpha} = (\alpha_0, \ldots, \alpha_{n-1})$ be a finite sequence of ordinal numbers. A function $f : \mathrm{Ord}^k \to \mathrm{Ord}$ is *ordinal λ-definable in parameters $\vec{\alpha}$* if there is an ordinal λ-term T in which all applications are of the form ${}^{\beta}(\cdot,\cdot)$ where $\beta \in \omega \cup \{\alpha_0, \ldots, \alpha_{n-1}\}$ such that for all $(\gamma_0, \ldots, \gamma_{k-1}) \in \mathrm{Ord}^k$ we have

$$\overline{(\ldots(T, \underline{\gamma_0}), \underline{\gamma_1}), \ldots), \underline{\gamma_{k-1}})} \simeq_v \underline{f(\gamma_0, \ldots, \gamma_{k-1})}.$$

If f is a partial function, we call f ordinal λ-definable in $\vec{\alpha}$ if $f\restriction\operatorname{dom} f$ is ordinal λ-definable in $\vec{\alpha}$ and $\overline{(T, \underline{\gamma})}\uparrow$ on $\gamma \notin \operatorname{dom} f$. If $\vec{\alpha} = \varnothing$, we simply speak of *ordinal λ-definability.*

As explained in Remark 5.3, we would like to add the capability of defining equality on ordinals to our calculus:

Definition 5.6. Let us add a constant symbol E to our alphabet and consider the terms formed over $\Sigma_{\mathrm{Ord}} \cup \{E\}$ with the additional rule ‘E is a term’ as the *ordinal $\lambda + E$-terms.* We extend Definition 5.4 by a case for terms of the form ${}^{1}({}^{1}(E, \underline{\alpha}), \underline{\beta})$: The algorithm is to replace ${}^{1}({}^{1}(E, \underline{\alpha}), \underline{\beta})$ with the normal form $\mathbf{T}_I = \lambda xy.(((y, \mathbf{I}), \mathbf{I}), x)$ if $\alpha = \beta$ and with the normal form $\mathbf{F}_I = \lambda x.(((x, \mathbf{I}), \mathbf{I}), \mathbf{I})$ else. In all other cases, E is to be treated like a variable symbol. The resulting notion of definability for functions on the ordinals is that of *ordinal $\lambda + E$-definable* functions.

The terms $\mathbf{T}_I$ and $\mathbf{F}_I$ will be used in the following manner:

Suppose $\overline{((P,\mathbf{I}),\mathbf{I})} \simeq_v \overline{((Q,\mathbf{I}),\mathbf{I})} \simeq_v \mathbf{I}$, which is the case, e.g., for P, Q Church numerals.

$$\text{Then } \overline{((B,P),Q)} \simeq_v \begin{cases} \overline{P} & \text{, if } \overline{B} = \mathbf{T}_I \\ \overline{Q} & \text{, if } \overline{B} = \mathbf{F}_I. \end{cases}$$

The term $((B,P),Q)$ hence may be read as `if` B `then` P `else` Q. We shall also consider variations of $\mathbf{T}_I$ and $\mathbf{F}_I$: The terms $\mathbf{T}_J$ and $\mathbf{F}_J$ as well as $\mathbf{T}_3$ and $\mathbf{F}_3$ are defined later on and behave similarly, for terms P and Q vanishing under different conditions.

Some examples are in order now. We take the liberty of abbreviating some terms in these examples, numerals for instance. Additional steps are added to write them out whenever necessary. We may also use extra lines to carry out substitutions etc. Apart from this, there is a line for every step in the normal form derivation where the stack has odd height, i.e., the topmost element is of the form $(N', \varnothing, 0, 0, \delta)$. Every line contains the first entry of the topmost stack element. Stack height is suggested by indentation. We mark the approximation steps by leading numbers to facilitate reading. We highlight the leftmost redex by overlining. We deliberately use our normal form notation here: the algorithm will recursively determine the normal form of the overlined term and, once found, replace the term with its normal form.

Example 5.7. Let us count up to ω to illustrate the desired lim inf behavior. The term $S_c^+ = \lambda nfx.(f,((n,f),x))$ classically defines the successor function of a given numeral. We show that the normal form of ${}^{\omega}(S_c^+, \underline{0})$ is $\underline{\omega}$. Let us run our algorithm:

${}^{\omega}(S_c^+, \underline{0})$	write out S_c^+
$\overline{{}^{\omega}(\lambda nfx.(f,((n,f),x)),\underline{0})}$	the entire term is the leftmost redex
1: $\lambda fx.(f,((n,f),x))\frac{\underline{0}}{n}$	start the approximation of the ω redex
$\lambda fx.(f,((\underline{0},f),x))$	write out $\underline{0}$
$\lambda fx.(f,(\overline{(\lambda gy.{}^{0}(g,y),f)},x))$	identify leftmost redex
1: $\lambda y.{}^{0}(g,y)\frac{f}{g}$	first and only approximation step
$\lambda y.{}^{0}(f,y)$	return redex-free term
$\lambda fx.(f,\overline{(\lambda y.{}^{0}(f,y),x)})$	identify leftmost redex
1: ${}^{0}(f,y)\frac{x}{y}$	first and only approximation step
${}^{0}(f,x)$	return redex-free term

	$\lambda fx.(f, {}^{0}(f,x))$	contract applications to obtain an a-minimal term
	$\lambda fx.(f,x)$	
	$\underline{1}$	supply redex-free term to next approximation step
2:	$\lambda fx.(f,((n,f),x))\frac{\underline{1}}{n}$	second approximation step
	$\vdots$	analogously to first approximation step
	$\underline{2}$	hand redex-free term to next approximation step
$\vdots$		analogously for all finite steps
ω:	$\liminf_{n\to\omega} \underline{n}$	take lim inf at limit step
	$\liminf_{n\to\omega} \lambda fx.{}^{n}(f,x)$	evaluate syntactical lim inf
	$\lambda fx.{}^{\omega}(f,x)$	last approximation step; return redex-free form
$\lambda fx.{}^{\omega}(f,x)$		normal form reached
$\underline{\omega}$		

The example above can be used to prove that the successor function on ordinals is ordinal λ-definable. The following example is given to show some nested limits, suggesting that recursively defined ordinal arithmetic can be implemented in a straightforward manner.

Example 5.8. If $\alpha, \beta > 0$ then $\lambda fx.((\underline{\beta}, (\underline{\alpha}, f)), x)$ defines the product $\alpha \cdot \beta$.

$\lambda fx.((\underline{\beta}, (\underline{\alpha}, f)), x)$		write out numerals
$\lambda fx.(\overline{((\lambda gy.{}^{\beta}(g,y), (\lambda hz.{}^{\alpha}(h,z), f))}, x)$		identify leftmost redex
1:	$\lambda y.{}^{\beta}(g,y)\frac{(\lambda hz.{}^{\alpha}(h,z), f)}{g}$	first and only approximation step
	$\lambda y.{}^{\beta}(\overline{(\lambda hz.{}^{\alpha}(h,z), f)}, y)$	identify leftmost redex
	1: $\lambda z.{}^{\alpha}(h,z)\frac{f}{h}$	first and only approximation step
	$\lambda z.{}^{\alpha}(f,z)$	return redex free term
	$\lambda y.{}^{\beta}\overline{(\lambda z.{}^{\alpha}(f,z), y)}$	identify leftmost redex
	1: ${}^{\alpha}(f,z)\frac{y}{z}$	first approximation step of the β redex
	${}^{\alpha}(f,y)$	hand redex-free term to next approximation step
	2: ${}^{\alpha}(f,z)\frac{{}^{\alpha}(f,y)}{z}$	second approximation step
	${}^{\alpha}(f, {}^{\alpha}(f,y))$	contract applications
	${}^{\alpha\cdot 2}(f,y)$	hand redex-free term to next approximation step
	$\vdots$	analogously for all finite steps
	ω: $\liminf_{n\to\omega} {}^{\alpha\cdot n}(f,y)$	take lim inf at limit step

$\quad\quad\quad {}^{\alpha\cdot\omega}(f,y)$	hand redex-free term to next approximation step
$\quad\quad\vdots$	analogously for all further steps
$\quad\beta\colon\ \liminf_{\gamma\to\beta} {}^{\alpha\cdot\gamma}(f,y)$	take lim inf at limit step
$\quad\quad\quad {}^{\alpha\cdot\beta}(f,y)$	last approximation step; return redex-free term
$\quad\lambda y.{}^{\alpha\cdot\beta}(f,y)$	return redex-free term
$\lambda fx.\overline{(\lambda y.{}^{\alpha\cdot\beta}(f,y),x)}$	identify leftmost redex
$\quad 1\colon\ \lambda y.{}^{\alpha\cdot\beta}(f,y)\frac{x}{y}$	first and only approximation step
$\quad\quad {}^{\alpha\cdot\beta}(f,x)$	return redex-free term
$\lambda fx.{}^{\alpha\cdot\beta}(f,x)$	normal form reached
$\underline{\alpha\cdot\beta}$	

The two examples just given only illustrate the limit case (c)(i). For an example for case (c)(ii), cf. § 7.2.

Classical $\lambda\mathbf{I}$-terms are ordinal λ-terms. On the other hand, if we have an ordinal λ-term where for all applications ${}^{\alpha}(M,N)$ we have that α is finite, we can convert it to a classical $\lambda\mathbf{I}$-term by a map φ given by replacing any subterm of the form ${}^{n}(M,N)$ by $\underbrace{(M,(M,(\ldots(M}_{n\text{-times}},N)\ldots)$ if $n>0$, and replacing any subterm of the form ${}^{0}(M,N)$ by $(((M,\mathbf{I}),\mathbf{I}),N)$. In particular, this maps $\underline{0}$ to $\underline{\underline{0}}' = \lambda fx.(((f,\mathbf{I}),\mathbf{I}),x)$, the term [1] uses in the treatment of $\lambda\mathbf{I}$-calculus.

Proposition 5.9. If M is a classical $\lambda\mathbf{I}$-term with classical normal form M' and M'' is the output of our algorithm on input M, then $\varphi(M'') \simeq_v M'_a$.

Proof. In $\lambda\mathbf{I}$-calculus, every reduction strategy (pattern of applying the $\boldsymbol{\beta}$-rule to various subterms until no redex is left) is normalizing, i.e., eventually yields normal forms. Our algorithm, although working with a-minimal terms, will simply run a finite number of applications of the $\boldsymbol{\beta}$-rule before halting with a term M'' without redexes. Converting this term to a classical $\lambda\mathbf{I}$-term does not introduce any redexes, so the resulting term is also classically in normal form. Since classically normal forms are unique up to renaming of variables, we have indeed found a term $\simeq_v$-equivalent to M'. Q.E.D.

6 A confluence property for our algorithm

The classical Church-Rosser result establishes that for any two terms Q and Q' that are obtained from the same term P via $\boldsymbol{\beta}$-reduction, there is a term R that can be obtained from Q and Q' by $\boldsymbol{\beta}$-reduction. This is known as the Church-Rosser property. It ensures that normal forms are unique. In the

$\lambda\mathbf{K}$-calculus, a term's unique normal form is obtained by a certain pattern of applications of the β-rule, whereas in $\lambda\mathbf{I}$, any pattern of applications of the β-rule leads to the term's normal form, given that one exists. In our situation, where we restrict ourselves from applying the β-rule freely for the sake of convergence at limits, we propose the following as the correct lifting of the Church-Rosser theorem:

Conjecture 6.1 (Seyfferth). Let T be a term with normal form and let $S \subseteq T$ be a subterm with normal form. Then $\overline{T} \simeq_v \overline{T\left[\frac{\overline{S}}{S}\right]}$.

For the purposes of this paper, the following result is sufficient, as it will establish that the composition of two λ-definable functions is λ-definable (results for $\lambda + E$ follow analogously).

Theorem 6.2. Let T be a term with normal form and let $S \subseteq T$ be a subterm that has a normal form and is of the form $S = {}^{\alpha}(M, N)$ such that all free variables in S are not bound in T. Then $\overline{T} \simeq_v \overline{T\left[\frac{\overline{S}}{S}\right]}$.

Proof. Fix some term S of the form ${}^{\alpha}(M, N)$. We can assume that S is not in normal form, otherwise we would be done. Let T be some term with S as a subterm, such that all free variables in S are not bound in T. Assume that T has a normal form. We shall compare the normal form derivations t of T and t' of $T\left[\frac{\overline{S}}{S}\right]$. It will be evident that t and t' have basically the same steps, except for subsequences of t that parallel the normal form derivation of S.

Let ϑ be the length of t and let ϑ' be the length of t'. We define recursively an injective and weakly monotonous map $f : \vartheta' \to \vartheta$ such that for every $\gamma < \vartheta'$ we have that every stack element of $t'(\gamma)$ can be obtained from $t(f(\gamma))$ by replacing copies of S by copies of $\overline{S}$ and the stacks $t'(\gamma)$ and $t(\gamma)$ have the same height. We define this map 'the other way round', by defining a surjective and increasing but not injective partial map g from t to t' such that our condition holds: At any time δ, we have that all stack elements of $t'(g(\delta))$ arise from $t(\delta)$ by replacing copies of S by copies of $\overline{S}$ (and possibly some variable renaming, which we shall surpress for the rest of this argument). We then can choose as $f(\gamma)$ the largest δ with $g(\delta) = \gamma$ (it will be clear from the construction that there always will be a largest pre-image). Let $\vartheta_0 + 1 = \vartheta$ and $\vartheta'_0 + 1 = \vartheta'$ (normal form derivations always have successor length). The construction below makes sure that g is defined on ϑ_0. Since g is surjective and increasing, $g(\vartheta_0) = \vartheta'_0$.

Then $t'(\vartheta'_0)$ is a stack of height 1 and its top element is in normal form. Therefore $t(f(\vartheta'_0))$ is also also a stack of height 1 with top element in normal form and we have $f(\vartheta'_0) = \vartheta_0$ and $\overline{T} \simeq_v \overline{T\left[\frac{\overline{S}}{S}\right]}$.

Throughout the recursive definition of g, we shall keep track of the *residuals* of S along the normal form derivation t. The term residual was coined by Church and Rosser in their paper proving confluence properties for the classical $\lambda\mathbf{I}$-calculus [2]. We use it in the following way: During a normal form derivation, the subterm S of the original term T may be moved around or duplicated due to other redexes being resolved. The resulting subterms ${}^{\alpha}(M', N')$ are called residuals of S. Residuals of S vanish as soon as the application term ${}^{\alpha}(M', N')$ is put on the stack and evaluated. Note that the free variables of residuals of S remain unbound in the surrounding term. We shall make only informal use of the notion of residuals, as all the details will be made clear in the definition of g.

At time 0, we set $g(0) = 0$. Our condition holds. So let g be defined up to δ and let $t(\delta)$ be of odd height. Consider the term T' that is the topmost element of $t(\delta)$ (more precisely, the first component of the topmost element, which is the term in whose normal form we are interested at time δ). Assume that our condition holds up to δ and let $S_0, S_1, \ldots, S_{k-1}$ denote those copies of S that are residuals of S in T'. Note that, as with S in T, all free variables of the S_i are unbound in T'.

Case 0. If T' is not a-minimal, consider the leftmost term of the form ${}^{\vartheta}(Q, {}^{\eta}(Q, R))$. If the second part ${}^{\eta}(Q, R)$ is not one of the S_i, set $g(\delta+1) = g(\delta) + 1$ as in both t and t' the next step simply is the contraction of applications. If ${}^{\eta}(Q, R) = S_i$, observe that the next step in t is the contraction of applications; followed by the first η-many approximation steps, i.e., the entire normal form derivation of S_i minus the last step that returns the result; again followed by the next ϑ-many approximation steps. Let γ be the length of the first η-many approximation steps. Let β be the length of the next ϑ-many approximation steps. Set $g(\delta + 1 + \gamma + i) = g(\delta) + 1 + i$ for $i < \beta$. Then our condition holds for $t(\delta + 1 + \gamma)$ and $t'(g(\delta) + 1)$ as in both stacks we are just at the beginning of the ϑ-many approximation steps of ${}^{\vartheta}(Q, \overline{S})$. The condition carries over to the next $\boldsymbol{\beta}$-many steps that are carried out in the same fashion in t and t'.

Case 1. If T' is in normal form, then the next two steps of t will be to substitute T' into a term immediately below T' on the stack. The topmost element of $t'(g(\delta))$ is identical to T' since there cannot be any copies of S around (we assumed S not to be in normal form). Set $g(\delta+1)$ and $g(\delta+2)$ to $g(\delta)+1$ and $g(\delta)+2$ respectively. Our condition, that all stack elements of $t'(g(\delta + j))$ arise from $t(\delta + j)$ by replacing copies of S by copies of $\overline{S}$, is retained (for $j = 1, 2$).

Case 2. So let T' not be in normal form. Let the leftmost redex be $P = {}^{\beta}(\lambda x.Q, R)$.

Case 2.1 P lies to the left of every S_i. Let γ be the length of the normal form derivation and set $g(\delta+j) = g(\delta)+j$ for $j < \delta$. Clearly, our condition holds for all these $\delta + i$. The residuals of S in the topmost element of the $t(\delta + i)$ are the same as those of T'.

Case 2.2. There are some S_j, for $j \in J \subseteq k$, that are subterms of P. Each of those is a subterm either of Q or of R. Let J_Q and J_R be the subsets of J containing the indices of subterms of Q and R respectively. We set $g(\delta + i) = g(\delta) + i$ for $i = 1, 2$. At $t(\delta + 1)$ the evaluation of ${}^{\beta}(\lambda x.Q, R)$ is prepared, so our condition holds. So consider $t(\delta + 2)$, i.e., the relevant next step in the normal form development of T. First consider those S_q with $q \in J_Q$. Since we assumed that all free variables of S are unbound in T and we took precautions to not accidentally bind variables, we know that S does not contain the variable x. Therefore, the topmost stack element $Q\frac{R}{x}$ contains these S_q as subterms. The replacement does not depend on the structure of S_q, so $t'(g(\delta + 2))$ contains $\overline{S}$ at exactly the same positions as $t(\delta+2)$ contains the S_q as subterms. Let us turn to the S_r for $r \in J_R$. In $t(\delta+2)$'s topmost element $Q\frac{R}{x}$, R has been substituted for all occurrences of x in Q and with it all the S_r. Since this substitution is also independent from the syntactic structure of the S_r, again the topmost element of $t'(g(\delta + 2))$ can be obtained from replacing all copies of all the S_r by $\overline{S}$.

Case 2.3 There is some $i < k$ such that P is S_i or a subterm of S_i. Let γ be the length of the normal form derivation of S. We set $g(\delta + \gamma)$ to $g(\delta)$. Observe that, by the definition of the normal form derivation, the steps of t between δ and $\delta + \gamma$ are used to determine the normal form of S_i and replace S_i in T' by $\overline{S}$. So the stacks $t(\gamma)$ and $t(\gamma+\delta)$ are of the same height and identical on all but the top layer which differs by the replacement of S_i by $\overline{S}$. Then, obviously our condition holds for $t(\delta + \gamma)$ and $t'(g(\delta))$.

We still need to define g at limits. Suppose κ is a limit and let g be defined on an unbounded subset $D \subseteq \kappa$. Assume without loss of generality, that $\kappa \setminus D$ is unbounded in κ, i.e., cofinally many steps from t are left out in $t{\upharpoonright}D$. All 'gaps' in D stem from Case 2.3. If there is a stack height σ that appears cofinally often in $t{\upharpoonright}D$, we know that σ also appears cofinally as stack height in $t{\upharpoonright}\kappa$. Since the gaps according to Case 2.3 each start and end with the same stack heights and only increase stack height in between, they cannot factor in the lim inf of $t{\upharpoonright}\kappa$. So we are safe to set $g(\kappa) = \lim_{\gamma<\kappa} g(\gamma)$. Now suppose the stack heights increase unboundedly in $t{\upharpoonright}D$. Since the gaps only make the stack higher, also the stack heights of $t{\upharpoonright}\kappa$ increase unboundedly. But in fact, the stack height at the beginning and the end of each gap is the same. Every gap ends before κ (since D is unbounded), so the part of the stack that gets built up and torn down within a gap cannot factor in to the 'limit stack' (as per Definition 5.4, (c), (ii)). Hence, the first skeleton on the limit stack that appears cofinally often in $t{\upharpoonright}D$ is the same as in $t{\upharpoonright}\kappa$. So we are safe to set $g(\kappa) = \lim_{\gamma<\kappa} g(\gamma)$. In both cases the condition holds and residuals are inherited from below the limit. Q.E.D.

7 Ordinal λ-definable functions and ordinal computability

We shall now explore which functions on the ordinals are ordinal λ-definable. In his Diploma thesis [4], the first author showed how various existing notions of ordinal computability coincide in strength. We tie in our proposed model of ordinal λ-definability into this framework to state our main result at the end of this section.

7.1 Primitive recursive set and ordinal functions

A generalization of primitive recursive functions on natural numbers, operating on the universe of sets, has been used in the study of the constructible hierarchy [7, 3]. In [7], Jensen and Karp gave a definition for $\mathrm{Prim}_{\mathrm{O}}$, the class of primitive recursive functions mapping ordinals to ordinals, which is compatible with their notion Prim of primitive recursiveness of functions mapping sets to sets defined alongside in their paper:

Definition 7.1. Let $b_0, \ldots, b_{n-1}$ be unary ordinal functions, i.e., $b_i : \mathrm{Ord} \to \mathrm{Ord}$ for $0 \leq i < k$. The symbol $\mathrm{Prim}_{\mathrm{O}}(b_0, \ldots, b_{k-1})$ (*primitive recursive ordinal functions in* $b_0, \ldots, b_{k-1}$) denotes the collection of all functions of type (1) to (5) closed under the schemes for substitution (a) and (b) and recursion (R).

(1) $f(\xi) = b_i(\xi)$ for $0 \leq i < k$

(2) $\mathrm{pr}_{n,i}(\vec{\xi}) = \xi_i$, for all $n \in \omega$, $\vec{\xi} = (\xi_1, \ldots, \xi_n)$ and $1 \leq i < n$.

(3) $f(\xi) = 0$

(4) $f(\xi) = \xi + 1$

(5) $c(\xi, \zeta, \gamma, \delta) = \begin{cases} \xi, \text{ if } \gamma < \delta \\ \zeta, \text{ else} \end{cases}$

(a) $f(\vec{\xi}, \vec{\zeta}) = g(\vec{\xi}, h(\vec{\xi}), \vec{\zeta})$

(b) $f(\vec{\xi}, \vec{\zeta}) = g(h(\vec{\xi}), \vec{\zeta})$

(R) $f(\xi, \vec{\zeta}) = g(\sup_{\eta<\xi} f(\eta, \vec{\zeta}), \xi, \vec{\zeta})$

We write $\mathrm{Prim}_{\mathrm{O}}$ for $\mathrm{Prim}_{\mathrm{O}}(\varnothing)$. Within this paper, we are interested in the case where the b_i are constant ordinal functions with value α_i. We therefore shall write $\mathrm{Prim}_{\mathrm{O}}(\alpha_0, \ldots, \alpha_{k-1})$ in these cases. A relation on ordinals is $\mathrm{Prim}_{\mathrm{O}}$ if its characteristic function is $\mathrm{Prim}_{\mathrm{O}}$. [7]

Let G_1, G_2 denote the inverses of the Gödel pairing function, mapping an ordinal α to the first or second coordinate of the αth Gödel pair. Based on [7, Theorem 4.4] one readily proves:

Lemma 7.2. Let α be admissible. There is a $\mathrm{Prim}_{\mathrm{O}}$ relation T such that for any partial $\mathbf{\Sigma}_1(L_\alpha)$-definable function $F : \alpha \rightharpoonup \alpha$ there is a bounded formula φ and an ordinal number β such that for all $\gamma \in \alpha$ we have $F(\gamma) = G_1 \min_{\xi \in \alpha} T(\xi, \ulcorner\varphi\urcorner, \gamma, \beta)$.

Proof. Let F be $\mathbf{\Sigma_1}$ via $\exists z \varphi$ and the parameter w, i.e.,

$$L(\alpha) \models F(x) = y \leftrightarrow \exists z \varphi(x, y, z, w).$$

Assume $x \in \operatorname{dom} F$. Then $F(x) = G_1 \min_{\xi \in \alpha} \varphi(x, G_1(\xi), G_2(\xi), w)$. Use the Prim enumeration N of all constructible sets from [7] to find a β such that $F(x) = G_1 \min_{\xi \in \alpha} \varphi(x, G_1(\xi), G_2(\xi), N(\beta))$. Since φ is is Δ_0 and truth of Δ_0 relations is Prim, we get a desired relation T as Prim and via [7, 3.5] as $\mathrm{Prim}_{\mathrm{O}}$. Q.E.D.

We obtain a different characterization of primitive recursive functions on ordinals by replacing rules (5) and (R) by:

(5') $$e(\xi, \zeta) = \begin{cases} 1 \text{ if } \xi = \zeta \\ 0 \text{ else} \end{cases}$$

(R') If g and h are given, define f by

$$\begin{aligned} f(0, \vec{\zeta}) &= g(\vec{\zeta}) \\ f(\xi + 1, \vec{\zeta}) &= h(f(\xi), \xi, \vec{\zeta}) \\ f(\xi, \vec{\zeta}) &= \liminf_{\eta < \xi} f(\eta, \vec{\zeta}) \text{ if } \xi \text{ is a limit ordinal} \end{aligned}$$

It was proved in Tim Fischbach's Diploma thesis [4, Appendix A] that these two schemes are equivalent.

Lemma 7.3. The class of $\mathrm{Prim}_{\mathrm{O}}$ functions and the class obtained from $\mathrm{Prim}_{\mathrm{O}}$ by replacing (5) by (5') and (R) by (R') are the same. Q.E.D.

We can now show that the $\mathrm{Prim}_{\mathrm{O}}$ functions are $\lambda + E$-definable, the first major step towards our main theorem.

Theorem 7.4. Every $\mathrm{Prim}_{\mathrm{O}}(\vec{\alpha})$ function is ordinal $\lambda + E$-definable in $\vec{\alpha}$.

Proof. A useful device in this proof is the following: Although we cannot forget arguments in the fashion of $\lambda xy.x$ due to the limitations of $\lambda\mathbf{I}$-calculus, we can do so if the argument is an ordinal. With $\mathbf{J} = \lambda z.(((S_c^+, z), \mathbf{I}), \mathbf{I})$ the term $(\mathbf{J}, \underline{\alpha})$ has normal form $\mathbf{I}$ for any α. Define $\mathbf{T}_J = \lambda xy.((\mathbf{J}, y), x)$ and $\mathbf{F}_J = \lambda xy.((\mathbf{J}, x), y)$.

(1) If $\vec{\alpha} = (\alpha_0, \ldots, \alpha_{k-1})$, then every constant function with value α_i is ordinal λ-definable in α_i as $\lambda x.((\mathbf{J}, x), \underline{\alpha_i})$.

(2) To retrieve the ith input ordinal, consider the following term:

$$P_i = \lambda x_0 x_1 \ldots x_{k-1}.((\ldots(\mathbf{J}, x_0)), (\mathbf{J}, x_1)), \ldots), (\mathbf{J}, x_{k-1})), x_i)$$

The term $(\ldots(P_i, \underline{\alpha}_0), \underline{\alpha}_1), \ldots), \underline{\alpha}_{k-1})$ has normal form $\underline{\alpha}_i$.

(3) As above, with

$$Z = \lambda x_0 x_1 \ldots x_{k-1}.((\ldots(\mathbf{J}, x_0)), (\mathbf{J}, x_1)), \ldots), (\mathbf{J}, x_{k-1})), \underline{0})$$

the term $(\ldots(Z, \underline{\alpha}_0), \underline{\alpha}_1), \ldots), \underline{\alpha}_{k-1})$ has $\underline{0}$ as normal form.

(4) From Example 5.7 it is evident that S_c^+ indeed defines the successor function on ordinals.

(5') The term $\lambda xy.((((((E, x), y), \mathbf{T}_J), \mathbf{F}_J), \underline{0}), \underline{1})$ defines the desired function e.

(a) Let g be defined by a term G and h by H. Define the term $(((G, \underline{\xi}), (H, \underline{\xi})), \underline{\zeta})$. Thanks to our Weak Church-Rosser Theorem, this term's normal form is $\simeq_v$-equivalent to the normal form of $(((G, \underline{\xi}), \underline{h(\xi)}), \underline{\zeta})$ and therefore the term defines the composition function $\overline{f(\xi, \zeta)} = g(\xi, h(\xi), \zeta)$. The term is straightforward to adapt for more than one parameter ζ. Similarly: (b).

(R') First, let us define the ordered pair of two ordinals α and β as the term $[\underline{\alpha}, \underline{\beta}] = \lambda y.((y, \underline{\alpha}), \underline{\beta})$. Then we have $\overline{([\underline{\alpha}, \underline{\beta}], \mathbf{T}_J)} = \underline{\alpha}$ and $\overline{([\underline{\alpha}, \underline{\beta}], \mathbf{F}_J)} = \underline{\beta}$. Let g be defined by G and h defined by H. Define

$$F_\xi = ((\underline{\xi}, \lambda x. \left[(((H, (x, \mathbf{T}_J)), (x, \mathbf{F}_J)), \underline{\zeta}), (S_c^+, (x, \mathbf{F}_J))\right]), \left[(G, \underline{\zeta}), \underline{0}\right]),$$

i.e., the ξ-fold application of some term

$$H^* = \left[(((H, (x, \mathbf{T}_J)), (x, \mathbf{F}_J)), \underline{\zeta}), \ (x, \mathbf{F}_J)\right]$$

to the term $G^* = [(G, \underline{\zeta}), \underline{0}]$. Inductively, all approximations ${}^{\eta}(H^*, G^*)$ for $0 < \eta < \xi$ have as normal form an ordered pair of two ordinals namely $[\underline{f(\eta, \zeta)}, \underline{\eta}]$, and the construction with $\mathbf{T}_J$ and $\mathbf{F}_J$ works at every approximation step. At limits, the pointwise lim inf ensures that limits are taken in both coordinates of the ordered pair. So $\overline{(F_\xi, \mathbf{T}_J)} = \underline{f(\xi, \zeta)}$ and, by parametrizing ξ and ζ, we can easily give a term F such that $\overline{((F, \xi), \zeta)} =$

$f(\xi,\zeta)$. The terms are straightforward to adapt for more than one parameter ζ.

This works fine for every $\xi > 0$. For the case of $\xi = 0$ we need to modify F_ξ to include a test for zero: Define

$$F'_\xi = ((((E,\underline{\xi}),\underline{0}),(G,\underline{\zeta})),F''_\xi)$$
$$F''_\xi = ((((((E,\underline{\xi}),\underline{0}),\underline{1}),\underline{\xi}),\lambda x.$$
$$[(((H,(x,\mathbf{T}_J)),(x,\mathbf{F}_J)),\underline{\zeta}),((S_c^+,x),\mathbf{F}_J)]),[(G,\underline{\zeta}),\underline{0}])$$

The modifications to F''_ξ make sure that its normal form is always an ordinal, even if $\xi = 0$. That way the case distinction in F'_ξ works as intended. Q.E.D.

7.2 Minimization

An *ordinal λ-definable predicate* on the ordinals is given by a term P such that $(P,\underline{\alpha})$ takes $\mathbf{T}_I$ as normal form for α in some subset or subclass of the ordinals and $\mathbf{F}_I$ for α in the complement.

In this section, we shall see that for every ordinal λ-definable predicate, there is a function defining its least witness. The proof is a generalization of [1, Chapter 9, §2] and Barendregt credits Kleene for the construction. In [1], for any classically λ-definable predicate P on ω, a term H_P is given that has the least witness of the predicate P as normal form. It turns out that the same term yields least witnesses for ordinal λ-definable predicates on Ord under our algorithm. The following definitions are direct adaptations.

Definition 7.5. Following [1, Chapter 9, §2] we define the following terms:

(a) $A_0 = \lambda xwt.((((((((w,\mathbf{T}_I),\mathbf{I}),\mathbf{I}),\mathbf{I}),(t,x)),\mathbf{I}),\mathbf{I}),x)$
(this term is used to escape the recursion)

(b) $A_1 = \lambda xwt.((((w,(t,(S_c^+,x))),(S_c^+,x)),w),t)$
(this term is used to continue the recursion)

(c) $\mathbf{T}_3 = \lambda xy.((((y,\mathbf{I}),\mathbf{I}),\mathbf{I}),x)$
(this term is used to forget an argument of the form A_i)

(d) $\mathbf{F}_3 = \lambda xy.((((x,\mathbf{I}),\mathbf{I}),\mathbf{I}),y)$
(this term is used to forget an argument of the form A_i)

(e) $W = \lambda x.((((x,\mathbf{T}_3),\mathbf{F}_3),A_0),A_1)$
(this term switches between A_0 and A_1 depending on the truth value of x)

The following facts from [1, Chapter 9, §2] hold true for our algorithm. We shall use these as macros in the upcoming proof of Theorem 7.8.

Lemma 7.6. (i) $(\mathbf{T}_I, (\mathbf{T}_3, \mathbf{F}_3)) \simeq_v \mathbf{T}_3$ and $(\mathbf{F}_I, (\mathbf{T}_3, \mathbf{F}_3)) \simeq_v \mathbf{F}_3$.

(ii) $\overline{(((A_0, \mathbf{I}), \mathbf{I}), \mathbf{I})} \simeq_v \mathbf{I}$

(iii) $\overline{(((A_1, \mathbf{I}), \mathbf{I}), \mathbf{I})} \simeq_v \mathbf{I}$

(iv) $\overline{((\mathbf{T}_3, A_i), A_j)} \simeq_v A_i$, for $i, j \in \{0, 1\}$

(v) $\overline{((\mathbf{F}_3, A_i), A_j)} \simeq_v A_j$, for $i, j \in \{0, 1\}$

(vi) $\overline{(W, \mathbf{T}_I)} \simeq_v A_0$

(vii) $\overline{(W, \mathbf{F}_I)} \simeq_v A_1$

Proof. Easily verified by running the algorithm. Q.E.D.

Now we can define the term H_P that has the least witness for P as normal form for any ordinal λ-definable predicate P.

Now we can define the term defining witnesses for predicates.

Definition 7.7. Let P be an ordinal λ-definable predicate. Then define $H_P = \lambda x.((((W, (P, x)), x), W), P)$.

We give the central result of this subsection, the second ingredient to our main result:

Theorem 7.8. Let P be an ordinal λ-definable predicate. Then $\overline{(H_P, \underline{0})} \simeq_v \underline{\gamma}$ where $\gamma = \min_{\gamma \in \mathrm{Ord}}(\overline{(P, \gamma)} \simeq_v \mathbf{T}_I)$ if such a γ exists. Otherwise $\overline{(H_P, \underline{0})} \uparrow$.

Proof. We demonstrate that the algorithm works correctly. The computation can be analyzed into the following stages:

Lines (1) to (8) contain some preliminary setup.

Lines (9) to (21) form one iteration of the main loop.

Line (18) marks the time after which the second stack level stabilizes: Through the following iterations, the stack remains stable up to its second level. In fact, at the first limit in time, every finite stack level reached so far contains a term of the same skeleton, cf. line (24).

Therefore, the same skeleton is assumed at limits and the following successor levels, until eventually level γ is reached in line (26).

Lines (26) to (42) then reduce the topmost stack level, from a complicated term still containing the subterms used to continue the recursion, to the bare Church numeral $\underline{\gamma}$.

In the steps abbreviated in line (43), the stack is torn down. Note that at all stack levels, the entire term was pushed onto the next stack level to be evaluated (e.g. lines (18), (24), (26)). Therefore, $\underline{\gamma}$ is handed down directly to the previous stack level without being inserted into some surrounding term. As soon as the first limit level is reached, the stack is immediately pruned to stack height 2 (cf. line (18)) as per Definition 5.4(b)(v).

In line (44), the stack has height 1 and contains $\underline{\gamma}$ as the desired normal form.

$$
\begin{array}{lr}
(H_P, \underline{0}) & \text{write out } H_P \ (1) \\
\overline{(\lambda x.((((W,(P,x)),x),W),P),\underline{0})} & \text{redex } (2) \\
\quad 1{:}\quad ((((W,(P,\underline{0})),\underline{0}),W),P) & \text{write out } W \ (3) \\
\quad\quad (((\overline{(\lambda x.((((x,\mathbf{T}_3),\mathbf{F}_3),A_0),A_1),(P,\underline{0}))},\underline{0}),W),P) & \text{redex } (4) \\
\quad\quad 1{:}\quad ((((\overline{(P,\underline{0})},\mathbf{T}_3),\mathbf{F}_3),A_0),A_1) & \text{WLOG } \overline{(P,\underline{0})} \simeq_v \mathbf{F}_I \ (5) \\
\quad\quad\quad\quad (((\overline{((\mathbf{F}_I,\mathbf{T}_3),\mathbf{F}_3)},A_0),A_1) & \text{Lemma 7.6 } (6) \\
\quad\quad\quad\quad \overline{((\mathbf{F}_3,A_0),A_1)} & \text{Lemma 7.6 } (7) \\
\quad\quad\quad\quad A_1 & \text{return } (8) \\
\quad\quad (((A_1,\underline{0}),W),P) & \text{write out } A_1 \ (9) \\
\quad\quad (((\overline{(\lambda xwt.((((w,(t,(S_c^+,x))),(S_c^+,x)),w),t),\underline{0})},W),P) & \text{redex } (10) \\
\quad\quad 1{:}\quad \lambda wt.((((w,(t,\overline{(S_c^+,\underline{0})})),\overline{(S_c^+,\underline{0})}),w),t) & \text{property of } S_c^+ \ (11) \\
\quad\quad\quad\quad \lambda wt.((((w,(t,\underline{1})),\underline{1}),w),t) & \text{return } (12) \\
\quad\quad \overline{((\lambda wt.((((w,(t,\underline{1})),\underline{1}),w),t),W)},P) & \text{redex } (13) \\
\quad\quad 1{:}\quad \lambda t.((((W,(t,\underline{1})),\underline{1}),W),t) & \text{write out } W \ (14) \\
\quad\quad\quad\quad \lambda t.(((\overline{(\lambda x.((((x,\mathbf{T}_3),\mathbf{F}_3),A_0),A_1),(t,\underline{1}))},\underline{1}),W),t) & \text{redex } (15) \\
\quad\quad\quad\quad 1{:}\quad (((((t,\underline{1}),\mathbf{T}_3),\mathbf{F}_3),A_0),A_1) & \text{return } (16) \\
\quad\quad\quad\quad \lambda t.(((((((t,\underline{1}),\mathbf{T}_3),\mathbf{F}_3),A_0),A_1),\underline{1}),W),t) & \text{return } (17) \\
\quad\quad \overline{(\lambda t.(((((((t,\underline{1}),\mathbf{T}_3),\mathbf{F}_3),A_0),A_1),\underline{1}),W),t),P)} & \text{cofinal skeleton } (18) \\
\quad\quad 1{:}\quad ((((((\overline{(P,\underline{1})},\mathbf{T}_3),\mathbf{F}_3),A_0),A_1),\underline{1}),W),P) & \text{WLOG } (19) \\
\quad\quad\quad\quad (((((\overline{((\mathbf{F}_I,\mathbf{T}_3),\mathbf{F}_3)},A_0),A_1),\underline{1}),W),P) & \text{Lemma 7.6 } (20) \\
\quad\quad\quad\quad ((((\overline{((\mathbf{F}_3,A_0),A_1)},\underline{1}),W),P) & \text{Lemma 7.6 } (21) \\
\quad\quad\quad\quad (((A_1,\underline{1}),W),P) & \text{write out } A_1 \ (22) \\
\quad\quad\quad\quad \vdots & \text{similarly } (23) \\
\quad\quad\quad\quad \overline{(\lambda t.(((((((t,\underline{2}),\mathbf{T}_3),\mathbf{F}_3),A_0),A_1),\underline{2}),W),t),P)} & \text{cofinal skeleton } (24)
\end{array}
$$

$\ddots$ eventually (else obvious divergence) (25)

$\overline{(\lambda t.((((((((t,\underline{\gamma}),\mathbf{T}_3),\mathbf{F}_3),A_0),A_1),\underline{\gamma}),W),t),P)}$

where $\overline{(P,\underline{\gamma})} \simeq_v \mathbf{T}_I$ (26)

1: $(((((((\overline{(P,\underline{\gamma})},\mathbf{T}_3),\mathbf{F}_3),A_0),A_1),\underline{\gamma}),W),P)$

assumption (27)

$(((((\overline{((\mathbf{T}_I,\mathbf{T}_3),\mathbf{F}_3)},A_0),A_1),\underline{\gamma}),W),P)$

Lemma 7.6 (28)

$(((\overline{((\mathbf{T}_3,A_0),A_1)},\underline{\gamma}),W),P)$ Lemma 7.6 (29)

$(((A_0,\underline{\gamma}),W),P)$ write out A_0 (30)

$((\overline{(\lambda xwt.((((((((w,\mathbf{T}_I),\mathbf{I}),\mathbf{I}),\mathbf{I}),(t,x)),\mathbf{I}),\mathbf{I}),x),\underline{\gamma})},W),P)$

redex (31)

1: $\lambda wt.((((((((w,\mathbf{T}_I),\mathbf{I}),\mathbf{I}),\mathbf{I}),(t,\underline{\gamma})),\mathbf{I}),\mathbf{I}),\underline{\gamma})$

return (32)

$(\overline{((\lambda wt.((((((((w,\mathbf{T}_I),\mathbf{I}),\mathbf{I}),\mathbf{I}),(t,\underline{\gamma})),\mathbf{I}),\mathbf{I}),\underline{\gamma}),W)},P)$

redex (33)

1: $\lambda t.(((((((\overline{(W,\mathbf{T}_I)},\mathbf{I}),\mathbf{I}),\mathbf{I}),(t,\underline{\gamma})),\mathbf{I}),\mathbf{I}),\underline{\gamma})$

Lemma 7.6 (34)

$\lambda t.((((\overline{(((A_0,\mathbf{I}),\mathbf{I}),\mathbf{I})},(t,\underline{\gamma})),\mathbf{I}),\mathbf{I}),\underline{\gamma})$

Lemma 7.6 (35)

$\lambda t.(((\overline{(\mathbf{I},(t,\underline{\gamma}))},\mathbf{I}),\mathbf{I}),\underline{\gamma})$ definition $\mathbf{I}$ (36)

$\lambda t.((((t,\underline{\gamma}),\mathbf{I}),\mathbf{I}),\underline{\gamma})$ return (37)

$\overline{(\lambda t.((((t,\underline{\gamma}),\mathbf{I}),\mathbf{I}),\underline{\gamma}),P)}$ redex (38)

1: $(((\overline{(P,\underline{\gamma})},\mathbf{I}),\mathbf{I}),\underline{\gamma})$ assumption (39)

$(\overline{((\mathbf{T}_I,\mathbf{I}),\mathbf{I})},\underline{\gamma})$ Lemma 7.6 (40)

$(\mathbf{I},\underline{\gamma})$ definition $\mathbf{I}$ (41)

$\underline{\gamma}$ return (42)

$\iddots$ note that when going through the stack ...

$\iddots$ ... also over limits in stack height ...

$\iddots$ ... no new redexes appear (43)

$\underline{\gamma}$ (44)

Q.E.D.

7.3 Main result

Theorem 7.9. A partial function $F : \mathrm{Ord} \rightharpoonup \mathrm{Ord}$ on the ordinals is $\lambda + E$-definable in finitely many ordinal parameters if and only if it is $\mathbf{\Sigma}_1$-definable over $\mathbf{L}$.

Proof. By Lemma 7.2, every ordinal $\mathbf{\Sigma}_1(L)$-definable function can be obtained by one minimization over a $\mathrm{Prim}_{\mathrm{O}}$ relation in an ordinal parameter (and by evaluating this result by another $\mathrm{Prim}_{\mathrm{O}}$ function). Theorem 7.8 shows that minimization over ordinal $\lambda + E$-definable relations is ordinal $\lambda + E$-definable. Since by Theorem 7.4 all $\mathrm{Prim}_{\mathrm{O}}$ functions are ordinal $\lambda + E$-definable, it follows that every $\mathbf{\Sigma}_1(L)$-definable function is already ordinal $\lambda + E$-definable. It remains to show that every ordinal $\lambda + E$-definable function is $\mathbf{\Sigma}_1(L)$-definable. We use the established equivalence of $\mathbf{\Sigma}_1(L)$-definable functions and α-computable functions [9] and define an OTM program which computes the normal form derivation of any $\lambda + E$-term:

At any point in the normal form derivation, we would like the current stack content to be coded on our Turing tape. Each stack element is composed of a finite number of terms plus some ordinals. Every term can be represented as the pair of its flesh (a tuple of ordinals) and its skeleton (a finite string over a countable alphabet). So every stack element can be coded into a finite sequence of ordinals. This can be coded into our tape via Gödel pairing $\langle\cdot,\cdot\rangle$: Cell number $\langle\langle n, \gamma\rangle, \xi\rangle = 1$ if and only if the nth ordinal of stack element γ is greater than ξ. All ordinals coded in this way are preserved as lim inf's across limits, due to the properties of OTMs. The operations on the stack that happen at successor times are all of syntactical nature on terms and can be carried out by an OTM.

At limits in the normal form derivation, our simulation thereof at first will set all ordinals involved to their lim inf:

First, consider the case where some stack height appears cofinal below the limit. We thus are in the case where some term ${}^{\beta}(M, N)$ is approximated. Then all stack levels up to the lim inf of stack heights will have stabilized and the trivial lim inf's are as desired. On the top stack level, however, the OTM lim inf-rule will produce garbage, as both the non-normalized form and the normal form of the approximations ${}^{\gamma}(M, N)$ appear on the cofinal stack height. But OTMs can recognize limits (by the flag-flashing technique introduced in [6]) and we can keep track of the stack height on some separate space on the tape (imagine, for simplicity, an extra tape for this task). Whenever a limit is reached, we can check whether there is a cofinally assumed stack height: Simply simulate the computation up to that point time and again, trying out all values from 0 to the number of the current iteration, again flashing a flag whenever the stack has the suspected height to see whether it appears cofinally. This means a vast increase of running time over the length of the normal form derivation to be simulated, but really poses no problem since, in any case, we shall be done eventually; nevertheless, time improvements are possible via diagonal enumeration. As soon as the cofinally assumed stack height is known, the desired $\liminf$'s can be found by one additional simulation up to the current step.

In the second case, where no stack height appears cofinally, we are interested in the first (with respect to stack height) skeleton that appears cofinally often in the ever increasing stack. We can search through the limit stack, checking for every possible skeleton whether it appears cofinally and, among those that do, choose the one that appears on the stack first. Once we identified this skeleton, we can search through the limit stack again to determine the desired $\liminf$. Q.E.D.

We can restrict both the length of the normal form derivation and the stack height in Definition 5.4 to some admissible ordinal α: If either reaches α, we say that the normal form derivation diverges. With the resulting notion of *α-normal form derivation* we can define the α-$\lambda + E$-definable (partial) functions on α (possibly in parameters $< \alpha$).

Corollary 7.10. Let α be admissible. A function partial $F : \alpha \rightharpoonup \alpha$ is α-$\lambda + E$-definable in a finite set of parameters $< \alpha$ if and only if it is $\mathbf{\Sigma}_1$-definable over L_α.

Proof. As above, Lemma 7.2 and Theorems 7.4 and 7.8 ensure that every $\mathbf{\Sigma}_1(L_\alpha)$-definable function is α-λ-definable in parameters $< \alpha$. For the converse, the admissibility of α ensures that the above construction can be carried out on an α-Turing machine (as defined, e.g., in [9]). Q.E.D.

8 Open questions

8.1 Strong Church-Rosser Theorem

We already stated Conjecture 6.1, that a strong version of the Church-Rosser Theorem holds for our calculus.

Conjecture 1. Let T be a term with normal form and let $S \subseteq T$ be a subterm. Then $\overline{T} \simeq_v \overline{T\left[\frac{\overline{S}}{S}\right]}$.

8.2 $\lambda + Z$-definability

Another open problem is whether we can prove our main result for calculi seemingly weaker than $\lambda + E$. We added a predicate for equality of ordinals to our calculus and defined the $\lambda + E$-definable functions to obtain the initial function (5') in the definition of primitive recursive functions. All other arguments from § 7 go through also for λ-definable functions. We conjecture that we can replace the predicate E by a predicate Z that tests numerals for being zero, in the same fashion as E tests for equality.

Conjecture 2. Theorem 7.9 holds for $\lambda + Z$-definable functions.

Proof idea. Equality of ordinals can be defined recursively from a test for 0:

$$\alpha = \beta \leftrightarrow (\alpha = 0 \wedge \beta = 0) \vee (\alpha \leq \beta \wedge \beta \leq \alpha)$$
$$\alpha \leq \beta \leftrightarrow \forall \gamma \in \alpha \, \exists \delta \in \beta \, \gamma = \delta$$

One should be able to carry out this recursion in our ordinal λ-calculus. The $\exists$ and $\forall$ quantifiers can be modeled in the following way: Let P be a predicate on the ordinals and assume we want to decide, e.g., whether P holds for some $\gamma < \alpha$. We define a function $f : \mathrm{Ord} \mapsto \{0, 1\}$ such that $f(\alpha) = 1 \leftrightarrow \exists \gamma < \alpha \, P(\alpha)$.

$$\begin{aligned} f(0) &= 0 \\ f(\gamma + 1) &= 1 \leftrightarrow P(\gamma) \vee f(\gamma) = 1 \\ f(\mu) &= \liminf_{\nu < \mu} f(\nu) && \text{if } \mu \text{ is a limit} \end{aligned}$$

This primitive recursion is $\lambda + Z$-definable if P is so.

Due to Remark 5.3, it seems unlikely that we are able to go even weaker than $\lambda + Z$. Since, syntactically, the numeral for 0 is indistinguishable from the numerals for non-zero ordinals, it appears doubtful to obtain a test for zero by syntactical tricks. Also, its arithmetical properties cannot be validated without a means to talk about equality of ordinals.

8.3 Variations of the model

As with the other models of ordinal computation, there are interesting variations imaginable. While λ-calculus does not come with a canonical distinction between time and space, our normal form algorithm can easily be restricted in both runtime or stack height. Asymmetric models, such as ITTMs or the restriction of OTMs in [11], have interesting theories so these two paths, i.e., restricting stack height but not runtime and restricting input complexity but neither runtime nor stack height, should be explored. In an early stage of the development, the authors conjectured that the present calculus restricted to finite stacks (and consequently without part (c)(ii) of Definition 5.4) would be equivalent in strength to the $\mathrm{Prim}_{\mathrm{O}}$ functions. While plausible from interpreting terms of the form ${}^{\alpha}(M, N)$ as some kind of `for`-loops, it turned out that this is false: Due to its un-typed nature, the thus defined generalization of λ-calculus is capable of giving 'primitive recursive' definitions for functionals such as the Ackermann function, while the calculus of primitive recursive functions is limited to defining only functions on ordinals in a recursive manner.

Moving away from looking at the calculus solely as means of defining functions on the ordinals, the work done to generalize λ-calculus may perhaps be used to extend other calculi that are centered on the re-writing of terms to the transfinite. The authors hope for the present work to be helpful in further studies in this direction.

Bibliography

[1] H. P. Barendregt. *The lambda calculus. Its syntax and semantics.*, volume 103 of *Studies in Logic and the Foundations of Mathematics*. North Holland, 1981.

[2] A. Church and J. B. Rosser. Some properties of conversion. *Transactions of the American Mathematical Society*, 39(3):472–482, 1936.

[3] K. J. Devlin. *Aspects of constructibility*, volume 354 of *Lecture Notes in Mathematics*. Springer-Verlag, Heidelberg, 1973.

[4] T. Fischbach. The Church-Turing thesis for ordinal computable functions. Master's thesis, Rheinische Friedrich-Wilhelms-Universität Bonn, 2010.

[5] T. Fischbach and B. Seyfferth. On λ-definable functions on ordinals. In P. Bonizzoni, V. Brattka, and B. Löwe, editors, *The Nature of Computation. Logic, Algorithms, Applications. 9th Conference on Computability in Europe, CiE 2013, Milan, Italy, July 1-5, 2013. Proceedings*, volume 7921 of *Lecture Notes in Computer Science*, pages 135–146. Springer-Verlag, 2013.

[6] J. D. Hamkins and A. Lewis. Infinite time Turing machines. *Journal of Symbolic Logic*, 65(2):567–604, 2000.

[7] R. B. Jensen and C. Karp. Primitive recursive set functions. In D. S. Scott, editor, *Axiomatic Set Theory*, volume XIII (Part 1) of *Proceedings of Symposia in Pure Mathematics*, pages 143–176. American Mathematical Society, 1971.

[8] P. Koepke. Turing computations on ordinals. *Bulletin of Symbolic Logic*, 11:377–397, 2005.

[9] P. Koepke and B. Seyfferth. Ordinal machines and admissible recursion theory. *Annals of Pure and Applied Logic*, 160(3):310–318, 2009.

[10] P. Koepke and R. Siders. Register computations on ordinals, 2006. Submitted.

[11] P. Schlicht and B. Seyfferth. Tree representations and ordinal machines. *Computability*, 1(1):45–57, 2012.

[12] B. Seyfferth. *Three models of ordinal computability*. PhD thesis, Rheinische Friedrich-Wilhelms-Universität Bonn, 2012.

Coding over core models

Sy-David Friedman[1*], Ralf Schindler[2†], and David Schrittesser[1‡]

[1] Kurt Gödel Research Center for Mathematical Logic, Universität Wien, Vienna, Austria
[2] Institut für Mathematische Logik und Grundlagenforschung, Westfälische Wilhelms-Universität Münster, Münster, Germany

Early in their careers, both Peter Koepke and Philip Welch made major contributions to two important areas of set theory, core model theory (cf. [10]) and coding (cf. [1]), respectively. In this article we aim to survey some of the work that has been done which combines these two themes, extending Jensen's original Coding Theorem from **L** to core models witnessing large cardinal properties.

The original result of Jensen can be stated as follows.

Theorem 1. (Jensen, cf. [1]) Suppose that (V, A) is a transitive model of ZFC + GCH (i.e., V is a transitive model of ZFC + GCH and replacement holds in V for formulas mentioning A as an additional unary predicate). Then there is a (V, A)-definable, cofinality-preserving class forcing P such that if G is P-generic over (V, A) we have:

(a) For some real R, $(V[G], A) \models$ ZFC + the universe is $\mathbf{L}[R]$ and A is definable with parameter R.

(b) The typical large cardinals properties consistent with **V**=**L** are preserved from V to $V[R]$: inaccessible, Mahlo, weak compact, Π^1_n indescribable, subtle, ineffable, α-Erdős for countable α.

Corollary 2. It is consistent to have a real R such that **L** and $\mathbf{L}[R]$ have the same cofinalities but R belongs to no set-generic extension of **L**.

The theme of this article is to consider the following question: To what extent is it possible to establish an analagous result when **L** is replaced by a core model **K** and the large cardinal properties in (b) are strengthened to those consistent with **V**=**K** (measurable, hypermeasurable, strong, Woodin)?

*The first author wishes to thank the FWF for its support through Einzelprojekt P25671. He sees in Peter Koepke and Philip Welch fellow disciples of our common mentor, Ronald Jensen.

†The result on pp. 176f. was produced while the second author was visiting the Erwin Schrödinger Institut, Vienna, in September 2013. He would like to thank Sy Friedman and the other organizers of the ESI Set Theory Program for their warm hospitality.

‡The third author wishes to thank Ralf Schindler for his support through SFB 878. He also wants to thank Sy Friedman and everyone at the KGRC for their hospitality.

A brief summary of the situation is as follows. Coding up to one measurable cardinal is unproblematic (cf. [4]), although already in this case there are some issues with condensation and the interesting new phenomenon of "ultrapower codings" arises. At the level of hypermeasurable cardinals there are serious condensation issues which obstruct a fully general result; nevertheless variants of Corollary 2 can be established and very special predicates A as in Theorem 1 can be coded (such as a generic for a Příkrý product, cf. [7]). In addition, although one is able to lift enough of the total extenders on the hierarchy of a core model witnessing hypermeasurability, it requires extra effort to lift more than one total extender for the same critical point (and it is not in general possible to lift all of the extenders (partial and total) on a fixed critical point κ satisfying $o(\kappa) = \kappa^{+++}$; we conjecture that this can be improved to $o(\kappa) = \kappa^{++}$). At the level of Woodin cardinals, even Corollary 2 is not possible if the aim is to lift all total extenders in a witness to Woodinness via the "A-strong" definition of this notion; however this obstacle is removed by instead considering witnesses to the definition of Woodinness in terms of "$j(f)(\kappa)$ strength" (cf. [6]).

There are a number of applications of coding over core models. In addition to those found in [5] based on Jensen's original method, we mention two other examples.

Theorem 3. (Friedman-Schrittesser, [9]) Relative to a Mahlo cardinal it is consistent that every set of reals in $L(\mathbb{R})$ is Lebesgue measurable but some projective (indeed lightface Δ^1_3) set of reals does not have the Baire property.

Theorem 4. (Friedman-Golshani, [7]) Relative to a strong cardinal (indeed relative to a cardinal κ that is $\mathbf{H}(\kappa^{+++})$-strong) it is consistent to have transitive models $V \subseteq V[R]$ of ZFC where R is a real, GCH holds in V and GCH fails at every infinite cardinal in $V[R]$. One can further require that V, $V[R]$ have the same cardinals.

1 About Jensen coding

To make what follows more intelligible it is worthwhile to first review the case of Jensen coding. No matter how you look at it, even this argument is complicated, although major simplifications can be made if one assumes the nonexistence of $0^\#$ in the ground model V. Our aim here however is not to delve into the fine points of the proof (and in particular we shall not reveal how the nonexistence of $0^\#$ can be exploited), but rather to give the architecture of the argument in order to facilitate a later discussion of generalisations.

For simplicity consider the special case where the cardinals of the ground model V are the same as those in $\mathbf{L}$ and the ground model is $(\mathbf{L}[A], A)$ where

A is a class of ordinals such that $\mathbf{H}(\alpha) = \mathbf{L}_\alpha[A]$ for each infinite cardinal α (the latter can be arranged using the fact that the GCH holds in V).

Coding is based on the method of almost disjoint forcing. Suppose that A is a subset of ω_1. Then we can code A into a real as follows: For each countable ordinal ξ attach a subset b_ξ of ω (so that the b_ξ's are almost disjoint) and force a real R such that R is almost disjoint from b_ξ iff ξ belongs to A. Actually it is convenient to modify this to: R almost contains b_ξ iff ξ belongs to A (where *almost contains* means contains with only finitely many exceptions). The conditions to achieve this are pairs (s, s^*) where s is an ω-Cohen condition (i.e., element of ${}^{<\omega}2$) and s^* is a finite subset of A; when extending to (t, t^*) we extend s to t, enlarge s^* to t^* and insist that if $s(n)$ is undefined but $t(n)$ equals 0 then n does not belong to b_ξ for any ξ in s^*. Then the generic G is determined by the union G_0 of the s for (s, s^*) in G and we can take R to be the set of n such that $G_0(n)$ equals 1. The forcing has the c.c.c. and ensures that A belongs to $\mathbf{L}[R]$ using the hypothesis $\omega_1 = \omega_1^{\mathbf{L}}$ to produce the b_ξ's in $\mathbf{L}$ (and therefore also in $\mathbf{L}[R]$).[1]

There is nothing to stop us from coding a subset A of ω_2 into a real in a similar fashion: First we use the hypothesis $\omega_2 = \omega_2^{\mathbf{L}}$ to choose subsets b_ξ of ω_1 to set up a forcing to code A into a subset B of ω_1 via the equivalence $\xi \in A$ iff B almost contains b_ξ, and then we code B into a real as in the previous paragraph. It is pretty clear how to do this for a subset of any ω_n, n finite.

If we have a subset A of $\aleph_\omega$ then we have to force subsets A_n of ω_n for each n so that A_n codes both A_{n+1} and $A \cap \omega_n$. At first this is confusing because there is no "top", i.e., no largest n to begin with, but further reflection reveals that there is no problem at all, as we don't need to know all of A_{n+1} to talk about conditions to add A_n. More precisely, a condition p will assign to each n a pair (s_n, s_n^*) so that s_n is an ω_n-Cohen condition and s_n^* is a size less than ω_n subset of the set of ξ such that $s_{n+1}(\xi)$ is defined with value 1. This makes sense even though s_{n+1} is not defined on all of ω_{n+1}. We also insist that all of the b_ξ's consist of even ordinals and that each $A \cap \omega_n$ is coded into the union of the s_n's using its values at odd ordinals. In the end A gets coded into a real and cofinalities are preserved since for any n the forcing factors into an ω_n-closed forcing (the nth upper part) followed by an ω_n-c.c. forcing (the nth lower part).

[1] In a more general setting we have to worry about how to find the b_ξ's in $\mathbf{L}[R]$. Jensen's trick to achieve this is to "reshape" A into a stronger predicate A' with the property that any countable ordinal ξ is in fact countable in $\mathbf{L}[A' \cap \xi]$; then after R decodes $A' \cap \xi$ it can find b_ξ and continue the decoding. A clever argument shows that such an A' can be added over $\mathbf{L}[A]$ by an ω-distributive forcing; when A is not just a subset of ω_1 but a subset of some larger cardinal or even a proper class of ordinals, then the "reshaping" forcing must be woven into the coding forcing itself.

Coding a subset A of $\aleph_{\omega+1}$ into a real requires a new idea. Actually by the previous paragraph it's enough to see how to code A into a subset of $\aleph_\omega$. Again we would like to assign a subset b_ξ of $\aleph_\omega$ to each $\xi < \aleph_{\omega+1}$ and then hope to force a subset B of $\aleph_\omega$ which almost contains b_ξ iff ξ belongs to A; how are we going to do that? The conditions to add B cannot be built from "$\aleph_\omega$-Cohen conditions" as this makes no sense for the singular cardinal $\aleph_\omega$. Instead they should look like conditions in the product of the ω_n-Cohen forcings, i.e., of the form $(s_n \mid n \in \omega)$ where each s_n is an ω_n-Cohen condition (as in the previous paragraph but without the "restraints" s_n^*). Actually it is very convenient to instead write $(s_{\omega_n} \mid n \in \{-1\} \cup \omega)$ where $\omega_{-1} = 0$ and s_{ω_n} is an ω_{n+1}-Cohen condition for each $n \geq -1$, and to think of s_{ω_n} as an ω_{n+1}-Cohen condition on the interval $[\omega_n, \omega_{n+1})$ rather than on ω_{n+1}, to separate the domains of the different s_{ω_n}'s. Thus the characteristic function of the generic subset B of $\aleph_\omega$ is the union of all of the s_{ω_n}'s which appear in the generic.

As said above we'd like to choose the b_ξ's so that ξ belongs to A iff the generic subset B of $\aleph_\omega$ almost contains b_ξ (i.e., contains b_ξ with a set of exceptions which is bounded in $\aleph_\omega$). This is done using a *scale*, i.e., a sequence $(f_\xi \mid \xi < \aleph_{\omega+1})$ of functions in $\prod_{n\geq -1}[\omega_n, \omega_{n+1})$ which is cofinal mod finite. Then we take b_ξ to be the range of f_ξ. Again it is convenient to change notation: instead of writing $f_\xi(n)$ we write $f_\xi(\omega_n)$. So the coding is: ξ belongs to A iff $G_{\omega_n}(f_\xi(\omega_n)) = 1$ for sufficiently large n, where G_{ω_n} denotes the union of the s_n's which appear in the generic.

As we are using a scale we can arrange the following: if $p = (s_{\omega_n} \mid -1 \leq n < \omega)$ is a condition then for some ordinal $|p| < \aleph_{\omega+1}$ called the *height* of p, if ξ is less than $|p|$ then $\xi \in A$ iff $s_{\omega_n}(f_\xi(\omega_n)) = 1$ for sufficiently large n and if ξ is at least $|p|$ then $f_\xi(\omega_n)$ is not in the domain of s_{ω_n} for sufficiently large n. In other words, p already codes A below $|p|$ but provides no information about future coding on the interval $[|p|, \aleph_{\omega+1})$. Notice the difference from the successor coding case: a single condition will definitively code an initial segment of A, in the sense that its values on a final segment of b_ξ for ξ in an initial segment of $\aleph_{\omega+1}$ have already been fixed (restraints are not needed). Of course no condition will code all of A, so this initial segment of A is proper.

But how do we know that this coding of $A \subseteq \aleph_{\omega+1}$ into a subset of $\aleph_\omega$ preserves the cardinal $\aleph_{\omega+1}$? For each n we can factor the forcing as the part $\geq \omega_n$ followed by the part below ω_n, and as the latter is a small forcing it causes no problems with cardinal-preservation; so we want to show that the forcing $\geq \omega_n$ (using conditions $p = (s_{\omega_k} \mid k \geq n)$) is ω_{n+1}-distributive, i.e., does not add new ω_n-sequences. For simplicity suppose that n is 0, so we want to hit ω-many open dense sets below any condition $p = (s_{\omega_k} \mid k \geq 0)$. Here is the worry: maybe things are going fine with the sequence $p = p_0 \geq p_1 \geq \cdots$ with corresponding heights $|p_0| \leq |p_1| \leq \cdots$ so we can conclude that the limit p_ω of the p_n's will code A up to the limit $|p_\omega|$ of the $|p_n|$'s. But there is the danger that p_ω "overspills" in the sense that it already has assigned cofinally many values on b_ξ for some $\xi \geq |p_\omega|$. This unintended assignment may conflict with the desired coding of A at the ordinal ξ.

The solution is to guide the construction using sufficiently elementary submodels and to refine our concept of scale. Namely, when we build the p_n's we also build a definable ω-chain of size $\aleph_\omega$ sufficiently elementary submodels $M_0 \prec M_1 \prec \cdots$ of the universe which are transitive below $\aleph_{\omega+1}$; we ensure that the p_n's are chosen from the M_n's and have heights $|p_n|$ which interleave with the ordinals $\gamma_n = M_n \cap \aleph_{\omega+1}$. The result is that the supremum of the $|p_n|$'s is exactly $\gamma_\omega = M_\omega \cap \aleph_{\omega+1}$, where M_ω is the union of the M_n's. Now how does this help? The point is that we can arrange for p_ω, the limit of the p_n's, to be definable over M_ω and therefore also over its transitive collapse $\overline{M}_\omega$; if we can also arrange our scale so that f_{γ_ω} eventually dominates any function in $\prod_n [\omega_n, \omega_{n+1})$ which is definable over $\overline{M}_\omega$, then p_ω will leave a final segment of the range of f_{γ_ω} untouched, as the sequence $(|p_\omega(\omega_n)| \mid n \in \omega)$ is indeed definable over $\overline{M}_\omega$. Finally, arranging our scale in this way is not a problem, as $\overline{M}_\omega$ is an initial segment of $\mathbf{L}$ which is so short that it still thinks that γ_ω is a cardinal (it is the image of $\aleph_{\omega+1}$ under the transitive collapse of M_ω) and we can define f_ξ to eventually dominate any function in $\prod_n [\omega_n, \omega_{n+1})$ which belongs to a model which still thinks that ξ is a cardinal (f_ξ is defined using Skolem hulls inside some big initial segment which sees that ξ is not a cardinal).

The reason we discussed the fine point above about the coding of a subset of $\aleph_{\omega+1}$ into $\aleph_\omega$ is to note that there is some condensation involved (we needed that $\overline{M}_\omega$ is an initial segment of our hierarchy). This is unproblematic for $\mathbf{L}$ (and even for $\mathbf{L}[U]$ where U is a single normal measure) but is a serious problem for large core models. The use of condensation is even more substantial when looking at $\aleph_{\omega^2}$, where one needs to simultaneously consider transitive collapses of unions of chains of sufficiently elementary submodels of any fixed size $\aleph_{\omega \cdot n}$ and worry about their transitive collapses being initial segments of the hierarchy. Indeed it is this issue with condensation which obstructs a fully general coding result over core models as in Theorem 1. Nearly all of the successes with coding over core models are variants of the weaker Corollary 2.

Now the fact that the strategy to code a subset of $\aleph_{\omega+1}$ into $\aleph_\omega$ fits so nicely with the strategy to code a subset of $\aleph_\omega$ into a real means that we can combine the two codings into a single coding of a subset of $\aleph_{\omega+1}$ into a real. Thus a condition is a function p that for each finite n assigns a pair (s_n, s_n^*) as in the latter coding so that in addition the sequence of s_n's is a condition in the former coding. For later use we change notation slightly: the domain of p consists of 0 together with the ω_n's and for each α in the domain of p, $p(\alpha) = (p_\alpha, p^*_\alpha)$ where p_α is an α^+-Cohen condition on the interval $[\alpha, \alpha^+)$ (0^+ is taken to be ω). And of course the restraint p^*_α is a size at less than α^+ subset of the set of ξ such that $p_{\alpha^+}(\xi)$ is defined with value 1. We also require that p_α codes $A \cap |p_\alpha|$ where the domain of p_α is $[\alpha, |p_\alpha|)$, using its values at odd ordinals. Finally, for some $|p| < \aleph_{\omega+1}$, if ξ is less than $|p|$ then ξ belongs to A iff $p_\alpha(\eta) = 1$ for sufficiently large η in $b_\xi = \text{ran}(f_\xi)$ and when ξ is at least $|p|$, sufficiently large η in b_ξ lie outside the domain of the p_α's.

This ends our introduction to Jensen coding. For arbitrary infinite cardinals α, the coding from a subset of α^{++} into a subset of α^+ is similar to the coding of a subset of ω_1 into a real and for arbitrary singular cardinals α, the coding of a subset of α^+ into a subset of α is similar to the above coding of a subset of $\aleph_{\omega+1}$ into a subset of $\aleph_\omega$. The final case of the coding of a subset of α^+ into a subset of α for inaccessible α uses either *full support* and thereby resembles the singular coding, or uses *Easton support* and thereby resembles the successor coding. In nearly all cases (including [1]) full support is used (it faciliates the preservation of large cardinals); Easton support coding is however needed in [9]. The reason is that in [9], we iterate Jensen coding to length κ, at the same time collapsing everything below κ; but we want to preserve κ itself. The usual strategy of "reducing to the lower part" fails below κ, since as we keep coding into ω, there are κ-many lower parts. Instead, a much more complex argument is needed, in which there is no fixed height where we cut into "upper" and "lower part": intuitively, we capture a given name by deciding it in different ways using larger and larger lower parts, catching our tail at an inaccessible below κ, where we shall have looked at all the relevant lower parts. For this to work, supports must be bounded below inaccessibles (we also have to assume κ is Mahlo).

2 One measurable cardinal

Suppose that there is a measurable cardinal κ in V. Can we code V into a real R preserving the measurability of κ?

Of course the model that results after coding into R cannot be $\mathbf{L}[R]$, but it could be $\mathbf{L}[U^R, R]$ where U^R is a normal measure on κ extending a given normal measure U on κ in V. As alluded to above there are serious issues with condensation when coding over core models and for this reason we shall only discuss here how to establish a version of Corollary 2: It is possible to force a real R over $\mathbf{L}[U]$ which preserves cofinalities, is not set-generic over $\mathbf{L}[U]$ and preserves the measurability of κ. Even in this special situation it is very helpful (and essential for further generalisations) to use a hierarchy for $\mathbf{L}[U]$ with good condensation properties, which we write as $\mathbf{L}[E]$. Note that the $\mathbf{L}[U]$-hierarchy does not obey even the weakest of consequences of condensation, the property that subsets of an infinite cardinal α appear in the hierarchy at a stage before α^+. The $\mathbf{L}[E]$ hierarchy inserts "partial measures" which ensure this property and more without altering the model: $\mathbf{L}[E] = \mathbf{L}[U]$. The measure U (or something very close to it) is placed on the $\mathbf{L}[E]$ hierarchy at an appropriate stage between κ^+ and κ^{++}, its *index* on the $\mathbf{L}[E]$-hierarchy, and there will be many approximations to it placed on the hierarchy at indices cofinal in any uncountable cardinal up to and including κ^+.

So proceed now to form conditions p in $\mathbf{L}[E]$ which resemble the coding conditions from Jensen coding: For α either 0 or an infinite cardinal, $p(\alpha)$ is a pair (p_α, p^*_α) where p_α is an α^+-Cohen condition on $[\alpha, \alpha^+)$ and p^*_α is a size at most α set of ξ such that $p_{\alpha^+}(\xi) = 1$. Also for limit cardinals λ we have a scale $(f_\xi \mid \xi \in [\lambda, \lambda^+))$ of functions in $\prod_{\alpha^+ < \lambda}[\alpha^+, \alpha^{++})$ and for $\xi < |p_\lambda|$, $p_\lambda(\xi) = 1$ iff $p_{\alpha^+}(f_\xi(\alpha^+)) = 1$ for sufficiently large $\alpha^+ < \lambda$. And $p \restriction \lambda$ does not interfere with future coding on $[|p_\lambda|, \lambda^+)$ in the sense that for $\xi \geq |p_\lambda|$, $p_{\alpha^+}(f_\xi(\alpha^+))$ is not defined for sufficiently large $\alpha^+ < \lambda$. The previous applies both to inaccessible and singular limit cardinals λ.

Now we need a strategy for showing that this forcing preserves the measurability of κ. It is best to think of measurability in terms of embeddings: In the ground model $V = \mathbf{L}[U] = \mathbf{L}[E]$ there is an elementary embedding $j : V \to M = \mathrm{Ult}_U$ with critical point κ, derived from the ultrapower given by U. The hierarchy provided by E is defined so that we have $j : \mathbf{L}[E] \to \mathbf{L}[E^*]$ where E, E^* agree up to the index of U (an ordinal between κ^+ and κ^{++}); for the present discussion we only need to know that this agreement persists at least up to the κ^{++} of M, the ultrapower of V by U. This has the important consequence that our coding forcing P agrees with $P^* = j(P)$, the coding forcing of the ultrapower M, up to the κ^{++} of M. More precisely, a function p defined at 0 together with the infinite cardinals $\leq \kappa^+$ such that $p(\alpha) = (p_\alpha, p^*_\alpha)$ for each α and $p^*_{\kappa^+} = \varnothing$ belongs to P^* iff it belongs to P, $|p_{\kappa^+}|$ is less $(\kappa^{++})^M$ and p^*_κ is a subset of $(\kappa^{++})^M$.[2]

Now Silver taught us that if we want to preserve the measurability of κ we should lift the embedding $j : V \to M$ to an embedding $j^* : V[G] \to M[G^*]$ where G^* is generic over M for $P^* = j(P)$. The key is to choose G^* to contain the pointwise image $j[G]$ of G as a subclass. There are many examples of such liftings in the context of reverse Easton forcing, where there are typically many choices for G^*. But notice that with coding there is only one candidate for G^*, the P^*-generic coded into the same real R

[2]This may not be entirely clear, as V has more subsets of κ^+ than M. However the coding is defined so that p_{κ^+} will belong to M provided its length is less than the κ^{++} of M.

that codes G. This is because $j^*(R)$ will equal R for any possible lifting j^* of j to $V[G]$.

Of course our desired generic G^* must include the image $j(p)$ of any condition p in G; it would be ideal if G^* were simply generated by these conditions in the sense that G^* is obtained as the class of all conditions extended by a condition in $j[G]$. This will however not be the case and it is instructive to see why not.

For G^* to be generic it must intersect all $\mathbf{L}[E^*]$-definable dense classes D on the forcing P^*. As $\mathbf{L}[E^*]$ is the ultrapower of $\mathbf{L}[E]$ by the measure U we can write D as $j(f)(\kappa)$ for some definable function f with domain κ in $\mathbf{L}[E]$ so that $f(\alpha)$ is dense on P for each α. Now our coding forcing P satisfies the following useful form of "diagonal distributivity": We say that a subclass D of P is *γ-dense* for a cardinal γ if any condition in P can be extended into D without changing its values below γ. Now suppose that $f(\alpha)$ is α^+-dense for each cardinal $\alpha < \kappa$ and p is a condition. Then p has an extension q which meets (i.e., extends an element of) each $f(\alpha)$. It follows that some condition p in G meets each $f(\alpha)$ and therefore on the ultrapower side, $j(p)$ will meet $j(f)(\kappa) = D$ provided D is κ^+-dense on P^*. In particular this means that the $j(p)$ for p in G will indeed provide us with a generic for the forcing P^* above κ^+, i.e., a generic subset of the κ^{++} of $\mathbf{L}[E]$ that in turn codes an entire generic class for the forcing P^* above κ^+. As the embedding j is the identity below κ, $j[G]$ also provides us with a generic below κ and indeed a generic subset G_κ of κ^+, as this is coded in both $\mathbf{L}[E]$ and $\mathbf{L}[E^*]$ into the generic below κ in the same way.

So $j[G]$ in fact gives us a subset $G^*_{\kappa^+}$ of $(\kappa^{++})^M$ which codes an entire P^*-generic above $(\kappa^{++})^M$, as well as a subset $G_\kappa = G^*_\kappa$ of κ^+ which is generically coded (in both the P and P^* forcings) into a real; what is missing is to ensure that G_κ, which generically codes G_{κ^+} over $V[G_{\kappa^+}]$, also generically codes $G^*_{\kappa^+}$ over $M[G^*_{\kappa^+}]$. We have to fit the "ultrapower coding" of $G^*_{\kappa^+}$ into G_κ together with the "V-coding" of G_{κ^+} into G_κ, in order to produce the desired P^*-generic G^*.

It is tempting now to make use of the fact that $V = \mathbf{L}[E]$ and $M = \mathbf{L}[E^*]$ actually agree up to $(\kappa^{++})^M$ in the sense that the hierarchies given by E and E^* are the same up to that point. Indeed it is natural to expect that G_κ will generically code $G^*_{\kappa^+}$ using $E \restriction (\kappa^{++})^M$, since it generically codes G_{κ^+} using E and $E^* \restriction (\kappa^{++})^M$ is an initial segment of E. This is encouraging, however it leads to a contradiction, as what G_κ codes below the ordinal $(\kappa^{++})^M$ using E is G_{κ^+} restricted to this ordinal, an element of V, whereas what we want G_κ to code over M, namely $G^*_{\kappa^+}$, cannot be an element of V (else both $G^*_{j(\kappa)}$ and its preimage G_κ would belong to V, reducing our class-forcing to a set-forcing).

Thus we need a different approach, in which the codings over $\mathbf{L}[E]$ and $\mathbf{L}[E^*]$ do not agree at κ^+, in the sense that the generic subset G_κ of κ^+ codes the generic subset G_{κ^+} of κ^{++} using E in a way which accomodates, but differs from, the way it codes the generic subset $G^*_{\kappa^+}$ using E^*. The solution is this: When defining conditions $p(\kappa) = (p_\kappa, p^*_\kappa)$ to almost disjoint code $p_{\kappa^+} : [\kappa^+, |p_{\kappa^+}|) \to 2$ we use sets b_ξ for $\xi < \kappa^{++}$ as before to ensure that $p_{\kappa^+}(\xi) = 1$ iff $p_\kappa(\delta) = 1$ for sufficiently large $\delta \in b_\xi$; however we additionally have sets b^*_ξ for $\xi < (\kappa^{++})^M$ to ensure that for $\xi < (\kappa^{++})^M$, $j(p)_{\kappa^+}(\xi) = 1$ iff $p_\kappa(\delta) = 1$ for sufficiently large $\delta \in b^*_\xi$. Thus there are two codings taking place simultaneously, one of p_{κ^+} and the other of $j(p)_{\kappa^+}$, with two different forms of restraint. To avoid conflicts between these codings we choose the b_ξ's to be very "thin" making use of the measure U. We choose a scale $(f_\xi \mid \xi \in [\kappa^+, \kappa^{++}))$ of functions from κ^+ to κ^+ so that the least function f_{κ^+} of this scale eventually dominates all functions from κ^+ to κ^+ in $M = \mathbf{L}[E^*]$; this is possible as there are only κ^+-many such functions in M. The net effect is that the resulting subset G_κ of κ^+ which is generic over $\mathbf{L}[E]$ will also be generic over $\mathbf{L}[E^*]$, as the thinness of the sets b_ξ allows us to show that conditions can be extended to meet the necessary dense sets from the $\mathbf{L}[E^*]$ coding without conflicting with the restraint imposed by the b_ξ for ξ in $p^*_{\kappa^+}$.

3 Measures of higher order

Suppose now that we are in a *Mitchell model* $\mathbf{L}[E]$ where we now have two normal measures U_0, U_1 on κ with U_0 below U_1 in the Mitchell order. Thus U_0 belongs to the ultrapower of V by the measure U_1. Can we create a real which is class-generic but not set-generic lifting both of the measures U_0 and U_1?

It is convenient to reformulate the situation of the last section (with a single measure U) as follows. Recall that at κ^+ we have two codings, that of $j(p)_{\kappa^+}$ into $G_\kappa \subseteq \kappa^+$ over the ultrapower Ult_U of V by U, and the other of p_{κ^+} into G_κ over V. As the latter coding takes place "above" the former (ultrapower) coding, it is natural to think of the κ^{++}-Cohen condition $p_{\kappa^+} : [\kappa^+, |p_{\kappa^+}|) \to 2$ in two parts: there is p_{κ^+} on $[\kappa^+, |j(p)_{\kappa^+}|)$ coinciding with $j(p)_{\kappa^+}$ and then p_{κ^+} on $[(\kappa^{++})^{\mathrm{Ult}_U}, |p_{\kappa^+}|)$, which is coded using the b_ξ's which "lie above" the ultrapower Ult_U. In this way there is in a sense just one coding, which uses restraints from Ult_U below the κ^{++} of Ult_U and restraints from V between the κ^{++} of Ult_U and the real κ^{++}. In fact $p \restriction \kappa$ is responsible for the coding below κ^{++} of Ult_U (via the embedding j) and p_{κ^+} is responsible for the coding above. But notice that viewed this way, the domain of p_{κ^+} is no longer an interval, but the union of two intervals, namely $[\kappa^+, |j(p)_{\kappa^+}|)$ and $[(\kappa^{++})^{\mathrm{Ult}_U}, |p_{\kappa^+}|)$. So p_{κ^+} is what one might call a "perforated string".

Now let us return to the more complex case of two measures U_0, U_1. At κ^+ the strings are doubly-perforated, since their domains consist of the union of three intervals: $[\kappa^+, |j_{U_0}(p)_{\kappa^+}|)$, $[(\kappa^{++})^{\mathrm{Ult}_{U_0}}, |j_{U_1}(p)_{\kappa^+}|)$ and $[(\kappa^{++})^{\mathrm{Ult}_{U_1}}, |p_{\kappa^+}|)$. For cardinals $\bar{\kappa}$ of Mitchell order 0 (i.e., carrying only normal measures concentrating on non-measurables), strings at $\bar{\kappa}^+$ will only be singly-perforated and at non-measurables we return to non-perforated strings. The situation is similar, but more complicated, when dealing with measurable κ of Mitchell order less than κ^{++} (the "real" coding takes place above the supremum of the $(\kappa^{++})^{\mathrm{Ult}_U}$ for U a normal measure on κ on the $\mathbf{L}[E]$ hierarchy).

But if we go as far as $o(\kappa) = \kappa^{++}$, where $(\kappa^{++})^U$ can be arbitrarily large in κ^{++} for measures U on κ, and wish to lift all of these measures, then we have a problem, as it seems that there is no longer room to code, as the entire interval $[\kappa^+, \kappa^{++})$ has been covered with ultrapower codings which must be respected. In fact, we cannot expect to add a class-generic real which is not set-generic but lifts all extenders, partial and total, when $o(\kappa) = \kappa^{+++}$:

To see this, work with $K = \mathbf{L}[E]$ and assume that $o(\kappa) = \kappa^{+++}$, κ is the largest measurable cardinal, but $\mathbf{L}[E]$ is also closed under sharps. Let $\Gamma(\lambda)$ denote the theory "ZF^-, λ is the largest measurable cardinal, λ^{+++} exists, and $o(\lambda) = \lambda^{+++}$," so that in particular $K \models \Gamma(\kappa)$. If M is a premouse, $M \models \Gamma(\lambda)$, and $M \models S = \{\xi < \lambda^{+++} \mid \mathrm{cf}(\xi) = \lambda^{++}\}$, then we may canonically and uniformly split S into two sets which are both stationary from the point of view of M; say $S = S^0_M \cup S^1_M$, where $M \models$ "S^0_M, S^1_M are both stationary," and S^0_M, S^1_M are both Δ_1-definable over $M|\lambda^{+++M}$ via defining formulas which are independent of M. For $i \in \{0,1\}$ we shall also write $T^i_M = S^i_M \cup \{\xi < \lambda^{+++M} \mid \mathrm{cf}^M(\xi) < \kappa^{++M}\}$ and $T^i = T^i_K$.

Assume now that for $i \in \{0,1\}$ there is a forcing P^i in K such that in any P^i-generic extension there exists a real r^i with the following property:

> For all ξ such that $K|\xi \models \Gamma(\lambda)$, there is a club through $T^i_{K|\xi}$ in $K|\xi[r^i]$. Moreover every (partial or total) extender on E lifts to $K[r^i]$. (*)

By the last sentence we mean that every iteration tree on K can be lifted to one on $K[r]$. Now force with $P = P^0 \times P^1$, obtaining reals r^0, r^1 as above (note that P collapses κ^{+++}, but this is irrelevant), and let g be generic for the collapse of κ^{++K} to ω over $K[r^0, r^1]$. By an unpublished result of Woodin (cf. [2]), $K[g]$ is Σ^1_4-correct in $K[g, r^i]$; this makes vital use of the fact that enough extenders from K lift to $K[r^i]$ (cofinally many total extenders suffice here). The following Π^1_3 statement $\Psi^i(r^i)$ holds in $K[g, r^i]$:

> Every countable mouse M_0 such that $M_0 \models \Gamma(\lambda)$ has a simple countable iterate M_1 such that r^i lifts all extenders on the M_1-sequence, and if M_0 is a countable mouse such that $M_0 \models \Gamma(\lambda)$ and r^i lifts all extenders on the M_0-sequence, then there is a club through $T^i_{M_0}$ in $M_0[r^i]$. $(**)$

That this statement is Π^1_3 boils down to the fact that being a mouse, in our setting, is Π^1_2. By a simple iterate we mean one without any drops. To see that the first part of $(**)$ holds of r^i, let M_0 be a countable mouse. Co-iterate M_0 with K until you reach $M_1 \lhd K'$, where K' is an iterate of K. Since r^i lifts all extenders on E, we can push forward $(*)$ to K' in the sense that $(*)$ holds with initial segments of K replaced by those of K'. If M_1 is not countable, then by taking a countable hull we shall obtain a *countable* iterate of M_0 which is as desired. To see the second part of $(**)$, we may argue in a similar fashion, this time co-iterating $M_0[r^i]$ with $K[r^i]$.

By Σ^1_4-correctness, we find s^0, s^1 in $K[g]$ such that $K[g] \models \Psi^0(s^0) \wedge \Psi^1(s^1)$. We claim the following:[3]

> For $i \in \{0,1\}$, K has a simple iterate W such that s^i lifts all extenders on the W-sequence and there is a club through T^i_W in $W[s^i]$. $(***)$

If $(***)$ is false for some $K|\vartheta \models \Gamma(\lambda)$, then we may pick some cardinal $\Omega > \vartheta$ and some $\sigma\colon \overline{K}[s^i] \to K|\Omega[s^i]$ with $\vartheta \in \operatorname{ran}(\sigma)$ such that $\overline{K}[s^i] \in K[g]$ is transitive and countable in $K[g]$. Write $\overline{\vartheta} = \sigma^{-1}(\vartheta)$, and let $h \in K[g]$ be $\operatorname{Col}(\omega, \overline{\vartheta})$-generic over $\overline{K}[s^i]$. By our hypothesis that $\mathbf{L}[E]$ be closed under sharps, $\overline{K}[s^i][h]$ will be Σ^1_2-correct in $K[g]$. This means that if we look at the family $\mathcal{F}$ of all $M \in \overline{K}[s^i][h]$ which in $\overline{K}[s^i][h]$ are countable simple iterates of $\overline{K}|\overline{\vartheta}$, then by $(**)$ densely many $M \in \mathcal{F}$ will be such that r^i lifts all extenders on the M-sequence and there is a club through T^i_M in $M[r^i]$. But then if $\tilde{M}$ is the direct limit of all mice in $\mathcal{F}$, then $\tilde{M} \in \overline{K}[s^i]$ by the homogeneity of $\operatorname{Col}(\omega, \overline{\vartheta})$, r^i lifts all extenders on the $\tilde{M}$-sequence, and there is a club through $T^i_{\tilde{M}}$ in $\tilde{M}[r^i]$. Moreover, in $\overline{K}[s^i]$, $\tilde{M}$ can be absorbed by a simple iterate of $\overline{K}|\overline{\vartheta}$ which has the same properties. But the elementarity of σ then yields a contradiction. We have verified $(***)$.

We may now use $(***)$ to find a simple iterate W of K such that for both $i = 0$ and $i = 1$, s^i lifts all extenders on the W-sequence and there is a club through T^i_W. Let $D^i \in W[s^i]$ be a club through T^i_W, and let $j\colon K \to W$ be the iteration map. Let D denote the limit points of

[3] We wish to thank the referee for pointing out a gap in our earlier presentation of this argument.

$j^{-1}[D^0 \cap D^1]$. Obviously, D is club in κ^{+++K} and $D \in K[g]$. Let $S \in K$ be such that $K \models S = \{\xi < \kappa^{+++} \mid \mathrm{cf}(\xi) = \kappa^{++}\}$. As S remains stationary in $K[g]$, we can find $\eta \in S \cap D$. Since j is continuous at points whose K-cofinality is bigger than κ, we have $j(\eta) \in D^0 \cap D^1$, and by elementarity $j(\eta) \in j(S)$. This is a contradiction, since $j(S) \cap D^0 \cap D^1 = \varnothing$. This finishes off the argument that no real r^i as in $(*)$ can exist.

We now argue that the problem with $(*)$ lies not in its first sentence, the coding part, but in its second sentence, the lifting of extenders. Consider the following weakening of $(*)$:

> For all ξ, *if* $K|\xi \models$ "λ is the greatest measurable and $\mathrm{o}(\lambda) = \lambda^{+++}$" and $K|\xi[r^i] \models$ "ZF^- and $\lambda^{+++} = (\lambda^{+++})^{K|\xi}$", then there is a club through $(T^i_\lambda)^{K|\xi}$ in $K|\xi[r^i]$.

We can produce a real r^i satisfying this by first shooting a club through T^i and then forcing to code it with localization (using a core model analogue of David's trick; cf. [5, Theorem 6.18]). The club is added by a κ^{+++} distributive forcing (a forcing adding no new κ^{++}-sequences) of K; then, the condensation provided by K suffices for the distributivity of the second forcing, as we can take Skolem hulls in K (as in [7]). In fact, if we weaken $(*)$ by just dropping the last requirement (i.e., that all extenders lift), we obtain a statement which should be forceable using a core model analogue of *strong coding* (cf. [3]). For these reasons we believe that the problem with $(*)$ lies with its second assertion, that all extenders lift.

In fact we conjecture that it is not possible to add a class-generic real which is not set-generic while lifting all normal measures on a measurable κ of order κ^{++}; in fact we conjecture that it is not even possible to do this while lifting all normal measures in a "cofinal" collection $\mathcal{S} \in \mathbf{L}[E]$ of such measures. (By "cofinal" we mean that the ordinals $(\kappa^{++})^{\mathrm{Ult}_U}$ for U in $\mathcal{S}$ are cofinal in κ^{++}.) But this does *not* mean that we cannot preserve the property $\mathrm{o}(\kappa) = \kappa^{++}$! As we shall see in the next section, it is possible to preserve cases of hypermeasurability, which in turn implies that the set of normal measures that are lifted is cofinal; it is not however clear that this set can contain a cofinal subset in the ground model.

4 Hypermeasurables

Can we preserve stronger forms of measurability? Suppose that κ is hypermeasurable in $V = \mathbf{L}[E]$ in the sense that some total extender F on the E-sequence with critical point κ witnesses that κ is $\mathbf{H}(\kappa^{++})$-strong, i.e., the ultrapower $j_F : V \to M$ has the property that $\mathbf{H}(\kappa^{++})$ is contained in M. (This is the same as saying that F is indexed past κ^{++} in the E-hierarchy.) Can we add a real which is class-generic but not set-generic and lifts F?

Again we want to set up our conditions so that the embedding $j = j_F$ can be lifted to $V[G]$. This time we have that the union of the $j(p)_{\kappa^{++}}$ is not in V yet like $p_{\kappa^{++}}$ must be coded into the same subset G_{κ^+} of κ^{++}. As in the one measure case this can be resolved by starting the coding of $p_{\kappa^{++}}$ above $(\kappa^{+++})^{\mathrm{Ult}_F}$, below which the former coding takes place. But we have a new problem: the union of the $j(p)_{\kappa^+}$ would appear to not belong to V and as V and Ult_F completely agree below κ^{++}, the set G_{κ^+} will code it in exactly the same way as it codes $G_{\kappa^{++}}$. This is a serious obstacle and the only way around it is to thin out the coding conditions to ensure that in fact the $j(p)_{\kappa^+}$ will be empty for each condition p.

To ensure the latter we require that for any condition p there is a closed unbounded subset C of κ such that for inaccessible α in C, p_{α^+} is the empty string. This ensures that $j(p)_{\kappa^+}$ will also be the empty string. The price one pays for this is that we only have a weaker form of diagonal distributivity: If $f(\alpha)$ is α^{++}-dense for each cardinal $\alpha < \kappa$ then any condition can be extended to meet each $f(\alpha)$. This only ensures that the pointwise image $j[G]$ will generate a generic over the ultrapower Ult_F above κ^{++}, coded into the subset $G^*_{\kappa^{++}}$ of $(\kappa^{+++})^M$ consisting of the union of the $j(p)_{\kappa^{++}}$ for p in G. G provides a generic below κ^{++} and now the task is to ensure that G_{κ^+} will code not only $G_{\kappa^{++}}$ but also $G^*_{\kappa^{++}}$. This is dealt with as in the one measure case, by starting the former coding "above" the latter, making use of an appropriate scale.

For a stronger total extender (of successor cardinal strength) the pattern is similar: Thin out the conditions to guarantee that $j(p)_\alpha$ is empty for cardinals α strictly between κ and the strength of the total extender. At the strength there are two codings which must be performed simultaneously, one over V and the other over the ultrapower. Conflicts between these codings are avoided by allowing the V-coding to make use of the total extender F when defining the coding sets b_ξ via an appropriate scale.

The above ideas are sufficient to lift a class $\mathcal{S}$ of total extenders (each of successor cardinal strength) which is *bounded* (the set of $(\alpha^+)^{\mathrm{Ult}_F}$ for total extenders F in $\mathcal{S}$ of strength exactly α is bounded in α^+ for each cardinal α) and *uniform (or coherent)* (if F belongs to $\mathcal{S}$ then $j_F(\mathcal{S})$ agrees with $\mathcal{S}$ below the index of F in the $\mathbf{L}[E]$-hierarchy), provided that in $\mathbf{L}[E]$ no inaccessible α is the stationary limit of cardinals which are strong up to α. This yields a version of Corollary 2 up to the level of a proper class of strong cardinals, but handling a stationary-limit of strong cardinals requires new ideas.

5 Woodin cardinals

As coding makes heavy use of condensation it is only reasonable to consider ground models for which a suitable core model theory is available, currently up to the level of Woodin cardinals. Recall that δ is *Woodin* if for each $A \subseteq \delta$ there is a $\kappa < \delta$ which is *A-strong in δ*, i.e., the critical point of embeddings $j : V \to M$ such that $j(A)$ agrees with A up to γ, for each $\gamma < \delta$. At first it appears that this indicates the end of the coding method, as Woodin proved the following (cf. [6] and [11, Theorem 7.14]): If $\mathcal{S}$ is a set of total extenders in V sufficient to witness Woodinness in this sense and R is a real such that each total extender in $\mathcal{S}$ lifts to $V[R]$, then in fact R is generic over V for a (δ-c.c.) forcing of size δ. So there appears to be no version of Corollary 2 in the context of a Woodin cardinal.

But actually there is another definition of Woodin cardinal with a different notion of witness: δ is Woodin if for each $f : \delta \to \delta$ there is a $\kappa < \delta$ closed under f which is *f-strong*, meaning that some embedding $j : V \to M$ with critical point κ is $j(f)(\kappa)$-strong (i.e., $\mathbf{H}(j(f)(\kappa))$ is contained in M). It is shown in [6] that if δ is Woodin in $V = \mathbf{L}[E]$ then in $\mathbf{L}[E]$ there is a witness T to Woodinness in this latter sense which can be lifted by a non-set-generic real R, thereby preserving the Woodinness of δ. And indeed this can be done simultaneously for all Woodin cardinals in $\mathbf{L}[E]$.

The proof of the latter result is much more involved than in the case of nonstationary limits of strongs. In that simpler setting, one can use the strength function $\alpha \mapsto$ (sup of the strengths of total extenders with critical point α) to thin out the codings uniformly below each inaccessible cardinal. In the Woodin cardinal setting one must instead use a uniform witness $\mathcal{T}$ to the Woodinness of each Woodin cardinal whose total extenders have non-Woodin critical point, and then thin out the codings using functions which witness the failure of these critical points to be Woodin. A major difference from the easier setting is that for total extenders F that are to be lifted and conditions p, it is no longer the case that $j_F(p)$ will be trivial between the critical point and strength of F; instead one must deal with this extra information at a cardinal α between the critical point and strength of F until reaching a condition which "recognises" that each of the finitely-many total extenders in $\mathcal{T}$ overlapping α has non-Woodin critical point; this is essential for showing that this extra information stabilises to a set in V.

6 Future work

The story is far from over regarding coding over core models. In terms of versions of Corollary 2, the current frontier is the preservation of measurable Woodinness, which will need a technique beyond what is sketched above for plain Woodinness. Going back all the way to hypermeasurable cardinals, there remains the difficult problem of condensation, which obstructs a satisfying version of Theorem 1. As mentioned, the special case of coding a generic for a Příkrý product is handled in [7], but this is an extremely special case and it is quite possible that there is a counterexample for the coding of more general predicates while preserving hypermeasurability. And of course it will be worthwhile to look at generalisations to the large cardinal setting of the many applications of Jensen coding (and its iterations), as found in [5, 9]. Finally, can one do something with coding at the level of supercompact cardinals? Of course the core model theory is not yet available there, but there has been considerable progress in showing that many of the nice features of $\mathbf{L}[E]$ models can be forced consistently with the strongest of large cardinal properties (cf., e.g., [8]). Are there coding theorems to be proved over such "pseudo" core models? A positive answer may have very interesting consequences.

References

[1] A. Beller, R. Jensen, and P. Welch. *Coding the universe*, volume 47 of *London Mathematical Society Lecture Note Series*. Cambridge University Press, 1982.

[2] P. Doebler. The 12th Delfino problem and universally Baire sets of reals. Master's thesis, Westfälische Wilhelms-Universität Münster, 2006.

[3] S. D. Friedman. Strong coding. *Annals of Pure and Applied Logic*, 35(1):1–98, 1987.

[4] S. D. Friedman. Coding over a measurable cardinal. *Journal of Symbolic Logic*, 54(4):1145–1159, 1989.

[5] S. D. Friedman. *Fine structure and class forcing*, volume 3 of *de Gruyter Series in Logic and its Applications*. Walter de Gruyter & Co., Berlin, 2000.

[6] S. D. Friedman. Genericity and large cardinals. *Journal of Mathematical Logic*, 5(2):149–166, 2005.

[7] S.-D. Friedman and M. Golshani. Killing the GCH everywhere with a single real. *Journal of Symbolic Logic*, 78(3):803–823, 2013.

[8] S. D. Friedman and P. Holy. A quasi-lower bound on the consistency strength of PFA. *Transactions of the American Mathematical Society*, to appear.

[9] S. D. Friedman and D. Schrittesser. Projective measure without projective Baire. Preprint.

[10] P. Koepke. Finestructure for inner models with strong cardinals, 1989. Habilitationsschrift, Albrecht-Ludwigs-Universität Freiburg.

[11] J. R. Steel. An outline of inner model theory. In M. Foreman and A. Kanamori, editors, *Handbook of set theory. Vol. 3*, pages 1595–1684. Springer-Verlag, 2010.

Extender based forcings, fresh sets and Aronszajn trees

Moti Gitik*

School of Mathematical Sciences, Tel Aviv University, Tel Aviv, Israel

Abstract

Extender based forcings are studied with respect of adding branches to Aronszajn trees. We construct a model with no Aronszajn tree over $\aleph_{\omega+2}$ from the optimal assumptions. This answers a question of Friedman and Halilović [1].

The reader interested only in the question of Friedman and Halilović may skip the first section and go directly to the second.

1 No new branches in κ^+-Aronszajn trees

We deal here with extender based Příkrý forcing, long and short extender Příkrý forcing. Let us refer for the definitions and basic properties to [5] or to [3, § 3] for extender based Příkrý forcing; to [3, § 2] or to [6] for long extender Příkrý forcing; and to [2, 6] for short extender Příkrý forcing.

Theorem 1.1. Let κ be a regular cardinal. Extender based Příkrý forcing over κ (i.e., extender based Příkrý forcing for adding λ-many ω-sequences to κ, for some regular $\lambda \geq \kappa^+$) cannot add a cofinal branch to a κ^+-Aronszajn tree.

Proof. Let $\langle T, \leq_T\rangle$ be a κ^+-Aronszajn tree. Denote by $\mathcal{P}$ the extender based Příkrý forcing over κ. Suppose that $\mathcal{P}$ adds a cofinal branch through T. Let $\underset{\sim}{b}$ be a name of such a branch and $0_{\mathcal{P}} \Vdash \underset{\sim}{b}$ is a κ^+-branch through T.

Let $p, q \in \mathcal{P}$ and $n < \omega$. We say that q is an n-extension of p iff $q \geq p$ and q is obtained from p by taking an n-element sequence $\langle \eta_1, ..., \eta_n\rangle$ from the first n-levels of the tree of sets of measures one over the maximal coordinate of p, adding it to p and projecting to all permitted coordinates of p. Denote such q by $p^\frown\langle \eta_1, ..., \eta_n\rangle$.

For each $\alpha < \kappa^+$ and $p \in \mathcal{P}$ there are $n < \omega$ and $p_\alpha \geq^* p$ such that any n-extension of p_α decides $\underset{\sim}{b}(\alpha)$, as was shown in [5]. Note that the branch $\underset{\sim}{b}\restriction\alpha + 1$ is decided as well, since $T \in V$ and so the value at the level α determines uniquely the branch to it below. Denote by $n(p, \alpha)$ the least such n.

*We are grateful to the referee of the paper for his or her remarks and corrections.

Stefan Geschke, Benedikt Löwe, Philipp Schlicht (*eds.*).
Infinity, computability, and metamathematics: Festschrift celebrating the 60th birthdays of Peter Koepke and Philip Welch. College Publications, London, 2014. Tributes, Volume 23.

Lemma 1.2. For each $p \in \mathcal{P}$ there are $p^* \geq^* p$ and $n^* < \omega$ such that for every $q \geq^* p^*$ and for every large enough $\alpha < \kappa^+$ we have $n(q, \alpha) = n^*$.

Proof. Suppose otherwise. Define by induction a $\geq^*$-increasing sequence $\langle p_k \mid k < \omega \rangle$ of direct extensions of p and an increasing sequence $\langle \alpha_k \mid k < \omega \rangle$ of ordinals below κ^+ such that $n(p_k, \alpha_k) < n(p_{k+1}, \alpha_{k+1})$, for every $k < \omega$.

Find $p_\omega \in \mathcal{P}$ which is $\leq^*$-stronger than every p_k. Extend it to a condition q that decides $\underset{\sim}{b}(\alpha_\omega)$, where $\alpha_\omega = \bigcup_{k<\omega} \alpha_k$. Let m be the length of the normal sequence of q. Denote the decided value by t. Pick k with $n_k > m$. Then there are two n_k-extensions q_1, q_2 of q which decide $\underset{\sim}{b}(\alpha_k)$ differently, say t_1 and t_2. But this is impossible since t must be above both t_1 and t_2. Q.E.D. (Lemma 1.2)

Suppose for simplicity that p^* is the empty condition and $n^* = 1$.

We define by induction a sequence of conditions $\langle p_\alpha \mid \alpha < \kappa^+ \rangle$ such that

1. p_α decides $\underset{\sim}{b}(\alpha)$ to be some $t_\alpha \in T$,
2. $p_\alpha, p_{\alpha'}$ are compatible, for every α, α' in some $S \subseteq \kappa^+$ of cardinality κ^+.

Then $\{t \in T \mid \exists \alpha \in S, t \leq_T t_\alpha\}$ is a cofinal branch in T, since $\alpha < \alpha', \alpha, \alpha' \in S$ implies by (2) that $t_\alpha <_T t_{\alpha'}$. This is a contradiction because T is a κ^+-Aronszajn tree.

Let q_0 be a direct extension of $0_{\mathcal{P}}$ such that every 1-extension of it decides $\underset{\sim}{b}(0)$. Set p_0 to be a 1-extension of q_0 by some ν_0 from the measure one set of its maximal coordinate. Consider a condition p'_0 which is obtained from p_0 by removing the projection $(\nu_0)^0$ of ν_0 to the normal measure of the extender, by creating a new maximal coordinate and moving to it the tree of p_0 putting the first level to be κ.

Let q_1 be a direct extension of p'_0 such that every 1-extension of it decides $\underset{\sim}{b}(1)$. Set p_1 to be a 1-extension of q_1 by some ν_1 from the measure one set of its maximal coordinate.

Continue further by induction. Suppose that $\langle p_\beta \mid \beta < \alpha \rangle$ is defined. Define p_α.

If α is a successor ordinal then proceed as above. Suppose that α is a limit ordinal. If $\operatorname{cof}(\alpha) < \kappa$, then pick a cofinal in α sequence $\langle \alpha_i \mid i < \operatorname{cof}(\alpha) \rangle$ and combine conditions $\langle p'_{\alpha_i} \mid i < \operatorname{cof}(\alpha) \rangle$ into one condition q'_α. Then pick a direct extension q_α of q'_α such that every 1-extension of it decides $\underset{\sim}{b}(\alpha)$. Set p_α to be a 1-extension of q_α by some ν_α from the measure one set of its maximal coordinate.

Suppose that $\operatorname{cof}(\alpha) = \kappa$. Pick a cofinal in α sequence $\langle \alpha_i \mid i < \kappa \rangle$. Combine conditions $\langle p'_{\alpha_i} \mid i < \kappa \rangle$ into a single condition q'_α as follows. For each $\eta \in \mathrm{supp}(p'_{\alpha_i})$ put a barrier i, add a new maximal coordinate, move trees of p'_{α_i} to it, take the diagonal intersection of them leaving the first level to be κ. Now proceed as before and define q_α, p'_α, ν_α and p_α. This completes the construction.

There are a stationary $S \subseteq \kappa^+$ and $\nu^* < \kappa$ such that for every $\alpha \in S$ we have $(\nu_\alpha)^0 = \nu^*$. Q.E.D. (Theorem 1.1)

In [7], Hamkins defined the following two useful notions: Let $V \subseteq V_1$. We say that the *δ-approximation property holds between V and V_1* if for every set A of ordinals in V_1, if $A \cap a \in V$ for all $a \in V$ with $V \models |a| < \delta$, then $A \in V$. If $A \subseteq \lambda$, $A \in V_1$, we say that A is *fresh* if for each $\alpha < \lambda$, $A \cap \alpha \in V$.

Theorem 1.3. Let κ be the limit of increasing sequence of regular cardinals $\langle \kappa_n \mid n < \omega \rangle$. Let Cohen($\omega$) be the Cohen real forcing, $\langle \mathcal{P}, \leq, \leq^* \rangle$ be long or short extender Příkrý or extender based Příkrý forcing over κ over $V^{\mathrm{Cohen}(\omega)}$. Then in $V^{\mathrm{Cohen}(\omega) * \mathcal{P}}$ there is no new fresh subsets of ordinals of cofinality bigger than κ.

Proof. We refer to [2, §§1&2] for definitions and basic properties of long (and short) extender forcing, a more detailed account may be found in [6]. Let us give here only a brief description. Conditions in this forcings are of the form $p = \langle p_n \mid n < \omega \rangle$ such that there is $\ell(p) < \omega$ with p_n being a Cohen condition in $\mathrm{Cohen}(\lambda, \kappa_n) = \{f \mid |f| \leq \kappa, f : \lambda \to \kappa_n\}$, for each $n < \ell(p)$. If $n \geq \ell(p)$, then $p_n = \langle a_n(p), f_n(p), A_n(p) \rangle$, where $a_n(p)$ is a partial order preserving function from λ to κ_n (or just a subset of λ in long extender forcing) of cardinality $< \kappa_n$, $A_n(p)$ is a set of measure one for the measure of E_n which corresponds to the maximal element of $\mathrm{rng}(a_n(p))$ and $f_n(p)$ is in $\mathrm{Cohen}(\lambda, \kappa_n)$.

Let $\mathcal{P}$ be the long extender Příkrý forcing over $\kappa = \bigcup_{n<\omega} \kappa_n$. The other two forcing notions are treated similar. Assume for simplicity that $\lambda = \kappa^+$. Let r be a Cohen real and $\underset{\sim}{X}$ be a $\mathcal{P}$-name of a subset of κ^+ with every initial segment in V. We shall show that then $\underset{\sim}{X}$ is in V as well.

Let us work in $V[r]$. Consider the following $\mathcal{P}$-name:

$$\underset{\sim}{Y} := \{\langle p, \check{\alpha} \rangle \mid p \Vdash \check{\alpha} \in \underset{\sim}{X}\}.$$

Then for every G generic for the forcing $\langle \mathcal{P}, \leq^* \rangle$ and every $a \subseteq \kappa^+, a \in V, |a| < \kappa_0$ (one may take $|a| < \aleph_2$ instead) we have $a \cap \underset{\sim}{Y}_G \in V$. This holds, since for every $q \in \mathcal{P}$ there is $p \geq^* q$ such that for all $\alpha \in a$, $p \| \alpha \in \underset{\sim}{X}$. Hence there is such $p \in G$. Then

$$a \cap \underset{\sim}{Y}_G = \{\alpha \in a \mid p \Vdash \alpha \in \underset{\sim}{X}\}.$$

So, $a \cap \underset{\sim}{Y}_G \in V[r]$. Remember that $p \Vdash$ "$\underset{\sim}{X}$ is fresh", hence $p \Vdash \underset{\sim}{X} \cap a \in V$. But, for all $\alpha \in a$, $p || \alpha \in \underset{\sim}{X}$. So, whenever H is $\langle \mathcal{P}, \leq \rangle$-generic with $p \in H$,

$$\{\alpha \in a \mid p \Vdash \alpha \in \underset{\sim}{X}\} = \underset{\sim}{X}_H \cap a \in V.$$

Hence, $a \cap \underset{\sim}{Y}_G \in V$.

Now the forcing $\langle \mathcal{P}, \leq^* \rangle$ is κ_0-closed, hence by Hamkins, the κ_0-approximation property holds between V and $V[r, G]$. In particular, $\underset{\sim}{Y}_G \in V$.

We apply this observation to conditions of $\mathcal{P}$ with different lengths of trunks. Let us start with $0_{\mathcal{P}}$. Pick $p_0 \geq^* 0_{\mathcal{P}}$ and $Y_0 \in V$ such that

$$p_0 \Vdash_{\langle \mathcal{P}, \leq^* \rangle} \underset{\sim}{Y} = \check{Y}_0.$$

Now let us construct by induction on $\nu \in A_0(p_0)$ (recall that $A_0(p_0)$ is the set of measure one of the first level of p_0) a sequence $\langle p_0(\nu) \mid \nu \in A_0(p_0) \rangle$ of extensions of p_0 and a sequence $\langle Y_0(\nu) \mid \nu \in A_0(p_0) \rangle$ of subsets of κ^+ such that

1. $Y_0(\nu) \in V$,
2. $p_0{}^\frown \nu \leq^* p_0(\nu)$,
3. $p_0(\nu) \Vdash_{\langle \mathcal{P}, \leq^* \rangle} \underset{\sim}{Y} = \check{Y}_0(\nu)$,
4. the sequence $\langle p_0(\nu) \setminus \nu \mid \nu \in A_0(p_0) \rangle$ is $\leq^*$-increasing, where $p_0(\nu) \setminus \nu \geq^* p_0$ is the condition obtained from $p_0(\nu)$ by removing ν and its projections to $a_0(p_0)$ but leaving all the rest.

Let p_1 be a $\leq^*$-upper bound of $\langle p_0(\nu) \setminus \nu \mid \nu \in A_0(p_0) \rangle$.

Continue similar (starting with p_1) to the second level (dealing with pairs) and define p_2. Proceed further and define p_3, p_4, etc. Let p_ω be a $\leq^*$-upper bound of $\langle p_n \mid n < \omega \rangle$.

Claim 1.4. Suppose that $q \geq p_\omega$ and for some $\alpha < \kappa^+$,

$$q \Vdash \alpha \in \underset{\sim}{X} \quad (\text{or } q \Vdash \alpha \notin \underset{\sim}{X}).$$

Then

$$p_\omega{}^\frown \langle \nu_1, ..., \nu_n \rangle \Vdash \alpha \in \underset{\sim}{X} \quad (\text{or } p_\omega{}^\frown \langle \nu_1, ..., \nu_n \rangle \Vdash \alpha \notin \underset{\sim}{X}),$$

where $\langle \nu_1, ..., \nu_n \rangle$ are such that

$$p_\omega \leq p_\omega{}^\frown \langle \nu_1, ..., \nu_n \rangle \leq^* q$$

and the measure one sets of p_ω are intersected with projections of the measure one sets of q.

Proof. Assume that $q \Vdash \alpha \in \underset{\sim}{X}$. Suppose for simplicity that $n = 1$. We have $p_\omega \leq p_\omega{}^\frown\nu \leq^* q$. First we show that for any $s \geq^* p_\omega{}^\frown\nu$, if $s \| \alpha \in \underset{\sim}{X}$, then $s \Vdash \alpha \in \underset{\sim}{X}$.

Suppose otherwise, i.e., $s \Vdash \alpha \notin \underset{\sim}{X}$. Recall that $p_\omega{}^\frown\nu \geq^* p_0(\nu)$ and

$$p_0(\nu) \Vdash_{\langle \mathcal{P}, \leq^* \rangle} \underset{\sim}{Y} = \check{Y}_0(\nu).$$

We have $q \Vdash \alpha \in \underset{\sim}{X}$. Then $\langle q, \check{\alpha} \rangle \in \underset{\sim}{Y}$. But $q \geq^* p_\omega{}^\frown\nu \geq^* p_0(\nu)$, hence

$$p_0(\nu) \Vdash_{\langle \mathcal{P}, \leq^* \rangle} \alpha \in \underset{\sim}{Y}.$$

Let now G be a generic set for $\langle \mathcal{P}, \leq^* \rangle$ with $s \in G$. Then in $V[r, G]$ we have $\alpha \in Y$. This means that there is $t \in G$ with $\langle t, \check{\alpha} \rangle \in \underset{\sim}{Y}$. So, by the definition of $\underset{\sim}{Y}$, $t \Vdash \alpha \in \underset{\sim}{X}$. But $s, t \in G$ so they are compatible also as conditions of $\langle \mathcal{P}, \leq \rangle$, which is impossible, since they force contradictory information about α.

Assume now that $p_\omega{}^\frown\nu \nVdash \alpha \in \underset{\sim}{X}$. There is $t \geq p_\omega{}^\frown\nu$ such that $t \Vdash \alpha \notin \underset{\sim}{X}$. Find $\eta_1, \ldots, \eta_k$ such that

$$t \geq^* p_\omega{}^\frown\nu^\frown\langle \eta_1, \ldots, \eta_k \rangle \geq p_\omega{}^\frown\nu.$$

The argument above shows that any $\leq^*$-extension of $p_\omega{}^\frown\nu^\frown\langle \eta_1, \ldots, \eta_k \rangle$ that decides the statement "$\alpha \in \underset{\sim}{X}$" does so negatively, i.e., it forces "$\alpha \notin \underset{\sim}{X}$". But take $q_1 \geq q$ be so that $q_1 \geq^* p_\omega{}^\frown\nu^\frown\langle \eta_1, \ldots, \eta_k \rangle$. Note that $\eta_1, \ldots, \eta_k$ have pre-images in measure one sets of q so such q_1 exists. We have $q_1 \Vdash \alpha \in \underset{\sim}{X}$, as $q_1 \geq q$. Contradiction. Q.E.D. (Claim 1.4)

Now let G be a generic for $\langle \mathcal{P}, \leq \rangle$ with $p_\omega \in G$. Clearly for every $\alpha < \kappa^+$ there is $q \in G, q \geq p_\omega$ which decides "$\alpha \in \underset{\sim}{X}$". Then, by the claim, an extension of the form $p_\omega{}^\frown\langle \eta_1, \ldots, \eta_n \rangle$ of p_ω which is $\leq^*$-weaker than q already decides the statement and does it the same way. We have then $p_\omega{}^\frown\langle \eta_1, \ldots, \eta_n \rangle \in G$ and the coordinates of measure one sets of $p_\omega{}^\frown\langle \eta_1, \ldots, \eta_n \rangle$ are the same (above the trunk) as those of p_ω.

The forcing $\langle \mathcal{P}, \leq \rangle$ preserves κ^+. Then for κ^+-many α's we shall have q's in G which decide "$\alpha \in \underset{\sim}{X}$" and have trunks of the same length. Then there will be a single sequence $\langle \eta_1, \ldots, \eta_n \rangle$ with $p_\omega{}^\frown\langle \eta_1, \ldots, \eta_n \rangle$ making the same decisions.

Consider now the subforcing

$$\mathcal{P}/p_\omega := \{ s \in \mathcal{P} \mid s \geq p_\omega, \forall n \geq \ell(p_\omega), a_n(s) = a_n(p_\omega) \}.$$

Then $\mathcal{P}/p_\omega$ is just equivalent to the usual tree Příkrý forcing. We can deal with coordinates of sets of measure one of p_ω and ignore the rest.

The argument of the previous paragraph implies that only $G{\restriction}\mathcal{P}/p_\omega$ is needed in order to decide $\underset{\sim}{X}$ completely.

It is easy to finish now. Recall that every initial segment of X is in V. Let $H \subseteq \mathcal{P}/p_\omega$ be generic. Find $n < \omega$ such that for κ^+-many β's there is a condition in H with a trunk of the length n which decides $\underset{\sim}{X}{\restriction}\beta$. Let $\langle \eta_1, ..., \eta_n\rangle$ be this trunk (it should be the same due to compatibility of members of H). Then

$$X = \bigcup\{Z \subseteq \kappa^+ \mid \exists \beta < \kappa^+ \exists q \geq^* p_\omega{}^\frown\langle \eta_1, ..., \eta_n\rangle \quad q \Vdash \underset{\sim}{X} \cap \beta = Z\},$$

since any two conditions with the same trunk are compatible in $\mathcal{P}/p_\omega$. The right side of the equality is obviously in V, and thus $X \in V$.

Q.E.D. (Theorem 1.3)

Note that the forcing $\langle \mathcal{P}, \leq, \leq^*\rangle$ over V adds fresh subsets to κ^+. Thus let $G \subseteq \mathcal{P}$ be a generic subset. Consider a set

$$A := \{\alpha < \kappa^+ \mid \exists p \in G, \ell(p) > 0, p = \langle p_n \mid n < \omega\rangle, p_0(\alpha) = 0\}.$$

Then for each $\beta < \kappa^+$ the set $A \cap \beta$ is in V, since a single extension decides it completely.

By [4], the Příkrý forcing does not add new fresh subsets to κ^+ (or to ordinals of cofinality $\geq \kappa^+$).

Theorem 1.5. Let Q be a forcing of cardinality $\leq \kappa$ and $\mathcal{P}$ be a forcing in V^Q that preserves κ^+ and does not add new fresh (relatively to V) subsets to κ^+. Suppose $\langle T, \leq_T\rangle$ is a κ^+-Aronszajn tree in V^Q. Then $\mathcal{P}$ does not add κ^+-branches to T.

Proof. Suppose otherwise. Assume without loss of generality that $\mathrm{Lev}_\alpha(T) = [\kappa \cdot \alpha, \kappa \cdot \alpha + \kappa)$. Let $\underset{\sim}{b}$ be a $Q * \mathcal{P}$ name such that

$$\langle 0_Q, 0_\mathcal{P}\rangle \Vdash \underset{\sim}{b} \text{ is a } \kappa^+\text{-branch through } \underset{\sim}{T}.$$

Let $G * H$ be a generic subset of $Q * \mathcal{P}$. There are $s \in G$, $\nu < \kappa$ and $S \subseteq \kappa^+$ stationary such that for every $\alpha \in S$ there is $p \in H$ with

$$\langle s, p\rangle \Vdash \kappa \cdot \alpha + \nu \in \underset{\sim}{b}.$$

Then for every $\alpha < \beta, \alpha, \beta \in S$ we have

$$s \Vdash_Q \kappa \cdot \alpha + \nu \leq_T \kappa \cdot \beta + \nu,$$

since otherwise there will be some $s' \geq s$ which forces "$\kappa \cdot \alpha + \nu, \kappa \cdot \beta + \nu$ are incompatible in T". Pick $p \in H$ such that

$$\langle s, \underset{\sim}{p} \rangle \Vdash \kappa \cdot \alpha + \nu \in \underset{\sim}{b}$$

and

$$\langle s, \underset{\sim}{p} \rangle \Vdash \kappa \cdot \beta + \nu \in \underset{\sim}{b}.$$

But then $\langle s', \underset{\sim}{p} \rangle \geq \langle s, \underset{\sim}{p} \rangle \Vdash \kappa \cdot \alpha + \nu \in \underset{\sim}{b}, \kappa \cdot \beta + \nu \in \underset{\sim}{b}$ and this is impossible.

Consider now the set T_* which consists of all ordinals $\kappa \cdot \alpha + \nu$ such that $\langle s, 0_P \rangle \not\Vdash \kappa \cdot \alpha + \nu \notin \underset{\sim}{b}$. Define an order $\leq_*$ on T_* (in V) as follows:

$$\kappa \cdot \alpha + \nu \leq_* \kappa \cdot \beta + \nu$$

if and only of

$$s \Vdash_Q \kappa \cdot \alpha + \nu \leq_T \kappa \cdot \beta + \nu.$$

The tree T_* need not be a κ^+-Aronszajn tree since its levels may have cardinality κ^+, but still its height is κ^+ and it has no κ^+-branches. Thus for a given $\delta < \kappa^+$ pick $\beta < \kappa^+$ such that $otp(\beta \cap S) \geq \delta$ and for some $p \in H$,

$$\langle s, p \rangle \Vdash \kappa \cdot \beta + \nu \in \underset{\sim}{b}.$$

Then $\kappa \cdot \beta + \nu \in T_*$ and for each $\alpha \in S \cap \beta$ we have

$$s \Vdash_Q \kappa \cdot \alpha + \nu \leq_T \kappa \cdot \beta + \nu,$$

as was observed above. Hence the level of $\kappa \cdot \beta + \nu$ in T_* is at least δ.

The tree T_* cannot have κ^+-branches, since any such branch will generate a κ^+-branch in T.

Now, a branch b translates easily into κ^+-branch c of T_*. Just set

$$c = \{\xi \in T_* \mid \exists \beta \in S, \kappa \cdot \beta + \nu \in b, \xi \leq_* \kappa \cdot \beta + \nu\}.$$

Then c will be a new fresh subset of κ^+. Contradiction. Q.E.D. (Theorem 1.5)

Corollary 1.6. Let Cohen(ω) be the Cohen real forcing, $\langle \mathcal{P}, \leq, \leq^* \rangle$ be long or short extender Příkrý or extender based Příkrý forcing over κ over $V^{\text{Cohen}(\omega)}$. Then $\langle \mathcal{P}, \leq, \leq^* \rangle$ does not add κ^+-branches to κ^+-Aronszajn trees in $V^{\text{Cohen}(\omega)}$.

Let us show directly that extender based forcings for a singular κ do not add κ^+-branches to κ^+-Aronszajn trees.

Theorem 1.7. Extender based forcings for a singular κ do not add κ^+-branches to κ^+-Aronszajn trees.

Proof. Let $\langle \mathcal{P}, \leq, \leq^* \rangle$ be such a forcing, $\langle T, \leq_T \rangle$ a κ^+-Aronszajn tree and $\underset{\sim}{b}$ a name of a κ^+-branch. The idea will be to find ω-levels $\langle \alpha_n \mid n < \omega \rangle$ and for every $n < \omega$ some δ_n points over the level α_n which are potential elements of $\underset{\sim}{b}$ with $\{\delta_n \mid n < \omega\}$ cofinal in κ. Then $|\prod_{n<\omega} \delta_n| = \kappa^+$ which allows to argue that the level $\bigcup_{n<\omega} \alpha_n$ has κ^+ many points.

Recall that each $p \in \mathcal{P}$ is of the form $\langle p_n \mid n < \omega \rangle$ and there is $\ell(p) < \omega$ such that p_n's with $n < \ell(p)$ are Cohen conditions which are κ^+-closed. For every $n \geq \ell(p)$, p_n is of the form $\langle a_n, A_n, f_n \rangle$ with the A_n's being sets of measure one for the measure of the extender corresponding to $\max(\mathrm{rng}(a_n))$.

Denote by $A_{\leq m}(p)$ the product of the first m measure one sets of p. Given $m < \omega$, by *an m-extension of p* we mean an extension of p obtained by choosing some $\langle \eta_1, ..., \eta_m \rangle \in A_{\leq m}(p)$ and extending p by adding it. Let us denote such an extension by $p^\frown \langle \eta_1, ..., \eta_m \rangle$.

Use Lemma 1.2 and fix $n^* < \omega$, p^* such that for every $q \geq^* p^*$ and $\alpha < \kappa^+$ there is $q_\alpha \geq^* q$ with any n^*-extension of it deciding $\underset{\sim}{b} \restriction \alpha$.

For $k, 1 \leq k < \omega$, let $r \geq^{*k} q$ mean that $r \geq^* q$ and for every level $i \leq k$ the conditions r, q have the same maximal coordinate at level i.

Now by the standard argument for each q, $k, 1 \leq k < \omega$, and $\alpha < \kappa^+$ there are $n < \omega$ and $q_{\alpha k} \geq^{*k} q$ such that any n-extension of $q_{\alpha k}$ decides $\underset{\sim}{b} \restriction \alpha$. Denote by $n(q, k, \alpha)$ the least such n.

Lemma 1.8. For any $k, 1 \leq k < \omega$ and $q \geq^* p^*$, we have $n^* = n(q, k, \alpha)$.

Proof. Clearly $n^* \leq n(q, k, \alpha)$. Let us show the equality. Run the standard argument trying to decide $\underset{\sim}{b} \restriction \alpha$ starting with q for each of its k-extensions, i.e., for any choice of $\langle \eta_1, ..., \eta_k \rangle$ in sets of measure one of the first k-levels of q. Then shrink these first k sets in order to have the same conclusion (n-extension decides or not $\underset{\sim}{b} \restriction \alpha$). Finally we shall have $r \geq^{*k} q$ and $n < \omega$ such that

1. every n-extension of r decides $\underset{\sim}{b} \restriction \alpha$.

2. for every $\langle \eta_1, ..., \eta_k \rangle, \langle \eta'_1, ..., \eta'_k \rangle$ two sequences from the first k sets of measure one of r and any $m < \omega$, if there is an m-extension of some $t \geq^* r^\frown \langle \eta_1, ..., \eta_k \rangle$ decides $\underset{\sim}{b} \restriction \alpha$, then any m-extension of $r^\frown \langle \eta'_1, ..., \eta'_k \rangle$ decides $\underset{\sim}{b} \restriction \alpha$.

Suppose for a moment that $n > n^*$. Pick some $s \geq^* r$ such that every n^*-extension of s decides $\underset{\sim}{b} \restriction \alpha$. Clearly, s need not be a $*k$-extension, i.e., maximal coordinates of levels $\leq k$ may increase. Let $\langle \varrho_1, ..., \varrho_k \rangle$ be a sequence from the first k sets of measure one of s. Denote its projection to the first k maximal coordinates of r by $\langle \eta_1, ..., \eta_k \rangle$. Then $s^\frown \langle \varrho_1, ..., \varrho_k \rangle \geq^* r^\frown \langle \eta_1, ... \eta_k \rangle$ and a $n^* - k$-extension of $s^\frown \langle \varrho_1, ..., \varrho_k \rangle$ decides $\underset{\sim}{b} \restriction \alpha$. So already a $n^* - k$-extension of $r^\frown \langle \eta_1, ..., \eta_k \rangle$ decides $\underset{\sim}{b} \restriction \alpha$ and then the same holds if we replace $\langle \eta_1, ..., \eta_k \rangle$ by any other sequence from the first k-sets of measures one of r. But this means that any n^*-extension of r already decides $\underset{\sim}{b} \restriction \alpha$. Q.E.D. (Theorem 1.7)

Suppose for simplicity that p^* is just $0_{\mathcal{P}}$. Fix a cofinal in κ sequence of regular cardinals $\langle \delta_n \mid n < \omega \rangle$ so that for some increasing sequence $\langle k_n \mid n < \omega \rangle$ of numbers above n^* we have $2^{\kappa_{k_n}} < \delta_n < \kappa_{k_n + 1}$.

Let $p \in \mathcal{P}$, $\delta = \delta_m$, for some $m < \omega$ and $k = k_m$.

Define by induction a continuous $\in$-chain of elementary submodels $\langle M_i \mid i \leq \delta \rangle$ and a sequence of conditions $\langle q_i \mid i \leq \delta \rangle$ such that the following conditions hold:

1. $p \in M_0$.
2. $\tilde{q}_i \geq^{*k} p$ where $\tilde{q}_i$ is obtained from q_i by intersecting its first n^* sets of measure one with those of p. This means that the first n^*-sets of measure one of q_i may be different (larger) than those of p.
3. If $i \leq j$, then $\tilde{q}_j \geq^{*k} q_i$, where $\tilde{q}_j$ is obtained from q_j by intersecting its first k sets of measure one with those of q_i.
4. $|M_i| < \delta$, for every $i < \delta$.
5. ${}^{\kappa_k} M_i \subseteq M_i$, for $i = 0$ or i a successor ordinal.
6. Each n^*-extension of q_i decides $\underset{\sim}{b} \restriction \mu_i$, where $\mu_i = \sup(M_i \cap \kappa^+)$.
7. (Maximality condition) If for some n^*-sequence $\langle \eta_1, ..., \eta_{n^*} \rangle$ we have
   (a) $q_i{}^\frown \langle \eta_1, ..., \eta_{n^*} \rangle \in \mathcal{P}$,
   (b) some $t \geq^{*k} q_i{}^\frown \langle \eta_1, ..., \eta_{n^*} \rangle$ decides $\underset{\sim}{b} \restriction \mu_i$,

   then $\langle \eta_1, ..., \eta_{n^*} \rangle \in A_{\leq n^*}(q_i)$ and already $q_i{}^\frown \langle \eta_1, ..., \eta_{n^*} \rangle$ decides $\underset{\sim}{b} \restriction \mu_i$
8. $q_i \in M_{i+1}$ and it is the least possible in some well order fixed in advance.

Note that the sequence $\langle q_i \mid i \leq \delta \rangle$ remains unaffected once we replace in p its first n^* sets of measure one but keep all the rest unchanged.

Denote the final M_δ by M, μ_δ by μ and q_δ by q. Note that $\delta > 2^{\kappa_k}$, so the sets of measures one of the first k-levels of δ many of q_i's are the same. Assume then that for all i's they are the same. Just shrink to i's with a constant value otherwise. Assume that the same sets stand in q as well.

Then each n^*-extension of q decides $\underset{\sim}{b}\restriction\mu$. We denote the set of all such decisions by $X = \{t_\xi \mid \xi < \kappa_{n^*}\}$. Note that $|X| \leq \kappa_{n^*} < \delta$. The set of decisions of n^*-extensions of q_i will be $X_i = \{t_\xi\restriction\mu_i \mid \xi < \kappa_{n^*}\}$.

Note that X and each particular t_ξ need not be in M, but X_i's and the initial segments of t_ξ are in M.

Let us fix $i^* < \delta$ such that all the branches in X already split before the level μ_{i^*}. This is possible since $\delta > \kappa_{n^*}$. Suppose that

> there are $i, i^* < i < \delta$ such that for every $\xi \in \kappa^+ \cap M$ and $r \geq^{*k} q_i, r \in M$ all of whose n^*-direct extensions decide $\underset{\sim}{b}\restriction\xi$ we have the following:
>
> for all $\langle \eta_1, ..., \eta_{n^*} \rangle$ an n^*-sequence from the first n^* measure one sets of r, $r^\frown\langle \eta_1, ..., \eta_{n^*} \rangle \Vdash \underset{\sim}{b}\restriction\xi = s$, and s is an initial segment of $t_\zeta \in X$, where the n^*-extension of q_{i^*} $q_{i^*}{}^\frown\langle \eta_1, ..., \eta_{n^*} \rangle \Vdash \underset{\sim}{b}\restriction\mu_{i^*} = t_\zeta\restriction\mu_{i^*}$. (*)

Let us choose (in M) for each $\xi \in \kappa^+$ one $r_\xi \geq^{*k} q_i$ such that all its n^*-direct extensions decide $\underset{\sim}{b}\restriction\xi$.

Then there are $\zeta < \kappa_{n^*}$, an unbounded $Z \subseteq \kappa^+, Z \in M$ and $\langle \eta_1^*, ..., \eta_{n^*}^* \rangle$ an n^*-sequence in M such that for every $\xi \in Z \cap M$, we have that $r_\xi \leq r_\xi{}^\frown\langle \eta_1^*, ..., \eta_{n^*}^* \rangle$ forces "$\underset{\sim}{b}\restriction\mu_{i^*} = t_\zeta\restriction\mu_{i^*}$". Set

$$e := \{s \mid \exists \xi \in Z, r_\xi{}^\frown\langle \eta_1^*, ..., \eta_{n^*}^* \rangle \Vdash s = \underset{\sim}{b}\restriction\xi\}.$$

Then e will be a κ^+-branch in T which is impossible. Hence (*) fails and therefore the following holds:

> For every $i, i^* < i < \delta$ there will be $\xi \in \kappa^+ \cap M$ and $r \geq^{*k} q_i, r \in M$ all of whose n^*-direct extensions decide $\underset{\sim}{b}\restriction\xi$, such that for some $\langle \eta_1, ..., \eta_{n^*} \rangle$ an n^*-sequence from the first n^* measure one sets of r (or equivalently from q_i or from p), let $r^\frown\langle \eta_1, ..., \eta_{n^*} \rangle \Vdash \underset{\sim}{b}\restriction\xi = s_{\langle \eta_1, ..., \eta_{n^*} \rangle}$. (**)

Then $s_{\langle \eta_1, ..., \eta_{n^*} \rangle}$ is not an initial segment of $t_\zeta \in X$, where the condition $q_{i^*}{}^\frown\langle \eta_1, ..., \eta_{n^*} \rangle$ forces "$\underset{\sim}{b}\restriction\mu_{i^*} = t_\zeta\restriction\mu_{i^*}$". By the choice of i^*, hence $s_{\langle \eta_1, ..., \eta_{n^*} \rangle}$ is not an initial segment of any other member of X as well.

We define a partition $F : A_{\leq n^*}(q) \to 2$ as follows: Set $F(\vec{\eta}) = 0$ iff there is $i(\vec{\eta}), i^* < i(\vec{\eta}) < \delta$, such for every $i, i(\vec{\eta}) \leq i < \delta$, there are $\xi \in \kappa^+ \cap M, \xi > \mu_i$ and $r \geq^{*k} q_i, r \in M$ all of whose n^*-direct extensions decide $\underset{\sim}{b}\restriction\xi$, $\vec{\eta} \in A_{\leq n^*}(r)$, such that

$$\text{if } r^\frown\vec{\eta} \Vdash \underset{\sim}{b}\restriction\xi = s_{\vec{\eta}}, \text{ then } s_{\vec{\eta}} \text{ is not an initial segment of } t_\zeta \in X,$$
$$\text{where } q_{i^*}{}^\frown\vec{\eta} \text{ forces "} \underset{\sim}{b}\restriction\mu_{i^*} = t_\zeta\restriction\mu_{i^*} \text{"}.$$

Set $X_0(p,\delta) = \{\vec{\eta} \in A_{\leq n^*}(p) \mid F(\vec{\eta}) = 0\}, X_1(p,\delta) = \{\vec{\eta} \in A_{\leq n^*}(p) \mid F(\vec{\eta}) = 1\}$. Once δ is fixed let us omit it.

Lemma 1.9. The set $X_0(p)$ is of measure one (relatively to the first n^* measures of p).

Proof. Otherwise $X_1(p)$ is of measure one. By the definition of F we have the following: for every $\vec{\eta} \in X_1(p)$, for every j with $i^* < j < \delta$, there is $i(j,\vec{\eta}), j \leq i(j,\vec{\eta}) < \delta$, for every $\xi \in \kappa^+ \cap M, \xi > \mu_{i(j,\vec{\eta})}$ and every $r \geq^{*k} q_{i(j,\vec{\eta})}, r \in M$ all of which n^*-direct extensions decide $\underset{\sim}{b}\restriction\xi$

$$\text{if } \vec{\eta} \in A_{\leq n^*}(r) \text{ and } r^\frown\vec{\eta} \Vdash \underset{\sim}{b}\restriction\xi = s_{\vec{\eta}}, \text{ then } s_{\vec{\eta}} \text{ is an initial segment}$$
$$\text{of } t_\zeta \in X, \text{ where } q_{i^*}{}^\frown\vec{\eta} \text{ forces "} \underset{\sim}{b}\restriction\mu_{i^*} = t_\zeta\restriction\mu_{i^*} \text{"}.$$

Let $\langle \vec{\eta}_\tau \mid \tau < \kappa_{n^*}\rangle$ be an enumeration of $X_1(p)$. Define by induction an increasing continuous sequence $\langle j_\tau \mid \tau < \kappa_{n^*}\rangle$: $j_0 = i^* + 1, j_1 = i(j_0, \vec{\eta}_0), \ldots, j_{\tau+1} = i(j_\tau, \vec{\eta}_\tau), \ldots$. Let $i^{**} = \bigcup_{\tau<\kappa_{n^*}} j_\tau$. Then for every $\vec{\eta} \in X_1(p)$, for every $\xi \in \kappa^+ \cap M$ and every $r \geq^{*k} q_{i^{**}}, r \in M$ all of which n^*-direct extensions decide $\underset{\sim}{b}\restriction\xi$

$$\text{if } \vec{\eta} \in A_{\leq n^*}(r) \text{ and } r^\frown\vec{\eta} \Vdash \underset{\sim}{b}\restriction\xi = s_{\vec{\eta}}, \text{ then } s_{\vec{\eta}} \text{ is an initial segment}$$
$$\text{of } t_\zeta \in X, \text{ where } q_{i^*}{}^\frown\vec{\eta} \text{ forces "} \underset{\sim}{b}\restriction\mu_{i^*} = t_\zeta\restriction\mu_{i^*} \text{"}.$$

Now the contradiction follows similar to (*) above.

Let us choose (in M, using elementarity) for each $\xi \in \kappa^+$ one $r_\xi \geq^{*k} q_{i^{**}}$ all of which n^*-direct extensions decide $\underset{\sim}{b}\restriction\xi$.

Then there are $\zeta < \kappa_{n^*}$, an unbounded $Z \subseteq \kappa^+, Z \in M$ and $\langle \eta_1^*, \ldots, \eta_{n^*}^*\rangle$ an n^*-sequence in $X_1(p)$ such that for every $\xi \in Z \cap M$ we have $\langle \eta_1^*, \ldots, \eta_{n^*}^*\rangle \in A_{\leq n^*}(r_\xi)$ and $r_\xi \leq r_\xi{}^\frown\langle \eta_1^*, \ldots, \eta_{n^*}^*\rangle$ forces "$\underset{\sim}{b}\restriction\mu_{i^*} = t_\zeta\restriction\mu_{i^*}$".

Set

$$e := \{s \mid \exists \xi \in Z, r_\xi{}^\frown\langle \eta_1^*, \ldots, \eta_{n^*}^*\rangle \Vdash s = \underset{\sim}{b}\restriction\xi\}.$$

Then e will be a κ^+-branch in T. Which is impossible. Q.E.D. (Lemma 1.9)

Set $X_0(p) = \bigcap_{m<\omega} X_0(p, \delta_m)$. Let us define $r \succeq^{n^*k} s$ if and only if $\tilde{r} \geq^{*k} s$, where $\tilde{r}$ is the condition obtained from r be intersecting its first n^* sets of measure one with those of s.

Lemma 1.10. For every $p \in \mathcal{P}$ there is $p^* \geq^{*k} p$ such that for every $s \succeq^{n^*k} p^*$ there is $r \succeq^{n^*k} s$ with $X_0(r) = X_0(p^*)$.

Proof. Suppose otherwise. Then

$$\exists p \forall p^* \geq^{*k} p \exists s(p^*) \succeq^{n^*k} p^* \forall r \succeq^{n^*k} s(p^*), X_0(r) \neq X_0(p^*).$$

Use completeness of $\langle \mathcal{P}, \succeq^{n^*k} \rangle$ to construct an $\succeq^{n^*k}$-increasing sequence $\langle p_\alpha \mid \alpha < (2^{\kappa_{n^*}})^+ \rangle$ such that $p_{\alpha+1} = s(\tilde{p}_\alpha)$, where $\tilde{p}_\alpha$ is obtained from p_α by intersecting its first n^* sets of measure one with those of p. In particular, $\tilde{p}_\alpha \geq^{*k} p$ and so $s(\tilde{p}_\alpha)$ is defined. The sequence of q_i's which corresponds to p_α is not affected by replacing p_α with $\tilde{p}_\alpha$. Hence $X_0(p_\alpha) = X_0(\tilde{p}_\alpha)$. Then there will be $\alpha < \beta < (2^{\kappa_{n^*}})^+$ with $X_0(p_{\alpha+1}) = X_0(p_{\beta+1})$. But this is impossible since $p_{\beta+1} \succeq^{n^*k} p_{\alpha+2} = s(\tilde{p}_{\alpha+1})$. Q.E.D. (Lemma 1.10)

Replace now the original p by p^*. Still denote it further by p. For each $\vec{\eta} \in A_{\leq n^*}(p) \cap X_0(p^*)$ and $i \geq i(\vec{\eta})$ pick $\xi(i, \vec{\eta})$ and $r(i, \vec{\eta})$ as in the definition of $F(\vec{\eta})$.

Build by induction sequences $\langle i_j \mid j < \delta \rangle$, $\langle \xi_j \mid j < \delta \rangle$, $\langle r_j(\vec{\eta}) \mid j < \delta \rangle$ such that

1. $r_j(\vec{\eta}) \geq^{*k} q_{i_j}$,
2. all n^*-direct extensions of $r_j(\vec{\eta})$ decide $\underset{\sim}{b} \restriction \xi_j$,
3. $i_j, \xi_j, r_j(\vec{\eta})$ satisfy the definition of $F(\vec{\eta})$,
4. $\mu_{i_j} > \xi_j$.

Finally we find a sequence $\langle p_j(\vec{\eta}) \mid j < \delta \rangle$, such that

1. $p_j(\vec{\eta}) \geq^{*k} r_j(\vec{\eta})$,
2. $X_0(p_j(\vec{\eta})) = X_0(p^*)$,
3. $\vec{\eta} \in A_{\leq n^*}(p_j(\vec{\eta}))$,
4. all of n^*-direct extensions of $p_j(\vec{\eta})$ decide $\underset{\sim}{b} \restriction \mu_\delta$.

This is possible by Lemma 1.10 (for item 3., use the maximality condition (7) of the definition of M_i's and q_i's). Now, using the above, we can find an increasing sequence $\langle \alpha_n \mid n < \omega \rangle$ of levels of T and $\langle p_u^{\vec{\eta}} \mid u \in \prod_{n<k} \kappa_n, k < \omega, \vec{\eta} \in A_{\leq n^*}(p) \cap X_0(p^*) \rangle$ such that

1. if u is an initial segment of v, then $p_u^{\vec{\eta}} \leq^* p_{u'}^{\vec{\eta}}$, for every $\vec{\eta} \in A_{\leq n^*}(p) \cap X_0(p^*)$,

2. all n^*-direct extensions of $p_u^{\vec{\eta}}$ decide $\underset{\sim}{b}\lceil\alpha_{|u|}$,

3. if u, v are incompatible then $p_u^{\vec{\eta}\frown}\vec{\eta}$ and $p_v^{\vec{\eta}\frown}\vec{\eta}$ provide incompatible decisions of $\underset{\sim}{b}\lceil\alpha_{|u|}$, $\underset{\sim}{b}\lceil\alpha_{|v|}$.

Fix some $\vec{\eta} \in A_{\leq n^*}(p) \cap X_0(p^*)$. Let $f \in \prod_{n<\omega} \kappa_n$. Let p_f be a $\leq^*$-upper bound of $\langle p_{f\lceil n}^{\vec{\eta}\frown}\vec{\eta} \mid n < \omega\rangle$. Then p_f will decide $\underset{\sim}{b}\lceil\alpha_\omega + 1$, where $\alpha_\omega := \bigcup_{n<\omega} \alpha_n$. Different f's will give different decisions, but the number of f's is κ^+. So the level α_ω of T will have cardinality κ^+, which is impossible.

Q.E.D. (Theorem 1.7)

2 A model with no Aronszajn tree over κ^{++}, for singular κ.

Our aim in this section will be to prove the following theorem:

Theorem 2.1. Assume GCH. Let $\kappa = \bigcup_{n<\omega} \kappa_n$ with $o(\kappa_n) = \kappa_n^{+n+2}$ and $\lambda > \kappa$ is a weakly compact. Then there is a generic extension $V^{\mathcal{P}}$ such that

1. GCH holds below κ,

2. κ^+ is preserved,

3. $\lambda = \kappa^{++}$,

4. $2^\kappa = \lambda$,

5. there are no κ^{++}-Aronszajn trees.

Proof. Suppose $\kappa = \bigcup_{n<\omega} \kappa_n$, $\langle \kappa_n \mid n < \omega\rangle$ increasing, $\lambda > \kappa$ and each of κ_n's carries an extender E_n (like in the short extender based forcings).

We refer to [2, §§1&2] for definitions and basic properties of short extender forcing, a more detailed account may be found in [6]. Let us give here only a brief description. Conditions in this forcings are of the form $p = \langle p_n \mid n < \omega\rangle$ such that there is $\ell(p) < \omega$ with p_n being a Cohen condition in $\mathrm{Cohen}(\lambda, \kappa_n) = \{f \mid |f| \leq \kappa, f : \lambda \to \kappa_n\}$, for each $n < \ell(p)$. If $n \geq \ell(p)$, then $p_n = \langle a_n, f_n, A_n\rangle$, where a_n is a partial order preserving function from λ to κ_n (or just a subset of λ in long extender forcing) of cardinality $< \kappa_n$, A_n is a set of measure one for the measure of E_n which corresponds to the maximal element of $\mathrm{rng}(a_n)$ and f_n is in $\mathrm{Cohen}(\lambda, \kappa_n)$.

We would like to make a small change here and allow Cohen parts of a condition to be names which depend on Příkrý sequences added before.

Namely, if $\alpha \in \mathrm{dom}(a_n)$, for each $n > n_0$, then Cohen functions on $\lambda \setminus \alpha+1$ may depend on the Příkrý sequence of α. Still we require that $\langle \mathrm{dom}(f_n) \mid n < \omega\rangle \in V$ as well as $\langle a_n \mid n \geq \ell(p)\rangle$. Also for each $\alpha < \lambda$, $\langle f_n(\alpha) \mid n < \omega\rangle$ should be in V, where α is a common point of domains of functions f_n (it is possible to present this forcing in a way that there is no such α's at all).

Let us denote this forcing by $\langle \mathcal{P}, \leq, \leq^*\rangle$. We define the following refinement $\leq^{*k}$ of $\leq^*$, for $k < \omega$: set $p \leq^{*k} q$ iff $p \leq^* q$ and $\mathrm{dom}(a_i(p)) = \mathrm{dom}(a_i(q))$, for every i with $\ell(p) \leq i \leq k$.

The next lemma is similar to the standard Příkrý condition lemma for these types of forcings:

Lemma 2.2. Let $p \in \mathcal{P}, k < \omega$ and suppose that σ is a statement of the forcing language. Then there is $q \geq^{*k} p$ which decides σ.

Suppose now $\nu < \lambda$ is an ordinal of cofinality $\geq \kappa^{++}$. Let $\mathcal{P}\lceil\nu$ be the natural restriction of $\mathcal{P}$ to ν. Note that we need first to restrict ourself to a dense subset of $\mathcal{P}$ which consists of conditions p such that $a_n(p) \cap \nu$ has a maximal element which is also a maximal in the extender (E_n) order, and then restrict everything to ν.

Let $\langle \underset{\sim}{T}, \leq_{\underset{\sim}{T}}\rangle$ be a $\mathcal{P}\lceil\nu$-name of a $(\kappa^{++})^{V^{\mathcal{P}\lceil\nu}}$-Aronszajn tree. Note that $(\kappa^{++})^{V^{\mathcal{P}\lceil\nu}} = \nu$. Suppose that the rest of the forcing, i.e., $\mathcal{P}/\mathcal{P}\lceil\nu$, adds a κ^{++}-branch with $\mathcal{P}$-name $\underset{\sim}{b}$. Assume without loss of generality that the αth level of T is just $[\kappa^+ \cdot \alpha, \kappa^+ \cdot \alpha + \kappa^+)$.

Lemma 2.3. Let $\alpha < \nu, p \in \mathcal{P}$. Then there are $q_\alpha \geq^* p$ and $n_\alpha < \omega$ such that every n_α-extension $q_\alpha{}^\frown\langle \eta_1, \dots, \eta_{n_\alpha}\rangle$ of q_α forces "$\underset{\sim}{b}\lceil\alpha = \underset{\sim}{t}_{\alpha,\langle \eta_1,\dots,\eta_{n_\alpha}\rangle}$," for some $\mathcal{P}\lceil\nu$-name $\underset{\sim}{t}_{\alpha,\langle \eta_1,\dots,\eta_{n_\alpha}\rangle}$.

Similarly, the following slight strengthening of the previous lemma holds:

Lemma 2.4. Let $\alpha < \nu, p \in \mathcal{P}, k < \omega$. Then there are $q_{\alpha k} \geq^{*k} p$ and $n_{\alpha k} < \omega$ such that every $n_{\alpha k}$-extension $q_{\alpha k}{}^\frown\langle \eta_1, \dots, \eta_{n_{\alpha k}}\rangle$ of $q_{\alpha k}$ forces "$\underset{\sim}{b}\lceil\alpha = \underset{\sim}{t}_{\alpha k,\langle \eta_1,\dots,\eta_{n_{\alpha k}}\rangle}$," for some $\mathcal{P}\lceil\nu$-name $\underset{\sim}{t}_{\alpha k,\langle \eta_1,\dots,\eta_{n_{\alpha k}}\rangle}$.

Let us denote the least n_α for which there is q_α as in Lemma 2.3 by $n(p,\alpha)$ and the least $n_{\alpha k}$ for which there is $q_{\alpha k}$ as in Lemma 2.4 by $n(p,\alpha,k)$.

Lemma 2.5. For each $p \in \mathcal{P}$ there are $p^* \geq^* p$ and $n^* < \omega$ such that for every $q \geq^* p^*$ and for every large enough $\alpha < \kappa^{++}$ we have $n(q,\alpha) = n^*$.

Proof. Suppose otherwise. Define by induction a $\geq^*$-increasing sequence $\langle p_k \mid k < \omega\rangle$ of direct extensions of p and an increasing sequence $\langle \alpha_k \mid k < \omega\rangle$ of ordinals below κ^{++} such that $n(p_k, \alpha_k) < n(p_{k+1}, \alpha_{k+1})$, for every $k < \omega$.

Find $p_\omega \in \mathcal{P}$ which is $\leq^*$-stronger than every p_k. Extend it to a condition q that decides $\underset{\sim}{b}\restriction\alpha_\omega$, where $\alpha_\omega = \bigcup_{k<\omega} \alpha_k$. Let $m = \ell(q)$. Denote the decided value by $\underset{\sim}{t}$. Pick k with $n(p_k, \alpha_k) > m$. Then this contradicts the definition of $n(p_k, \alpha_k)$ being the least possible. Q.E.D. (Theorem 2.1)

Suppose that $p \in \mathcal{P}$ and p^*, n^* are given by the lemma.

Lemma 2.6. For every natural number k, we have $n^* = n(p^*, k, \alpha)$

Proof. Clearly $n^* \leq n(p^*, k, \alpha)$. Let us show the equality. We first run the standard argument trying to decide $\underset{\sim}{b}\restriction\alpha$ starting with p^* for each of its k-extensions, i.e., for any choice of $\langle \eta_1, ..., \eta_k \rangle$ in sets of measure one of the first k-levels of p^*. Then shrink this first k sets in order to have the same conclusion (whether n-extensions decide $\underset{\sim}{b}\restriction\alpha$ or not). Finally we shall have $r \geq^{*k} p^*$ and $n < \omega$ such that

1. every n-extension of r decides $\underset{\sim}{b}\restriction\alpha$,
2. for every $\langle \eta_1, ..., \eta_k \rangle, \langle \eta'_1, ..., \eta'_k \rangle$ two sequences from the first k sets of measure one of r and any $m < \omega$, if there is an m-extension of some $t \geq^* r^\frown\langle \eta_1, ..., \eta_k \rangle$ which decides $\underset{\sim}{b}\restriction\alpha$, then any m-extension of $r^\frown\langle \eta'_1, ..., \eta'_k \rangle$ decides $\underset{\sim}{b}\restriction\alpha$.

Suppose for a moment that $n > n^*$. Pick some $s \geq^* r$ such that every n^*-extension of s decides $\underset{\sim}{b}\restriction\alpha$. Clearly, s need not be a $*k$-extension, i.e., maximal coordinates of levels $\leq k$ may increase. Let $\langle \varrho_1, ..., \varrho_k \rangle$ be a sequence from the first k sets of measure one of s. Denote its projection to the first k maximal coordinates of r by $\langle \eta_1, ..., \eta_k \rangle$. Then $s^\frown\langle \varrho_1, ..., \varrho_k \rangle \geq^* r^\frown\langle \eta_1, ...\eta_k \rangle$ and an $(n^* - k)$-extension of $s^\frown\langle \varrho_1, ..., \varrho_k \rangle$ decides $\underset{\sim}{b}\restriction\alpha$. So already an $n^* - k$-extension of $r^\frown\langle \eta_1, ..., \eta_k \rangle$ decides $\underset{\sim}{b}\restriction\alpha$ and then the same holds if we replace $\langle \eta_1, ..., \eta_k \rangle$ by any other sequence from the first k-sets of measures one of r. But this means that any n^*-extension of r already decides $\underset{\sim}{b}\restriction\alpha$. Q.E.D. (Lemma 2.6)

Suppose for simplicity that p^* is just $0_\mathcal{P}$.

Lemma 2.7. Let $p \in \mathcal{P}$ and let δ be a regular cardinal above $2^{\kappa_{n^*}}$ and $2^{\kappa_k} < \delta < \kappa_{k+1}$ for some $k < \omega$. Then there are $\alpha < \nu$, $q \geq^{*k} p\restriction\nu$ and a sequence of $(*, k)$-direct extensions $\langle p_\xi \mid \xi < \delta \rangle$ of p with the same sequence of sets of measures one up to the level k such that

1. $p_\xi\restriction\nu = q$,

2. every n^*-extension of p_ξ decides $\underset{\sim}{b}\restriction\alpha+1$,

3. if $\xi \neq \xi'$, $p_\xi{}^\frown\langle\eta_1, ..., \eta_{n^*}\rangle$ an n^*-extension of p_ξ and $p_{\xi'}{}^\frown\langle\eta'_1, ..., \eta'_{n^*}\rangle$ an n^*-extension of $p_{\xi'}$ then q forces (in $\mathcal{P}\restriction\nu$) that decisions made by $p_\xi{}^\frown\langle\eta_1, ..., \eta_{n^*}\rangle$ and $p_{\xi'}{}^\frown\langle\eta'_1, ..., \eta'_{n^*}\rangle$ are different (incompatible).

Proof. Suppose otherwise. Let p and δ be a counterexample. Assume δ is a regular cardinal above κ_{n^*} and $2^{\kappa_k} < \delta < \kappa_{k+1}$ for some $k < \omega$.

Define by induction a continuous $\in$-chain of elementary submodels $\langle M_i \mid i \leq \delta\rangle$ and a sequence of direct extensions $\langle q_i \mid i \leq \delta\rangle$ of p so that

1. $q_i \geq^{*k} p$,

2. if $i \leq j$, then $\tilde{q}_j \geq^{*k} q_i$, where $\tilde{q}_j$ is obtained from q_j by intersecting its first k sets of measure one with those of q_i,

3. $|M_i| < \delta$, for every $i < \delta$,

4. ${}^{\kappa_k}M_i \subseteq M_i$, for $i = 0$ or i a successor ordinal,

5. each n^*-extension of q_i decides $\underset{\sim}{b}\restriction\mu_i + 1$, i.e., for some $\mathcal{P}\restriction\nu$-name $\underset{\sim}{t}$ of $\underset{\sim}{T}\restriction\mu_{i+1}+1$ and $\gamma < \kappa^+$ we have $q_i \Vdash \underset{\sim}{b}\restriction\mu_i+1 = \underset{\sim}{t}$ and $\underset{\sim}{b}(\mu_i) = \kappa^+\cdot\mu_i+\gamma$, where $\mu_i = \sup(M_i \cap \nu)$, and

6. $q_i \in M_{i+1}$.

Denote the final M_δ by M, μ_δ by μ and q_δ by q. Note that $\delta > 2^{\kappa_k}$ so sets of measures one of the first k-levels of δ many of q_i's are the same. Assume then that for all i's they are the same in order to insure $q \geq^{*k} q_i$ for every i. Then each n^*-extension of q decides $\underset{\sim}{b}\restriction\mu$. Denote by $X = \{t_\xi \mid \xi < \kappa_{n^*}\}$ the set of all such decisions. Note that $|X| \leq \kappa_{n^*} < \delta$. Shrink the sets of measure one of the maximal coordinates of each $q_i, i < \delta$ for every level $\leq n^*$ to the projection of the corresponding set of measure one of q. Denote still the resulting conditions by q_i. We have M_{i+1} is closed under κ_{n^*}-sequences, so such new q_i will be in M_{i+1}.

Then the set of decisions of n^*-extensions of q_i will be $X_i = \{t_\xi\restriction\mu_i \mid \xi < \kappa_{n^*}\}$. We have that X and each particular t_ξ need not be in M, but X_i's and the initial segments of t_ξ are in M. Note that for each $i \leq \delta$ any n^*-extension $q_i{}^\frown\langle\eta_1, ..., \eta_{n^*}\rangle$ of q_i gives the values $\langle\underset{\sim}{b}(\mu_j) \mid j < i\rangle$, by item 5. above. Denote the sequence of the values by $\langle\gamma_{j\langle\eta_1,...,\eta_{n^*}\rangle} \mid j < i\rangle$. Then $j' < j$ implies

$$q_i\restriction\nu \Vdash \gamma_{j'\langle\eta_1,...,\eta_{n^*}\rangle} \leq_{\underset{\sim}{T}} \gamma_{j\langle\eta_1,...,\eta_{n^*}\rangle}.$$

Recall that $\delta > 2^{\kappa_{n^*}}$. Hence there will be $i^* < \delta$ such that all the branches in X already split before the level μ_{i^*}. Actually for every $j \leq \delta$ of cofinality

$\geq \kappa_{n^*}$ the same is true, i.e., all the branches of X_j split already at some $j^* < j$.

Assume that we are at a stage $i+1$ of the construction. So $q_i \in M_{i+1}$ and we have q_{i+1} and the list $X_{i+1} = \{\underset{\sim}{t}_{i+1\xi} \mid \xi < \kappa_{n^*}\}$.

Suppose for a moment that the following holds:

$$\text{there are } i, i^* \leq i < \delta \text{ and } r \in \mathcal{P}\lceil\nu, r \geq q_i\lceil\nu, r \in M_{i+1} \text{ such that for all } \xi \in \nu \cap M_{i+1}, \text{ for all } \tilde{r} \text{ if } \tilde{r} \geq q_i, \tilde{r}\lceil\nu \geq r \text{ and } \tilde{r} \text{ decides } \underset{\sim}{b}\lceil\xi \text{ then } \tilde{r} \Vdash \underset{\sim}{b}\lceil\xi \text{ is an initial segment of one of } \underset{\sim}{t}_{i+1\zeta}, \text{ for some } \zeta < \kappa_{n^*}. \tag{$\dagger$}$$

By extending r if necessary we may assume that $\ell(r) \geq n^*$.

Let us choose (in M) for each $\xi \in \kappa^+$ one r_ξ that witnesses (†) for ξ.

Then there are an unbounded $Z \subseteq \nu, Z \in M$ and $\langle \eta_1^*, ..., \eta_{n^*}^* \rangle$ an n^*-sequence in M such that for every $\xi \in Z \cap M$, the sequence $\langle \eta_1^*, ..., \eta_{n^*}^* \rangle$ is the projection to the first n^*-maximal coordinates of q_i, i.e., $q_i{}^\frown\langle \eta_1^*, ..., \eta_{n^*}^* \rangle \leq r_\xi$. Then $r_\xi \Vdash \underset{\sim}{b}\lceil\xi$ is an initial segment of $\underset{\sim}{t}_{i+1\zeta}$, where ζ corresponds to the μ_{i+1}-branch determined by $q_i{}^\frown\langle \eta_1^*, ..., \eta_{n^*}^* \rangle$ (remember that we are above i^* so it does not split further in q_{i+2}, q_{i+3}, etc.). Set

$$e := \{\underset{\sim}{s} \mid \underset{\sim}{s}\text{is a } \mathcal{P}\lceil\nu \text{ name and } \exists\xi \in Z, r_\xi \Vdash \underset{\sim}{s} = \underset{\sim}{b}\lceil\xi\}.$$

Then e will be forced by r to be a κ^{++}-branch in $\underset{\sim}{T}$. Which is impossible. Hence (†) fails.

So,

$$\text{for every } i, i^* \leq i < \delta \text{ and } r \in \mathcal{P}\lceil\nu, r \geq q_i\lceil\nu, r \in M_{i+1} \text{ there will be } \xi \in \nu \cap M_{i+1} \text{ and } \tilde{r} \text{ such that } \tilde{r} \geq q_i, \tilde{r}\lceil\nu \geq r, \tilde{r} \text{ decides } \underset{\sim}{b}\lceil\xi \text{ and } \tilde{r} \Vdash \underset{\sim}{b}\lceil\xi \text{ is not an initial segment of none of } \underset{\sim}{t}_{i+1\zeta}, \text{ for } \zeta < \kappa_{n^*}. \tag{$\ddagger$}$$

Start with $i = i^*$. Our next task will be to extend $q_i\lceil\nu$ by finding a $(*k)$-direct extension $q_i^* \in \mathcal{P}\lceil\nu$ such that

$$\text{for some } n_i, n^* \leq n_i < \omega \text{ for every } n_i\text{-extension } q_i^{*\frown}\langle \eta_1, ..., \eta_{n_i} \rangle \text{ of } q_i^* \text{ there is some } r_{i\langle \eta_1, ..., \eta_{n_i} \rangle} \in \mathcal{P}\lceil\lambda \setminus \nu \text{ such that } q_i^{*\frown}\langle \eta_1, ..., \eta_{n_i} \rangle^\frown r_{i\langle \eta_1, ..., \eta_{n_i} \rangle} \in \mathcal{P}\lceil\nu, r_{i\langle \eta_1, ..., \eta_{n_i} \rangle} \text{ decides } \underset{\sim}{b}\lceil\xi \text{ and } q_i^{*\frown}\langle \eta_1, ..., \eta_{n_i} \rangle^\frown r_{i\langle \eta_1, ..., \eta_{n_i} \rangle} \Vdash \underset{\sim}{b}\lceil\xi \text{ is not an initial segment of none of } \underset{\sim}{t}_{i+1\zeta}, \text{ for } \zeta < \kappa_{n^*}. \tag{$\ddagger_i$}$$

Let us combine such $r_{i\langle \eta_1, ..., \eta_{n_i} \rangle}$'s into a $\mathcal{P}\lceil\nu$ name $\underset{\sim}{r_i}$. Note that

$$q_i^{*\frown}\langle \eta_1, ..., \eta_{n_i} \rangle^\frown r_{i\langle \eta_1, ..., \eta_{n_i} \rangle} \Vdash \underset{\sim}{b}\lceil\xi$$

is an end extension of $\underset{\sim}{t}_{i\zeta}$ which corresponds to $\langle\eta_1, ..., \eta_{n^*}\rangle$, since the last condition extends $q_i{}^\frown\langle\eta_1, ..., \eta_{n^*}\rangle$.

We run a standard Příkrý type argument running over all the possibilities over the levels, shrinking each of them in order to have same decisions.

It is easy to combine q_i^* and the part of q_i above ν into one condition. Just increase maximal coordinates of the part of q_i above ν and shrink the resulting measure one sets. Denote such combination by q_i^{**}. Also this definition can be carried out inside M_{i+1}.

Note that q_i^{**} need not be compatible with q_{i+1}. So we just replace q_i by q_i^{**} and define the new q_j's, $(i < j \leq \delta)$ and the new i^* (however, we use the same notation).

At the next stage, we go to the next i above the previous one and $\geq$ the new i^* and define its q_i^{**} and so on. At limit stages there may be not enough completeness in order to intersect sets of measure one at certain levels. In this case we put this sets to be as large as possible (i.e., over the level n just κ_n) and then continue the process.

Denote the final q_δ^{**} by q.

Finally let p_i be a $(*k)$-extension of $q{}^\frown \underset{\sim}{r}_i$ which decides $\underset{\sim}{b}\restriction\mu_\delta$.

The choice of $\underset{\sim}{r}_i$'s provides the desired conclusion. Q.E.D. (Lemma 2.7)

Fix now a cofinal in κ sequence $\langle\delta_n \mid n < \omega\rangle$ of measurable cardinals with $2^{\kappa_n} < \delta_n < \kappa_{n+1}$. Let us use Lemma 2.7 to find an increasing sequence $\langle\alpha_n \mid n < \omega\rangle$ of levels of $\underset{\sim}{T}$, $\langle Y_n \mid n < \omega\rangle$ and sequences of conditions $\langle q_n \mid n < \omega\rangle$, $\langle p_u \mid u \in \prod_{n<k} Y_n, k < \omega\rangle$ such that

1. $Y_n \subseteq \delta_n$ of cardinality δ_n,
2. $q_n \in \mathcal{P}\restriction\nu$,
3. $q_n \leq^{*n} q_{n+1}$,
4. $p_u\restriction\nu = q_n$, for each u with $|u| = n$,
5. if u is an initial segment of v, then $p_u \leq^* p_v$,
6. all n^*-direct extensions of p_u decide $\underset{\sim}{b}\restriction\alpha_{|u|}$,
7. if u, v are incompatible then any n^*-extensions of p_u and of p_v provide incompatible decisions of $\underset{\sim}{b}\restriction\alpha_{|u|}$, $\underset{\sim}{b}\restriction\alpha_{|v|}$.

Apply Lemma 2.7 to δ_0. This will produce α_0, q_0 and $\langle p_{\langle\xi\rangle} \mid \xi < \delta_0\rangle$. Now let $\xi < \delta_0$. Apply Lemma 2.7 to δ_1 and $p_{\langle\xi\rangle}$. We shall obtain $\alpha_{\langle\xi\rangle}$, $q'_{\langle\xi\rangle} \geq^{*1} q_0$ and $\langle p'_{\langle\xi\zeta\rangle} \mid \zeta < \delta_1\rangle$.

Set $\alpha_1 = \max(\alpha_0 + 1, \bigcup_{\xi<\delta_0} \alpha_{\langle\xi\rangle})$. We stretch each of $p'_{\langle\xi\zeta\rangle}$ to its $(*1)$-extension $p''_{\langle\xi\zeta\rangle}$ with all n^*-extensions deciding $\underset{\sim}{b}\restriction\alpha_1$. Do this by induction on ξ keeping the parts in $\mathcal{P}\restriction\nu$ increasing, except possibly for the first 2 sets of measures one (we do not have enough completeness in order to intersect this sets). At the final stage let us use the measurability of δ_1. Let U_1 be a normal measure on it. Pick a set $Z_\xi \in U_1$ on which all first 2 sets of measures one are the same. Use this constant value in the final condition and restrict ourself only to conditions with indexes in Z_ξ. Denote the result by $\langle p'''_{\langle\xi\zeta\rangle} \mid \zeta \in Z_\xi\rangle$ and let $q'''_{\langle\xi\rangle}$ be the restriction of this conditions to $\mathcal{P}\restriction\nu$.

We perform the process described above by induction on $\xi < \delta_0$ and require that the sequence $\langle q'''_{\langle\xi\rangle} \mid \xi < \delta_0\rangle$ be $(*1)$-increasing, again except possibly for the first set of measures one. Set $Z'_1 = \bigcap_{\xi<\delta_0} Z_\xi$. Then $Z_1 \in U_1$. Combine $\langle q'''_{\langle\xi\rangle} \mid \xi < \delta_0\rangle$ into one condition. We need to stabilize the first set of measure one. Pick a normal ultrafilter U_0 on δ_0. There is $Z_0 \in U_0$ on which we shall have this sets the same. The result taking this common set of the measure one on the first level will be our q_1. Shrink to Z_0 and deal further only with $\langle p_{\langle\xi\rangle} \mid \xi \in Z_0\rangle$. Then let $p_{\xi\zeta}$ be the final condition after stabilizing the set of measure over the first level. So we constructed $\langle p_{\langle\xi\zeta\rangle} \mid \xi \in Z_0, \zeta \in Z_1\rangle$.

Continue further in same fashion. At the final stage we shall take Y_n to be intersections of ω-many sets Z corresponding to u's of the length n. Such Y_n will be still in U_n and in particular will have cardinality δ_n.

Let $f \in \prod_{n<\omega} Y_n$. Let p_f be a $\leq^*$-upper bound of $\langle p_{f\restriction n} \mid n < \omega\rangle$. Let us pick an n^*-sequence $\langle \eta_1^f, ..., \eta_{n^*}^f\rangle$ in the first n^* sets of measure one of p_f. Then $p_f{}^\frown\langle \eta_1^f, ..., \eta_{n^*}^f\rangle$ will decide $\underset{\sim}{b}\restriction\alpha_\omega + 1$, where $\alpha_\omega := \bigcup_{n<\omega} \alpha_n$. Different f's will give different decisions. But the number of f's in $V^{\mathcal{P}\restriction\nu}$ is $\nu = (\kappa^{++})^{V^{\mathcal{P}\restriction\nu}}$. So the level α_ω of T will have cardinality κ^{++}, which is impossible since we assumed that T is a κ^{++}-Aronszajn tree in $V^{\mathcal{P}\restriction\nu}$. Contradiction. Q.E.D. (Theorem 2.1)

3 Down to $\aleph_{\omega+2}$.

We shall move κ of the previous section to $\aleph_\omega$. The process will be rather standard and so we will concentrate only on few new points. In general the idea in this sort of constructions is to use collapses together with main things that are done and not afterwards.

We work here entirely with short extender forcing. Long extender forcing does not allow us to move down to $\aleph_\omega$.

In our setting there will be two sorts of collapses: $\mathrm{Col}(\varrho_n^{+n+4}, < \kappa_n)$ and $\mathrm{Col}(\kappa_n^{+n+8}, < \varrho_{n+1})$, for each $n < \omega$, where ϱ_n denotes a generic one element Příkrý sequence for the normal measure of the extender E_n. The first collapse will be guided by a function F_n defined on the projection of the maximal coordinate of a level n of a condition to one corresponding to the normal measure of E_n, i.e., κ_n. For each $\nu \in \mathrm{dom}(F_n)$, we require that $F_n(\nu) \in \mathrm{Col}(\nu^{+n+4}, <\kappa_n)$.

Let now $\nu < \lambda$ be as in the previous section. The collapses will not involve names and so they will be in $\mathcal{P}\lceil\nu$.

The analogue of Lemma 2.7 will deal with δ such that $2^{\kappa_k} = \kappa_k^+ < \delta < \kappa_k^{+k+8}$. Note that the collapse over κ_k starts only further up from κ_k^{+8}, which provides enough completeness for running the argument.

Let us give the definition of the forcing used here.

Definition 3.1. $\mathcal{P}$ consists of sequences $p = \langle p_n \mid n < \ell(p)\rangle^\frown\langle p_n \mid \ell(p) \leq n < \omega\rangle$ such that

1. $\ell(p) < \omega$,
2. for every $n < \ell(p), p_n$ is of the form $\langle \varrho_n, h_{<n}, h_{>n}, f_n\rangle$ where
   (a) ϱ_n is the one element Příkrý sequence for the normal measure of E_n (i.e., an indiscernible for it),
   (b) $h_{<n} \in \mathrm{Col}(\varrho_n^{+n+4}, <\kappa_n)$,
   (c) $h_{>n} \in \mathrm{Col}(\kappa_n^{+n+8}, <\varrho_{n+1})$, if $n+1 < \ell(p)$, and $h_{>n} \in \mathrm{Col}(\kappa_n^{+n+8}, <\kappa_{n+1})$, if $n+1 = \ell(p)$,
   (d) f_n is a partial function of cardinality at most κ from λ to κ_n.
3. For every $n \geq \ell(p)$, p_n is of the form $\langle a_n, A_n, S_n, h_{>n}, f_n\rangle$ where
   (a) a_n, A_n and f_n are as in § 2,
   (b) S_n is a function with domain the projection of A_n to the normal measure of E_n such that for each $\nu \in \mathrm{dom}(S_n)$ we have $S_n(\nu) \in \mathrm{Col}(\nu^{+n+4}, <\kappa_n)$,
   (c) $\min(\mathrm{dom}(S_n)) > \sup(\mathrm{rng}(h_{>n-1}))$,
   (d) $h_{>n} \in \mathrm{Col}(\kappa_n^{+n+8}, <\kappa_{n+1})$.

At the final step of the argument of Theorem 2.1, we cannot use measurable δ_n's which differ from κ_n's. The use of κ_n's seems problematic, since the forcing $\mathrm{Col}(\kappa_n^{+n+8}, <\kappa_{n+1})$ of cardinality κ_{n+1} is involved and its degree of completeness is only κ_n^{+n+8}.

Let us overcome the difficulty by taking in advance (before the forcing with short extenders) δ_n's to be successor cardinals but carrying precipitous ideals I_n for which the forcing with positive sets, i.e., $\mathcal{P}(\delta_n)/I_n$, being δ_n^--strategically closed.

Let us construct such I_n's in advance by collapsing measurables. Namely, we fix, for every $n < \omega$, a measurable cardinal δ_{n+1} with $\kappa_n < \delta_{n+1} < \kappa_{n+1}$

and a normal ultrafilter W_{n+1} over it. Then force with the full support iteration of $\mathrm{Col}(\kappa_n^{+n+5}, < \delta_{n+1})$. The filter W_{n+1} will be as desired in this generic extension, i.e., the forcing with its positive sets will be isomorphic to $\mathrm{Col}(\kappa_n^{+n+5}, < i_{W_{n+1}}(\delta_{n+1}))$ which is κ_n^{+n+5}-closed and κ_n^{+n+5} is the immediate predecessor of δ_{n+1} in this generic extension, where $i_{W_{n+1}}$ is the elementary embedding corresponding to W_{n+1}.

References

[1] S.-D. Friedman and A. Halilović. The tree property at $\aleph_{\omega+2}$. *Journal of Symbolic Logic*, 76(2):477–490, 2011.

[2] M. Gitik. Blowing up power of a singular cardinal—wider gaps. *Annals of Pure and Applied Logic*, 116(1-3):1–38, 2002.

[3] M. Gitik. Prikry-type forcings. In M. Foreman and A. Kanamori, editors, *Handbook of set theory. Vol. 2*, pages 1351–1447. Springer-Verlag, 2010.

[4] M. Gitik, V. Kanovei, and P. Koepke. Intermediate models of Prikry generic extensions, 2010. Preprint.

[5] M. Gitik and M. Magidor. The singular cardinal hypothesis revisited. In H. Judah, W. Just, and H. Woodin, editors, *Set theory of the continuum, Papers from the workshop held in Berkeley, California, October 16–20, 1989*, volume 26 of *Mathematical Sciences Research Institute Publications*, pages 243–280. Springer-Verlag, 1992.

[6] M. Gitik and S. Unger. Adding many ω-sequences to a singular cardinal. In J. Cummings and E. Schimmerling, editors, *Appalachian Set Theory 2006–2012*, volume 406 of *London Mathematical Society Lecture Notes Series*, pages 245–264. Cambridge University Press, 2012.

[7] J. D. Hamkins. Extensions with the approximation and cover properties have no new large cardinals. *Fundamenta Mathematicae*, 180(3):257–277, 2003.

On ground model definability

Victoria Gitman and Thomas A. Johnstone

Department of Mathematics, New York City College of Technology, New York NY, United States of America

Abstract

Laver, and Woodin independently, showed that models of ZFC are uniformly definable in their set-forcing extensions, using a ground model parameter [9, 16]. We investigate ground model definability for models of fragments of ZFC, particularly of $\mathsf{ZF} + \mathsf{DC}_\delta$ and of ZFC^-, and we obtain both positive and negative results. Generalizing the results of [9], we show that models of $\mathsf{ZF}+\mathsf{DC}_\delta$ are uniformly definable in their set-forcing extensions by posets admitting a gap at δ, using a ground model parameter. In particular, this means that models of $\mathsf{ZF}+\mathsf{DC}_\delta$ are uniformly definable in their forcing extensions by posets of size less than δ. We also show that it is consistent for ground model definability to fail for canonical ZFC^- models $\mathbf{H}_{\kappa^+}$. Using forcing, we produce a ZFC universe in which there is a cardinal $\kappa >> \omega$ such that $\mathbf{H}_{\kappa^+}$ is not definable in its Cohen forcing extension. As a corollary, we show that there is always a countable transitive model of ZFC^- violating ground model definability. These results turn out to have a bearing on ground model definability for models of ZFC. It follows from our proof methods that the hereditary size of the parameter that Woodin used in [16] to define a ZFC model in its set-forcing extension is best possible.

1 Introduction

It took four decades since the invention of forcing for set theorists to ask (and answer) what post factum seems as one of the most natural questions regarding forcing. Is the ground model a definable class of its set-forcing extensions? Laver published the positive answer in a paper mainly concerned with whether rank-into-rank cardinals can be created by small forcing [9]. Woodin obtained the same result independently, and it appeared in the appendix of [16].

Theorem 1.1 (Laver, Woodin)**.** Suppose V is a model of ZFC, $\mathbb{P} \in V$ is a forcing notion, and $G \subseteq \mathbb{P}$ is V-generic. Then in $V[G]$, the ground model V is definable from the parameter $\wp(\gamma)^V$, where $\gamma = |\mathbb{P}|^V$.

Indeed, it follows from the proof of Theorem 1.1 that this definition of the ground model is uniform across all its set-forcing extensions. There is a first-order formula which, using the ground model parameter $\wp(\gamma)^V$ where

Stefan Geschke, Benedikt Löwe, Philipp Schlicht (*eds.*).
Infinity, computability, and metamathematics: Festschrift celebrating the 60th birthdays of Peter Koepke and Philip Welch. College Publications, London, 2014. Tributes, Volume 23.

$\gamma = |\mathbb{P}|^V$,[1] defines the ground model in any set-forcing extension[2]. Before Theorem 1.1, properties of the forcing extension in relation to the ground model could be expressed in the forcing language using the predicate $\check{V}$ for the ground model sets. But having a uniform definition of ground models in their set-forcing extensions was an immensely more powerful result that opened up rich new avenues of research. Hamkins and Reitz used it to introduce the *Ground Axiom*, a first-order assertion that a universe is not a nontrivial set-forcing extension [12]. Research on the Ground Axiom in turn grew into the set-theoretic geology project that reverses the forcing construction by studying what remains from a model of set theory once the layers created by forcing are removed [1]. Woodin made use of Theorem 1.1 in studying generic multiverses—collections of set-theoretic universes that are generated from a given universe by closing under generic extensions and ground models [16]. In addition, Theorem 1.1 proved crucial to Woodin's pioneering work on suitable extender models, a potential approach to constructing the canonical inner model for a supercompact cardinal [15].

In this article we investigate ground model definability for models of fragments of ZFC, particularly of $\mathsf{ZF} + \mathsf{DC}_\delta$ and of ZFC^-, and we obtain both positive and negative results.

Laver's proof [9] that ground models of ZFC are definable in their set-forcing extensions uses Hamkins's techniques and results on pairs of models with the δ-cover and δ-approximation properties.

Definition 1.2 (Hamkins [6]). Suppose $V \subseteq W$ are transitive models of (some fragment of) ZFC and δ is a cardinal in W.

(1) The pair $V \subseteq W$ satisfies the *δ-cover property* if for every $A \in W$ with $A \subseteq V$ and $|A|^W < \delta$, there is $B \in V$ with $A \subseteq B$ and $|B|^V < \delta$.

(2) The pair $V \subseteq W$ satisfies the *δ-approximation property* if whenever $A \in W$ with $A \subseteq V$ and $A \cap a \in V$ for every a of size less than δ in V, then $A \in V$.

Pairs of the form the ground model with its forcing extension, $V \subseteq V[G]$, satisfy the δ-cover and δ-approximation properties for any cardinal $\delta \geq \gamma^+$, where γ is the size of the forcing poset. This fact is proved in [9], and it is an immediate corollary of Lemma 13 of [6], which easily generalizes to the following theorem.

[1]The parameter $\wp(\gamma)^V$ for $\gamma = |\mathbb{P}|^V$ appeared in Woodin's statement of Theorem 1.1, while Laver's statement of it used the less optimal parameter $\mathbf{V}_{\delta+1}$ for $\delta = \gamma^+$.

[2]It is known that the ground model may not be definable in a class-forcing extension satisfying ZFC. A counterexample to definability, attributed to Sy-David Friedman, is the forcing extension by the class Easton product adding a Cohen subset to every regular cardinal [3].

Theorem 1.3 (Hamkins). Suppose δ is a cardinal and $\mathbb{P}$ is a poset which factors as $\mathbb{R} * \dot{\mathbb{Q}}$, where $\mathbb{R}$ is nontrivial[3] of size less than δ and $\Vdash_{\mathbb{R}} \dot{\mathbb{Q}}$ is strategically $<\delta$-closed. Then the pair $V \subseteq V[G]$ satisfies the δ-cover and δ-approximation properties for any forcing extension $V[G]$ by $\mathbb{P}$.

Theorem 1.4 (Hamkins, cf. [9]). Suppose V, V' and W are transitive models of ZFC, δ is a regular cardinal in W, the pairs $V \subseteq W$ and $V' \subseteq W$ have the δ-cover and δ-approximation properties, $\wp(\delta)^V = \wp(\delta)^{V'}$, and $(\delta^+)^V = (\delta^+)^W$. Then $V = V'$.

Laver's proof of Theorem 1.1 proceeds by combining his weak version of Theorem 1.3 with Hamkins's uniqueness Theorem 1.4 as follows. A forcing extension $V[G]$ by a poset $\mathbb{P}$ of size γ has the δ-cover and δ-approximation properties for $\delta = \gamma^+$, and moreover it holds that $(\delta^+)^V = (\delta^+)^{V[G]}$. It is not difficult to see that there is an unbounded definable class C of ordinals such that for every $\lambda \in C$, the δ-cover and δ-approximation properties reflect down to the pair $\mathbf{V}_\lambda \subseteq V[G]_\lambda$ and both $\mathbf{V}_\lambda$ and $V[G]_\lambda$ satisfy a large enough fragment of ZFC, call it ZFC*, for the proof of Theorem 1.4 to go through. Letting $s = \wp(\delta)^V$, the sets $\mathbf{V}_\lambda$, for $\lambda \in C$, are then defined in $V[G]$ as the unique transitive models $M \models \mathsf{ZFC}^*$ of height λ, having $\wp(\delta)^M = s$ such that the pair $M \subseteq V[G]_\lambda$ has the δ-cover and δ-approximation properties. Finally, we can replace the parameter $s = \wp(\delta)^V$ with $\wp(\gamma)^V$ by observing that $\wp(\gamma)^V$ is definable from $\wp(\delta)^V$ in $V[G]$ using the δ-approximation property (cf. § 4 in the paragraph before Theorem 4.1 for the argument).

Forcing constructions over models of ZF can be carried out in some overarching ZFC context because the essential properties of forcing such as the definability of the forcing relation and the Truth Lemma do not require choice. Also, forcing over models of ZF preserves ZF to the forcing extension.[4] Is every model of ZF definable in its set-forcing extensions? Although at the outset, it might appear that the δ-cover and δ-approximation properties machinery, used to prove the definability of ZFC-ground models, isn't applicable to models without full choice, we shall show that much of it can be salvaged with only a small fragment of choice. In § 3, we prove an analogue of Theorem 1.4 for models of $\mathsf{ZF}+\mathsf{DC}_\delta$ (Theorem 3.2) and derive from it a partial definability result for ground models of $\mathsf{ZF}+\mathsf{DC}_\delta$ and forcing extensions by posets admitting a gap at δ. Posets admitting a gap at δ are particularly suited to forcing over models of $\mathsf{ZF}+\mathsf{DC}_\delta$ because they also preserve DC_δ to the forcing extension (Theorem 2.3).

---

[3] Here, and elsewhere in this article, a poset is *nontrivial* if it necessarily adds a new set if used as a forcing poset.

[4] All these facts follow by examining Shoenfield's proofs of them in [14] for the ZFC context. Since maximal antichains need not exist without choice, generic filters must meet all dense subsets.

Lévy [10] introduced the dependent choice axiom variant DC_δ, for an ordinal δ, asserting that for any nonempty set S and any binary relation R, if for each sequence $s \in S^{<\delta}$ there is a $y \in S$ such that s is R-related to y, then there is a function $f : \delta \to S$ such that $f \upharpoonright \alpha \, R f(\alpha)$ for each $\alpha < \delta$. It is easy to see that DC_δ implies the choice principle AC_δ, the assertion that indexed families $\{A_\xi \mid \xi < \delta\}$ of nonempty sets have choice functions. The full AC is clearly equivalent to the assertion $\forall\delta\ \mathsf{DC}_\delta$, while AC_δ is much weaker than DC_δ, as it provides choice functions only for already well-ordered families of nonempty sets.[5] Some of the natural models of $\mathsf{ZF}+\mathsf{DC}_\delta$ arise as symmetric inner models of forcing extensions and models of the form $\mathbf{L}(\mathbf{V}_{\delta+1})$. In [5], Hamkins defined that a poset $\mathbb{P}$ *admits a gap* at a cardinal δ if it factors as $\mathbb{R} * \dot{\mathbb{Q}}$, where $\mathbb{R}$ is nontrivial forcing of size less than δ, and it is forced by $\mathbb{R}$ that $\dot{\mathbb{Q}}$ is strategically $\leq\delta$-closed. By Theorem 1.3, a ZFC ground model with a forcing extension by a poset admitting a gap at a cardinal δ satisfy the δ-cover and δ-approximation properties. Indeed the analogous result for $\mathsf{ZF} + \mathsf{DC}_\delta$ holds as well (Theorem 3.1).

Main Theorem 1. Suppose V is a model of $\mathsf{ZF} + \mathsf{DC}_\delta$, $\mathbb{P} \in V$ is a forcing notion admitting a gap at δ, and $G \subseteq \mathbb{P}$ is V-generic. Then in $V[G]$, the ground model V is definable from the parameter $\wp(\delta)^V$.

Models of the theory ZFC^-, known as set theory without powerset, are used widely throughout set theory. Typically, but not necessarily, these have a largest cardinal κ. The canonical ones are models $\mathbf{H}_{\kappa^+}$, which are collections of all sets of hereditary size at most κ for some cardinal κ. Models of ZFC^- also play a prominent role in the theory of smaller large cardinals, many of which, such as weakly compact, remarkable, unfoldable, and Ramsey cardinals, are characterized by the existence of elementary embeddings of ZFC^- models. While set theorists often think of ZFC^- as simply the axioms of ZFC with the powerset axiom removed, the situation is more complex. Indeed, removing the powerset axiom from ZFC has many surprising consequences as illustrated by the work of Zarach. Zarach showed that without the powerset axiom, the collection and replacement schemes are not equivalent, and that the axiom of choice does not imply that every set can be well-ordered [19, 18]. Together with Hamkins, we continued Zarach's project in our article [2], whose theme was the importance of including collection and not just replacement in what we understand to be set theory without powerset. We showed that a number of crucial set-theoretic results, such as the Łoś Theorem for ultrapowers, or Gaifman's theorem that a Σ_1-elementary cofinal embedding is fully elementary, may fail for models of replacement but not collection in the absence of powerset. In light of these

[5]Indeed, for any fixed δ, the principle AC_δ does not imply DC_ω, while the assertion $\forall\delta\mathsf{AC}_\delta$ does imply DC_ω but not DC_{ω_1} (cf. [8, Chapter 8]).

facts, we define ZFC^- as in [2] to mean the theory ZFC without the powerset axiom, with the replacement scheme replaced by the collection scheme and with the axiom of choice replaced by the assertion that every set can be well-ordered.

Forcing over models of ZFC^- preserves ZFC^- to the forcing extension and the rest of standard forcing machinery carries over as well.[6] However, in § 4, we show that Laver's ground model definability result cannot be generalized to ZFC^- ground models. Using forcing, we produce a ZFC universe with a cardinal κ such that ground model definability fails for $\mathbf{H}_{\kappa^+}$. In this case, ground model definability is violated in the strongest possible sense because $\mathbf{H}_{\kappa^+}$ has a set-forcing extension in which it is not definable even using a parameter from the extension. We can set up the preparatory forcing so that κ is any ground model cardinal and so that the forcing extension violating ground model definability is by a poset of the form $\mathrm{Add}(\delta, 1)$ for some regular cardinal $\delta \ll \kappa$.[7] It will follow from our arguments that there is always a countable transitive model of ZFC^- violating ground model definability.

Main Theorem 2. Assume that δ, κ are cardinals such that δ is regular and either $2^{<\delta} \ll \kappa$ or $\delta = \kappa = 2^{<\kappa}$ holds. If $V[G]$ is a forcing extension by $\mathrm{Add}(\delta, \kappa^+)$, then $\mathbf{H}_{\kappa^+}^{V[G]}$ is not definable in its forcing extension by $\mathrm{Add}(\delta, 1)$. It follows that there is a countable transitive model of ZFC^- that is not definable in its Cohen forcing extension.

For instance, it follows that it is consistent for $\mathbf{H}_{\omega_2}$ to fail to be definable in its Cohen forcing extension $\mathbf{H}_{\omega_2}[g]$.

The proof method of Main Theorem 2 has an interesting consequence for ground model definability of ZFC models. We noted earlier that the natural parameter $\wp(|\mathbb{P}|^+)$ can be improved to $\wp(|\mathbb{P}|)^V$, but we shall show that it cannot be improved any further.

Theorem 1.5. It is consistent that a ground model $V \models \mathsf{ZFC}$ cannot be defined in its Cohen extension $V[g]$ using any parameter of hereditary size less than 2^ω.

2 ZF-Forcing preliminaries

Many of the concepts from the standard forcing toolbox use the axiom of choice in both obvious and subtle ways. For instance, nice names, which play a crucial role in forcing constructions, need not exist in choiceless models. Also, without choice, the concept of closure of a forcing notion loses

[6] It is an open question whether forcing extensions of models of set theory without powerset where replacement is used in place of collection continue to satisfy replacement.

[7] We denote by $\mathrm{Add}(\delta, \gamma)$, where δ is an infinite cardinal and γ is any cardinal, the poset which adds γ-many Cohen subsets to δ, using conditions of size less than δ.

much of its potency because a nontrivial infinite poset may, for instance, be vacuously countably closed simply because there are no infinite descending chains. Indeed, it is not difficult to see that the assertion that $\leq\delta$-closed forcing does not add new δ-sequences of ground model sets is actually equivalent to DC_δ. Another issue which arises when forcing over models satisfying only a fragment of choice is that this fragment need not be preserved to the forcing extension. For instance, Monro showed that it is possible to have a model of $\mathsf{ZF} + \mathsf{DC}_\delta$ for a cardinal $\delta >> \omega$ that has a forcing extension that does not even satisfy AC_ω [11]. In this section, we shall briefly discuss how to adapt certain forcing related concepts, such as nice names and full names, to the choiceless setting. We shall also show that posets admitting a gap at δ preserve DC_δ to the forcing extension. This is critical for our results because the generalized uniqueness theorem (Theorem 3.2), which we prove in § 3, requires all three models to satisfy DC_δ, and so it can only apply to models $V, V' \subseteq V[G]$ provided that $V[G] \models \mathsf{DC}_\delta$.

Suppose $\mathbb{P}$ is a poset in a model $V \models \mathsf{ZF}$ and σ is a $\mathbb{P}$-name. As a natural replacement for nice names, we define that a *good name* for a subset of σ is any $\mathbb{P}$-name τ such that $\tau \subseteq \mathrm{dom}(\sigma) \times \mathbb{P}$. It is easy to see that good names share the defining property of nice names, namely that if σ, μ are $\mathbb{P}$-names, then there is a good $\mathbb{P}$-name τ for a subset of σ such that $\mathbb{1} \Vdash (\mu \subseteq \sigma \to \mu = \tau)$. Therefore, good names can be used instead of nice names in the construction of the canonical names for the $\mathbf{V}_\alpha$-hierarchy of the forcing extension. We define these by recursion as follows: $\sigma_0 = \varnothing$, $\sigma_{\alpha+1} = \{\langle\tau, \mathbb{1}\rangle \mid \tau \text{ is a good name for a subset of } \sigma_\alpha\}$, and $\sigma_\lambda = \bigcup_{\alpha<\lambda} \sigma_\alpha$ for limit ordinals λ. It then follows that $(\sigma_\alpha)_G = V[G]_\alpha$ for every V-generic filter G. Moreover, assuming we used a flat pairing function for constructing $\mathbb{P}$-names,[8] we get that $\sigma_\alpha \subseteq \mathbf{V}_\alpha$ for every $\alpha \geq \gamma\cdot\omega$, where γ is the rank of $\mathbb{P}$, and hence $\mathbf{V}_\alpha[G] = V[G]_\alpha$ for all sufficiently large α. Having the canonical names σ_α and knowing that good names suffice to represent all subsets of σ_α, we get that for any $\mathbb{P}$-name σ, there is an ordinal γ such that whenever $p \in \mathbb{P}$ is a condition and μ is a $\mathbb{P}$-name such that $p \Vdash \mu \in \sigma$, then there is another $\mathbb{P}$-name $\tau \in \mathbf{V}_\gamma$ such that $p \Vdash \mu = \tau$.

There are a few different approaches to defining a two-step forcing iteration $\mathbb{P} * \dot{\mathbb{Q}}$ in ZFC, all of which can be shown to have the desired properties in the absence of choice as well. For concreteness, we use full names. For a poset $\mathbb{P}$, a $\mathbb{P}$-name τ is called *full* if $\tau = \mathrm{dom}(\tau) \times \{\mathbb{1}\}$ and whenever $p \in \mathbb{P}$ and σ is a $\mathbb{P}$-name such that $p \Vdash \sigma \in \tau$, then there is a $\sigma' \in \mathrm{dom}(\tau)$ such that $p \Vdash \sigma = \sigma'$. In models of ZFC, restricting to full names comes without a loss, since for every $\mathbb{P}$-name τ such that $\mathbb{1} \Vdash \tau \neq \varnothing$, there is a full name τ' such that $\mathbb{1} \Vdash \tau = \tau'$. The argument to see this uses nice names, the

[8] A *flat pairing function* is a way of defining ordered pairs which ensures that if $a, b \in \mathbf{V}_\alpha$ and $\alpha \geq \omega$, then so does the ordered pair of a and b.

canonical names σ_α, and the technique of *mixing* to verify that whenever $p \Vdash \sigma \in \tau$, then there is a another name σ' such that $p \Vdash \sigma = \sigma'$ and $\mathbb{1} \Vdash \sigma' \in \tau$. In models of ZF, good names again take on the role of nice names in this argument and, provided that for some fixed name τ_0, we have that $\mathbb{1} \Vdash \tau_0 \in \tau$, we can mix τ_0 and σ to create the required name σ' without any need for maximal antichains. Thus, in ZF, whenever a $\mathbb{P}$-name τ has an element τ_0 such that $\mathbb{1} \Vdash \tau_0 \in \tau$, it follows that a full name for τ exists. Since we can insist that any $\mathbb{P}$-name that is forced by $\mathbb{1}$ to be a poset comes with the top element $\dot{\mathbb{1}}$, it follows that, in models of ZF, every $\mathbb{P}$-name for a poset is forced by $\mathbb{1}$ to be equal to a full name. Thus, using full names, we define that $\mathbb{P} * \dot{\mathbb{Q}}$, where $\mathbb{P}$ is a poset and $\dot{\mathbb{Q}}$ is a full name for a poset, is the partial order consisting of conditions $(p, \dot{q})$, where $p \in \mathbb{P}$ and $q \in \text{dom}(\dot{\mathbb{Q}})$, ordered so that $(p, \dot{q}) \leq (p', \dot{q}')$ whenever $p \leq p'$ and $p \Vdash \dot{q} \leq \dot{q}'$.[9]

With the technical preliminaries out of the way, we now proceed to argue that a general class of posets, extending those admitting a gap at δ, preserve DC_δ to the forcing extension.

Theorem 2.1. Suppose that $V \models \mathsf{ZF} + \mathsf{DC}_\delta$ for an ordinal δ and $\mathbb{P}$ is well-orderable of order type at most δ. Then every forcing extension $V[G]$ by $\mathbb{P}$ is a model of DC_δ.[10]

Proof. In $V[G]$, suppose R is a relation and A is a set such that for all $s \in A^{<\delta}$ there is a $y \in A$ with sRy. Fix $\mathbb{P}$-names $\dot{A}$ and $\dot{R}$ such that $\dot{A}_G = A$ and $\dot{R}_G = R$, and also fix a condition $p \in G$ forcing the hypothesis of DC_δ for $\dot{A}$ and $\dot{R}$. By the previous remarks, we can find an ordinal γ such that whenever $p \Vdash \sigma \in \dot{A}$ for some $\mathbb{P}$-name σ, then there is another $\mathbb{P}$-name $\mu \in \mathbf{V}_\gamma$ such that $p \Vdash \sigma = \mu$. Let $B = \{\sigma \in \mathbf{V}_\gamma \mid p \Vdash \sigma \in \dot{A}\}$. Recall also that if $s = \langle \sigma_\xi \mid \xi < \alpha \rangle \in V$ is any sequence of $\mathbb{P}$-names, then there is a canonical $\mathbb{P}$-name τ_s such that $\mathbb{1} \Vdash$ "τ_s is an α-sequence" and $\mathbb{1} \Vdash \tau_s(\xi) = \sigma_\xi$ for all $\xi < \alpha$. We shall define a binary relation R^* on $B^{<\delta} \times B$ as follows. For $s \in B^{<\delta}$ and $\sigma \in B$, we define that $sR^*\sigma$ whenever $p \Vdash \tau_s \dot{R} \sigma$. We now argue that the hypothesis of DC_δ is satisfied for B and R^*. Towards this end, we suppose that $s \in B^\alpha$ for some $\alpha < \delta$. By the DC_δ hypothesis for $\dot{A}$ and $\dot{R}$ forced by p, it follows that $p \Vdash \exists x \in \dot{A}\ \tau_s \dot{R} x$. Since $\mathbb{P}$ is well-orderable, there exists a maximal antichain below p of conditions

[9] An alternative approach to defining a two-step iteration that does not require $\dot{\mathbb{Q}}$ to be a full name, starts with a proper class of conditions $(p, \dot{q})$, where $p \in \mathbb{P}$ and $\mathbf{1} \Vdash \dot{q} \in \dot{\mathbb{Q}}$, that is later cut down to a set using good names together with the names σ_α. Although this approach appears to skirt the need for a top element, most arguments involving iterations rely on the ability to turn a name $\dot{q}$ forced by some condition p to be an element of $\dot{\mathbb{Q}}$ into a name $\dot{q}'$ such that $p \Vdash \dot{q} = \dot{q}'$ and $(p, \dot{q}') \in \mathbb{P} * \dot{\mathbb{Q}}$, which needs mixing or alternatively a top element.

[10] In the case when $\mathbb{P}$ collapses the cardinality of δ, what really is preserved is DC_γ for $\gamma = |\delta|^{V[G]}$.

q such that for some name σ, we have $q \Vdash$ "$\sigma \in \dot{A}$ and $\tau_s \dot{R} \sigma$". Now using AC_δ, for each such q we choose a name σ_q and mix to obtain a single $\mathbb{P}$-name σ such that $p \Vdash$ "$\sigma \in \dot{A}$ and $\tau_s \dot{R} \sigma$". Without loss of generality, we may assume that $\sigma \in \mathbf{V}_\gamma$, and so $sR^*\sigma$ as desired. Now applying DC_δ, there is a sequence $s = \langle \sigma_\xi \mid \xi < \delta \rangle$ such that $s \upharpoonright \alpha \; R^* \sigma_\alpha$ for each $\alpha < \delta$. By the definition of R^*, it follows that $p \Vdash \tau_s \upharpoonright \alpha \; \dot{R} \; \tau_s(\alpha)$ for all $\alpha < \delta$. The interpretation $(\tau_s)_G$ is thus the desired δ-sequence witnessing DC_δ for A and R in $V[G]$. Q.E.D.

Recall that a poset is *strategically* $<\!\gamma$*-closed* if in the game of ordinal length γ in which two players alternatively select conditions from it to construct a descending γ-sequence with the second player playing at limit stages, the second player has a strategy that allows her to always continue playing; a poset is *strategically* $\leq\!\gamma$*-closed* if the corresponding game has length $\gamma + 1$.

Theorem 2.2. Suppose that $V \models \mathsf{ZF} + \mathsf{DC}_\delta$ for an ordinal δ and $\mathbb{P}$ is a strategically $\leq\!\delta$-closed poset in V. Then every forcing extension $V[G]$ by $\mathbb{P}$ is a model of DC_δ.

Proof. We shall follow the proof of Theorem 2.1, while avoiding the need for mixing (the only place where the well-orderability of $\mathbb{P}$ was used) by using the strategic closure property instead. To simplify the presentation, let us assume first that $\mathbb{P}$ is $\leq\!\delta$-closed and outline at the end of the proof how the argument can be modified in the case when $\mathbb{P}$ is merely strategically $\leq\!\delta$-closed. In $V[G]$, suppose R is a relation and A is a set such that for all $s \in A^{<\delta}$ there is a $y \in A$ with sRy. Fix a $\mathbb{P}$-name $\dot{A}$ and $\dot{R}$ such that $\dot{A}_G = A$ and $\dot{R}_G = R$, and also fix a condition $p \in G$ forcing the hypothesis of DC_δ for $\dot{A}$ and $\dot{R}$. Next, we find an ordinal γ such that whenever a condition $q \Vdash \sigma \in \dot{A}$ for some $\mathbb{P}$-name σ, then there is another $\mathbb{P}$-name $\mu \in \mathbf{V}_\gamma$ such that $q \Vdash \sigma = \mu$. We shall argue that it is dense below p to have conditions q forcing the existence of a sequence witnessing DC_δ for $\dot{A}$ and $\dot{R}$. Towards this end, we fix some $q \leq p$. Let $B = \{\langle r, \sigma \rangle \mid r \leq q, \sigma \in \mathbf{V}_\gamma, r \Vdash \sigma \in \dot{A}\}$. We shall define a binary relation R^* on $B^{<\delta} \times B$ as follows. Suppose $z = \langle \langle r_\xi, \sigma_\xi \rangle : \xi < \alpha \rangle$ is a sequence of elements of B for some $\alpha < \delta$ and let $s = \langle \sigma_\xi : \xi < \alpha \rangle$. If $\langle r_\xi : \xi < \alpha \rangle$ is a descending sequence of conditions, then we define that $zR^*\langle r, \sigma \rangle$ whenever r is below all the r_ξ and $r \Vdash \tau_s \dot{R} \sigma$. Otherwise, we define that $zR^*\langle r, \sigma \rangle$ for every $\langle r, \sigma \rangle \in B$. We now argue that the hypothesis of DC_δ is satisfied for B and R^*. If $z = \langle \langle r_\xi, \sigma_\xi \rangle : \xi < \alpha \rangle$ with $\langle r_\xi \mid \xi < \alpha \rangle$ descending, then since $\mathbb{P}$ is (much more than) $\leq\!\alpha$-closed, there is a condition $r^* \in \mathbb{P}$ below all the r_ξ. It is clear that $r^* \Vdash \sigma_\xi \in \dot{A}$ for all $\xi < \alpha$ and thus, $r^* \Vdash \exists x \in \dot{A} \; \tau_s \dot{R} x$. In contrast to the proof of Theorem 2.1, we cannot use mixing to obtain a witnessing name, but instead we strengthen r^* to a condition r for which there exists a $\mathbb{P}$-name σ such that $r \Vdash$ "$\sigma \in \dot{A}$ and $\tau_s \dot{R} \sigma$". Without loss of generality, we may assume that $\sigma \in \mathbf{V}_\gamma$, and so $z\ R^* \langle r, \sigma \rangle$, as desired.

Now applying DC_δ in V, there is a sequence $z = \langle \langle r_\xi, \sigma_\xi \rangle : \xi < \delta \rangle$ of elements of B such that $z \upharpoonright \xi\ R^*\ \langle r_\xi, \sigma_\xi \rangle$ for each $\xi < \delta$. We let $s = \langle \sigma_\xi \mid \xi < \delta \rangle$ and consider τ_s. By induction on ξ, it is easy to see that $\langle r_\xi \mid \xi < \delta \rangle$ is a descending sequence of conditions in $\mathbb{P}$ and each $r_\xi \Vdash \tau_s \upharpoonright \xi \dot{R} \tau_s(\xi)$. Since $\mathbb{P}$ is $\leq\delta$-closed, there is a condition $r \in \mathbb{P}$ below all the r_ξ. By the definition of R^*, it follows that $r \Vdash \tau_s \upharpoonright \xi\ \dot{R}\ \tau_s(\xi)$ for all $\xi < \delta$. Thus, $r \leq q$ forces that there is a sequence witnessing DC_δ for $\dot{A}$ and $\dot{R}$. This proves the theorem in the case when $\mathbb{P}$ is $\leq\delta$-closed.

It is straightforward to modify the argument for the case when $\mathbb{P}$ is merely strategically $\leq\delta$-closed, say with winning strategy Σ for player II. The definition of R^* has to be modified to insist that the descending sequence $\langle r_\xi : \xi < \alpha \rangle$ is built according to Σ, and when arguing that R^* satisfies the hypothesis for DC_δ, the condition r below all the r_ξ must be chosen according to the strategy Σ. Q.E.D.

Theorem 2.1 and 2.2 have the immediate corollary:

Theorem 2.3. Suppose that $V \models \mathsf{ZF} + \mathsf{DC}_\delta$ for a cardinal δ and $\mathbb{P} \in V$ is a poset which factors as $\mathbb{R} * \dot{\mathbb{Q}}$, where $|\mathbb{R}| \leq \delta$ and $\Vdash_{\mathbb{R}}$ "$\dot{\mathbb{Q}}$ is strategically $\leq\delta$-closed. Then every forcing extension $V[G]$ by $\mathbb{P}$ is a model of DC_δ. In particular, posets admitting a gap at δ preserve DC_δ to the forcing extension.

Posets described in the hypothesis of Theorem 2.3, with the additional assumption that $\mathbb{R}$ is nontrivial, are said to admit a *closure point* at δ. These forcing notions were introduced by Hamkins in [6] as a significant generalization of posets admitting a gap at δ. As we already noted, it follows from Theorem 1.3 that the ground model and its forcing extension by a poset admitting a gap at δ have the δ-cover and δ-approximation properties, but if the forcing extension is by a poset admitting a closure point at δ, then we are only guaranteed to have the δ^+-cover and δ^+-approximation properties. Chiefly because of this difference, we succeed in showing that every ground model of $\mathsf{ZF} + \mathsf{DC}_\delta$ is definable in its set-forcing extensions by posets admitting a gap at δ, while the analogous fact about closure point forcing is an open question (cf. § 5).

3 Definable ZF-ground models

In this section, we show that models of $\mathsf{ZF} + \mathsf{DC}_\delta$ are uniformly definable in their set-forcing extensions by posets admitting a gap at δ. Recall that Laver's proof of ZFC ground model definability combined Hamkins's uniqueness theorem (Theorem 1.4) with the fact that the ground model and its forcing extension by $\mathbb{P}$ always have the δ-cover and δ-approximation property for $\delta > |\mathbb{P}|$. Following this strategy, we proceed by first extending the uniqueness theorem to pairs of models of (a fragment of) $\mathsf{ZF}+\mathsf{DC}_\delta$ with the δ-cover and δ-approximation properties. Continuing to follow Laver, this would yield only that a ground model of $\mathsf{ZF}+\mathsf{DC}_\delta$ is definable its forcing extensions by well-ordered posets of size less than δ. But with the help of the following $\mathsf{ZF}+\mathsf{DC}_\delta$ analogue of Theorem 1.3, we are able to significantly expand the class of posets for which $\mathsf{ZF}+\mathsf{DC}_\delta$ ground models are definable, to non-well-orderable posets also.

Theorem 3.1. Suppose $V \models \mathsf{ZF}+\mathsf{DC}_\delta$ for a cardinal δ and $\mathbb{P}$ is a poset which factors as $\mathbb{R} * \dot{\mathbb{Q}}$, where $\mathbb{R}$ is nontrivial of size less than δ and $\Vdash_{\mathbb{R}} \dot{\mathbb{Q}}$ is strategically $<\delta$-closed. Then the pair $V \subseteq V[G]$ satisfies the δ-cover and δ-approximation properties for any forcing extension $V[G]$ by $\mathbb{R}*\dot{\mathbb{Q}}$. Indeed, if $\delta = \gamma^+$, then DC_γ suffices.

Proof. The proof is essentially identical to that of [7, Lemma 12], except that one needs to exercise care wherever the full axiom of choice is used. For instance, the δ-cover property is verified by observing that it holds for each step of the forcing. For the second step, we use that $<\delta$-closed forcing does not add new $<\delta$-sequences of ground model elements, which in our case relies on the preservation of DC_δ by the first step of the forcing (Theorem 2.1). Another key step of that proof uses mixing in $\mathbb{R}$, which we are able to do as well because $\mathbb{R}$ is well-ordered of size less than δ and AC_δ holds in V. Q.E.D.

Thus, in particular, the pair consisting of a model $V \models \mathsf{ZF}+\mathsf{DC}_\delta$ and its forcing extension $V[G]$ by a poset admitting a gap at δ has the δ-cover and δ-approximation properties. Crucially, it also follows that such a pair has the δ^+-cover and δ^+-approximation properties. The additional cover and approximation properties will allow us to fulfill a hypothesis of the generalized uniqueness theorem that we are about to state and prove.

As in the proof of ZFC ground model definability (sketched in the introduction), we shall eventually need that the uniqueness theorem holds for some fragment of $\mathsf{ZF}+\mathsf{DC}_\delta$ such that there is a proper definable class of ordinals α for which $\mathbf{V}_\alpha$ is a model of this fragment. We shall denote by Z^* the fragment of ZF consisting of Zermelo set theory (without choice) together with the axiom asserting that the universe is the union of the von Neumann hierarchy.[11] Observe that if $V \models \mathsf{ZF}+\mathsf{DC}_\delta$, then for every limit ordinal λ,

[11] Interestingly, over Zermelo set theory, the axiom asserting that for every ordinal α, $\mathbf{V}_\alpha$ exists does not imply that the universe is the union of the von Neumann hierarchy. A counterexample model was constructed by Sam Roberts in response to a MathOverflow question [13]: Recall that the *Zermelo ordinals* are defined by $Z(\varnothing)=\varnothing$, $Z(\alpha+1)=\{Z(\alpha)\}$, and $Z(\lambda)=\{Z(\alpha) \mid \alpha<\lambda\}$. Now consider the model M obtained by starting with $\mathbf{V}_{\omega+\omega}$, adding $Z(\omega+\omega)$ and closing under pairing, union, subsets, and powersets in

$\mathbf{V}_\lambda \models \mathsf{Z}^*$, and if moreover $\lambda > \delta$, then $\mathbf{V}_\lambda \models \mathsf{DC}_\delta$ as well. Previous versions of the uniqueness theorem have used different fragments of ZFC. In his proof of ground model definability, Woodin argued that the uniqueness theorem holds for models of the theory $\mathrm{ZC}^{(\mathrm{VN})} + \Sigma_1$-replacement, where $\mathrm{ZC}^{(\mathrm{VN})}$ consists of Zermelo set theory with choice and the additional axiom that $\mathbf{V}_\alpha$ exists for every ordinal α, and Σ_1-replacement is the replacement axiom for Σ_1-definable functions [16]. Reitz, in his ground axiom paper [12], argued that the uniqueness theorem holds for models of the theory ZFC_δ (for a regular cardinal δ), consisting of Zermelo set theory with choice, replacement for definable functions with domain $\leq\delta$, and the additional axiom asserting that every set is coded by a set of ordinals[12] below, replacement for functions with domain $\leq\delta$ turns out to superfluous because the ranges of the functions in question are contained in a set and therefore DC_δ (which follows from choice) suffices to argue that their ranges are themselves sets.[13] Over Zermelo set theory, the Σ_1-replacement axiom implies Reitz's coding axiom and strengthens the assertion that $\mathbf{V}_\alpha$ exists for every α to our assertion that the universe is the union of the von Neumann hierarchy. Because our proof combines a weak version of coding which already follows from $\mathsf{Z}^* + \mathsf{DC}_\delta$ with an induction that relies on the fact that the universe is the union of the von Neumann hierarchy, we avoid the need for any additional replacement or coding axioms.

Theorem 3.2. Suppose that V, V', and W are transitive models of $\mathsf{Z}^* + \mathsf{DC}_\delta$, for some regular cardinal δ of W. Suppose that the pairs $V \subseteq W$ and $V' \subseteq W$ have the δ-cover and δ-approximation properties, $\wp(\delta)^V = \wp(\delta)^{V'}$, and $\mathbf{H}^W_{\delta^+} \cap V = \mathbf{H}^V_{\delta^+}$. Then $V = V'$.

Proof. We follow the main ideas of the proof for the ZFC context as presented in [9, Theorem 1, Lemmas 1.1 & 1.2] closely, but make a few significant changes to adapt the arguments to the ZF case. The proof of [9] proceeds, for instance, by arguing that V and V' have the same sets of ordinals, a condition which suffices to conclude that $V = V'$ only if both are models of ZFC. Our changes to that argument are designed to overcome this and other uses of full choice.

ω-many steps. It is not difficult to see that M is a model of Zermelo set theory of height $\omega + \omega$. Thus, M satisfies that $\mathbf{V}_\alpha$ exists for every ordinal α, but it is not a model of Z^*.

[12]In [12], this coding axiom is given formally as $\forall A \exists \alpha \in \mathrm{ORD}\, \exists E \subseteq \alpha \times \alpha\, \langle \alpha, E\rangle \cong \langle \mathrm{tc}(\{A\}), \in\rangle$, but this formalization appears to be slightly too weak. In order to truly code every set by a set of ordinals, Reitz's arguments use the ability to *decode* sets of ordinals into transitive sets, but without the replacement axiom, this may not be possible. It thus appears that Reitz's ZFC_δ should include an additional requirement, namely that for all ordinals α and every well-founded extensional relation $E \subseteq \alpha \times \alpha$, the Mostowski collapsing map of $\langle \alpha, E\rangle$ exists.

[13]Reitz's theory would still need replacement for functions on ω because the proof uses the existence of transitive closures, which does not follow from Zermelo set theory.

First, we make a general observation about the Mostowski collapse in models of Z^* that will be used throughout the proof. Suppose that E is a well-founded extensional relation on a set A. Even though replacement may fail to hold, because E is a relation on a set, we can define the E-rank function $e : \mathrm{ORD} \to \wp(A)$. In particular, if the E-ranks are bounded, that is, there is an α such that $e(\alpha) = e(\alpha + 1)$, then the Mostowski collapse maps into $\mathbf{V}_\alpha$, and therefore exists as a set. Thus, in models of Z^*, any extensional well-founded set relation with bounded ranks is isomorphic to the $\in$-relation on a transitive set. For instance, it follows from this that $\mathbf{H}^V_{\delta+} = \mathbf{H}^{V'}_{\delta+}$. To see this, suppose that $A \in V$ is a transitive set having a bijection to some ordinal $\delta' \leq \delta$. This bijection imposes a relation E on δ' corresponding to the $\in$-relation on A. Since $\wp(\delta)^V = \wp(\delta)^{V'}$, we have that $E \in V'$. Clearly, since E codes A, the E-ranks are bounded and hence V' can Mostowski collapse E to recover A.

Note that, using the δ-cover and δ-approximation properties, it follows that V, V', and W all have the same ordinals. Next, we observe that a set $A \in V$ has size less than δ in V if and only if it has size less than δ in W. Clearly, since $V \subseteq W$, then $|A|^W \leq |A|^V$. For the other direction, it suffices to observe that, since V satisfies DC_δ, it follows that either a set there has size less than δ or there is an injection of δ into it. The same statement obviously holds for V' and W as well, and thus when dealing with sets of size less than δ, it does not matter in which of the three models that size is computed. Using the hypothesis that $\mathbf{H}^W_{\delta+} \cap V = \mathbf{H}^V_{\delta+}$, we can extend this conclusion to transitive sets of size δ. This additional assumption replaces an analogous hypothesis of Theorem 1.4 that $(\delta^+)^V = (\delta^+)^W$, which appears insufficient without full choice, since a set in V of size δ in W may not even be well-orderable in V.

Claim 3.3. Suppose that $T \in V \cap V'$ is a transitive set and $A \in W$ is any subset of T of size less than δ. Then there exist a common cover $B \in V \cap V'$ with $A \subseteq B$ and a common bijection $f : B \to \delta'$ with $f \in V \cap V'$ for some $\delta' \leq \delta$.

Proof. Fix $A \in W$ with $A \subseteq T$ and $|A| < \delta$. Using the δ-cover property of the pair $V \subseteq W$, there is a cover $B_0 \in V$ such that $A \subseteq B_0 \subseteq T$ and $|B_0| < \delta$. Now we make the key observation that, since T is transitive, we can, working in V and using DC_δ, extend B_0 to an $\in$-extensional cover of size less than δ. Thus, we may assume without loss of generality that B_0 is already extensional. The set B_0 may in turn be covered by an extensional $B_1 \in V'$ with $B_1 \subseteq T$ and $|B_1| < \delta$, this time using the δ-cover property of the pair $V' \subseteq W$. Now we observe that W can tell which subsets of T are elements of V or V' by consulting $\wp(T)^V$ and $\wp(T)^{V'}$, both of which are elements of W. Thus, working in W and using DC_δ together with the regularity of δ to get through limit stages, we may obtain a sequence

$$A \subseteq B_0 \subseteq B_1 \subseteq \cdots \subseteq B_\xi \subseteq \cdots \text{ for } \xi < \delta$$

with cofinally many $B_\xi \in V$ and cofinally many in V' such that each $B_\xi \subseteq T$ is extensional and $|B_\xi| < \delta$. Let $B = \bigcup_{\xi<\delta} B_\xi$ and note that it has size at most δ in W by DC_δ. Since δ is regular, it follows that if $a \in W$ is any set of size less than δ, then $B \cap a = B_\xi \cap a$ for a sufficiently large ξ, and so $B \in V \cap V'$ by the δ-approximation property. Thus, it remains to demonstrate the existence of the required bijection f. Since $\in$ is clearly extensional on B, we let $\pi : \langle B, \in \rangle \to \langle b, \in \rangle$ be the Mostowski collapse of $\langle B, \in \rangle$. Note that π exists both in V and in V', by the uniqueness of the collapsing map, and so b is a transitive set in $V \cap V'$ with $|b|^W \leq \delta$. By our earlier remarks, it follows that $|b|^V \leq \delta$ as well, and thus, we let $g \in V$ be a bijection from b onto δ' for some $\delta' \leq \delta$. Since the bijection g can be coded by a subset of δ', and $\wp(\delta)^V = \wp(\delta)^{V'}$ by assumption, it follows that $g \in V'$ also. Let $f = g \circ \pi$ be the composition map. Then $f \in V \cap V'$, and $f : B \to \delta'$ is the desired a bijection that exists in $V \cap V'$. Q.E.D. (Claim 3.3)

Claim 3.4. Suppose that $T \in V \cap V'$ is a transitive set and $A \in W$ is any subset of T of size less than δ. Then $A \in V$ if and only if $A \in V'$.

Proof. We assume that $A \in V$. By Claim 3.3, there is a cover $B \in V \cap V'$ with $A \subseteq B$ and a bijection $f \in V \cap V'$ with $f : B \to \delta'$ for some $\delta' \leq \delta$. Since V and V' have the same subsets of δ by assumption and $f\text{"}A \subseteq \delta' \subseteq \delta$, it follows that $f\text{"}A \in V'$ and hence $A \in V'$. Q.E.D. (Claim 3.4)

Remarkably, the δ-approximation property now allows us to strengthen Claim 3.4 to apply to all $A \subseteq T$, whether well-orderable or not.

Claim 3.5. Suppose that $T \in V \cap V'$ is a transitive set and $A \in W$ is any subset of T. Then $A \in V$ if and only if $A \in V'$.

Proof. We assume that $A \in V$. Since $T \in V \cap V'$, it follows that $A \subseteq V'$, and so we may apply the δ-approximation property of the pair $V' \subseteq W$ to argue that $A \in V'$. Toward this end, we fix some $a \subseteq T$ in V' of size less than δ, and proceed to show that $a \cap A \in V'$. By Claim 3.4, it follows that a is an element of V. Thus, $a \cap A$ is an element of V of size less than δ. Using Claim 3.4 once again, we have that $a \cap A$ is an element of V', which completes the argument that $A \in V'$. Q.E.D. (Claim 3.5)

Finally, to prove the theorem, we first argue by induction that $\mathbf{V}_\alpha^V = \mathbf{V}_\alpha^{V'}$ for all ordinals α. The limit step is trivial, and for the successor step, we assume inductively that $\mathbf{V}_\xi^V = \mathbf{V}_\xi^{V'}$ and apply Claim 3.5 to conclude that $\mathbf{V}_{\xi+1}^V = \wp(\mathbf{V}_\xi^V) = \wp(\mathbf{V}_\xi^{V'}) = \mathbf{V}_{\xi+1}^{V'}$. It follows that $V = V'$ since the Z^*-axioms include the assertion that the universe the union of the von Neumann hierarchy. Q.E.D.

As we already noted in the proof of Theorem 3.2, its hypothesis that $\mathbf{H}^W_{\delta^+} \cap V = \mathbf{H}^V_{\delta^+}$ replaces the analogous hypothesis of Theorem 1.4 that $(\delta^+)^V = (\delta^+)^W$, which might be weaker in the choiceless context. However, in the presence of slightly more choice, namely DC_{δ^+} rather than DC_δ, we claim that $(\delta^+)^V = (\delta^+)^W$ implies $\mathbf{H}^W_{\delta^+} \cap V = \mathbf{H}^V_{\delta^+}$. If $V \models \mathsf{Z}^* + \mathsf{DC}_{\delta^+}$ and $A \in V$, then either A has size $\leq \delta$ in V, or there is an injection from $(\delta^+)^V$ into A. Thus, if $V \subseteq W$ with $(\delta^+)^V = (\delta^+)^W$ and $A \in \mathbf{H}^W_{\delta^+} \cap V$, then such an injection cannot exist and so A has hereditary size at most δ in V, as desired.

The proof of Main Theorem 1 now follows by combining Theorem 3.2 together with Theorem 2.3 and the proof of Theorem 1.1 from [9] sketched in the introduction.

Main Theorem 1. Suppose V is a model of $\mathsf{ZF} + \mathsf{DC}_\delta$, $\mathbb{P} \in V$ is a forcing notion admitting a gap at δ, and $G \subseteq \mathbb{P}$ is V-generic. Then in $V[G]$, the ground model V is definable from the parameter $\wp(\delta)^V$.

Proof. First, observe that we can assume without loss of generality that δ is regular. If δ was singular, then we could replace it by the regular cardinal $\delta' = \gamma^+$, where $\mathbb{P}$ factors as $\mathbb{R} * \dot{\mathbb{Q}}$ with $|\mathbb{R}| = \gamma < \delta$, witnessing that it admits a gap at δ.

By Theorem 3.1, the pair $V \subseteq V[G]$ has the δ-cover and δ-approximation properties. Moreover, as we observed earlier the pair $V \subseteq V[G]$ has the δ^+-cover and δ^+-approximation properties. It follows that any set in V that has size $\leq\delta$ in $V[G]$ also has size $\leq\delta$ in V, and so in particular we have $\mathbf{H}^{V[G]}_{\delta^+} \cap V = \mathbf{H}^V_{\delta^+}$. Finally, by Theorem 2.3, the forcing extension $V[G] \models \mathsf{DC}_\delta$.

It is easy to see that the δ-cover and δ-approximation properties reflect down to pairs $\mathbf{V}_\lambda \subseteq \mathbf{V}_\lambda[G]$ for λ of cofinality $\geq \delta$. Moreover, any such $\mathbf{V}_\lambda \models \mathsf{Z}^* + \mathsf{DC}_\delta$. Thus, the sets $\mathbf{V}_\lambda$, for ordinals $\lambda > \delta^+$ of cofinality $\geq \delta$ in $V[G]$, are now defined in $V[G]$ as the unique transitive models $M \models \mathsf{Z}^* + \mathsf{DC}_\delta$ of height λ, having $\wp(\delta)^M = \wp(\delta)^V$ such that the pair $M \subseteq V[G]_\lambda$ has the δ-cover and δ-approximation properties. Q.E.D.

4 Undefinable ZFC^- ground models

In this section, we show that it is consistent for Laver's and Woodin's ground model definability result to fail for the canonical ZFC^- models $\mathbf{H}_{\kappa^+}$. Starting with a universe V and a cardinal $\kappa \in V$, we produce a forcing extension $V[G]$ in which $\mathbf{H}^{V[G]}_{\kappa^+}$ is not definable in its forcing extension by a poset of the form $\mathrm{Add}(\delta, 1)$. The same forcing construction carried out over the

Mostowski collapse of a countable elementary submodel of a sufficiently large $\mathbf{H}_{\vartheta^+}$ shows the existence of countable transitive models of ZFC^- that violate ground model definability. All our counterexample models violate ground model definability in a strong sense, because they have set-forcing extensions in which they are not definable, not even when using parameters from the extension.

Main Theorem 2. Assume that δ, κ are cardinals such that δ is regular and either $2^{<\delta} \ll \kappa$ or $\delta = \kappa = 2^{<\kappa}$ holds. If $V[G]$ is a forcing extension by $\mathrm{Add}(\delta, \kappa^+)$, then $\mathbf{H}_{\kappa^+}^{V[G]}$ is not definable in its forcing extension by $\mathrm{Add}(\delta, 1)$. It follows that there exists a countable transitive model of ZFC^- that is not definable in its Cohen forcing extension.

Proof. For ease of presentation we shall only prove the specific case of the theorem when $\delta = \omega$ and $\kappa = \omega_1$, so that $V[G]$ is the forcing extension by $\mathrm{Add}(\omega, \omega_2)$. We shall say a few words about the proof of the general case at the end of this proof.

Fix any $V[G]$-generic $g \subseteq \mathrm{Add}(\omega, 1)$. We shall show that $\mathbf{H}_{\omega_2}^{V[G]}$ is not definable in its $\mathrm{Add}(\omega, 1)$-forcing extension $\mathbf{H}_{\omega_2}^{V[G]}[g]$, not even when using parameters from the forcing extension $\mathbf{H}_{\omega_2}^{V[G]}[g]$. In $V[G]$, every nice $\mathrm{Add}(\omega, 1)$-name for a subset of ω_1 has hereditary size at most ω_1, and since every element of $\mathbf{H}_{\omega_2}^{V[G]}$ can be coded by a subset of ω_1 via the Mostowski collapse, it follows that $\mathbf{H}_{\omega_2}^{V[G][g]} = \mathbf{H}_{\omega_2}^{V[G]}[g]$. Using this equality, we now suppose towards a contradiction that $\mathbf{H}_{\omega_2}^{V[G]}$ is definable in $\mathbf{H}_{\omega_2}^{V[G][g]}$ by some formula $\varphi(x, a)$ for some parameter $a \in \mathbf{H}_{\omega_2}^{V[G][g]}$. Without loss of generality, assume that $a \subseteq \omega_1$. Let $\dot{a}$ be a nice $\mathrm{Add}(\omega, \omega_2) \times \mathrm{Add}(\omega, 1)$-name for a subset of ω_1 such that $(\dot{a})_{G \times g} = a$, and let $\dot{G}$ be the canonical $\mathrm{Add}(\omega, \omega_2) \times \mathrm{Add}(\omega, 1)$-name for a generic filter on the $\mathrm{Add}(\omega, \omega_2)$ part of the product. Let g_α denote the Cohen subset on coordinate α of G. Since each g_α is an element of $\mathbf{H}_{\omega_2}^{V[G]}$, we have that $\mathbf{H}_{\omega_2}^{V[G][g]} \models \varphi(g_\alpha, a)$ for all $\alpha < \omega_2$. Thus, we proceed to fix a condition $(p, q) \in G \times g$ forcing that for all $\alpha < \omega_2$, $\mathbf{H}_{\omega_2} \models \varphi(x, \dot{a})$ of the Cohen subset x on coordinate α of $\dot{G}$. Since $\mathrm{Add}(\omega, \omega_2)$ has the ccc, there is $\beta < \omega_2$ such that $(\dot{a})_{G \times g} = (\dot{a})_{G_\beta \times g}$, where G_β is the restriction of G to the first β-many coordinates of $\mathrm{Add}(\omega, \omega_2)$. We may choose β large enough so that $p \in G_\beta$.

Fix any ordinal γ in ω_2 above β. A standard approach to take towards obtaining a contradiction would be to try to interchange g with g_γ. This doesn't quite work because a is an element of the extension $\mathbf{H}_{\omega_2}^{V[G][g]}$ and therefore g may be necessary to interpret $\dot{a}$ correctly. Instead, our strategy will be to use an automorphism π of $\mathrm{Add}(\omega, 1)$ in $V[g]$ whose point-wise image of g_γ produces a filter $\overline{g}_\gamma$ that together with g_γ codes in g. For instance, we can take π to be the function mapping a sequence $s \in \mathrm{Add}(\omega, 1)$

to the sequence of the same length with the value of a bit of s flipped whenever the value of that bit in g is 1. Viewing π as an automorphism of $\mathrm{Add}(\omega, \omega_2)$ which acts only on coordinate γ, we obtain, by applying it to G, the filter $\overline{G}$, where g_γ is replaced by $\overline{g}_\gamma$ on coordinate γ. Since π is an automorphism of $\mathrm{Add}(\omega, \omega_2)$ in $V[g]$, the filter $\overline{G}$ is $V[g]$-generic and $V[g][G] = V[g][\overline{G}]$. Because we are using product forcing, it follows that $V[G][g] = V[\overline{G}][g]$. The condition (p, q) is an element of $\overline{G} \times g$, since G_β is an initial segment of $\overline{G}$, and it forces that $\mathbf{H}_{\omega_2} \models \varphi(x, \dot{a})$ of the Cohen subset x on coordinate γ of $\dot{G}$. Since $(\dot{G})_{\overline{G}\times g} = \overline{G}$ and $(\dot{a})_{\overline{G}\times g} = a$, it follows that $\mathbf{H}_{\omega_2}^{V[\overline{G}][g]} = \mathbf{H}_{\omega_2}^{V[G][g]} \models \varphi(\overline{g}_\gamma, a)$, and hence $\overline{g}_\gamma$ is an element of $\mathbf{H}_{\omega_2}^{V[G]}$. Now we have g_γ and $\overline{g}_\gamma$ both in $\mathbf{H}_{\omega_2}^{V[G]}$, from which it follows by the definition of π that the filter g is in $\mathbf{H}_{\omega_2}^{V[G]}$ as well. Thus, we have reached a contradiction, showing that $\mathbf{H}_{\omega_2}^{V[G]}$ could not have been definable in $\mathbf{H}_{\omega_2}^{V[G][g]}$.

To see that there exists a transitive model of ZFC^- that is not definable in its Cohen extension, we carry out the above forcing construction over the collapse of a countable elementary submodel of some sufficiently large $\mathbf{H}_{\vartheta^+}$. For instance, suppose ϑ is large enough that $\mathbf{H}_{\omega_2} \in \mathbf{H}_{\vartheta^+}$. Let M be a transitive model of ZFC^- that is the collapse of some countable elementary submodel of $\mathbf{H}_{\vartheta^+}$. Since M is countable, there is, in V, an M-generic filter for $\mathrm{Add}(\omega, \omega_2)^M$. Consider the countable transitive ZFC^--model $N = \mathbf{H}_{\omega_2}^{M[G]}$ and use the above argument to show that N is not definable in its Cohen forcing extension.

The proof of the general case for arbitrary cardinals δ, κ is essentially the same. In the case when $2^{<\delta} < \kappa$, the poset $\mathrm{Add}(\delta, \kappa^+)$ has the κ-cc and is $<\delta$-closed, and it thus preserves all cardinals $\leq\delta$ and all cardinals $\geq\kappa$. When $\delta = \kappa = 2^{<\kappa}$, the poset $\mathrm{Add}(\delta, \kappa^+)$ has the κ^+-cc and is $<\kappa$-closed, and it thus preserves all cardinals. In both cases, κ is preserved as a cardinal in $V[G]$, and so it makes sense to consider $\mathbf{H}_{\kappa^+}^{V[G]}$. The κ^+-cc allows for the analogous nice name argument. Q.E.D.

In particular, it follows from the theorem that it is consistent to have cardinals $\kappa >> \omega$ such that $\mathbf{H}_{\kappa^+}$ is not definable in its Cohen forcing extension.

Recall from the introduction that following Laver's proof [9] a natural parameter to use when defining V in a forcing extension $V[G]$ by a poset $\mathbb{P}$ is the parameter $\wp(\delta)^V$, where $\delta = |\mathbb{P}|^+$ in V (cf. also [12]). The second author and Hamkins observed in 2012 how this parameter can easily be reduced to the parameter $\wp(|\mathbb{P}|)^V$, the parameter that Woodin had used in his ZFC ground model definition [16]. Namely, if $V \subseteq V[G]$ has the δ-approximation property, then $A \subseteq \delta$ is an element of $\wp(\delta)^V$ if and only if for every bounded subset $a \in \wp_\delta(\delta)^V$ the intersection $a \cap A$ is an element of $\wp_\delta(\delta)^V$.[14] Thus, $\wp(\delta)^V$ is definable in $V[G]$ from $\wp_\delta(\delta)^V$. But since every element of $\wp_\delta(\delta)^V$ is coded by an element of $\wp(|\mathbb{P}|)^V$, it follows that $\wp(\delta)^V$ is definable in $V[G]$ from $\wp(|\mathbb{P}|)^V$.

Theorem 4.1. It is consistent that a ground model V cannot be defined in its Cohen forcing extension $V[g]$ with a parameter of hereditary size less than 2^ω.

Proof. Suppose that CH holds in V and $V[G]$ is a forcing extension by $\mathrm{Add}(\omega, \omega_1)$. Clearly CH continues to hold in $V[G]$. Let $V[G][g]$ be the forcing extension of $V[G]$ by $\mathrm{Add}(\omega, 1)$. The proof of Main Theorem 2 easily modifies to show that $\wp(\omega)^{V[G]}$, and thus, $V[G]$ itself, cannot be defined in $V[G][g]$ using a parameter from $\mathbf{H}_{\omega_1}$. Q.E.D.

The pair $\mathbf{H}_{\omega_2}^{V[G]}$ and $\mathbf{H}_{\omega_2}^{V[G][g]}$, from the proof of Main Theorem 2, witnesses another violation of a standard ZFC-result about ground models and their forcing extensions. A model of ZFC can never be an elementary submodel of its set-forcing extension since, using the ground model powerset of the poset as a parameter, it is expressible that the forcing extension contains a filter meeting all the ground model dense sets. But in set theory without powerset, forcing extensions can be elementary.

Theorem 4.2 (Hamkins [4]). It is consistent that there is a model $M = \mathbf{H}_{\kappa^+}$ and a poset $\mathbb{P} \in M$ such that M is an elementary submodel of its forcing extensions by $\mathbb{P}$.

Proof. For instance, consider the models $M = \mathbf{H}_{\omega_2}^{V[G]}$ and $M[g] = \mathbf{H}_{\omega_2}^{V[G][g]}$ from the proof of Main Theorem 2. We argue that $M \prec M[g]$. Suppose $M \models \varphi(a)$ and assume that $a \subseteq \omega_1$. Fix an $\mathrm{Add}(\omega, \omega_2)$-name $\dot{a}$ such that $(\dot{a})_G = a$ and a condition $p \in \mathrm{Add}(\omega, \omega_2)$ forcing that $\mathbf{H}_{\check{\omega}_2} \models \varphi(\dot{a})$. Now observe that the poset $\mathrm{Add}(\omega, \omega_2) \times \mathrm{Add}(\omega, 1)$ is isomorphic to $\mathrm{Add}(\omega, \omega_2)$, and such an isomorphism may be chosen to fix any initial segment of the product $\mathrm{Add}(\omega, \omega_2)$. Thus, we let $F : \mathrm{Add}(\omega, \omega_2) \times \mathrm{Add}(\omega, 1) \to \mathrm{Add}(\omega, \omega_2)$ be an isomorphism that fixes p and the name $\dot{a}$. Next, we let $\overline{G}$ be the image of $G \times g$ under F and observe that $V[G][g] = V[\overline{G}]$ and $p \in \overline{G}$. Since $p \in \overline{G}$ and $(\dot{a})_{\overline{G}} = (\dot{a})_G$, it follows that $\mathbf{H}_{\omega_2}^{V[\overline{G}]} = \mathbf{H}_{\omega_2}^{V[G][g]} = M[g]$ satisfies $\varphi(a)$. Since we chose an arbitrary formula $\varphi(x)$ with a an arbitrary element of M, this concludes the proof that $M \prec M[g]$. Q.E.D.

All our models violating ground model definability share the feature that the powerset of the poset in whose forcing extension they are not definable is

[14] As is standard, we let $\wp_\alpha(\delta)$ denote the collection of all subsets of δ of size less than α.

too large to be an element of the model. Indeed, it is by exploiting this very feature that the counterexample models are obtained. Is it possible that a model of ZFC^- is not definable in its forcing extension by a poset whose powerset is an element of the model? Is it possible that a model $\mathbf{H}_{\kappa^+}$ is not definable in its forcing extension by a poset whose powerset has size $\leq\kappa$? David Asperó communicated to the authors that, by a result of Woodin, such a model exists in a universe with an I_0-cardinal, one of the strongest known large cardinal notions. A cardinal $\kappa < \lambda$ is an I_0-cardinal if it is the critical point of an elementary embedding $j : \mathbf{L}(\mathbf{V}_{\lambda+1}) \to \mathbf{L}(\mathbf{V}_{\lambda+1})$. The I_0-cardinals were introduced by Woodin, and their existence pushes right up against the *Kunen Inconsistency*, the existence of a nontrivial elementary embedding $j : V \to V$.

Theorem 4.3 (Woodin [17]). If there is an elementary embedding $j : \mathbf{L}(\mathbf{V}_{\lambda+1}) \to \mathbf{L}(\mathbf{V}_{\lambda+1})$ with critical point $\kappa < \lambda$, then $\mathbf{H}_{\lambda^+}$ is not definable in its forcing extension by any poset $\mathbb{P} \in \mathbf{V}_\lambda$ adding a countable sequence of elements of λ. In particular, $\mathbf{H}_{\lambda^+}$ is not definable in its Cohen extensions.

Proof. Since $\mathbb{P} \in \mathbf{V}_\lambda$, we have that $\mathbf{V}_{\lambda+1}^{V[G]} = \mathbf{V}_{\lambda+1}[G]$ and $\mathbf{H}_{\lambda^+}^{V[G]} = \mathbf{H}_{\lambda^+}[G]$. Thus, whenever $\mathbf{H}_{\lambda^+}$ is definable in $\mathbf{H}_{\lambda^+}^{V[G]} \in \mathbf{L}(\mathbf{V}_{\lambda+1}^{V[G]})$, we must have $\mathbf{V}_{\lambda+1} \in \mathbf{L}(\mathbf{V}_{\lambda+1}^{V[G]})$. Indeed, in this case, $\mathbf{V}_{\lambda+1}$ already appears by some finite stage $\mathbf{L}_n(\mathbf{V}_{\lambda+1}^{V[G]})$ of the construction because $\mathbf{H}_{\lambda^+}^{V[G]}$ is isomorphic to the structure $\mathcal{H}$ built out of equivalence classes of subsets of λ coding extensional well-founded relations. The equivalence relation, as well as the membership relation, on the codes is definable from information in $\mathbf{V}_{\lambda+1}^{V[G]}$. Using the definition of $\mathbf{H}_{\lambda^+}$ in $\mathbf{H}_{\lambda^+}^{V[G]}$, we can recover elements of $\mathbf{V}_{\lambda+1}$ from $\mathcal{H}$ because their Mostowski collapses already exist in $\mathbf{V}_{\lambda+1}^{V[G]}$. The theorem now follows directly from a result of Woodin showing that if $j : \mathbf{L}(\mathbf{V}_{\lambda+1}) \to \mathbf{L}(\mathbf{V}_{\lambda+1})$ is an elementary embedding with critical point $\kappa < \lambda$, $\mathbb{P} \in \mathbf{V}_\lambda$ is a poset, and $({}^\omega\lambda)^V \neq ({}^\omega\lambda)^{V[G]}$ in a forcing extension $V[G]$ by $\mathbb{P}$, then $\mathbf{V}_{\lambda+1} \notin \mathbf{L}_\lambda(\mathbf{V}_{\lambda+1}^{V[G]})$. Q.E.D.

It appears that nothing else is currently known about whether large cardinals are needed for the existence of such counterexample models.

The next observation is intended as a road map of which paths to avoid when attempting to construct a counterexample model $\mathbf{H}_{\kappa^+}$ that is not definable in its forcing extension by some poset whose powerset has size $\leq\kappa$.

Theorem 4.4. Suppose $\mathbb{P} \in \mathbf{H}_{\kappa^+}$ is a forcing notion of size γ and $\kappa^\gamma = \kappa$. Then $\mathbf{H}_{\kappa^+}$ is definable in its forcing extensions by $\mathbb{P}$ using the ground model parameter $\wp_{\gamma^+}(\kappa)$. For instance, we have:

(1) If the GCH holds and $\mathbb{P} \in \mathbf{H}_{\kappa^+}$ with $|\mathbb{P}| < cf(\kappa)$, then $\mathbf{H}_{\kappa^+}$ is definable in all its forcing extensions by $\mathbb{P}$. In particular, for κ such that $\mathrm{cf}(\kappa) > \omega$, we then have that $\mathbf{H}_{\kappa^+}$ is definable in its Cohen extensions.

(2) If $\kappa^{<\kappa} = \kappa$, then $\mathbf{H}_{\kappa^+}$ is definable in all its forcing extensions by posets of size less than κ.

Proof. First, observe that if $\mathbb{P}$ is a poset of size γ in a model $M \models \mathsf{ZFC}^-$ and γ is not the largest cardinal of M, then the usual ZFC argument generalizes easily to show that the pair $M \subseteq M[G]$ has the δ-cover and δ-approximation properties for $\delta = \gamma^+$. Now let $G \subseteq \mathbb{P}$ be $\mathbf{H}_{\kappa^+}$-generic for a poset as in the statement of the theorem. To see that $\mathbf{H}_{\kappa^+}$ is definable in $\mathbf{H}_{\kappa^+}[G]$, it suffices to definably select those subsets of κ that belong to $\mathbf{H}_{\kappa^+}$. Thus, consider in $\mathbf{H}_{\kappa^+}$ the definable class $b = \wp_{\gamma^+}(\kappa)$. Since b has size κ by assumption, b exists as a set in $\mathbf{H}_{\kappa^+}$. Since $\mathbf{H}_{\kappa^+} \subseteq \mathbf{H}_{\kappa^+}[G]$ has the δ-approximation property, it follows that in $\mathbf{H}_{\kappa^+}[G]$ a set $A \subseteq \kappa$ is in $\mathbf{H}_{\kappa^+}$ if and only if for every $a \in b$ the set $A \cap a$ is an element of b. Assertions (1) and (2) are immediate consequences. Q.E.D.

Note that even if $\wp_{\gamma^+}(\kappa)$ is too large to be an element of $\mathbf{H}_{\kappa^+}$, it would suffice for the proof if it was definable in the forcing extension $\mathbf{H}_{\kappa^+}[G]$. Thus, for example, it is relatively consistent that $\mathbf{H}_{\aleph_\omega^+}$ is definable in its Cohen extensions, even if $2^{\omega_1} >> \aleph_\omega$, by starting in L and blowing up the powerset of ω_1. Note also that since any κ with $\kappa^{<\kappa} = \kappa$ retains this property after forcing with $\mathrm{Add}(\kappa, \kappa^+)$, it is relatively consistent by Main Theorem 2 that $\mathbf{H}_{\kappa^+}$, for such a cardinal κ, is not definable in its forcing extension by $\mathrm{Add}(\kappa, 1)$. Thus, the size requirement on the posets in assertion (2) of Theorem 4.4 is optimal.

The proof of Theorem 4.4 applies to arbitrary transitive ZFC^--models with a largest cardinal also. Indeed, if $M \models \mathsf{ZFC}^-$ with the largest cardinal $\kappa \in M$, and $\mathbb{P} \in M$ is a forcing notion of size γ with $\kappa^\gamma = \kappa$ in M, then M is definable in its forcing extensions by $\mathbb{P}$ using the parameter $\wp_{\gamma^+}(\kappa)^M$. If in addition $\kappa^{<\kappa} = \kappa$ holds in M, then M is definable in all its forcing extensions by posets of size less than κ, using the same parameter $\wp_{\gamma^+}(\kappa)^M$. However, M need not be definable in its forcing extensions by posets of size κ. Several large cardinal notions κ below a measurable cardinal are characterized by the existence of elementary embeddings of such ZFC^--models with κ as the largest cardinal and the critical point of the embedding. Consistency results concerning these large cardinals are obtained by forcing over such models and so it should be noted that the observations above completely characterize the ground definability situation for such models.

5 Questions

There remain several open questions surrounding the topics of this paper. In Main Theorem 1, we generalized ground model definability to models of $\mathsf{ZF}+\mathsf{DC}_\delta$ and posets admitting a gap at δ. We conjecture that in general a model of ZF need not be definable in its set-forcing extension.

Question 5.1. Is every model of ZF a definable class of its set-forcing extensions?

More specifically, since it follows from Main Theorem 1 that every model of $\mathsf{ZF}+\mathsf{DC}_{\omega_1}$ is definable its Cohen forcing extensions, we ask:

Question 5.2. Is every model $V \models \mathsf{ZF}+\mathsf{DC}_\omega$ definable in its Cohen extensions?

The class of posets admitting a closure point at δ, defined in § 2, is a natural extension of the class of posets admitting a gap at δ. What additional assumptions must be added to generalize Main Theorem 1 to posets admitting a closure point at δ? By Theorem 3.1, the pair consisting of a $\mathsf{ZF}+\mathsf{DC}_\delta$ ground model and its forcing extension by a poset admitting a closure point at δ has the δ^+-cover and δ^+-approximation properties. Since Theorem 3.2 requires DC_{δ^+} to conclude uniqueness for submodels with the δ^+-cover and δ^+-approximation properties, it seems reasonable to expect that the necessary choice principle will have to be strengthened to DC_{δ^+}.

Question 5.3. Is every model $V \models \mathsf{ZF}+\mathsf{DC}_{\delta^+}$ definable in its set-forcing extensions by posets admitting a closure point at δ?

There are two difficulties in answering this question using our methods. First, we do not know whether posets admitting a closure point at δ preserve DC_{δ^+} (by Theorem 2.3, we know only that they preserve DC_δ). Second, forcing extensions by posets admitting a closure point at δ may collapse δ^{++}, violating the $\mathbf{H}^{V[G]}_{\delta^{++}} \cap V = \mathbf{H}^V_{\delta^{++}}$ requirement, a condition that was crucially used in the proof of the uniqueness Theorem 3.2. A resolution of question 5.3 may come down to answering the following.

Question 5.4. Is the $\mathbf{H}^W_{\delta^+} \cap V = \mathbf{H}^V_{\delta^+}$ requirement necessary in Theorem 3.2?

A natural approach to answer this question may be to first address the analogous situation in the ZFC context and find out whether the $(\delta^+)^V = (\delta^+)^W$ requirement is necessary in the uniqueness Theorem 1.4. In regards to the preservation of the choice fragments DC_δ by forcing, it seems natural to ask whether Theorem 2.2 can be strengthened to show that DC_δ is preserved by $\leq \delta$-distributive forcing.

Question 5.5. Do $\leq\delta$-distributive posets preserve DC_δ to the forcing extension?

In Main Theorem 2, we showed that there always exist models of ZFC^- violating ground model definability, and also that it is consistent for this to be the case for canonical models $\mathbf{H}_{\kappa^+}$. But all our counterexample models had the feature that the powerset of the forcing poset was a proper class. By a result of Woodin, we know that in a universe with an I_0-cardinal, there is a model $\mathbf{H}_{\lambda^+}$ that is not definable in any of its forcing extensions by a poset of size less than λ that adds a countable sequence of elements of λ (Theorem 4.3). This introduces the exciting possibility that the existence of ZFC^--models violating ground model definability for such posets may carry large cardinal strength.

Question 5.6. What is the consistency strength of the existence of a model $\mathbf{H}_{\kappa^+}$ that is not definable in its forcing extension by a poset whose powerset has size $\leq\kappa$?

A more targeted approach would be to settle the situation with the definability of models $\mathbf{H}_{\kappa^+}$ in their Cohen extensions. In this case, Theorem 4.4 (1) suggests that such a κ should be a singular cardinal κ of cofinality ω. Thus, it is not coincidental that Woodin's λ is just such a cardinal.

Question 5.7. If $2^\omega \leq \kappa$, what is the consistency strength of having a model $\mathbf{H}_{\kappa^+}$ that is not definable in its Cohen extension?

References

[1] G. Fuchs, J. D. Hamkins, and J. Reitz. Set-theoretic geology. Submitted.

[2] V. Gitman, J. D. Hamkins, and T. A. Johnstone. What is the theory ZFC without power set? Submitted.

[3] J. D. Hamkins. Answer to "Definability of ground model". Posting on mathoverflow.net, 12 December 2011.

[4] J. D. Hamkins. Answer to "Elementary end extensions of a countable model for ZF". Posting on mathoverflow.net, 19 March 2012.

[5] J. D. Hamkins. Gap forcing. *Israel Journal of Mathematics*, 125:237–252, 2001.

[6] J. D. Hamkins. Extensions with the approximation and cover properties have no new large cardinals. *Fundamenta Mathematicae*, 180(3):257–277, 2003.

[7] J. D. Hamkins and T. A. Johnstone. Indestructible strong unfoldability. *Notre Dame Journal of Formal Logic*, 51(3):291–321, 2010.

[8] T. J. Jech. *The axiom of choice*, volume 75 of *Studies in Logic and the Foundations of Mathematics*. North-Holland, 1973.

[9] R. Laver. Certain very large cardinals are not created in small forcing extensions. *Annals of Pure and Applied Logic*, 149(1-3):1–6, 2007.

[10] A. Lévy. The interdependence of certain consequences of the axiom of choice. *Fundamenta Mathematicae*, 54:135–157, 1964.

[11] G. P. Monro. On generic extensions without the axiom of choice. *Journal of Symbolic Logic*, 48(1):39–52, 1983.

[12] J. Reitz. The ground axiom. *Journal of Symbolic Logic*, 72(4):1299–1317, 2007.

[13] S. Roberts. Answer to "Does the existence of the von Neumann hierarchy in models of Zermelo set theory with foundation imply that every set has ordinal rank?". Posting on `mathoverflow.net`, 1 September 2013.

[14] J. R. Shoenfield. Unramified forcing. In D. S. Scott, editor, *Axiomatic set theory, Proceedings of the Symposium in Pure Mathematics of the American Mathematical Society held at the University of California, Los Angeles, California, July 10-August 5, 1967*, pages 357–381. American Mathematical Society, 1971.

[15] W. H. Woodin. Suitable extender models I. *Journal of Mathematical Logic*, 10(1-2):101–339, 2010.

[16] W. H. Woodin. The continuum hypothesis, the generic-multiverse of sets, and the Ω conjecture. In J. Kennedy and R. Kossak, editors, *Set theory, arithmetic, and foundations of mathematics: theorems, philosophies*, volume 36 of *Lecture Notes in Logic*, pages 13–42. Association for Symbolic Logic, 2011.

[17] W. H. Woodin. Suitable extender models II: beyond ω-huge. *Journal of Mathematical Logic*, 11(2):115–436, 2011.

[18] A. Zarach. Unions of ZF$^-$-models which are themselves ZF$^-$-models. In D. van Dalen, D. Lascar, and T. J. Smiley, editors, *Logic Colloquium '80, Papers intended for the European Summer Meeting of the Association for Symbolic Logic to have been held in Prague, August 24–30, 1980*, volume 108 of *Studies in Logic and the Foundations of Mathematics*, pages 315–342. North-Holland, 1982.

[19] A. M. Zarach. Replacement $\nrightarrow$ collection. In P. Hájek, editor, *Gödel '96. Proceedings of the conference held in Brno, 1996, Logical foundations of mathematics, computer science, and physics—Kurt Gödel's legacy*, volume 6 of *Lecture Notes in Logic*, pages 306–322. Association for Symbolic Logic, 2001.

On the automorphisms in the Gitik-Koepke construction

Vladimir Kanovei*

Institute for Information Transmission Problems, Russian Academy of Sciences, Moscow, Russia
Department of Mathematics, Moscow State University of Railway Engineering, Moscow, Russia

Abstract

It is well known that in ZFC, the assumption that GCH first fails at $\aleph_\omega$ implies the existence of inner models with large cardinals. Gitik and Koepke [5] demonstrated that this is not the case in ZF. Namely, any model of ZFC + GCH has a cardinal-preserving symmetric extension in which all axioms of ZF hold, the axiom of choice fails, $\operatorname{card} 2^{\aleph_n} = \aleph_{n+1}$ for all natural numbers n, and there is a surjection from $2^{\aleph_\omega}$ onto λ, where $\lambda > \aleph_{\omega+1}$ is any previously chosen cardinal in the ground model. In other words, in such an extension GCH holds in the proper sense for all cardinals $\aleph_n$ but fails at $\aleph_\omega$ in Hartogs's sense. The goal of this paper is to analyse the system of automorphisms involved in the Gitik-Koepke construction.

1 Introduction

The continuum problem asks whether the *continuum hypothesis* $2^{\aleph_0} = \aleph_1$ (CH) holds. The problem was introduced by Cantor in the aftermath of his famous proof that the set of real numbers is uncountable and was included in the famous list of Hilbert's problems [7] in 1900. This has been one of the focal points in set theoretic studies since the very first years of set theory. As a matter of curiosity, a flawed attempt to prove its negation was presented at the third international congress of mathematicians in Heidelberg by Julius König [8, 9]. Hausdorff formulated the *generalized continuum hypothesis* (GCH), which asserts that $2^{\aleph_\alpha} = \aleph_{\alpha+1}$ for all ordinals α.

Gödel [6] proved that GCH, and hence CH as well, is consistent with the axioms of ZFC. Cohen [2] established that conversely, the negation of CH is consistent with ZFC. Soon after, Easton [3] proved that powers of regular cardinals can assume any values in suitable models of ZFC, restricted only by a few simple and well-known rules. Yet his method turned out to be not applicable to powers of singular cardinals. And in fact, Silver [10] demonstrated that a singular cardinal of uncountable cofinality, such as $\aleph_{\omega_1}$

*The author would like to acknowledge partial support in the form of RFFI grant 13-01-00006.

cannot be the smallest cardinal at which the GCH fails. Singular cardinals of countable cofinality such as $\aleph_\omega$ can be the first instances of failure of GCH, but the relevant models of ZFC require a large cardinal hypothesis stronger than measurability [4].

Apter and Koepke [1] demonstrated that the picture changes in the absence of the axiom of choice, if one considers the violation of GCH in a sense close to Hartogs's cardinal definition. Namely, assuming only the existence of a measurable cardinal, they construct a symmetric extension in which the axioms of ZF hold and GCH holds for all $\aleph_n$ but fails at $\aleph_\omega$ in the sense that there is a surjective map $f : [\aleph_\omega]^\omega \to \aleph_{\omega+\alpha}$ for any previously chosen ordinal α.

Furthermore, Gitik and Koepke [5] proved that one can avoid large cardinals at the cost of weakening the properties of the surjection. Namely, there is a cardinal-preserving symmetric extension of any model of ZFC + GCH in which all axioms of ZF hold, the axiom of choice fails, $\operatorname{card} 2^{\aleph_n} = \aleph_{n+1}$ for all natural numbers n, and there is a surjection from $2^{\aleph_\omega}$ onto λ, where $\lambda > \aleph_{\omega+1}$ is any previously chosen cardinal in the ground model. Thus in such an extension, GCH holds in the proper sense for all cardinals $\aleph_n$ but fails at $\aleph_\omega$ in a weak sense.

For convenience we formulate the main result of [5] as follows.

Theorem 1 (Gitik, Koepke [5]). *Suppose that $\lambda > \aleph_{\omega+1}$ is a cardinal in the constructible universe $\mathbf{L}$. There is a set-generic extension $\mathbf{L}[G]$ of $\mathbf{L}$ and a symmetric submodel $\mathbf{L}_{\text{sym}}[G] \subseteq \mathbf{L}[G]$ with the same cardinals as $\mathbf{L}$ such that in $\mathbf{L}_{\text{sym}}[G]$*

(i) *all axioms of* ZF *hold,*

(ii) $\operatorname{card} 2^{\aleph_n} = \aleph_{n+1}$ *for all natural numbers n, and*

(iii) *there is a surjection from $2^{\aleph_\omega}$ onto λ.*

The model in Theorem 1 is a *symmetric model* (or *permutation model*) and the construction involves a very interesting system of automorphisms consisting of three subsystems. The goal of this paper is to explicitly define and analyse these systems of automorphisms. Based on our analysis, we present the proof of Theorem 1 in a somewhat more pedestrian way than in [5].

The structure of the exposition is as follows. We define the forcing $\mathbb{T}$ in § 2. §§ 3, 4, and 5 introduce three systems of transformations of the forcing, which we call *permutations*, *swaps*, and *flips*. The symmetry lemma established in § 6 is a typical result in the theory of symmetric extensions. § 7 describes the structure of $\mathbb{T}$-generic extensions. § 8 is devoted to the key technical result, the definability lemma. Finally, § 9 accomplishes the proof of Theorem 1.

2 Basic definitions and the forcing

After an array of auxiliary definitions, we shall introduce a forcing notion $\mathbb{T}$. Let λ be a fixed cardinal everywhere; $\lambda > \aleph_\omega$.

To prove Theorem 1, we are going to force λ-many subsets of $\aleph_\omega$. This final goal is behind the lengthy definition of the forcing notion $\mathbb{T}$ given below, whose structure will lead to specific cardinality properties of the extension, and to a symmetric subextension satisfying Theorem 1.

2.1 Basic definitions

We define $\mathbb{D}[n]$ be the set of all sets $d \subseteq [\aleph_n, \aleph_{n+1})$ such that $\sup(d) < \aleph_{n+1}$; $\mathbb{P}^+[n]$ be the set of all functions $p : \mathrm{dom}(p) \to 2$ such that $\varnothing \neq \mathrm{dom}(p) \subseteq [\aleph_n, \aleph_{n+1})$; $\mathbb{P}[n]$ be the set of all functions $p \in \mathbb{P}^+[n]$, such that $\mathrm{dom}(p) \in \mathbb{D}[n]$; $\mathbb{D}$ be the set of all sets $d \subseteq [\omega, \aleph_\omega)$ such that $d \cap [\aleph_n, \aleph_{n+1}) \in \mathbb{D}[n]$ for all n; $\mathbb{D}^*$ be the set of all sets $d \subseteq [\omega, \aleph_\omega)$ such that $d \cap [\aleph_n, \aleph_{n+1}) \in \mathbb{D}[n]$ for all but finite n; $\mathbb{P}^+$ be the set of all functions $p : \mathrm{dom}(p) \to 2$ such that $\mathrm{dom}(p) \subseteq [\omega, \aleph_\omega)$; and, finally, $\mathbb{P}$ be the set of all functions $p \in \mathbb{P}^+$ such that $\mathrm{dom}(p) \in \mathbb{D}$. If $n \in \omega$, $d \in \mathbb{D}$ and $p \in \mathbb{P}^+$, we let $d[n] = d \cap [\aleph_n, \aleph_{n+1})$ and $p[n] = p \restriction [\aleph_n, \aleph_{n+1})$. Thus $d \in \mathbb{D}$ iff $d[n] \in \mathbb{D}[n]$ for all n, and $p \in \mathbb{P}$ iff $p[n] \in \mathbb{P}[n]$ for all n. We order $\mathbb{P}$ so that $p \leqslant q$ iff $\mathrm{dom}(q) \subseteq \mathrm{dom}(p)$ and $q = p \restriction \mathrm{dom}(q)$.

Remark 2.1. As a forcing notion, each $\mathbb{P}[n]$ adds (the characteristic function of) a new subset of the interval $[\aleph_n, \aleph_{n+1})$, but does not add any new subset of $\aleph_n$. Accordingly, one may view $\mathbb{P}$ as the product of all forcing notions $\mathbb{P}[n]$, so that $\mathbb{P}$ adds a subset of the interval $[\omega, \aleph_\omega)$.

Note that if $m \neq n$ then $\mathbb{P}^+[n] \cap \mathbb{P}^+[m] = \mathbb{P}[n] \cap \mathbb{P}[m] = \varnothing$.

2.2 Narrow subconditions

Let $\mathbb{H}^+$ consist of all indexed sets of the form $h = \{h_\xi\}_{\xi \in |h|}$, where $|h| \subseteq [\omega, \aleph_\omega)$ (the *domain* of h) and $h_\xi \in \mathbb{P}^+[n]$ for all n and all $\xi \in |h| \cap [\aleph_n, \aleph_{n+1})$.

We put $h[n] = h \restriction [\aleph_n, \aleph_{n+1})$ (restriction) for $h \in \mathbb{H}^+$ and any n. Thus still $h[n] \in \mathbb{H}^+$ and $|h[n]| = |h| \cap [\aleph_n, \aleph_{n+1})$. Let $\mathbb{H}$ consist of all $h \in \mathbb{H}^+$ such that

(h1) for all n, $h[n] \in \mathbb{D}[n]$, i.e. $\sup(|h[n]|) < \aleph_{n+1}$,

(h2) the set $\mathrm{bas}(h) = \{n : h[n] \neq \varnothing\}$ is finite,

(h3) $h_\xi \in \mathbb{P}[n]$ for all n and all indices $\xi \in |h[n]| = |h| \cap [\aleph_n, \aleph_{n+1})$.

We say that a condition $h \in \mathbb{H}$ is *regular* at some $n \in \mathrm{bas}(h)$, iff for every $\xi \in |h[n]| = |h| \cap [\aleph_n, \aleph_{n+1})$ the set $\{\eta \in |h[n]| : h_\eta = h_\xi\}$ has cardinality exactly $\aleph_n$, and we say that $h \in \mathbb{H}$ is *stronger* than another condition $g \in \mathbb{H}$, symbolically $h \leqslant g$, iff $|g| \subseteq |h|$, and $h_\xi \leqslant g_\xi$ for all $\xi \in |g|$. The empty condition $\varnothing \in \mathbb{H}$ ($|\varnothing| = \varnothing$) is $\leqslant$-largest in $\mathbb{H}$.

We further define $\mathbb{H}[n] = \{h \in \mathbb{H} : |h| \subseteq [\aleph_n, \aleph_{n+1})\}$; thus $\mathbb{H}[n]$ consists of all indexed sets $h = \{h_\xi\}_{\xi \in |h|}$, where $|h| \in \mathbb{D}[n]$ (i.e., by definition we have $|h| \subseteq [\aleph_n, \aleph_{n+1})$ and $\sup(|h|) < \aleph_{n+1}$), and $h_\xi \in \mathbb{P}[n]$ for all $\xi \in |h|$. It is clear that $h \in \mathbb{H}$ iff $h[n] \in \mathbb{H}[n]$ for all n and the set $\mathrm{bas}(h)$ is finite.

As a forcing notion, $\mathbb{H}[n]$ is equal to the product of $\aleph_{n+1}$ copies of the forcing $\mathbb{P}[n]$, with $\aleph_n$-support, and hence, by Remark 2.1, it adds $\aleph_{n+1}$-many subsets of the interval $[\aleph_n, \aleph_{n+1})$. Accordingly, $\mathbb{H}$ is the finite support product of all forcing notions $\mathbb{H}[n]$, and it adds $\prod_n \aleph_{n+1}$ subsets of the interval $[\omega, \aleph_\omega)$, each of which can be obtained by picking, for any n, one particular subset of $[\aleph_n, \aleph_{n+1})$ of those forced by $\mathbb{H}[n]$.

2.3 Wide subconditions

Let $\mathbb{Q}^+$ consist of all indexed sets $q = \{q_\gamma\}_{\gamma \in |q|}$, where $|q| \subseteq \lambda$ and $q_\gamma \in \mathbb{P}^+$ for all indices $\gamma \in |q|$. We define

$$\begin{aligned} \mathbb{Q}^* &= \text{all } q \in \mathbb{Q}^+ \text{ such that } |q| \text{ is finite,} \\ \mathbb{Q} &= \text{all } q \in \mathbb{Q}^+ \text{ such that } |q| \text{ is finite and } q_\gamma \in \mathbb{P} \text{ for all } \gamma \in |q|. \end{aligned}$$

We say that a condition $q \in \mathbb{Q}^+$ is: *uniform*, if $\mathrm{dom}(q_\gamma[n]) = \mathrm{dom}(q_\delta[n])$ for all $\gamma, \delta \in |q|$ and all $n \in \omega$; we say that it is *compatible* with an assignment $a \in \mathbb{A}$, iff we have $q_\gamma[n] = q_\delta[n]$ whenever $\gamma, \delta \in |q| \cap |a|$, $n \in \mathrm{bas}(a)$, and $a(n,\gamma) = a(n,\delta)$; we say that it is *equally shaped* as another condition $p \in \mathbb{Q}^+$, iff $|p| = |q|$, and we have $\mathrm{dom}(p_\gamma[n]) = \mathrm{dom}(q_\gamma[n])$ for all $\gamma \in |p|$ and all $n \in \omega$; and we say that it is *stronger* than another condition $p \in \mathbb{Q}^+$, symbolically $q \leqslant p$, iff $|p| \subseteq |q|$, and $p_\gamma \leqslant q_\gamma$ in $\mathbb{P}$ for all $\gamma \in |p|$. Once again, the empty condition $\varnothing \in \mathbb{Q}$ ($|\varnothing| = \varnothing$) is $\leqslant$-largest in $\mathbb{Q}$.

As a forcing notion, $\mathbb{Q}$ is equal to the finite support product of λ copies of the forcing $\mathbb{P}$, and hence, by Remark 2.1, it adds λ subsets of the interval $[\omega, \aleph_\omega)$.

2.4 Assignments

An *assignment* will be any function a such that

(a1) $\mathrm{dom}(a) = \mathrm{bas}(a) \times |a|$, where $\mathrm{bas}(a) \subseteq \omega$ and $|a| \subseteq \lambda$ are finite sets, and

(a2) if $\langle n, \gamma \rangle \in \mathrm{dom}(a)$ then $a(n,\gamma) \in [\aleph_n, \aleph_{n+1})$.

Let $\mathbb{A}$ be the set of all assignments. In particular, $\varnothing$ (the empty assignment) belongs to $\mathbb{A}$. We suppose that $\mathrm{bas}(\varnothing) = |\varnothing| = \varnothing$, but it can be consistently assumed that either $\mathrm{bas}(\varnothing) = \varnothing$ and $|\varnothing| = \Gamma \subseteq \lambda$ is any finite set, or $|\varnothing| = \varnothing$ and $\mathrm{bas}(\varnothing) = N \subseteq \omega$ is any finite set, depending on the context.

Meanwhile any assignment $a \neq \varnothing$ has definite values of $|a|$ and $\mathrm{bas}(a)$. The set $\mathbb{A}$ is *ordered* so that $a \leqslant b$ (a is stronger) iff

(a3) $\mathrm{bas}(b) \subseteq \mathrm{bas}(a)$ and $|b| \subseteq |a|$, and

(a4) if $n \in \mathrm{bas}(a) \setminus \mathrm{bas}(b)$ and $\gamma \neq \delta$ belong to $|b|$ then $a(n,\gamma) \neq a(n,\delta)$.

Clearly $\varnothing$ is the $\leqslant$-largest (weakest, as a condition) element in $\mathbb{A}$. If $n \in \mathrm{bas}(a)$, then we define a map $a[n]$ on the set $|a|$ by $a[n](\gamma) = a(n,\gamma)$. Assignments a, b are *coherent* iff $\mathrm{dom}(a) = \mathrm{dom}(b)$, and for any $n \in \mathrm{bas}(a) = \mathrm{bas}(b)$ and $\gamma, \delta \in |a| = |b|$ we have: $a(n,\gamma) = a(n,\delta)$ if and only if $b(n,\gamma) = b(n,\delta)$. A condition $q \in \mathbb{Q}^+$ is *compatible* with an assignment $a \in \mathbb{A}$, if and only if we have $q_\gamma[n] = q_\delta[n]$ whenever $\gamma, \delta \in |q| \cap |a|$, $n \in \mathrm{bas}(a)$, and $a(n,\gamma) = a(n,\delta)$. If $a \in \mathbb{A}$ and $\Delta \subseteq |a|$ then let $a \Uparrow \Delta$ be the restriction $a \upharpoonright (\mathrm{bas}(a) \times \Delta)$.

2.5 Conditions

The forcing $\mathbb{T}$ consists of all triples of the form $t = \langle q^t, a^t, h^t \rangle$, where $q^t \in \mathbb{Q}$, $a^t \in \mathbb{A}$, $h^t \in \mathbb{H}$, and

(t1) $|a^t| = |q^t|$ and $\mathrm{bas}(a^t) = \mathrm{bas}(h^t)$—we put $|t| := |a^t|$ and $\mathrm{bas}(t) := \mathrm{bas}(a^t)$,

(t2) $\mathrm{ran}(a^t) \subseteq |h^t|$ and $h^t_{a^t(n,\gamma)} = q^t_\gamma[n]$ for all $n \in \mathrm{bas}(t)$ and $\gamma \in |t|$, and

(t3) q^t is compatible with a^t in the sense above, i.e., if $\gamma, \delta \in |t|$, $n \in \mathrm{bas}(t)$, and $a^t(n,\gamma) = a^t(n,\delta)$, then $q^t_\gamma[n] = q^t_\delta[n]$.

Note that the last condition follows from the first two conditions.

The set $\mathbb{T}$ is ordered componentwise: a condition $t \in \mathbb{T}$ is *stronger than* $s \in \mathbb{T}$, symbolically $t \leqslant s$, iff $q^t \leqslant q^s$ in $\mathbb{Q}$, $a^t \leqslant a^s$ in $\mathbb{A}$, $h^t \leqslant h^s$ in $\mathbb{H}$. Clearly $t = \langle \varnothing, \varnothing, \varnothing \rangle$ is the largest condition in $\mathbb{T}$.

A condition $t \in \mathbb{T}$ is *uniform*, symbolically $t \in \mathbb{T}^{\mathrm{uni}}$, iff q^t is uniform.

The idea behind the forcing notion $\mathbb{T}$ can be explained as follows. Let $t \in \mathbb{T}$ and $q^t = \{q^t_\gamma\}_{\gamma \in |q^t|} \in \mathbb{Q}$, where $|q^t| \subseteq \lambda$ is finite.

Let $\gamma \in |q^t|$. Then $q^t_\gamma \in \mathbb{P}$, hence q^t_γ can be viewed as the sequence $\{q^t_\gamma[n]\}_{n \in \omega}$, where $q^t_\gamma[n] \in \mathbb{P}[n]$ for each n. And q^t_γ as a condition in $\mathbb{P}$ is responsible for adding a generic subset, say Y_γ, of the interval $[\omega, \aleph_\omega)$, while each $q^t_\gamma[n] \in \mathbb{P}[n]$ is responsible for adding a generic subset $Y_\gamma[n] = Y_\gamma \cap [\aleph_n, \aleph_{n+1}) \subseteq [\aleph_n, \aleph_{n+1})$, cf. Remark 2.1.

Let $n \in \omega$. If $n \notin \mathrm{bas}(a^t)$ then the remainder of this remark is vacuous. Now, suppose that $n \in \mathrm{bas}(a^t) = \mathrm{bas}(h^t)$. Then by (t2) we have $q^t_\gamma[n] = h^t_\xi$, where $\xi = a^t(n,\gamma)$, and moreover, the equality $q^{t'}_\gamma[n] = h^{t'}_\xi$ holds for any stronger condition $t' \in \mathbb{T}$. However h^t_ξ is a condition in $\mathbb{P}[n]$, hence, it is responsible for adding a generic subset, say $X_\xi \subseteq [\aleph_n, \aleph_{n+1})$ (still

Remark 2.1). And the assignment part of the forcing $\mathbb{T}$ is responsible for the equality $Y_\gamma[n] = X_\xi$.

In other words, the nth part $Y_\gamma[n]$ of Y_γ belongs to the set $\{X_\xi : \xi \in [\aleph_n, \aleph_{n+1})\}$ of cardinality $\aleph_{n+1}$, whenever $\gamma < \lambda$ and independently of the fixed size of λ. This will be the key property of this forcing.

3 Permutations

In this section and the following two sections, we consider three groups of full or partial order-preserving transformations of conditions. Let Π_{fin} be the group of all permutations of the set $[\omega, \aleph_\omega)$ such that

(A) for any n, the restriction $\pi[n] = \pi \restriction [\aleph_n, \aleph_{n+1})$ is a permutation of the interval $[\aleph_n, \aleph_{n+1})$, and

(B) the set $\text{bas}(\pi) = \{n : \pi[n] \neq \text{the identity}\}$ is finite.

Let $\Pi_{\text{fin}}[n]$ consist of all $\pi \in \Pi_{\text{fin}}$ equal to the identity outside of $[\aleph_n, \aleph_{n+1})$. Any $\pi \in \Pi_{\text{fin}}[n]$ is naturally identified with $\pi[n]$. We define that any transformation $\pi \in \Pi_{\text{fin}}$ acts on the set of assignments $\mathbb{A}$ by $\pi \cdot a = \pi \circ a$, so that if $a \in \mathbb{A}$ then $a' = \pi \cdot a \in \mathbb{A}$, $\text{dom}(a') = \text{dom}(a)$, and $a'(n,\gamma) = \pi(a(n,\gamma))$ for all $\langle n,\gamma\rangle \in \text{dom}(a)$; and acts on $\mathbb{H}^+$ (and on $\mathbb{H} \subseteq \mathbb{H}^+$) by $\pi \cdot h = h \circ \pi^{-1}$, so that if $h \in \mathbb{H}^+$ then $h' = \pi \cdot h \in \mathbb{H}^+$, $|h'| = \{\pi(\xi) : \xi \in |h|\}$, and $h'_{\pi(\xi)} = h_\xi$ for all $\xi \in |h|$.

Finally if $t = \langle q^t, a^t, h^t\rangle \in \mathbb{T}$ then put $\pi \cdot t = \langle q^t, \pi \cdot a^t, \pi \cdot h^t\rangle$. The following lemma is rather obvious.

Lemma 2. *Any $\pi \in \Pi_{\text{fin}}$ is an order-preserving automorphism of the ordered sets $\mathbb{A}$, $\mathbb{H}$, and $\mathbb{T}$. Moreover if $a \in \mathbb{A}$ and $n \in \text{bas}(a) \smallsetminus \text{bas}(\pi)$ then $(\pi \cdot a)[n] = a[n]$, and accordingly if $h \in \mathbb{H}$ and $n \notin \text{bas}(\pi)$ then $(\pi \cdot h)[n] = h[n]$.*

4 Swaps

Suppose that $a, b \in \mathbb{A}$, $\text{dom}(a) = \text{dom}(b) = D$, and $\text{ran}(a) = \text{ran}(b)$. Let $\mathbb{A}_a = \{c \in \mathbb{A} : c \leqslant a\}$, $\mathbb{Q}_a^+ = \{q \in \mathbb{Q}^+ : |a| \subseteq |q| \wedge q \text{ is compatible with } a\}$ and $\mathbb{Q}_a = \{q \in \mathbb{Q} : |a| \subseteq |q| \wedge q \text{ is compatible with } a\}$ (and similarly, $\mathbb{A}_b$, $\mathbb{Q}_b^+$, and $\mathbb{Q}_b$. The assignments a and b induce a *swap transformation* $\mathbf{S}_{ab}$, acting from $\mathbb{A}_a$ to $\mathbb{A}_b$, from $\mathbb{Q}_a^+$ to $\mathbb{Q}_b^+$, and from $\mathbb{Q}_a$ to $\mathbb{Q}_b$. Recall that $q \in \mathbb{Q}^+$ is compatible with $a \in \mathbb{A}$ iff $q_\gamma[n] = q_\delta[n]$ holds whenever $\gamma, \delta \in |a| \cap |q|$, $n \in \text{bas}(a)$, and $a(n,\gamma) = a(n,\delta)$. Obviously $\mathbb{Q}_a = \mathbb{Q}_a^+ \cap \mathbb{Q}$.

The action of $\mathbf{S}_{ab}$ on $\mathbb{A}_a$ is defined as follows:

(1) if $c \in \mathbb{A}_a$ then $c' = \mathbf{S}_{ab} \cdot c \in \mathbb{A}$, $\text{dom}(c') = \text{dom}(c)$, $c' \restriction D = b$ (where $D = \text{dom}(a) = \text{dom}(b)$), and $c' \restriction (\text{dom}(c) \smallsetminus D) = c \restriction (\text{dom}(c) \smallsetminus D)$.

The action of $\mathbf{S}_{ab}$ on $\mathbb{Q}_a^+$ is defined as follows. First of all, if $n \in \text{bas}(a)$ and $\gamma \in |a|$ then let $\mathbf{s}_{ab}^n(\gamma)$ be the least $\vartheta \in |a|$ satisfying $a(n, \vartheta) = b(n, \gamma)$; such ordinals ϑ exist because $\text{ran}(a) = \text{ran}(b)$. Thus $\mathbf{s}_{ab}^n : |a| \to |a|$. Now we define:

(2) if $q \in \mathbb{Q}_a^+$ then $q' = \mathbf{S}_{ab} \cdot q \in \mathbb{Q}^+$, $|q'| = |q|$, and for all $n \in \omega$ and $\gamma \in |q|$:

(a) if $\gamma \in |a|$ and $n \in \text{bas}(a)$ then $q'_\gamma[n] = q_\vartheta[n]$, where $\vartheta = \mathbf{s}_{ab}^n(\gamma)$,

(b) if either $\gamma \notin |a|$ or $n \notin \text{bas}(a)$ then $q'_\gamma[n] = q_\gamma[n]$.

Finally if $t \in \mathbb{T}_a = \{t \in \mathbb{T} : a^t \leqslant a\}$ (then $a^t \in \mathbb{A}_a$ and $q^t \in \mathbb{Q}_a$), then put

$$\mathbf{S}_{ab} \cdot t = \langle \mathbf{S}_{ab} \cdot q^t, \mathbf{S}_{ab} \cdot a^t, h^t \rangle.$$

Lemma 3. *Assume that* $a, b \in \mathbb{A}$, $\text{bas}(a) = \text{bas}(b) = B$, $|a| = |b| = \Delta$, *and* $\text{ran}(a) = \text{ran}(b)$. *Then* $\mathbf{S}_{ab}$ *is an order-preserving bijection* $\mathbb{A}_a \to \mathbb{A}_b$, $\mathbb{Q}_a \to \mathbb{Q}_b$, $\mathbb{T}_a \to \mathbb{T}_b$ *and* $\mathbf{S}_{ba}$ *is the inverse in each of the three cases.*

Let $t \in \mathbb{T}_a$. *Then* $t' = \mathbf{S}_{ab} \cdot t \in \mathbb{T}_b$, $|t| = |t'|$, $\text{bas}(t) = \text{bas}(t')$, *and:*

(i) *if t is uniform, then so is t' and $q^t, q^{t'}$ are equally shaped;*

(ii) *if* $n \in B$, $\gamma \in \Delta$, *and* $a(n, \gamma) = b(n, \gamma)$ *then* $a^t(n, \gamma) = a^{t'}(n, \gamma) = a(n, \gamma) = b(n, \gamma)$ *and* $q_\gamma^{t'}[n] = q_\gamma^t[n]$;

(iii) *if* $n \in |t|$ *then* $\{q_\gamma^{t'}[n] : \gamma \in |t'|\} = \{q_\gamma^t[n] : \gamma \in |t|\}$.

Proof. The first essential part of the lemma is to show that if $t \in \mathbb{T}_a$ then $t' = \mathbf{S}_{ab} \cdot t \in \mathbb{T}_b$. It is enough to show that $t' \in \mathbb{T}$. And here the only essential task is to prove condition (t2) of § 2.5, i.e., to establish the equality $q_\gamma^{t'}[n] = h_{a^{t'}(n,\gamma)}^{t'}$ for all $n \in \text{bas}(t')$ and $\gamma \in |t'|$.

We can assume that $n \in \text{bas}(a)$ and $\gamma \in |a|$, simply because $\mathbf{S}_{ab}$ is the identity outside of $\text{dom}(a) = \text{bas}(a) \times |a|$. We have $a^{t'}(n, \gamma) = b(n, \gamma)$ within this narrower domain, hence the result to prove is $q_\gamma^{t'}[n] = h_{b(n,\gamma)}^t$ for all $n \in \text{bas}(a)$ and $\gamma \in |a|$. (Recall that $\mathbf{S}_{ab}$ does not change h^t, so that $h^{t'} = h^t$.) However $q_\gamma^{t'}[n] = q_\vartheta^t[n]$ by (2)a, where $\vartheta = \mathbf{s}_{ab}^n(\gamma)$, so that, in particular, $a(n, \vartheta) = b(n, \gamma)$. Thus the equality required turns out to be $q_\vartheta^t[n] = h_{a(n,\vartheta)}^t$, which is true since t is a condition.

The other essential claim is that the action of $\mathbf{S}_{ba}$ is the inverse of the action of $\mathbf{S}_{ab}$. Suppose that $t \in \mathbb{T}_a$ and let $t' = \mathbf{S}_{ab} \cdot t$; $t \in \mathbb{T}_b$. Put $s = \mathbf{S}_{ba} \cdot t'$; $s \in \mathbb{T}_a$ once again. We have to show that $s = t$. The key fact is $q_\gamma^s[n] = q_\gamma^t[n]$ for all $n \in \text{bas}(a)$ and $\gamma \in |a|$. By definition $q_\gamma^s[n] = q_\zeta^{t'}[n]$, where $\zeta = \mathbf{s}_{ba}^n$, in particular, $b(n, \zeta) = a(n, \gamma)$. Still by definition, $q_\zeta^{t'}[n] = q_\vartheta^t[n]$, where $\vartheta = \mathbf{s}_{ab}^n(\zeta)$, so that $a(n, \vartheta) = b(n, \zeta)$. To conclude, $q^s_\gamma[n] = q^t_\vartheta[n]$, where $a(n,\gamma) = a(n,\vartheta)$. But then $q^t_\gamma[n] = q^t_\vartheta[n]$ by (t3) of § 2.5, and hence we have $q^s_\gamma[n] = q^t_\gamma[n]$, as required.

Claims (i), (ii) are rather obvious. It follows from (2)b that claim (iii) is trivial for $n \in |t| \smallsetminus B$, while in the case $n \in B$ it suffices to prove $\{q^{t'}_\gamma[n] : \gamma \in B\} = \{q^t_\gamma[n] : \gamma \in B\}$. The inclusion $\subseteq$ holds because $q^{t'}_\gamma[n] = q^t_\vartheta[n]$ by (2)a, where $\vartheta = \mathbf{s}^n_{ab}(\gamma)$. The inclusion $\supseteq$ holds by the same reason with respect to the inverse swap $\mathbf{S}_{ba}$. Q.E.D.

5 Flips

This is a somewhat more complicated type of transformations, not related to permutations of indices as the first two types, but rather connected with certain topological homeomorphisms of product spaces of the form 2^X. We define it by extension, beginning from most elementary conditions.

5.1 Simple flips

If $d \in \mathbb{D}$ and $p \in \mathbb{P}$, or generally even $d \in \mathbb{D}^*$ and $p \in \mathbb{P}^+$, then define $d \cdot p = p' : \mathrm{dom}(p') \to 2$ so that $\mathrm{dom}(p) = \mathrm{dom}(p')$ and

$$p'(\alpha) = \begin{cases} p(\alpha) & \text{whenever} \quad \alpha \in (\mathrm{dom}(p)) \smallsetminus d\,, \\ 1 - p(\alpha) & \text{whenever} \quad \alpha \in d \cap \mathrm{dom}(p)\,. \end{cases}$$

Clearly $p \mapsto d \cdot p$ is an order-preserving automorphism of $\mathbb{P}$ and of $\mathbb{P}^+$. Transformations of this type, as well as those based on them and defined below, will be called *flips*.

5.2 Flips for narrow subconditions

We define product flips acting on conditions in $\mathbb{H}^+$ and in $\mathbb{H} \subseteq \mathbb{H}^+$. Let $\mathbb{FN}$ (flips for narrow subconditions) consist of all indexed sets $\psi = \{\psi_\xi\}_{\xi \in |\psi|}$, where $|\psi| \subseteq [\omega, \aleph_\omega)$ is finite, and $\psi_\xi \in \mathbb{D}[n]$ for all $n \in \omega$ and $\xi \in |\psi| \cap [\aleph_n, \aleph_{n+1})$. If $\psi \in \mathbb{FN}$ and $h \in \mathbb{H}^+$ then define $h' = \psi \cdot h \in \mathbb{H}^+$ so that $|h'| = |h|$ and for all ξ:

$$h'_\xi = \begin{cases} h_\xi & & \text{whenever} \quad \xi \in |h| \smallsetminus |\psi|\,, \\ \psi_\xi \cdot h_\xi & \text{(in the sense of § 5.1)} & \text{whenever} \quad \xi \in |h| \cap |\psi|\,. \end{cases}$$

Let $\mathbb{FN}[n] = \{\psi \in \mathbb{FN} : |\psi| \subseteq [\aleph_n, \aleph_{n+1})\}$; and accordingly if $\psi \in \Psi$ then let $\psi[n] = \psi \restriction [\aleph_n, \aleph_{n+1})$; then $\psi[n] \in \mathbb{FN}[n]$. The next lemma is obvious.

Lemma 4. *If $\psi \in \mathbb{FN}$ then the map $h \mapsto \psi \cdot h$ is an order-preserving surjective action $\mathbb{H}^+ \to \mathbb{H}^+$ and $\mathbb{H} \to \mathbb{H}$.*

5.3 Flips for wide subconditions

Now define product flips acting on conditions in $\mathbb{Q}^+$ and $\mathbb{Q} \subseteq \mathbb{Q}^+$. Let $\mathbb{FW}$ (flips for wide subconditions) consist of all indexed sets $\varphi = \{\varphi_\xi\}_{\xi \in |\varphi|}$, where $|\varphi| \subseteq \lambda$ is a finite set and $\varphi_\gamma \in \mathbb{D}$ for all $\gamma \in |\varphi|$. If $\varphi \in \mathbb{FW}$ and $q \in \mathbb{Q}^+$ then define $q' = \varphi \cdot q \in \mathbb{Q}^+$ so that $|q'| = |q|$ and for all γ:

$$q'_\gamma = \begin{cases} q_\gamma & \text{whenever} \quad \gamma \in |q| \smallsetminus |\varphi|, \\ \varphi_\gamma \cdot q_\gamma \quad (\text{in the sense of } \S\, 5.1) & \text{whenever} \quad \gamma \in |\varphi| \cap |q|. \end{cases}$$

The lext elementary lemma is left to the reader.

Lemma 5. *If $\varphi \in \mathbb{FW}$ then the map $q \mapsto \varphi \cdot q$ is an order-preserving surjective action $\mathbb{Q}^+ \to \mathbb{Q}^+$ and $\mathbb{Q} \to \mathbb{Q}$. If $q \in \mathbb{Q}^+$ then q and $\varphi \cdot q$ are equally shaped.*

As above, say that a flip $\varphi \in \mathbb{FW}$ is: *compatible* with an assignment $a \in \mathbb{A}$, in symbol $\varphi \in \mathbb{FW}_a$, iff $\varphi_\gamma[n] = \varphi_\delta[n]$ holds whenever $\gamma, \delta \in |\varphi| \cap |a|$, $n \in \operatorname{bas}(a)$, and $a(n,\gamma) = a(n,\delta)$. In this case, if in addition $|\varphi| \subseteq |a|$ then we define:

(1) a flip $\psi = \varphi \downarrow a \in \mathbb{FN}$ (*a-projection*) so that

$$|\psi| = \{a(n,\gamma) : n \in \operatorname{bas}(a) \wedge \gamma \in |\varphi|\}$$

and if $n \in \operatorname{bas}(a)$, $\gamma \in |\varphi|$, and $\xi = a(n,\gamma)$ then $\psi_\xi = \varphi_\gamma[n]$;

(2) a flip $\varepsilon = \varphi^{\to} a \in \mathbb{FW}$ (*a-extension*) so that $|\varepsilon| = |a|$, $\varepsilon_\delta = \varphi_\delta$ for all $\delta \in |\varphi|$, and the following holds for all $\gamma \in |a| \smallsetminus |\varphi|$ and $n \in \omega$:

$$\varepsilon_\gamma[n] = \begin{cases} \varphi_\delta[n] & \text{iff} \quad n \in \operatorname{bas}(a) \wedge \delta \in |\varphi| \wedge a(n,\gamma) = a(n,\delta), \\ \varnothing & \text{iff} \quad n \notin \operatorname{bas}(a) \vee \neg \exists \delta \in |\varphi|\, (a(n,\gamma) = a(n,\delta)). \end{cases}$$

The consistency of both (1) and (2) follows from the compatibility assumption.

5.4 Flips for conditions

Finally we define how any $\varphi \in \mathbb{FW}$ acts on the set

$$\mathbb{T}_\varphi = \{t \in \mathbb{T} : |\varphi| \subseteq |t| \wedge \varphi \text{ is compatible with } a^t\}.$$

If $t \in \mathbb{T}_\varphi$ then let $\varphi \cdot t = t'$, where $q^{t'} = (\varphi^{\to} a^t) \cdot q^t$, $a^{t'} = a^t$, $h^{t'} = (\varphi \downarrow a^t) \cdot h^t$.

Lemma 6. *Suppose that $\varphi \in \mathbb{FW}$. Then the map $t \mapsto \varphi \cdot t$ is an order-preserving surjective action $\mathbb{T}_\varphi \to \mathbb{T}_\varphi$, with $t \mapsto \varphi^{-1} \cdot t$ being the inverse. If $t \in \mathbb{T}_\varphi$ is uniform then so is $t' = \varphi \cdot t$, and $q^t, q^{t'}$ are equally shaped.*

Proof. Assume that $t \in \mathbb{T}_\varphi$ and prove that $t' = \varphi \cdot t$ belongs to $\mathbb{T}_\varphi$ as well; this is the only part of the lemma not entirely trivial. We have to check (t2) of § 2.5, i.e., $h^{t'}_{a^{t'}(n,\gamma)} = q^{t'}_\gamma[n]$ for all $n \in \mathrm{bas}(t')$ and $\gamma \in |t'|$. By definition $a^{t'} = a^t$, $\mathrm{bas}(t') = \mathrm{bas}(t)$, and $|t'| = |t|$, hence we have to prove $q^{t'}_\gamma[n] = h^{t'}_{a^t(n,\gamma)}$, for all $n \in \mathrm{bas}(t) = \mathrm{bas}(t')$, $\gamma \in |t| = |t'|$. Note that $q^{t'} = (\varphi^{\rightarrow} a^t) \cdot q^t$ and $h^{t'} = \psi \cdot h^t$, where $\psi = \varphi \downarrow a^t \in \mathbb{FN}$.

Case 1: $\gamma \in |\varphi|$. Then $q^{t'}_\gamma[n] = \varphi_\gamma[n] \cdot q^t_\gamma[n]$. Let $\xi = a^t(n,\gamma)$. By definition $h^{t'}_\xi = \psi_\xi \cdot h^t_\xi$. On the other hand, $\psi_\xi = \varphi_\gamma[n]$ and $h^t_\xi = q^t_\gamma[n]$. Therefore $h^{t'}_\xi = \varphi_\gamma[n] \cdot q^t_\gamma[n] = q^{t'}_\gamma[n]$, as required.

Case 2: $\gamma \notin |\varphi|$, and there is an ordinal $\delta \in |\varphi|$ such that $a^t(n,\gamma) = a^t(n,\delta)$. Then the extended flip $\varepsilon = \varphi^{\rightarrow} a^t$ satisfies $\varepsilon_\gamma[n] = \varphi_\delta[n]$, and hence $q^{t'}_\gamma[n] = \varepsilon_\gamma[n] \cdot q^t_\gamma[n] = \varphi_\delta[n] \cdot q^t_\delta[n] = q^{t'}_\delta[n] = h^{t'}_\xi$, where $\xi = a^t(n,\gamma) = a^t(n,\delta)$ (we refer to Case 1), as required.

Case 3: $\gamma \notin |\varphi|$, but there is no ordinal $\delta \in |\varphi|$ such that $a^t(n,\gamma) = a^t(n,\delta)$. The extended flip $\varepsilon = \varphi^{\rightarrow} a^t$ satisfies $\varepsilon_\gamma[n] = \varnothing$ in this case, and hence $q^{t'}_\gamma[n] = q^t_\gamma[n]$. Moreover, the assumption in Case 3 means that $\xi = a^t(n,\gamma) \notin |\psi|$, and hence $h^{t'}_\xi = h^t_\xi$, and we are done. Q.E.D.

The action symbol $\cdot$ is used for several different actions in this paper, including three different kinds of transformations of conditions. Nevertheless there should be no confusion since the domains of actions do not intersect each other.

6 The symmetry lemma

We begin with auxiliary definitions. If $t \in \mathbb{T}$ then let $\mathbb{T}_{\leqslant t} = \{t' \in \mathbb{T} : t' \leqslant t\}$ be the set of all conditions in $\mathbb{T}$ stronger than t.

Definition 7. Suppose that $N \subseteq \omega$ and $\Gamma \subseteq \lambda$ are finite sets. Conditions $s, t \in \mathbb{T}$ are *similar on* $N \times \Gamma$ iff

(a) $\Gamma \subseteq |s| = |t|$, $N \subseteq \mathrm{bas}(s) = \mathrm{bas}(t)$,

(b) $q^s \restriction \Gamma = q^t \restriction \Gamma$ and the restricted assignments $a^s \Vert\!\restriction \Gamma$ and $a^t \Vert\!\restriction \Gamma$ are coherent (cf. § 2.4),

(c) if $n \in N$ then $h^s[n] = h^t[n]$, and $a^s(n,\gamma) = a^t(n,\gamma)$ for all $\gamma \in \Gamma$,

and *strongly similar on* $N \times \Gamma$ if in addition

(d) s, t are uniform conditions, and q^s, q^t are equally shaped (cf. § 2.3),

(e) $\mathrm{ran}(a^s) = \mathrm{ran}(a^t)$ and $|h^s| = |h^t|$,

(f) conditions h^s and h^t are regular at every $n \in \mathrm{bas}(s) \smallsetminus N$ (§ 2.2),

(g) $\{h^s_\xi : \xi \in |h^s|\} = \{h^t_\xi : \xi \in |h^t|\}$; then easily $\{h^s_\xi : \xi \in |h^s| \cap [\aleph_n, \aleph_{n+1})\}$ $= \{h^t_\xi : \xi \in |h^t| \cap [\aleph_n, \aleph_{n+1})\}$ for all n.

Theorem 8 (the symmetry lemma). *Suppose that $N \subseteq \omega$ and $\Gamma \subseteq \lambda$ are finite sets, conditions $s, t \in \mathbb{T}$ are strongly similar on $N \times \Gamma$, $B = \mathrm{bas}(s) = \mathrm{bas}(t)$, and $\Delta = |s| = |t|$. Then:*

(i) *there exists a transformation $\pi \in \Pi_{\mathrm{fin}}$ such that $\pi[n]$ is the identity for all $n \in N$, condition $u = \pi \cdot s$ is strongly similar to t on $N \times \Gamma$, and moreover $\pi \cdot h^s = h^u = h^t$, and $a^u \Uparrow \Gamma = a^t \Uparrow \Gamma$;*

(ii) *the condition $v = \mathbf{S}_{a^u a^t} \cdot u$ is strongly similar to t on $N \times \Gamma$, and moreover $h^v = h^u$ and $a^v = a^t$;*

(iii) *there is a flip $\varphi \in \mathbb{FW}_{a^v}$ i.e., φ is compatible with a^v) such that $|\varphi| = \Delta$, $\varphi_\gamma[n] = \varnothing$ for all $n \in B$ and $\gamma \in \Delta$* (then obviously φ is compatible with each of the assignments a^s, a^t, a^u, a^v), *and moreover $t = \varphi \cdot v$;*

(iv) *the superposition $\tau = \varphi \circ \mathbf{S}_{a^u a^t} \circ \pi$ is an order preserving bijection from $\mathbb{T}_{\leqslant s}$ onto $\mathbb{T}_{\leqslant t}$;*

(v) *any condition $s' \in \mathbb{T}_{\leqslant s}$ is similar to $t' = \tau \cdot s'$ on $N \times \Gamma$.*

Proof. (i) Let $\Xi = |h^s| = |h^t|$. Under our assumptions, obviously there is a transformation $\pi \in \Pi_{\mathrm{fin}}$ such that

(1) $\mathrm{bas}(\pi) = B$ and if $n \in N$ then $\pi[n]$ is the identity;

(2) $\pi(a^s(n, \gamma)) = a^t(n, \gamma)$ for all $n \in B$ and $\gamma \in \Gamma$—as s, t are similar on Γ, here we avoid a contradiction related to the possibility of equalities $a^t(n, \gamma) = a^t(n, \gamma')$ for $\gamma \neq \gamma'$ in Γ ;

(3) π maps the set Ξ onto itself, and π is the identity outside of Ξ,

(4) if $\xi \in \Xi = |h^s|$ then $h^s_\xi = h^t_{\pi(\xi)}$.

The only point of contention is whether (2) does not contradict to (4). That is, we have to check that $h^s_{a^s(n,\gamma)} = h^t_{a^t(n,\gamma)}$. Note that $h^s_{a^s(n,\gamma)} = q^s_\gamma[n]$ and $h^t_{a^t(n,\gamma)} = q^t_\gamma[n]$ by (t2) of § 2. On the other hand $q^s_\gamma = q^t_\gamma$ by (b) of Definition 7, as required.

Lemma 9. *The transformation π satisfies* (i) *of Theorem 8, and in addition if $s' \in \mathbb{T}_{\leqslant s}$ then s' is similar to $u' = \pi \cdot s'$ on $N \times \Gamma$.*

Proof. To begin with, we prove that $h^u = \pi \cdot h^s$ is equal to h^t. (This is a fragment of (i).) We have $|h^u| = \{\pi(\xi) : \xi \in |h^s|\} = \Xi$ by (3), and $|h^t| = \Xi$ as well. Thus it remains to prove that $h^u_\eta = h^t_\eta$ for any $\eta = \pi(\xi) \in \Xi$, where $\xi \in \Xi$. Yet by definition (§ 3) $h^u_\eta = h^s_\xi$, and $h^t_\eta = h^s_\xi$ by (4). The equality $a^u \Uparrow \Gamma = a^t \Uparrow \Gamma$ follows from (2) since $a^u(n, \gamma) = \pi(a^s(n, \gamma))$.

Let us prove that any condition $s' \in T$, $s' \leqslant s$, is similar to $u' = \pi \cdot s'$ on $N \times \Gamma$. Item (a) of Definition 7 holds for the pair of conditions s', u' simply because the action of any $\pi \in \Pi_{\text{fin}}$ preserves $|\cdot|$ and bas.

Now prove (b). We have $q^{s'} = q^{u'}$ because the action of π does not change $q^{s'}$ at all. To show the coherence of $a^{s'} \Uparrow \Gamma$ and $a^{u'} \Uparrow \Gamma$, suppose that $\gamma, \delta \in \Gamma$, $n \in \omega$, and $a^{s'}(n,\gamma) = a^{s'}(n,\delta)$, and prove that $a^{u'}(n,\gamma) = a^{u'}(n,\delta)$. (The inverse implication can be checked in the same way.) Suppose first that $n \in B$. Then $a^{s'}(n,\gamma) = a^{s}(n,\gamma)$ and $a^{s'}(n,\delta) = a^{s}(n,\gamma)$, therefore $a^{s}(n,\gamma) = a^{s}(n,\delta)$. It follows that $a^{t}(n,\gamma) = a^{t}(n,\delta)$ by the coherence in (b) for s,t, therefore $a^{u}(n,\gamma) = a^{u}(n,\delta)$ since $a^{u} \Uparrow \Gamma = a^{t} \Uparrow \Gamma$, and finally $a^{u'}(n,\gamma) = a^{u'}(n,\delta)$, as required. Now suppose that $n \notin B$. Then the equality $a^{s'}(n,\gamma) = a^{s'}(n,\delta)$ implies $\gamma = \delta$ by (a4) of § 2, so obviously $a^{u'}(n,\gamma) = a^{u'}(n,\delta)$.

To check (c), i.e., $h^{u'}[n] = h^{s'}[n]$ and $a^{u'}(n,\gamma) = a^{s'}(n,\gamma)$ for all $\gamma \in \Gamma$ and $n \in N$, we use the fact that $\pi[n]$ is the identity for any $n \in N$ by (1).

Finally prove that s is strongly similar to $u = \pi \cdot s$ on $N \times \gamma$. We have (d) of Definition 7 (for the pair of conditions s', u') for rather obvious reasons. The equalities $\operatorname{ran}(a^{u'}) = \operatorname{ran}(a^{s'})$ and $|h^{u'}| = |h^{s'}|$ in (e) hold by (3), as $\operatorname{ran}(a^{u'})$ is equal to the π-image of $\operatorname{ran}(a^{s'})$. Finally the equality $\{h^{u'}_{\xi} : \xi \in |h^{u'}|\} = \{h^{s'}_{\xi} : \xi \in |h^{s'}|\}$ in (g) holds whenever $u' = \pi \cdot s'$ for some π. This proves the strong similarity of s and u. We conclude that conditions u and t are strongly similar on $N \times \Gamma$. Q.E.D.

We return to the proof of Theorem 8.To prove the next claim (ii) of Theorem 8, let $a = a^{u}$ and $b = a^{t}$. Thus $a, b \in \mathbb{A}$, $\operatorname{dom}(a) = \operatorname{dom}(b) = B \times \Delta$, $\operatorname{ran}(a) = \operatorname{ran}(b)$, and $a \Uparrow \Gamma = b \Uparrow \Gamma$ by the above. Thus, as clearly $u \in \mathbb{T}^{\text{uni}}_{a}$, we define $v = \mathbf{S}_{ab} \cdot u \in \mathbb{T}^{\text{uni}}_{b}$.

Lemma 10. *Condition* (ii) *of Theorem 8 holds, and in addition if $u' \in \mathbb{T}_{\leqslant u}$ then u' is similar to $v' = \mathbf{S}_{ab} \cdot u'$ on $N \times \Gamma$.*

Proof. That equalities $h^{v} = h^{u}$ and $a^{v} = a^{t}$ in (ii) hold is clear by definition: for instance swaps do not change h^{u} at all.

Prove that any condition $u' \in T$, $u' \leqslant u$, is similar to $v' = \mathbf{S}_{ab} \cdot u'$ on $N \times \Gamma$. Indeed by definition (cf. § 4) v' and u' are equal outside of the domain $N \times \Delta$, and $h^{v'} = h^{u'}$. Therefore we can assume without loss of generality that $|v'| = |u'| = \Delta$ and $\operatorname{bas}(v') = \operatorname{bas}(u') = B$. Then $a^{v'} = b = a^{t}$ and $a^{u'} = a = a^{u}$, thus the restricted assignments $a^{v'} \Uparrow \Gamma = b \Uparrow \Gamma$ and $a^{u'} \Uparrow \Gamma = a \Uparrow \Gamma$ are not merely coherent (as required by (b) of Definition 7) but just equal by the above. The equality $q^{v'} \upharpoonright \Gamma = q^{u'} \upharpoonright \Gamma$ in (b) follows from $a \Uparrow \Gamma = b \Uparrow \Gamma$ as well. And finally we have $h^{v'} = h^{u'}$ ($\mathbf{S}_{ab}$ does not change this component), proving (c).

We now prove that any u is strongly similar to $v = \mathbf{S}_{ab} \cdot u$ on $N \times \gamma$. We skip (d) of Definition 7 as quite simple. Further, as $h^v = h^u$, we have $|h^v| = |h^u|$ in (e) and the whole of (g). It remains to show $\operatorname{ran}(a^v) = \operatorname{ran}(a^u)$ in (e). Recall that $a^v = a^t$ while conditions s, t, u are strongly similar, therefore $\operatorname{ran}(a^v) = \operatorname{ran}(a^t) = \operatorname{ran}(a^s) = \operatorname{ran}(a^u)$. We conclude that conditions v and t are strongly similar on $N \times \Gamma$. Q.E.D.

(iii) Thus v, t are uniform conditions, strongly similar on $N \times \Gamma$, and $a^v = a^t$. In particular q^v and q^t are equally shaped, i.e., in this case, $|q^v| = |q^t| = \Delta$ and $\operatorname{dom}(q^v_\gamma[n]) = \operatorname{dom}(q^t_\gamma[n])$ holds for all $\gamma \in \Delta$ and $n \in \omega$. Define a flip $\varphi \in \mathbb{FW}$ so that still $|\varphi| = \Delta$, and

$$\varphi_\gamma[n] = \{\alpha \in \operatorname{dom}(q^v_\gamma[n]) = \operatorname{dom}(q^t_\gamma[n]) : q^v_\gamma(\alpha) \neq q^t_\gamma(\alpha)\}$$

for all $\gamma \in \Delta$ and $n \in \omega$. Then clearly $\varphi \cdot q^v = q^t$. Moreover φ is compatible with $a^v = a^t$, because so are q^t and q^v in the sense of (t3) of § 2. Thus conditions v and t belong to $\mathbb{T}_\varphi$, so $\varphi \cdot v$ makes sense.

Lemma 11. *Assertion* (iii) *of Theorem 8 holds, and in addition if* $v' \in \mathbb{T}_{\leqslant v}$, *then* v' *is similar to* $t' = \varphi \cdot v'$ *on* $N \times \Gamma$.

Proof. Recall that $a^v = a^t$ and $h^v = h^u = h^t$ by (i), (ii). It follows by (t2) of § 2 that $q^v_\gamma[n] = q^t_\gamma[n]$, and hence $\varphi_\gamma[n] = \varnothing$, whenever $\gamma \in \Delta$ and $n \in B$. To accomplish the proof of (iii) we need to check that $\varphi \cdot v = t$. Indeed $a^v = a^t$, since φ does not change this component. Further, $q^t = \varphi \cdot q^v$ simply by the choice of φ. Let us show that $h^t = h^v$ as well. Indeed, since by definition $\operatorname{bas}(h^v) = B = \operatorname{bas}(t)$, any change in h^v by the action of φ can be only due to a component $\varphi_\gamma[n]$ for some $\gamma \in \Delta$ and $n \in B$—but this is the identity since $\varphi_\gamma[n] = \varnothing$ in this case.

We now prove that any $v' \in T$, $v' \leqslant v$, is similar to $t' = \varphi \cdot v'$ on $N \times \Gamma$. By definition $a^{v'} = a^{q'}$, covering the coherence in (b) of Definition 7. Further the extended flip $\varphi' = \varphi^{\rightarrow} a^{v'}$ obviously satisfies the same property $\varphi'_\gamma[n] = \varnothing$ for all $n \in B$ and $\gamma \in \Delta' = |v'|$. This implies $h^{t'}[n] = h^{v'}[n]$ even for all $n \in B$, so that (c) holds for v', t' for all $n \in B$. It only remains to prove that $q^{t'} \restriction \Gamma = q^{v'} \restriction \Gamma$ in (b) of Definition 7, i.e., $q^{t'}_\gamma = q^{v'}_\gamma$ for all $\gamma \in \Gamma$.

By definition it suffices to show that $\varphi_\gamma[n] = \varnothing$ for all $\gamma \in \Gamma$ and $n \in \omega$, or equivalently, $q^v \restriction \Gamma = q^t \restriction \Gamma$—yet this is the case since v and t are similar on $N \times \Gamma$ by the above. Q.E.D.

Finally, part (iv) of Theorem 8 is a consequence of Lemmas 2, 3, 6, while part (v) is a corollary of Lemmas 9, 10, 11. Q.E.D.

7 The extension and symmetric subextension

We consider $\mathbf{L}$, the constructible universe, as the ground model. As usual, $\aleph_\xi^{\mathbf{L}}$ will denote the ξth infinite cardinal in $\mathbf{L}$.[1] Accordingly, $[\aleph_n^{\mathbf{L}}, \aleph_{n+1}^{\mathbf{L}})$ and $[\omega, \aleph_\omega^{\mathbf{L}})$ denote intervals in the ordinals, while $\lambda > \aleph_\omega^{\mathbf{L}}$ continues to be any $\mathbf{L}$-cardinal.

Blanket agreement 12. Below, the objects $\mathbb{D}$, $\mathbb{D}[n]$, $\mathbb{P}$, $\mathbb{P}[n]$, $\mathbb{H}$, $\mathbb{Q}$, $\mathbb{A}$, $\mathbb{T}$, Π_{fin}, $\mathbb{FN}$, $\mathbb{FW}$ (and those related to them) mean objects defined in $\mathbf{L}$, by definitions in §§ 2–5 relativized to $\mathbf{L}$. All of them belong to $\mathbf{L}$, of course. Thus, for example, $\mathbb{D}[n] \in \mathbf{L}$ will be assumed from now on to consist of all sets $d \subseteq [\aleph_n^{\mathbf{L}}, \aleph_{n+1}^{\mathbf{L}})$ such that $\sup(d) < \aleph_{n+1}^{\mathbf{L}}$.

In particular, $\mathbb{T}$, the forcing, belongs to $\mathbf{L}$, and hence we can consider sets $\mathbb{T}$-generic over $\mathbf{L}$ and corresponding generic extensions of $\mathbf{L}$.

7.1 The extension

Let a set $G \subseteq \mathbb{T}$ be $\mathbb{T}$-generic over $\mathbf{L}$. This naturally produces:

1. for any n and $\xi \in [\aleph_n^{\mathbf{L}}, \aleph_{n+1}^{\mathbf{L}})$, a function $\mathbf{x}_\xi^G = \bigcup_{t \in G} h_\xi^t \in 2^{[\aleph_n^{\mathbf{L}}, \aleph_{n+1}^{\mathbf{L}})}$;
2. for every n, a function $\mathbf{x}^G[n] = \{\mathbf{x}_\xi^G\}_{\xi \in [\aleph_n^{\mathbf{L}}, \aleph_{n+1}^{\mathbf{L}})} : [\aleph_n^{\mathbf{L}}, \aleph_{n+1}^{\mathbf{L}}) \to 2^{[\aleph_n^{\mathbf{L}}, \aleph_{n+1}^{\mathbf{L}})}$;
3. for any $\gamma < \lambda$, a function $\mathbf{y}_\gamma^G = \bigcup_{t \in G} q_\gamma^t \in 2^{[\omega, \aleph_\omega^{\mathbf{L}})}$;
4. for any $\gamma < \lambda$ and n, a function $\mathbf{y}_\gamma^G[n] = \mathbf{y}_\gamma^G \restriction [\aleph_n^{\mathbf{L}}, \aleph_{n+1}^{\mathbf{L}}) \in 2^{[\aleph_n^{\mathbf{L}}, \aleph_{n+1}^{\mathbf{L}})}$;
5. $\vec{\mathbf{y}}[G] = \{\mathbf{y}_\gamma^G\}_{\gamma < \lambda}$, a λ-sequence of functions in $2^{[\omega, \aleph_\omega^{\mathbf{L}})}$;
6. a map $\mathbf{a}^G = \bigcup_{t \in G} a^t : \omega \times \lambda \to [\omega, \aleph_\omega^{\mathbf{L}})$ such that $\mathbf{a}^G(n, \gamma) \in [\aleph_n^{\mathbf{L}}, \aleph_{n+1}^{\mathbf{L}})$ for all $n \in \omega$ and $\gamma < \lambda$.

Lemma 13. *If a set $G \subseteq \mathbb{T}$ is $\mathbb{T}$-generic over L, then*

(i) *if $n < \omega$, $\gamma < \lambda$, and $\mathbf{a}^G(n, \gamma) = \xi$, then $\mathbf{y}_\gamma^G[n] = \mathbf{x}_\xi^G$;*

(ii) *if $n < \omega$, $\gamma, \delta < \lambda$, and $\mathbf{a}^G(n, \gamma) \ne \mathbf{a}^G(n, \delta)$, then $\mathbf{y}_\gamma^G[n] \ne \mathbf{y}_\delta^G[n]$;*

(iii) *if $\gamma \ne \delta < \lambda$ then there is a number $n_0 = n_0(\gamma, \delta)$ such that $\mathbf{a}^G(n, \gamma) \ne \mathbf{a}^G(n, \delta)$ for all $n \ge n_0$.*

---

[1] In fact, the generic extensions of $\mathbf{L}$ considered below will preserve cardinals (cf. Corollary 24) but it is not convenient to pursue this here.

Proof. (i) is obvious.

(ii) Suppose that a condition $t \in G$ forces otherwise, and $\gamma, \delta \in |t|$, $n \in \mathrm{bas}(t)$. Then $\xi = a^t(n, \gamma) \neq a^t(n, \delta) = \eta$; here ξ, η are ordinals in $[\aleph_n^{\mathbf{L}}, \aleph_{n+1}^{\mathbf{L}})$. Note that h_ξ^t and h_η^t are conditions in $\mathbb{P}[n]$. Let $w_\xi \leqslant h_\xi^t$ and $w_\eta \leqslant h_\eta^t$ be any pair of incompatible conditions in $\mathbb{P}[n]$. Let $t' \in T$ be a condition which differs from t only in the following: $q_\gamma^{t'}[n] = h_\xi^{t'} = w_\xi$ and $q_\delta^{t'}[n] = h_\eta^{t'} = w_\eta$. Obviously $t' \leqslant t$, and t' forces that $\mathbf{y}_\gamma^G[n] \neq \mathbf{y}_\delta^G[n]$.

(iii) Definitely there is a condition $t \in G$ such that $|t|$ contains both γ and δ. Let $B = \mathrm{bas}(t)$ (a finite subset of ω) and let n_0 be bigger than $\max B$. Now if $s \in G$, $s \leqslant t$, and $n \in \mathrm{bas}(s)$, $n \geq n_0$, then $a^s \leqslant a^t$, and hence $a^s(n, \gamma) \neq a^s(n, \delta)$. This implies $\mathbf{a}^G(n, \gamma) \neq \mathbf{a}^G(n, \delta)$. Q.E.D.

7.2 Symmetric subextension

Now let us define a symmetric subextension of $\mathbf{L}[G]$, on the base of certain symmetric hulls of sets $\mathbf{x}^G[n]$ and $\mathbf{y}_\gamma^G$. Arguing inside $\mathbf{L}[G]$, for every n, we let $\mathbf{X}^G[n]$ be the $(\Pi_{\text{fin}}, \mathbb{FN})$-hull of $\mathbf{x}^G[n]$, i.e., the set of all elements of the form $\pi \cdot (\psi \cdot \mathbf{x}^G[n])$, where $\pi \in \Pi_{\text{fin}}$ and $\psi \in \mathbb{FN}$. Finally, we let $\vec{\mathbf{X}}[G] = \{\mathbf{X}^G[n]\}_{n<\omega}$.

Recall that the actions of $\pi \in \Pi_{\text{fin}}$ and $\psi \in \mathbb{FN}$ are defined in §§3 & 5 above. In particular $\psi \cdot \mathbf{x}^G[n]$ and $\pi \cdot (\psi \cdot \mathbf{x}^G[n])$ are maps $[\aleph_n^{\mathbf{L}}, \aleph_{n+1}^{\mathbf{L}}) \to 2^{[\aleph_n^{\mathbf{L}}, \aleph_{n+1}^{\mathbf{L}})}$ in $\mathbf{L}[\mathbf{x}^G[n]]$. It is clear that $\mathbf{X}^G[n]$ is closed under further application of transformations in Π_{fin} and $\mathbb{FN}$, so there is no need to consider iterated actions.

It takes more time to define suitable hulls of elements $\mathbf{y}_\gamma^G$. First of all, for any n and $\gamma < \lambda$, let $\mathbf{Y}_\gamma^G[n] = \{d \cdot \mathbf{y}_\gamma^G[n] : d \in \mathbb{D}[n]\} \subseteq 2^{[\aleph_n^{\mathbf{L}}, \aleph_{n+1}^{\mathbf{L}})}$; for any n, let $\mathbf{Y}^G[n] = \bigcup_{\gamma<\lambda} \mathbf{Y}_\gamma^G[n]$—still $\mathbf{Y}^G[n] \subseteq 2^{[\aleph_n^{\mathbf{L}}, \aleph_{n+1}^{\mathbf{L}})}$, and obviously $\mathbf{Y}^G[n]$ is the $\mathbb{D}[n]$-hull of $\{\mathbf{y}_\gamma^G[n] : \gamma < \lambda\}$.

Finally, if $\gamma < \lambda$ then we let $\mathbf{Y}_\gamma^G$ be the set of all $z \in 2^{[\omega, \aleph_\omega^{\mathbf{L}})}$ in $\mathbf{L}[G]$ such that there exist a set $d \in \mathbb{D}$ and a number n_0 satisfying:

(1) $z[n] = d[n] \cdot \mathbf{y}_\gamma^G[n]$ for all $n \geq n_0$;[2]

(2) $z[n] \in \mathbf{Y}^G[n]$ for all $n < n_0$.

In other words, to obtain $\mathbf{Y}_\gamma^G$ we first define the $\mathbb{D}$-hull $\mathbb{D} \cdot \mathbf{y}_\gamma^G = \{d \cdot \mathbf{y}_\gamma^G : d \in D\}$ of $\mathbf{y}_\gamma^G$, and then allow to substitute sets in $\mathbf{Y}^G[n]$ for $y[n]$ for any $y \in \mathbb{D} \cdot \mathbf{y}_\gamma^G$ and finitely many n, so that

> $\mathbf{Y}_\gamma^G$ is the set of all $z \in 2^{[\omega, \aleph_\omega^{\mathbf{L}})}$ (in $\mathbf{L}[G]$) such that there exist an element $y \in \mathbb{D} \cdot \mathbf{y}_\gamma^G$ and a number n_0 satisfying: $z[n] = y[n]$ for all $n \geq n_0$, and $z[n] \in \mathbf{Y}^G[n]$ for all $n < n_0$. $(\star)$

[2] Regarding the action of $d \in \mathbb{D}$, cf. § 5.1.

Lemma 14. *If* $\gamma \neq \delta$ *then* $\mathbf{Y}^G_\gamma \cap \mathbf{Y}^G_\delta = \varnothing$.

Proof. Suppose to the contrary that $z \in \mathbf{Y}^G_\gamma \cap \mathbf{Y}^G_\delta$. Then by ($\star$) there exist flips $d', d'' \in \mathbb{D}$ and a number n_0 such that the elements $y' = d' \cdot \mathbf{y}^G_\gamma$ and $y'' = d'' \cdot \mathbf{y}^G_\delta$ satisfy $y[n] = y'[n]$ for all $n \geq n_0$. In other words, $\mathbf{y}^G_\gamma[n] = (d \cdot \mathbf{y}^G_\delta)[n]$ for all $n \geq n_0$, where $d = d' \mathbin{\triangle} d'' \in \mathbb{D}$ (symmetric difference). We now use Lemma 13(iii) to obtain a number $n \geq n_0$ such that $\mathbf{a}^G(n,\gamma) \neq \mathbf{a}^G(n,\delta)$; still we have $\mathbf{y}^G_\gamma[n] = d[n] \cdot \mathbf{y}^G_\delta[n]$. But this yields a contradiction similarly to the proof of Lemma 13(ii). Q.E.D.

Now we let, in $\mathbf{L}[G]$, $\vec{\mathbf{Y}}[G] = \{\mathbf{Y}^G_\gamma\}_{\gamma<\lambda}$, a function defined on λ. We finally define

$$W[G] = \bigcup_n \mathbf{X}^G[n] \cup \bigcup_{\gamma<\lambda} \mathbf{Y}^G_\gamma \cup \{\vec{\mathbf{X}}[G], \vec{\mathbf{Y}}[G]\}.$$

Definition 15. Let $\mathbf{L}_{\mathrm{sym}}[G]$ denote **HOD** in $\mathbf{L}[G]$ with parameters in $W[G]$. Thus, by definition, every set in $\mathbf{L}_{\mathrm{sym}}[G]$ is definable in $\mathbf{L}[G]$ by a formula with parameters in $\mathbf{L}$, two special parameters $\vec{\mathbf{X}}[G]$ and $\vec{\mathbf{Y}}[G]$, and parameters which belong to the sets $\mathbf{X}^G[n]$ and $\mathbf{Y}^G_\gamma$ for various numbers $n < \omega$ and ordinals $\gamma < \lambda$.

The next lemma allows to reduce the last category of parameters essentially to those in $\{\mathbf{x}^G[n] : n < \omega\} \cup \{\mathbf{y}^G_\gamma : \gamma < \lambda\}$.

Lemma 16. *If* $n < \omega$ *then every* $x \in \mathbf{X}^G[n]$ *belongs to* $\mathbf{L}[\mathbf{x}^G[n]]$. *If* $\gamma < \lambda$ *and* $z \in \mathbf{Y}^G_\gamma$ *then there is a finite set* $\Delta \subseteq \lambda$ *such that* $z \in \mathbf{L}[\{\mathbf{y}^G_\delta : \delta \in \Delta\}]$.

Proof. By definition x belongs to the $(\Pi_{\mathrm{fin}}, \mathbb{FN})$-hull of $\mathbf{x}^G[n]$. But Π_{fin} and $\mathbb{FN}$ belong to $\mathbf{L}$ (cf. Blanket Agreement 12). Regarding the claim for $z \in \mathbf{Y}^G_\gamma$, come back to ($\star$). Note that y as in ($\star$) belongs to $\mathbf{L}[\mathbf{y}^G_\gamma]$ (since $\mathbb{D} \in \mathbf{L}$). Then to obtain z from y we replace a finite number of intervals $y[n]$ in y by elements of sets $\mathbf{Y}^G[n]$. Thus suppose that $n < \omega$ and $w \in \mathbf{Y}^G[n]$, i.e., $w \in \mathbf{Y}^G_\delta[n]$, where $\delta < \lambda$. But then $w \in \mathbf{L}[\mathbf{y}^G_\delta]$ (since $\mathbb{D}[n] \in \mathbf{L}$), so that it suffices to define Δ as the (finite) set of all ordinals δ which appear in this argument for all intervals $y[n]$ to be replaced. Q.E.D.

8 Definability lemma

The next theorem plays the key role in the analysis of the symmetric subextension.

Theorem 17 (definability lemma). *Suppose that a set* $G \subseteq \mathbb{T}$ *is* $\mathbb{T}$*-generic over* L, *and* $N \subseteq \omega$, $\Gamma \subseteq \lambda$ *are finite sets. Let* $Z \in \mathbf{L}[G]$, $Z \subseteq L$, *be a set definable in* $\mathbf{L}[G]$ *by a formula with parameters in* $\mathbf{L}$ *and those in the list*

$$\{\vec{\mathbf{X}}[G], \vec{\mathbf{Y}}[G]\} \cup \{\mathbf{x}^G[n] : n \in N\} \cup \{\mathbf{y}^G_\gamma : \gamma \in \Gamma\}.$$

Then $Z \in \mathbf{L}[\{\mathbf{x}^G[n] : n \in N\}, \{\mathbf{y}^G_\gamma : \gamma \in \Gamma\}]$.

Beginning the proof of Theorem 17, we put $\vec{\mathbf{x}}_N[G] = \{\mathbf{x}^G[n]\}_{n\in N}$ and $\vec{\mathbf{y}}_\Gamma[G] = \{\mathbf{y}^G_\gamma\}_{\gamma\in\Gamma}$, and let

$$\vartheta(z) := \vartheta(z, \vec{\mathbf{X}}[G], \vec{\mathbf{Y}}[G], \vec{\mathbf{x}}_N[G], \vec{\mathbf{y}}_\Gamma[G])$$

be a formula such that $Z = \{z : \vartheta(z)\}$ in $\mathbf{L}[G]$. By Lemma 13(iii) there is n_0 such that $\mathbf{a}^G(n,\gamma) \neq \mathbf{a}^G(n,\delta)$ whenever $n > n_0$ and $\gamma \neq \delta$ belong to Γ.

Let $M = N \cup \{n : n \leqslant n_0\}$. We say that a condition $t \in \mathbb{T}$ *complies with* $\vec{\mathbf{x}}_N[G], \vec{\mathbf{y}}_\Gamma[G]$ if $M \subseteq \operatorname{bas}(t)$, $\Gamma \subseteq |t|$, and

(I) if $n \in N$ and $\xi \in |h^t| \cap [\aleph^{\mathbf{L}}_n, \aleph^{\mathbf{L}}_{n+1})$ then $h^t_\xi \subset \mathbf{x}^G_\xi$,

(II) if $\gamma \in \Gamma$ then $q^t_\gamma \subset \mathbf{y}^G_\gamma$,

(III) if $n \in \operatorname{bas}(t)$ and $\gamma \in \Gamma$ then $a^t(n,\gamma) = \mathbf{a}^G(n,\gamma)$.

For instance, any condition $t \in G$ with $M \subseteq \operatorname{bas}(t)$, $\Gamma \subseteq |t|$ complies with $\vec{\mathbf{x}}_N[G], \vec{\mathbf{y}}_\Gamma[G]$ by obvious reasons.

It is quite clear that the set $\mathbb{T}_{N\Gamma}[G]$ of all conditions $t \in \mathbb{T}$ which comply with $\vec{\mathbf{x}}_N[G], \vec{\mathbf{y}}_\Gamma[G]$ belongs to $\mathbf{L}[\vec{\mathbf{x}}_N[G], \vec{\mathbf{y}}_\Gamma[G]]$. Therefore to prove the theorem it suffices to verify the following assertion:

> if $z \in \mathbf{L}$, $s, t \in \mathbb{T}_{N\Gamma}[G]$, and s forces $\vartheta(z)$, then t does not force $\neg\,\vartheta(z)$.

Contrary assumption 18. Suppose to the contrary that this fails, so that there exist $z \in \mathbf{L}$ and $s, t \in \mathbb{T}_{N\Gamma}[G]$ such that s forces $\vartheta(z)$ and t forces $\neg\,\vartheta(z)$.

The proof of Theorem 17 continues in §§ 8.1 and 8.2.

8.1 Proof of the definability lemma, part 1

Our temporary goal here will be to strengthen conditions s, t towards the requirements of Theorem 8 (the symmetry lemma).

Lemma 19. *There exists a condition* $s' \in \mathbb{T}_{N\Gamma}[G]$ *such that* $|s'| = |s| \cup |t|$, $\operatorname{bas}(s') = \operatorname{bas}(s) \cup \operatorname{bas}(t)$, *and* $s' \leqslant s$. *Accordingly there is a condition* $t' \in \mathbb{T}_{N\Gamma}[G]$ *such that* $|t'| = |s| \cup |t|$, $\operatorname{bas}(t') = \operatorname{bas}(s) \cup \operatorname{bas}(t)$, *and* $t' \leqslant t$.

Proof. We define $a^{s'}$; the definition splits into subdomains.

subdomain $\operatorname{bas}(s) \times |s|$. If $n \in \operatorname{bas}(s)$ and $\gamma \in |s|$ then put $a^{s'}(n,\gamma) = a^s(n,\gamma)$ and $q^{s'}(n,\gamma) = q^s(n,\gamma)$.

subdomain $(\operatorname{bas}(t) \smallsetminus \operatorname{bas}(s)) \times \Gamma$. If $n \in \operatorname{bas}(t) \smallsetminus \operatorname{bas}(s)$ and $\gamma \in \Gamma$ then put $a^{s'}(n,\gamma) = \mathbf{a}^G(n,\gamma)$, and $q^{s'}(n,\gamma) = q^s(n,\gamma)$, as above.

subdomain $(\text{bas}(t) \smallsetminus \text{bas}(s)) \times (|s| \smallsetminus \Gamma)$. For any $n \in \text{bas}(t) \smallsetminus \text{bas}(s)$ fix a bijection $\delta \longmapsto \xi^n_\delta$ from $|s| \smallsetminus \Gamma$ to $[\aleph^{\mathbf{L}}_n, \aleph^{\mathbf{L}}_{n+1}) \smallsetminus \{a^{s'}(n,\gamma) : \gamma \in \Gamma\}$. If now $\delta \in |s| \smallsetminus \Gamma$ then put $a^{s'}(n,\delta) = \xi^n_\delta$ and $q^{s'}(n,\delta) = \varnothing$.

subdomain $(\text{bas}(t) \cup \text{bas}(s)) \times (|t| \smallsetminus |s|)$. Fix an ordinal $\delta^* \in |s|$. If $n \in \text{bas}(t) \cup \text{bas}(s)$ and $\delta \in |t| \smallsetminus |s|$ then put $a^{s'}(n,\delta) = a^{s'}(n,\delta^*)$ and $q^{s'}(n,\delta) = q^{s'}(n,\delta^*)$.

subdomain $(\omega \smallsetminus (\text{bas}(t) \cup \text{bas}(s))) \times (|t| \smallsetminus |s|)$. If $n \notin \text{bas}(t) \cup \text{bas}(s)$ and $\delta \in |t| \cup |s|$ then put $q^{s'}(n,\delta) = \varnothing$ and keep $a^{s'}(n,\delta)$ undefined.

On the top of the above definition, define $h^{s'}$ so that

$$|h^{s'}| = |h^s| \cup \{\xi^n_\delta : n \in \text{bas}(t) \smallsetminus \text{bas}(s) \wedge \delta \in |s| \smallsetminus \Gamma\},$$

$h^{s'}_\xi = h^s_\xi$ for all $\xi \in |h^s|$, and $h^{s'}_{\xi^n_\delta} = \varnothing$ for all $n \in \text{bas}(t) \cup \text{bas}(s)$ and $\delta \in |t| \smallsetminus |s|$.

We claim that s' is as required. The key issue is to prove $a^{s'} \leqslant a^s$, in particular, (a4) of § 2 for $a = a^{s'}$, $b = a^s$. Note that if $\gamma \neq \delta$ belong to Γ and $n \notin \text{bas}(s)$ then $\mathbf{a}^G(n,\gamma) \neq \mathbf{a}^G(n,\delta)$ by the choice of M and because $M \subseteq \text{bas}(s)$. Therefore if $n \in \text{bas}(t) \smallsetminus \text{bas}(s)$ and γ, δ as indicated then by definition $a^{s'}(n,\gamma) \neq a^{s'}(n,\delta)$, as required.

We have (I), (II), (III) by obvious reasons: in particular, $q^{s'}_\gamma = q^s_\gamma$ for all $\gamma \in \Gamma$, and if $n \in N$ then $n \in |s|$ and hence by construction $|h^{s'}| \cap [\aleph^{\mathbf{L}}_n, \aleph^{\mathbf{L}}_{n+1}) = |h^s| \cap [\aleph^{\mathbf{L}}_n, \aleph^{\mathbf{L}}_{n+1})$ and $h^{s'}_\xi = h^s_\xi$ for all $\xi \in |h^{s'}| \cap [\aleph^{\mathbf{L}}_n, \aleph^{\mathbf{L}}_{n+1})$.

Q.E.D.

By the lemma, we can assume without loss of generality in Contrary assumption 18 that

(1) conditions s, t satisfy $|s| = |t|$ and $\text{bas}(s) = \text{bas}(t)$.

Moreover we can assume without loss of generality that in addition:

(2) $|h^s| = |h^t|$, and if $n \in \text{bas}(s) = \text{bas}(t)$ then the set $|h^s| \cap [\aleph^{\mathbf{L}}_n, \aleph^{\mathbf{L}}_{n+1}) = |h^t| \cap [\aleph^{\mathbf{L}}_n, \aleph^{\mathbf{L}}_{n+1})$ is infinite.

This is rather elementary. If say $\xi \in |h^s| \smallsetminus |h^t|$ then simply add ξ to $|h^t|$ and define $h^t_\xi = \varnothing$.

Further, we can assume without loss of generality that, in addition to Contrary assumption 18 and (1), (2), the following holds:

(3) conditions s, t satisfy $\text{ran}(a^s) = \text{ran}(a^t)$.

Indeed, suppose that $n \in \text{bas}(s)$ and, say, $\xi \in (\text{ran}(a^t) \smallsetminus \text{ran}(a^s)) \cap [\aleph_n^{\mathbf{L}}, \aleph_{n+1}^{\mathbf{L}})$. Put $\xi_n = \xi$ and for any $m \in \text{bas}(s)$, $m \neq n$ pick an ordinal $\xi_m \in |h^s| \cap [\aleph_m^{\mathbf{L}}, \aleph_{m+1}^{\mathbf{L}})$, $\xi_m \notin \text{ran}(a^s) \cup \text{ran}(a^t)$ (this is possible by (2)). Add an ordinal $\gamma \notin |s| = |t|$ to $|s|$ and to $|t|$. If $m \in \text{bas}(s) = \text{bas}(t)$ then put $a^s(m,\gamma) = a^t(m,\gamma) = \xi_m$ and $q_\gamma^s[m] = q_\gamma^t[m] = h_{\xi_m}^t$, and in addition define $q_\gamma^s[m] = q_\gamma^t[m] = \varnothing$ for all $m \notin \text{bas}(s) = \text{bas}(t)$. Conditions s, t extended this way still satisfy conditions of Contrary assumption 18, together with (1), (2), but now $\xi \in \text{ran}(a^s)$. One has to maintain such an extension for all indices ξ in $\text{ran}(a^t) \smallsetminus \text{ran}(a^s)$ and $\text{ran}(a^s) \smallsetminus \text{ran}(a^t)$ one by one; the details are left to the reader.

After this step, the sets $\Delta = |s| = |t|$ and $B = \text{bas}(s) = \text{bas}(t)$ (finite subsets of resp. λ and ω) will not be changed, as well as the assignments $a = a^s$ and $b = a^t$ ($\text{dom}(a) = \text{dom}(b) = B \times \Delta$). We put $\Xi = |h^s| = |h^t|$.

Further we can assume without loss of generality that in addition:

(4) subconditions q^s, q^t are uniform and equally shaped.

It suffices to define a pair of stronger conditions $s', t' \in \mathbb{T}_{N\Gamma}[G]$ such that

$$|s'| = |t'| = \Delta\,,\ |h^{s'}| = |h^{t'}| = \Xi\,,\ \text{bas}(s') = \text{bas}(t') = B\,,\ a^{s'} = a\,,\ a^{t'} = b\,,$$

and in addition $q^{s'}, q^{t'}$ are uniform and equally shaped.

Consider any $n \in \omega$. Put $d[n] = \bigcup_{\delta \in \Delta}(\text{dom}(q_\delta^s[n]) \cup \text{dom}(q_\delta^t[n]))$, a set in $\mathbb{D}[n]$. If $\delta \in \Delta$, then we define extensions $q_\delta^{s'}[n], q_\delta^{t'}[n] \in \mathbb{P}[n]$ of resp. $q_\delta^s[n], q_\delta^t[n]$ so that

(i) $\text{dom}(q_\delta^{s'}[n]) = \text{dom}(q_\delta^{t'}[n]) = d[n]$,

(ii) if $n \in N$ and $\delta \in \Gamma$ then simply $q_\delta^{s'}[n] = q_\delta^{t'}[n] = \mathbf{y}_\delta^G \restriction d[n]$,

(iii) if $n \in B$ and $\gamma, \delta \in \Delta$ then: if $a(n,\delta) = a(n,\gamma)$ then $q_\delta^{s'}[n] = q_\gamma^{s'}[n]$, and if $b(n,\delta) = b(n,\gamma)$ then $q_\delta^{t'}[n] = q_\gamma^{t'}[n]$.

In addition to this, define $h_{a(n,\delta)}^{s'} = q_\delta^{s'}[n]$ and $h_{b(n,\delta)}^{t'} = q_\delta^{t'}[n]$ for all $n \in B$ and $\delta \in \Delta$. In the rest, put $|h^{s'}| = |h^{t'}| = \Xi$ (recall that $\Xi = |h^s| = |h^t|$), and $h_\xi^{s'} = h_\xi^s$, $h_\xi^{t'} = h_\xi^t$ for all $\xi \in \Xi$ *not* in $\text{ran}(a) = \text{ran}(b)$.

Further, we can assume without loss of generality that, in addition to assumptions (1)–(4):

(5) conditions s, t coincide on the domain $N \times \Gamma$, so that

(a) if $\gamma \in \Gamma$ then $q_\gamma^s = q_\gamma^t$,

(b) if $n \in N$ then $h^s[n] = h^t[n]$, i.e., $h_\xi^s = h_\xi^t$ for all $\xi \in |h^s| \cap [\aleph_n^{\mathbf{L}}, \aleph_{n+1}^{\mathbf{L}}) = |h^t| \cap [\aleph_n^{\mathbf{L}}, \aleph_{n+1}^{\mathbf{L}})$, and

(c) if $n \in N$ and $\gamma \in \Gamma$ then $a^s(n,\gamma) = a^t(n,\gamma) = \mathbf{a}^G(n,\gamma)$—but this already follows from the compliance assumption.

Regarding (5)a, note that this is already done. Indeed, q^s, q^t are equally shaped by (4), and satisfy $q^s_\gamma \subset \mathbf{y}^G_\gamma$ and $q^t_\gamma \subset \mathbf{y}^G_\gamma$ by Contrary assumption 18, therefore $q^s_\gamma = q^t_\gamma$. Now consider (5)b; suppose that $n \in N$. Let $\xi \in |h^s| \cap [\aleph^{\mathbf{L}}_n, \aleph^{\mathbf{L}}_{n+1})$.

If $\xi \in \operatorname{ran}(a^s) = \operatorname{ran}(a^t)$ then $\xi = a^s(n,\gamma) = a^t(n,\delta)$ for some $\gamma, \delta \in \Delta$, and then $h^s_\xi = q^s_\gamma[n]$ and $h^t_\xi = q^t_\gamma[n]$. It follows that $\operatorname{dom}(h^s_\xi) = \operatorname{dom}(h^t_\xi)$, by (4). Therefore $h^s_\xi = h^t_\xi$, because we have $h^s_\xi \subset \mathbf{x}^G_\xi$ and $h^t_\xi \subset \mathbf{x}^G_\xi$.

If $\xi \in \operatorname{ran}(a^s) = \operatorname{ran}(a^t)$ then still $h^s_\xi \subset \mathbf{x}^G_\xi$ and $h^t_\xi \subset \mathbf{x}^G_\xi$, thus h^s_ξ and h^t_ξ are compatible as conditions in $\mathbb{P}$, and we simply replace either of them by $h^s_\xi \cup h^t_\xi$.

Finally, we can assume without loss of generality that in addition to Contrary assumption 18 and (1)–(5) the following holds:

(6) we have $\{h^s_\xi : \xi \in |h^s|\} = \{h^t_\xi : \xi \in |h^t|\}$ as in (g) of Definition 7, and subconditions h^s, h^t are regular on every $n \in B \setminus N$ (§ 2.2).

The equality $\{h^s_\xi : \xi \in |h^s| \cap [\aleph^{\mathbf{L}}_n, \aleph^{\mathbf{L}}_{n+1})\} = \{h^t_\xi : \xi \in |h^t| \cap [\aleph^{\mathbf{L}}_n, \aleph^{\mathbf{L}}_{n+1})\}$ holds already for all $n \in N$ by (5).

Now suppose that $n \in B \setminus N$. The requirement of compliance with $\vec{\mathbf{x}}_N[G], \vec{\mathbf{y}}_\Gamma[G]$ is void for $n \notin N$, therefore we can simply extend $h^s[n]$ and $h^t[n]$ to a bigger domain and appropriately define h^s_ξ and h^t_ξ for all "new" elements ξ in these extended domains so that (6) holds, without changing q^s, q^t and a^s, a^t.

To conclude, we can assume without loss of generality in Contrary assumption 18 that (1)–(6) hold, i.e., in other words, conditions $s, t \in \mathbb{T}_{N\Gamma}[G]$ are strongly similar on $N \times \Gamma$ in the sense of Definition 7.

8.2 Proof of the definability lemma, part 2

We now continue the proof of Theorem 17. Our intermediate result and the starting point of the final part of the proof is Contrary assumption 18 with the additional assumption that conditions $s, t \in \mathbb{T}_{N\Gamma}[G]$ as in Contrary assumption 18 are strongly similar on $N \times \Gamma$, and to complete the proof of the theorem it suffices to derive a contradiction. This will be obtained by means of Theorem 8.

In accordance with Theorem 8, let $B = \operatorname{bas}(s) = \operatorname{bas}(t)$, $\Delta = |s| = |t|$, and let transformations π, $\mathbf{S}_{a^u a^t}$, φ and $\tau = \varphi \circ \mathbf{S}_{a^u a^t} \circ \pi$, and conditions $v, u \in \mathbb{T}$ satisfy $\operatorname{bas}(u) = \operatorname{bas}(v) = B$, $|u| = |v| = \Delta$, and the following:

(i) $\pi \in \Pi_{\text{fin}}$, $\pi[n]$ is the identity for all $n \in N$, $u = \pi \cdot s$, u is strongly similar to t on $N \times \Gamma$, and moreover $\pi \cdot h^s = h^u = h^t$, and $a^u {\upharpoonright\upharpoonright} \Gamma = a^t {\upharpoonright\upharpoonright} \Gamma$;

(ii) $v = \mathbf{S}_{a^u a^t} \cdot u$, v is strongly similar to t on $N \times \Gamma$, $h^v = h^u$, $a^v = a^t$;

(iii) $\varphi \in \mathbb{FW}_{a^v}$, $|\varphi| = \Delta$, $\varphi_\gamma[n] = \varnothing$ for all $n \in B$ and $\gamma \in \Delta$, and $t = \varphi \cdot v$;

(iv) $\tau = \varphi \circ \mathbf{S}_{a^u a^t} \circ \pi$ is an order preserving bijection from $\mathbb{T}_{\leqslant s}$ onto $\mathbb{T}_{\leqslant t}$;

(v) any condition $s' \in \mathbb{T}_{\leqslant s}$ is similar to $t' = \tau \cdot s'$ on $N \times \Gamma$.

(these are items (i)–(v) of Theorem 8).

Consider a set $G \subseteq \mathbb{T}$ generic over $\mathbf{L}$ and containing s. We assume that s is the largest (i.e., weakest) condition in G. Then, by (v), $H = \{\tau \cdot s' : s' \in G\} \subseteq \mathbb{T}$ is generic over $\mathbf{L}$ either, and $\mathbf{L}[H] = \mathbf{L}[G]$. Moreover $t = \tau \cdot s \in H$. Therefore it follows from Contrary assumption 18 that

$$\vartheta(z, \vec{\mathbf{X}}[G], \vec{\mathbf{Y}}[G], \vec{\mathbf{x}}_N[G], \vec{\mathbf{y}}_\Gamma[G]) \text{ is true in } \mathbf{L}[G], \text{ but} \qquad (\dagger)$$
$$\vartheta(z, \vec{\mathbf{X}}[H], \vec{\mathbf{Y}}[H], \vec{\mathbf{x}}_N[H], \vec{\mathbf{y}}_\Gamma[H]) \text{ is false in } \mathbf{L}[H] = \mathbf{L}[G].$$

Our strategy to derive a contradiction will be to show that the parameters in the formulas are pairwise equal, and hence one and the same formula is simultaneously true and false in one and the same class. This is the content of the following lemma.

Lemma 20. (i) $\vec{\mathbf{y}}_\Gamma[G] = \vec{\mathbf{y}}_\Gamma[H]$, *i.e., if* $\gamma \in \Gamma$ *then* $\mathbf{y}_\gamma^G = \mathbf{y}_\gamma^H$;

(ii) $\vec{\mathbf{x}}_N[G] = \vec{\mathbf{x}}_N[H]$, *i.e., if* $n \in N$ *and* $\xi \in [\aleph_n^{\mathbf{L}}, \aleph_{n+1}^{\mathbf{L}})$ *then* $\mathbf{x}_\xi^G = \mathbf{x}_\xi^H$;

(iii) $\vec{\mathbf{X}}[G] = \vec{\mathbf{X}}[H]$, *i.e.,* $\mathbf{X}^G[n] = \mathbf{X}^H[n]$ *for all* $n \in \omega$;

(iv) $\vec{\mathbf{Y}}[G] = \vec{\mathbf{Y}}[H]$, *i.e.,* $\mathbf{Y}_\gamma^G = \mathbf{Y}_\gamma^H$ *for all* $\gamma < \lambda$.

Proof. (i) If $\gamma \in \Gamma$ then by definition, $\mathbf{y}_\gamma^G = \bigcup_{s' \in G} q_\gamma^{s'}$ and $\mathbf{y}_\gamma^H = \bigcup_{t' \in H} q_\gamma^{t'} = \bigcup_{s' \in G} q_\gamma^{(\tau \cdot s')}$. Yet if $s' \in G$ then condition $t' = \tau \cdot s'$ satisfies $q_\gamma^{t'} = q_\gamma^{s'}$ by (v).

(ii) This is a similar argument. Suppose that $n \in N$ and $\xi \in [\aleph_n^{\mathbf{L}}, \aleph_{n+1}^{\mathbf{L}})$. By definition, $\mathbf{x}_\xi^G = \bigcup_{s' \in G} h_\xi^{s'}$ and $\mathbf{x}_\xi^H = \bigcup_{t' \in H} h_\xi^{t'} = \bigcup_{s' \in G} h_\xi^{(\tau \cdot s')}$. However if $s' \in G$ then condition $t' = \tau \cdot s'$ satisfies $h_\xi^{t'} = h_\xi^{s'}$ by (v).

(iii) By definition, $\mathbf{X}^G[n]$ and $\mathbf{X}^H[n]$ are the $(\Pi_{\text{fin}}, \mathbb{FN})$-hulls of resp.

$$\mathbf{x}^G[n] = \{\mathbf{x}_\xi^G\}_{\xi \in [\aleph_n^{\mathbf{L}}, \aleph_{n+1}^{\mathbf{L}})} \quad \text{and} \quad \mathbf{x}^H[n] = \{\mathbf{x}_\xi^H\}_{\xi \in [\aleph_n^{\mathbf{L}}, \aleph_{n+1}^{\mathbf{L}})}.$$

Thus it remains to prove that $\mathbf{x}^H[n]$ belongs to the $(\Pi_{\text{fin}}, \mathbb{FN})$-hull of $\mathbf{x}^G[n]$, and vice versa. Let $\psi = \varphi \downarrow a^v$ (a flip in $\mathbb{FN}$, cf. § 5). By definition, if $s' \in G$ and $t' = \tau \cdot s'$, then the subconditions $h^{s'}$ and $h^{t'}$ satisfy $h^{t'} = \psi \cdot (\pi \cdot h^{s'})$. (The middle transformation $\mathbf{S}_{a^u a^t}$ does not act on the h-components). It easily follows that $\mathbf{x}^H[n] = \psi \cdot (\pi \cdot \mathbf{x}^G[n])$, as required.

(iv) Note that the sequences $\vec{\mathbf{y}}_\Gamma[G] = \{\mathbf{y}^G_\gamma\}_{\gamma<\lambda}$ and $\vec{\mathbf{y}}_\Gamma[H] = \{\mathbf{y}^H_\gamma\}_{\gamma<\lambda}$ satisfy $\vec{\mathbf{y}}_\Gamma[H] = \varphi\cdot(\mathbf{S}_{a^u a^t}\cdot\vec{\mathbf{y}}_\Gamma[G])$ (since the permutation π does not act on wide subconditions). That is, the construction of $\vec{\mathbf{y}}_\Gamma[H]$ from $\vec{\mathbf{y}}_\Gamma[G]$) has two steps.

Step 1. We define $\vec{\mathbf{r}} = \{\mathbf{r}_\gamma\}_{\gamma<\lambda}$ by $\vec{\mathbf{r}} = \mathbf{S}_{a^u a^t}\cdot\vec{\mathbf{y}}_\Gamma[G]$. Thus, by definition,

(1) $\mathbf{r}_\gamma \in 2^{[\omega,\aleph^{\mathbf{L}}_\omega)}$ for all γ,

(2) if $n \notin B$ or $\gamma \notin \Delta$ then $\mathbf{r}_\gamma[n] = \mathbf{y}^G_\gamma[n]$, and

(3) if $n \in B$ and $\gamma \in \Delta$ then $\mathbf{r}_\gamma[n] = \mathbf{y}^G_\vartheta[n]$, where $\vartheta = \mathbf{s}^n_{a^u a^t}(\gamma)$.

Thus the difference between $\vec{\mathbf{r}}$ and $\vec{\mathbf{y}}_\Gamma[G]$ is located within the finite domain $B \times \Delta$. Moreover, as in Lemma 3(iii), we have

$$\{\mathbf{r}_\gamma[n] : \gamma < \lambda\} = \{\mathbf{y}^G_\gamma[n] : \gamma < \lambda\} \text{ for every } n. \tag{‡}$$

Step 2. We define $\vec{\mathbf{y}}_\Gamma[H] = \varphi\cdot\vec{\mathbf{r}}$. Thus, by definition,

(4) if $\gamma \in \Delta$ then directly $\mathbf{y}^H_\gamma = \varphi\cdot\mathbf{r}_\gamma$, i.e., $\mathbf{y}^H_\gamma[n] = \varphi[n]\cdot\mathbf{r}_\gamma[n]$, $\forall n$;

(5) if $\gamma \notin \Delta$ and $\exists\delta \in \Delta\,(\mathbf{a}^G(n,\gamma) = \mathbf{a}^G(n,\delta))$, then $\mathbf{y}^H_\gamma[n] = \varphi[n]\cdot\mathbf{r}_\gamma[n]$;

(6) if $\gamma \notin \Delta$ but $\neg\,\exists\delta \in \Delta\,(\mathbf{a}^G(n,\gamma) = \mathbf{a}^G(n,\delta))$, then $\mathbf{y}^H_\gamma[n] = \mathbf{r}_\gamma[n]$.

Now it immediately follows from (‡) that $\mathbf{Y}^G[n] = \mathbf{Y}^H[n]$ for every n: both sets are equal to the $\mathbb{D}[n]$-hull of one and the same set mentioned in (‡).

We are ready to prove that $\mathbf{Y}^G_\gamma = \mathbf{Y}^H_\gamma$ for every $\gamma < \lambda$.

We start with a couple of definitions. If $y, y' \in 2^{[\aleph^{\mathbf{L}}_n,\aleph^{\mathbf{L}}_{n+1})}$ and there exists a set $d \in \mathbf{Y}^G[n] = \mathbf{Y}^H[n]$ such that $y' = d\cdot y$ then write $y \equiv_n y'$. If $y, y' \in 2^{[\omega,\aleph^{\mathbf{L}}_\omega)}$ and there exists a number n_0 such that $y'[n] \equiv_n y[n]$ for all $n < n_0$ and $y'[n] = y[n]$ for all $n \ge n_0$ then write $y \equiv^* y'$. Then by $(\star)$ in § 7 we have:

$$\left.\begin{array}{rcl} \mathbf{Y}^G_\gamma &=& \{z \in 2^{[\omega,\aleph^{\mathbf{L}}_\omega)} : \exists y \in \mathbb{D}\cdot\mathbf{y}^G_\gamma\,(z \equiv^* y)\} \\ \mathbf{Y}^H_\gamma &=& \{z \in 2^{[\omega,\aleph^{\mathbf{L}}_\omega)} : \exists y \in \mathbb{D}\cdot\mathbf{y}^H_\gamma\,(z \equiv^* y)\} \end{array}\right\} \tag{**}$$

and hence to prove $\mathbf{Y}^G_\gamma = \mathbf{Y}^H_\gamma$ it suffices to check that $\mathbf{y}^H_\gamma \in \mathbf{Y}^G_\gamma$ and $\mathbf{y}^G_\gamma \in \mathbf{Y}^H_\gamma$.

Case 1: $\gamma \in \Delta$. It follows from (2) and (3) that $\mathbf{r}_\gamma \equiv^* \mathbf{y}^G_\gamma$ and hence $\mathbf{y}^H_\gamma \equiv^* y$ by (4), where $y = \varphi\cdot\mathbf{y}^G_\gamma$. Thus $\mathbf{y}^H_\gamma \in \mathbf{Y}^G_\gamma$ by the first part of (**). On the other hand, $\mathbf{r}_\gamma = \varphi^{-1}\cdot\mathbf{y}^H_\gamma$ still by 4), so that $\mathbf{y}^G_\gamma \in \mathbf{Y}^H_\gamma$ by the second part of (**).

Case 2: $\gamma \notin \Delta$. Note that for a given γ, (5) holds only for finitely many numbers n by Lemma 13(iii), so (6) holds for almost all n. Therefore $\mathbf{y}^H_\gamma \equiv^* \mathbf{r}_\gamma$. But $\mathbf{r}_\gamma = \mathbf{y}^G_\gamma$ in this case by (2). Thus $\mathbf{y}^H_\gamma \in \mathbf{Y}^G_\gamma$ by (**), the first line (with $y = \mathbf{y}^G_\gamma$). And $\mathbf{y}^G_\gamma \in \mathbf{Y}^H_\gamma$ holds by a similar argument. Q.E.D.

9 The structure of the symmetric subextension

Here we accomplish the proof of Theorem 1. We fix a set $G \subseteq \mathbb{T}$, $\mathbb{T}$-generic over $\mathbf{L}$, for this section. It will be shown that the symmetric subextension $\mathbf{L}_{\text{sym}}[G] = L(W[G])$ (cf. § 7.2) satisfies Theorem 1. The following is a key technical claim.

Theorem 21. *Suppose that $\nu < \omega$, and $Z \in \mathbf{L}_{\text{sym}}[G]$, $Z \subseteq [0, \aleph_{\nu+1})$. Then $Z \in \mathbf{L}[\{\mathbf{x}^G[n] : n \leqslant \nu\}]$.*

Proof. It follows from Lemma 16 and Theorem 17 that there exist finite sets $N \subseteq \omega$ and $\Gamma \subseteq \lambda$ such that $Z \in \mathbf{L}[\{\mathbf{x}^G[n] : n \in N\}, \{\mathbf{y}^G_\gamma : \gamma \in \Gamma\}]$. We can assume that

(1) $N = \{0, 1, 2, \ldots, \kappa\}$ for some $\kappa < \omega$, $\kappa \geq \nu$;

(2) if $\gamma \neq \delta$ belong to Γ and $n < \omega$ satisfies $\mathbf{a}^G(n, \gamma) = \mathbf{a}^G(n, \delta)$ then $n \leqslant \kappa$.

(Lemma 13(iii) is used to justify (2).) Define in $\mathbf{L}$

$$\mathbb{T}[N, \Gamma] = \{s \in \mathbb{T} : \text{bas}(s) = N \wedge |s| = \Gamma \wedge a^s = \mathbf{a}^G \restriction (N \times \Gamma)\}.$$

Lemma 22. *The set $G[N, \Gamma] = G \cap \mathbb{T}[N, \Gamma]$ is $\mathbb{T}[N, \Gamma]$-generic over $\mathbf{L}$ and $Z \in \mathbf{L}[G[N, \Gamma]]$.*

Proof. Suppose that $t \in \mathbb{T}$, $N \subseteq \text{bas}(t)$, $\Gamma \subseteq |t|$. Define the *projection* $s = t[N, \Gamma] \in \mathbb{T}[N, \Gamma]$ so that $q^s = q^t \restriction \Gamma$, $a^s = a^t \restriction (N \times \Gamma)$, and h^s is the restriction of h^t to the set $|h^t| \cap \bigcap_{n \leqslant \kappa} [\aleph_n, \aleph_{n+1})$ (it is not asserted that $t \leqslant s$). Given a condition $s' \in \mathbb{T}[N, \Gamma]$, $s' \leqslant s$, we have to accordingly find a condition $t' \leqslant t$ such that $t'[N, \Gamma] = s'$.

We define t' as follows. First of all, $\text{bas}(t)' = \text{bas}(t)$, $|t'| = |t|$, $a^{t'} = a^t$. Put $h^{t'}[n] = h^{s'}[n]$ for $n \in N$ but $h^{t'}[n] = h^t[n]$ for $n \in \text{bas}(t) \smallsetminus N$. If $n \notin N$ then put $q^{t'}_\gamma[n] = q^{t'}_\gamma[n]$ for all $\gamma \in |t'| = |t|$. If $n \in N$ and $\gamma \in |t'|$ then put $q^{t'}_\gamma[n] = h^{t'}_\xi = h^{s'}_\xi$, where $\xi = a^{t'}(n, \gamma)$. Q.E.D. (Lemma 22)

In continuation of the proof of Theorem 21, let us consider $\mathbb{T}[N, \Gamma]$ as the forcing notion. It appears to be quite similar to the product $\prod_{n=0}^{\kappa} \mathbb{H}[n] \times \mathbb{P}^\Gamma$: indeed, if $s \in \mathbb{T}[N, \Gamma]$ then h^s can be seen as an element of $\prod_{n=0}^{\kappa} \mathbb{H}[n]$, q^s can be seen as an element of $\mathbb{P}^\Gamma$ (the product of $\text{card}\,\Gamma$ copies of $\mathbb{P}$;

$\operatorname{card}\Gamma < \omega$), while $a^s = \mathbf{a}^G \restriction (N \times \Gamma)$ is a constant. However if $n \in N$ and $\gamma \in \Gamma$ then $q^s_\gamma[n] = q^t_{a^s(n,\gamma)}$, hence in fact $\mathbb{T}[N,\Gamma]$ can be identified with

$$\prod_{n=0}^{\kappa} \mathbb{H}[n] \times \Big(\prod_{n=\kappa+1}^{\infty} \mathbb{P}[n]\Big)^\Gamma = \prod_{n=0}^{\kappa} \mathbb{H}[n] \times \prod_{n=\kappa+1}^{\infty} (\mathbb{P}[n]^\Gamma). \qquad (\#)$$

However the sets $\mathbb{P}[n]$ and $\mathbb{H}[n]$ as forcing notions are $\aleph_{n+1}$-closed, meaning that any decreasing sequence of length $\leqslant \aleph_n$ has a lower bound in the same set. Therefore if we present $\mathbb{T}[N,\Gamma]$ as

$$\prod_{n=0}^{\nu} \mathbb{H}[n] \times \prod_{n=\nu+1}^{\kappa} \mathbb{H}[n] \times \prod_{n=\kappa+1}^{\infty} (\mathbb{P}[n]^\Gamma), \qquad (\P)$$

then it becomes clear that the second and third subproducts are $\aleph_{\nu+2}$-closed forcing notions (in $\mathbf{L}$). Hence, by basic results of forcing theory, the set $Z \subseteq [0, \aleph^{\mathbf{L}}_{\nu+1})$ belongs to the subextension corresponding to the first subproduct $\prod_{n=0}^{\nu} \mathbb{H}[n]$. That is, $Z \in \mathbf{L}[\{\mathbf{x}^G[n] : n \leqslant \nu\}]$, as required.

Q.E.D. (Theorem 21)

Corollary 23. *If $n < \omega$ then it is true in $\mathbf{L}_{\text{sym}}[G]$ that $\aleph^{\mathbf{L}}_n$ remains a cardinal, the power set $\wp(\aleph^{\mathbf{L}}_n)$ is wellorderable, and* $\operatorname{card}(\wp(\aleph^{\mathbf{L}}_n)) = \aleph^{\mathbf{L}}_{n+1}$.

Yet cardinal preservation holds for all cardinals.

Corollary 24. *Any cardinal in $\mathbf{L}$ remains a cardinal in $\mathbf{L}_{\text{sym}}[G]$.*

Proof. Indeed we have established (cf. the proof of Theorem 21) that any set $Z \in \mathbf{L}_{\text{sym}}[G]$, $Z \subseteq L$, belongs to a generic extension of $\mathbf{L}$ via a forcing as in (#) in the proof of Theorem 21. However any such a forcing is cardinal-preserving by a simple cardinality argument. Q.E.D. (Theorem 21)

To accomplish the proof of Theorem 1, it remains to check that the symmetric subextension $\mathbf{L}_{\text{sym}}[G]$ contains a surjection $\sigma : 2^{[\omega, \aleph^{\mathbf{L}}_\omega)} \to \lambda$. We define σ in $\mathbf{L}_{\text{sym}}[G]$ as follows. If $\gamma < \lambda$ and $z \in \mathbf{Y}^G_\gamma$ then put $\sigma(z) = \gamma$ (the definition is consistent by Lemma 14). If $z \in 2^{[\omega, \aleph^{\mathbf{L}}_\omega)}$ does not belong to $\bigcup_{\gamma<\lambda} \mathbf{Y}^G_\gamma$ then $\sigma(z) = 0$. As any set $\mathbf{Y}^G_\gamma$ definitely contains $\mathbf{y}^G_\gamma$, σ is a surjection onto λ, as required.

References

[1] A. W. Apter and P. Koepke. The consistency strength of choiceless failures of SCH. *Journal of Symbolic Logic*, 75(3):1066–1080, 2010.

[2] P. J. Cohen. *Set theory and the continuum hypothesis*. W. A. Benjamin, 1966.

[3] W. B. Easton. Powers of regular cardinals. *Annals of Pure and Applied Logic*, 1:139–178, 1970.

[4] M. Gitik. The strength of the failure of the singular cardinal hypothesis. *Annals of Pure and Applied Logic*, 51(3):215–240, 1991.

[5] M. Gitik and P. Koepke. Violating the singular cardinals hypothesis without large cardinals. *Israel Journal of Mathematics*, 191(2):901–922, 2012.

[6] K. Gödel. *The Consistency of the Continuum Hypothesis*, volume 3 of *Annals of Mathematics Studies*. Princeton University Press, 1940.

[7] D. Hilbert. Mathematical problems. Lecture delivered before the international congress of mathematicians at Paris in 1900. Translated by Mary Winston Newson. *Bulletin of the American Mathematical Society*, 8:437–479, 1902.

[8] S. Koppelberg. General theory of Boolean algebras. In J. D. Monk and R. Bonnet, editors, *Handbook of Boolean algebras. Vol. 1*, pages 741–773. North-Holland, 1989.

[9] G. H. Moore. Towards a history of cantor's continuum problem. In D. E. Rowe and J. McCleary, editors, *The history of modern mathematics. Vol. I, Proceedings of the symposium held at Vassar College, Poughkeepsie, New York, June 20–24, 1989, Ideas and their reception*, pages 79–121. Academic Press, 1989.

[10] J. Silver. On the singular cardinals problem. In *Proceedings of the International Congress of Mathematicians. Volume 1, Held in Vancouver, B.C., August 21–29, 1974*, pages 265–268. Canadian Mathematical Congress, 1975.

On compactness for being λ-collectionwise Hausdorff

Menachem Magidor*

Einstein Institute of Mathematics, Hebrew University of Jerusalem, Jerusalem, Israel

1 Introduction

Reflection principles (or dually compactness principles) are some of the most fruitful sources for set theoretical problems. A reflection principle is the statement that for a certain class of structures, if the the structure has a certain property then there is a smaller cardinality substructure with the same property. A compactness property is the statement for a structure in a given class, if every smaller cardinality substructure has a certain property then the whole structure has this property. Reflection properties and compactness properties come in dual pairs: A reflection principle for a certain property is equivalent to a compactness property for the negation of the property and vice versa.

In this paper we shall make some remarks on the property of a topological space being λ-collectionwise Hausdorff.

Definition 1.1.
1. A subset Y of a topological space X is separated if there is a family $\{U_y \mid y \in Y\}$ of mutually disjoint open sets such that for all $y \in Y$, $U_y \cap Y = \{y\}$.
2. The space X is said to be collectionwise Hausdorff (cwH) if every discrete closed subset of X can be separated.
3. The space X is said to be λ-collectionwise Hausdorff (λ-cwH) if every closed discrete subset of X of cardinality less than λ is separated.

The property of cwH was extensively studied as part of the study of conditions of metrizablity of the space X. Cf., e.g., [4, 5]. In particular there is interest in identifying situations in which λ-cwH implies cwH. Generalizing the famous counterexample of Bing [2] one can get (cf. [5]) for every λ a space X_λ such that X is λ-cwH but not λ^+-cwH. So if we want to have a compactness result of the form "λ-cwH implies λ^+-cwH" we need to restrict the spaces for which we can expect the compactness to hold.

*A paper dedicated to Peter Koepke and Philip Welch on their 60th birthdays: a tribute to dear friends and great colleagues. The research presented in this paper was supported by Israel Science Foundation Grant no. 817/11.

Stefan Geschke, Benedikt Löwe, Philipp Schlicht (*eds.*).
Infinity, computability, and metamathematics: Festschrift celebrating the 60th birthdays of Peter Koepke and Philip Welch. College Publications, London, 2014. Tributes, Volume 23.

Definition 1.2. 1. A topological space X is said to be locally of cardinality $< \lambda$ if every $x \in X$ has a neighborhood of cardinality less than λ. The cardinal $\lambda(X)$ is is the minimal cardinal λ such that X is locally of cardinality $< \lambda$.

2. By Λ_κ we denote the class of spaces X such that $\lambda(X) \leq \kappa$ and $\Lambda_{<\kappa} = \bigcup_{\eta<\kappa} \Lambda_\eta$.

3. The character of the space X $(\chi(X))$ is the minimal cardinal χ such that every point of X has a neighborhood base of cardinality less than χ

4. By Ω_κ we denote the class of spaces X such that $\chi(X) \leq \kappa$ and $\Omega_{<\kappa} = \bigcup_{\eta<\kappa} \Omega_\eta$.

Definition 1.3. 1. A cardinal κ is said to be cwH weakly compact for the the class of topological spaces $\mathcal{C}$ if every space $X \in \mathcal{C}$ which is κ-cwH is κ^+-cwH.

2. A cardinal κ is said to be cwH compact for the the class of topological spaces $\mathcal{C}$ if every space $X \in \mathcal{C}$ which is κ-cwH is cwH.

3. A cardinal κ is said to be cwH weakly compact if it is cwH weakly compact for $\Lambda_{<\kappa}$.

4. A cardinal κ is said to be cwH compact if it it is cwH compact for $\Lambda_{<\kappa}$.

It can be seen (cf. [4]) that if κ is weakly compact then it is is cwH weakly compact for both Λ_κ and Ω_κ. Similarly if κ is a strongly compact cardinal then it is cwH compact for both Λ_κ and Ω_κ. The following theorem shows that, as expected, in the constructible universe there is no compact cwH for similar classes of spaces.

Theorem 1.4 (Shelah [11]). Let κ be regular cardinal such that there is stationary set $S \subseteq \kappa$ which does not reflect and such that for $\alpha \in S$, $\mathrm{cf}(\alpha) = \omega$. Then there is a topological space X such that $\lambda(X) = \chi(X) = \omega_1$ such that X is κ-cwH but not κ^+-cwH. So if $V = L$ there is no cardinal which is cwH compact even for Λ_{ω_1} or for Ω_{ω_1}.

This theorem can be easily generalized as follows:

Theorem 1.5. Let $\lambda < \kappa$ be regular cardinals such that there is a stationary set $S \subseteq \kappa$ such that for $\alpha \in S$, $\mathrm{cf}(\alpha) = \lambda$ such that S does not reflect. Then there is a space X such that $\lambda(X) = \chi(X) = \lambda^+$ and such X is λ-cwH but not λ^+-cwH.

Another case of incompactness was proved in [5]. There the authors show that under appropriate set-theoretic assumptions (which hold for instance in the constructible universe L) one can get a singular cardinal λ which is not cwH-weakly compact for the class $\Omega_{(2^{\aleph_0})^+}$. By further forcing they get a singular cardinal which is not cwH for the class of first countable spaces. (i.e., $\Omega_{\aleph_1}$.) The space they construct has local cardinality λ so it is in Λ_{λ^+}. So it is not a counterexample for λ being cwH weakly compact.

On the other hand it is consistent (modulo the consistency of a supercompact cardinal) that a small cardinal can be compact for an interesting class of spaces.

Theorem 1.6 (Shelah [11]). If the existence of supercompact cardinal is consistent then there is a model of set theory such that ω_2 is cwH compact for Λ_{ω_1}. Also in the the same model ω_2 is cwH compact for the class of spaces X such that $X \in \Lambda_{\omega_2} \cap \Omega_{\omega_1}$.

In fact the model in the above theorem can be obtained by Lévy collapsing all the cardinals less than the supercompact cardinal κ to ω_1 (such that κ becomes ω_2). A natural question is whether the condition $\chi(X) = \omega_1$ can be omitted from the second clause of the last theorem. Also the question whether we can get the same kind of result if the parameters in the above theorem are moved up by one cardinal.

The purpose of this paper is to make some remarks about the cardinals that can be cwH weakly compact or cwH compact for Λ_κ for various cardinals κ. (In particular we show that the answer to the question in the previous paragraph is "No".) There is a substantial study of that cardinals that can be compact with respect to different properties. (Cf., e.g., [10, 9].) A representative example is compactness for the property of an Abelian group being free, namely for what cardinals κ is an Abelian group such that every smaller cardinality subgroup is free itself free. A typical theorem that holds for many properties is a *pump-up lemma* which states that if there is a counter example to the compactness property at a regular cardinal κ then this example can be pumped up to κ^+. For instance, the following theorem holds.

Theorem 1.7 (Eklof, [3]). Let κ be a regular cardinal such that there is a non free Abelian group G such that every smaller cardinality subgroup is free. Then there is a non free Abelian group of cardinality κ^+ such that every smaller cardinality subgroup is free.

This theorem is prototype of a much more general phenomenon. On the other hand for singular cardinals we have the famous Shelah singular cardinals compactness (again, it is much more general than the statement here):

Theorem 1.8 (Shelah, [10]). Let κ be a singular cardinal. Let G be an Abelian group of cardinality κ such that every smaller cardinality subgroup is free. Then G is free.

In § 2 we show that in the case of the property cwH a theorem corresponding to Theorem 1.7 can fail.

Theorem 1.9. 1. Let κ be a weakly compact cardinal and $\lambda < \kappa$ a regular cardinal.Let $V[G]$ be the model obtained by Lévy collapsing all the cardinals less than κ to λ, thus making $\kappa = \lambda^+$. Then in $V[G]$, κ is cwH weakly compact.

2. Suppose that it is consistent to assume the existence a proper class of supercompact cardinals Then it is consistent to assume that every successor of a regular cardinal is cwH weakly compact.

For instance it is consistent (assuming the consistency of a weakly compact cardinal) that ω_3 is cwH weakly compact. By Theorem 1.5 in every universe of set theory there is a space X with $\lambda(X) = \omega_2$ which is ω_2-cwH but not ω_3-cwH. But in the model obtained by Theorem 1.9, every space X with $\Lambda(X) = \omega_3$ which is ω_3-cwH is also ω_4-cwH. So in this case a theorem similar to Theorem 1.7 fails.

On the other hand for singular cardinals Shelah singular compactness applies, namely:

Theorem 1.10. Every singular cardinal κ is cwH weakly compact.

Note that in view of the results of [5] we can not expect to get a ZFC singular cardinals compactness for cwH for the class $\Omega_{(2^{\aleph_0})^+}$ or even for $\Omega_{\aleph_1}$. This leaves the problem of the possibility of cwH compactness at successors of singular cardinals. In § 3 following [9] we can show

Theorem 1.11. Let $\kappa < \lambda$ be two regular cardinals and let $\tau = \lambda^{+\kappa+1}$ then there is a space $X \in \Lambda_{\lambda^+}$ which is τ-cwH but not τ^+-cwH.

In particular we get

Corollary 1.12. There is a space X that $\lambda(X) = \omega_2$ such that X is $\aleph_{\omega+1}$-cwH but it is not $\aleph_{\omega+2}$-cwH. So $\aleph_{\omega+1}$ is not cwH weakly compact for Λ_{ω_2}.

Recall that a cardinal κ is called a cardinal fixed point if κ is the κ-th cardinal, i.e., $\kappa = \aleph_\kappa$.

Corollary 1.13. 1. No successor of a regular cardinal can be cwH compact.

2. No successor of singular less than $\aleph_{\omega^2+1}$ can be cwH weakly compact.

3. No cardinal less than the first cardinal fixed point can be cwH compact.

In § 5 we show that the above incompactness results are the best possible, at least if we assume the consistency of some large cardinals. The definition of the large cardinal we shall use is given by:

Definition 1.14. Let κ be a cardinal and α an ordinal. We say that κ is (n, α)-huge if there is a transitive class M and an elementary embedding $j : V \to M$ such that κ is the critical point of j and $M^{j^n(\kappa)^{+\alpha}} \subseteq M$.

Note that the usual definition of huge cardinal is the same as being $(1, 0)$ huge and that the usual definition of 2-huge implies that if κ is 2-huge witnessed by the embedding j then κ is $(1, \alpha)$ huge for every $\alpha < j(\kappa)$. In § 5 we prove

Theorem 1.15. Assuming the consistency of a $(1, \omega + 1)$-huge cardinal, there is a model of ZFC in which for every space X with $\lambda(X) < \aleph_\omega$, if X is $\aleph_{\omega \cdot 2}$-cwH, then it is $\aleph_{\omega \cdot 2+1}$-cwH.

Following [9] we show (in § 4) that (again assuming the consistency of large cardinals) we cannot improve Corollary 1.13.

Theorem 1.16. Assume that the existence of ω supercompact cardinals is consistent. Then the following hold:

1. It is consistent that $\aleph_{\omega^2+1}$ is cwH weakly compact.

2. It is consistent that the first cardinal fixed point is cwH-compact.

2 Some cases of compactness

Proof of Theorem 1.9. Let $\lambda < \kappa$ be like in the statement of the theorem, where κ is weakly compact and λ is regular. Recall the definitions of an element of $\wp_\lambda(\mathbf{H}_\kappa)$ being internally approachable (cf. [6]). (By $\wp_\lambda(\mathbf{H}_\kappa)$ we mean all subsets M of $\mathbf{H}_\kappa$ of cardinality less than λ such that $M \cap \lambda \in \lambda$.)

Definition 2.1. A is any set. A set $M \subseteq A$ is said to be internally approachable if there is an ordinal δ and a sequence $\langle M_\alpha \mid \alpha < \delta \rangle$ such that the sequence is increasing, continuous, for every $\beta < \delta$, $\langle M_\alpha \mid \alpha \leq \beta \rangle \in M_{\beta+1}$, and $M = \bigcup_{\alpha<\delta} M_\alpha$.

We denote the set of internally approachable subsets of $\mathbf{H}_\kappa$ by IA_κ. Note that for this definition to make sense we need that A contains as members the sets M_α for $\alpha < \delta$ which are also subsets of A. So typically when we use this definition for M of cardinality $< \lambda$ we assume that $\wp_\lambda(A) \subseteq A$. Combining [6, Lemma 28] and the argument of [1], using the fact that κ is weakly compact one can show:

Lemma 2.2. Let G be generic with respect to the Lévy collapse $\mathrm{Coll}(\lambda, <\kappa)$ of all cardinals less than κ to λ. Note that $\mathbf{H}_\kappa^{V[G]} = \mathbf{H}_\kappa^V[G]$. Then in $V[G]$ the following reflection principle holds: For every $S \subseteq \mathrm{IA}_\kappa$ which is stationary in $\wp_\lambda(\mathbf{H}_\kappa)$:

$$\{A \mid A \in \wp_\kappa(\mathbf{H}_\kappa), S \cap \wp_\lambda(A) \cap A \text{ is stationary in } \wp_\lambda(A)\}$$

is stationary in $\wp_\kappa(\mathbf{H}_\kappa)$.

The reflection principle given by Lemma 2.2 will be denoted by $\mathsf{REF}(\kappa, \lambda)$. The first clause of Theorem 1.9 will follow from the following lemma:

Lemma 2.3. Let $\lambda < \kappa$ be two regular cardinals. Assume $\mathsf{REF}(\kappa, \lambda)$, $\gamma^{<\lambda} < \kappa$ for all $\gamma < \kappa$, and $2^{<\kappa} = \kappa$. Then κ is cwH weakly compact for Λ_λ.

Proof of Lemma 2.3. Let X be a topological space such that $\lambda(X) \leq \lambda$ and such that X is κ-cwH. We have to show that X is κ^+-cwH. For every $y \in X$ let U_y be a an open neighborhood of y whose cardinality is less that λ. A subset of X, Z, is said to be full if for $y \in Z$, $U_y \subseteq Z$. Note that for every $Z \subseteq X$ there is a full set $Z \subseteq Z^* \subseteq X$ such that if $|Z| < \lambda$ then $|Z^*| < \lambda$ and if $|Z| \geq \lambda$, then $|Z^*| = |Z|$.

Let Z be a closed discrete subset of X of cardinality κ. We have to show that Z is separated. Let $Z \subseteq Z^*$ be a full subset such that $|Z^*| = \kappa$. It is enough to show that Z is separated in the subspace Z^*, so we can assume without loss generality that $X = Z^*$ and hence $|X| = \kappa$. So without loss of generality we can assume $X = \kappa$.

Lemma 2.4. Let $Z \subseteq X$ be as above.Then Z is separated iff $R_Z = \{M \in \wp_\lambda(\mathbf{H}_\kappa) \cap \mathrm{IA}_\kappa \mid \exists y \in Z - M(y \in \bar{M})\}$ is not stationary in $\wp_\lambda(\mathbf{H}_\kappa)$. ($\bar{M}$ is the closure of M in X.)

Proof of Lemma 2.4. Assume that Z is separated but R_Z is a stationary subset of $\wp_\lambda(\mathbf{H}_\kappa)$. Let $\{W_y \mid y \in Z\}$ be a family of mutually disjoint open sets such that $y \in W_y$. For $x \in X$ let $F(x)$ be the unique $y \in Z$ such that $x \in W_y$ if such $y \in Z$ exists. If not then $F(x)$ is any member of X. Since R_Z is stationary in $\wp_\lambda(\mathbf{H}_\kappa)$ there is $M \in R_Z$ such that M is closed under the function F. Since $M \in R_Z$ there is $y \in \bar{M} \cap (Z - M)$. The set W_y is an open neighborhood of y so there is $x \in M \cap W_y$. Hence $F(x) = y$ but $y \notin M$ and M is closed under F. We get a contradiction. Note that in this direction we did not use the assumptions that all the members of M are in IA_κ.

Now we prove that other direction. Assume that R_Z is not stationary in $\wp_\lambda(\mathbf{H}_\kappa)$. Hence there is an algebra $\mathcal{A}$ such that for every $M \in \wp_\lambda(\mathbf{H}_\kappa) \cap \mathrm{IA}_\kappa)$ which is a subalgebra of $\mathcal{A}$, $(Z - M) \cap \bar{M} = \varnothing$. Without loss of generality we can assume $\mathcal{A}$ contains Skolem functions for the structure $\mathcal{B} = \langle \mathbf{H}_\kappa, \varepsilon, \triangleleft, \mathcal{U}(x,y)\rangle$ where $\triangleleft$ is a well ordering of $\mathbf{H}_\kappa$ of order type κ and $\mathcal{U}(x,y)$ is the binary relation defined by $\mathcal{U}(x,y) \Leftrightarrow x \in U_y$. We can also assume that the set of operations of the algebra $\mathcal{A}$ is closed under composition. It easily seen that $M \subseteq \mathbf{H}_\kappa$ which is closed under the operations of $\mathcal{A}$ and such that $M \cap \lambda$ is an ordinal, is full.

Claim 2.5. The set D of $\beta \in \kappa$ such that β (as subset of X) is full and such that there is no $y \in (Z - \beta)$ which in the closure of β, contains a closed unbounded subset of κ.

Proof of the claim. The cardinal arithmetic assumptions in Lemma 2.3 imply that the cardinality of $\mathbf{H}_\kappa^{<\lambda}$ is κ, so in the structure $\langle \mathbf{H}_\kappa, \varepsilon, \triangleleft, \mathcal{U}\rangle$ one can define a bijection $G : \kappa \to \mathbf{H}_\kappa$. Note that the same cardinal arithmetic assumptions imply that the set

$$E = \{\beta < \kappa \mid \text{for every bounded subset } a \text{ of } \beta \text{ of cardinality } < \lambda, \quad a = G(\gamma) \text{ for some } \gamma < \beta\}$$

contains a closed unbounded subset of κ.

For $M \subseteq \mathbf{H}_\kappa$ let M^* be the minimal subset of $\mathbf{H}_\kappa$ which is closed under the operations of $\mathcal{A}$ and such that $M \cap \lambda$ is an ordinal. Note that if $|M| < \lambda$ then $M^* \in \wp_\lambda(\mathbf{H}_\kappa)$. Let K be the set of $\beta \in \kappa$ such that $\beta^* \cap \kappa = \beta$ (namely the set of ordinals such that their $\mathcal{A}$-closure does not increase them). Clearly K is closed unbounded subset of κ and every $\beta \in K$ is full. Define the required set $D = \{\beta \in E \mid \lambda < \beta, \beta \text{ is a limit point of } K\}$. The set D is clearly a closed unbounded subset of κ. We claim that D is the required set for the claim.

Suppose that $\beta \in D$ but that there is $y \in (Z - \beta) \cap \bar{\beta}$. Since $\lambda(X) \leq \lambda$ there is $W \subseteq \beta, |W| < \lambda$ such that $y \in \bar{W}$. We can write W as increasing continuous union of bounded subsets β of length less than λ, i.e., $W = \bigcup_{\varrho<\delta} W_\varrho$ where $\delta < \lambda$. Without loss of generality we can assume that $\delta \leq \mathrm{cf}(\beta)$. By induction on $\varrho < \delta$ we define a sequence of elements of $\wp_\lambda(\mathbf{H}_\kappa)$, $\langle M_\varrho \mid \varrho < \delta\rangle$. Inductively we assume that for $\varrho < \delta$, $M_\varrho = G(\eta_\varrho)$ for some $\eta_\varrho < \beta$ and that $W_\mu \subseteq M_\varrho$ for all $\mu < \varrho$.

We let $M_0 = \delta^*$. For limit ϱ we define $M_\varrho = \bigcup_{\mu<\varrho} M_\mu$. Note that since $\varrho < \mathrm{cf}(\beta)$ and since for $\mu < \varrho : G^{-1}(M_\varrho) < \beta$ we get that $J = \{\eta_\mu \mid \mu < \varrho\}$ is a bounded subset of β, so by $\beta \in D$ we get that $J = G(\nu)$ for some $\nu < \beta$. Since $M_\varrho = \bigcup\{G^{-1}(\xi) \mid \xi \in J\}$, we get that $G^{-1}(M_\varrho) < \beta$. Note also that the sequence $\langle M_\mu \mid \mu < \varrho\rangle$ is definable from ν.

Now we deal with the successor case: $\varrho = \sigma + 1$. As in the limit case we get that $\langle M_\mu \mid \mu \leq \sigma\rangle = G(\nu)$ for some $\nu < \beta$, so $M_\varrho = (\{\nu\} \cup W_\sigma)^*$. We can easily see that $\langle M_\mu \mid \mu \leq \sigma\rangle \in M_\varrho$ and also $M_\sigma \subseteq M_\varrho$. We still have to verify that $G^{-1}(M_\varrho) < \beta$. Since $\beta \in D$, β is a limit point of K, there is $\gamma \in K$ such that $\max(\nu, \sup(W_\sigma)) < \gamma$. Every member of M_ϱ is of the form $x = g(\nu, \vec{\eta}, \vec{(\tau)})$ where g is one of the functions of the algebra $\mathcal{A}$, $\vec{\eta}$ is a finite sequence of ordinals in λ and $\vec{(\tau)}$ is a finite sequence of ordinals in W_σ. By γ being relatively closed under the operations of $\mathcal{A}$ and by $\lambda \cup \{\nu\} \cup W_\sigma \subseteq \gamma$ we get for $x \in M_\varrho$ that $G^{-1}(x) < \gamma$. So the set $L = \{G^{-1}(x) \mid x \in M_\varrho\}$ is a subset of γ. So it is a bounded subset of β of cardinality less than λ. We get that $G^{-1}(L) < \beta$. But $G^{-1}(M_\varrho)$ is definable in $\mathcal{B}$ from $G(L)$, so by $\beta \in K$ we get that $G^{-1}(M_\varrho) < \beta$. $W_\sigma \subseteq M_\varrho$ is clear from the definition of M_ϱ.

Let $M = \bigcup_{\varrho<\delta} M_\varrho$. The construction yields immediately that $M \in \mathrm{IA}_\kappa$ as witnessed by the sequence $\langle M_\varrho \mid \varrho < \delta \rangle$. The set M is closed under the operations of the algebra $\mathcal{A}$, so $M \notin R_Z$. But $W \subseteq M \cap \kappa \subset \beta$, $y \in (Z - M)$ and it is in the closure of M. So $M \in R_Z$ which is clearly a contradiction. Q.E.D.

We resume the proof of Lemma 2.4. We want to show that Z is separated. Let D be the closed unbounded subset of κ satisfying the conditions in the claim. We shall define the separating family of open sets $\langle W_y \mid y \in Z \rangle$ such they are mutually disjoint and such that $y \in W_y$. We shall define the sequence by induction on $\beta \in D$. Namely at stage $\beta \in D$ we define $\langle W_y \mid y \in Z \cap \beta \rangle$ where the inductive assumption is of course that $\langle W_y \mid y \in Z \cap \beta \rangle$ separates $Z \cap \beta$ and that for $y \in Z \cap \beta$, $W_y \subseteq \beta$.

For $\beta = \min(D)$ pick any sequence $\langle W_y \mid y \in Z \cap \beta \rangle$ separating $Z \cap \beta$. It exists since X is κ-cwH and we can assume that for $y \in Z \cap \beta$, $W_y \subset \beta$, since β is full. For β which a limit point of D the definition of $\langle W_y \mid y \in Z \cap \beta \rangle$ is simply the union of $\langle W_y \mid y \in Z \cap \gamma \rangle$ for $\gamma \in D \cap \beta$.

So we are left with the case that β is a successor point of D. Let $\gamma \in D$ be $\max(D \cap \beta)$. Let $\langle T_y \mid y \in Z \cap (\beta - \gamma) \rangle$ be a family of mutually disjoint open sets separating $Z \cap (\beta - \gamma)$ such that for $y \in Z \cap (\beta - \gamma)$, $T_y \subseteq \beta$. Again we use the fact that X is κ-cwH and that β is full. Because $\gamma \in D$, we have that if $y \in Z \cap\ (\beta - \gamma)$, then Y is not in the closure of γ. In particular y is not in the closure of $\bigcup_{z \in Z \cap \gamma} W_z$ so we can shrink each T_y to W_y such that $W_y \cap \bigcup_{z \in Z \cap \gamma} W_z = \varnothing$. The sequence $\langle W_y \mid y \in Z \cap \beta \rangle$ obviously satisfies the inductive assumption for β. Q.E.D.

We finish the proof of Lemma 2.3. Assume that $X = \kappa$ and that $Z \subseteq X$ is a closed discrete set which not separated. By Lemma 2.4 the set R_Z is stationary in $\wp_\lambda(\mathbf{H}_\kappa)$. For $M \in R_Z$ let $F(M)$ be a member of $Z - M$ which is in the closure of M. Since $R_Z \subseteq \mathrm{IA}_\kappa$ we can use $\mathsf{REF}(\kappa, \lambda)$ to get $A \in \wp_\kappa$ such that $R_Z \cap \wp_\lambda(A)$ is stationary in $\wp_\lambda(A)$. We have stationarily many possible A, so we can assume that A is full and closed under F. Since $Z \cap A$ has cardinality $< \kappa$, $Z \cap A$ can be separated. Let $\langle W_y \mid y \in Z \cap A \rangle$ be a family of mutually disjoint open sets separating $Z \cap A$. Since A is full we can assume that for $y \in Z \cap A$, $W_y \subseteq A$. For very $\eta \in A$ let $J(\eta)$ be the unique member of $Z \cap A$ such that $\eta \in W_y$ if it exists. Otherwise $J(\eta)$ is any member of $Z \cap A$. Since $R_Z \cap A$ is stationary in $\wp_\lambda(A)$ there is $M \in R_Z \cap A$ which is closed under J. Then $y \in F(M)$ is in A, it is a member of $Z - M$ and it is in the closure of M. So there is $\eta \in M \cap W_y$. But also $\eta \in W_{J(\eta)}$. But $J(\eta) \in M$, hence $y \neq J(\eta)$. So $\eta \in W_y \cap W_{J(\eta)} = \varnothing$. Clearly a contradiction. Q.E.D.

We still have to prove the second clause of Theorem 1.9. Namely we assume that we have a model of ZFC with a proper class of supercompact cardinals.

Let us define by induction an increasing sequence of regular cardinals $\langle \kappa_\gamma \mid \gamma \in \mathrm{Ord} \rangle$ such that for successor γ, κ_γ is supercompact. We let $\kappa_0 = \omega$. For γ limit κ_γ is the minimal regular cardinal greater or equal to $\sup(\{\kappa_\beta \mid \beta < \gamma\})$. (Thus if $\eta = \sup\{\kappa_\beta \mid \beta < \gamma\}$ is regular $\kappa_\gamma = \eta$ otherwise $\kappa_\gamma = \eta^+$.) For $\gamma = \beta + 1$, κ_γ is the minimal supercompact greater than κ_β.

We force with the class forcing which is the Easton support iteration of the forcings of $\mathrm{Coll}(\kappa_\gamma, < \kappa_{\gamma+1})$ for $\gamma \subset \mathrm{Ord}$. Denote this (class) forcing by $\mathcal{P}$. For an ordinal γ there is a natural representation of $\mathcal{P}$ as $\mathcal{P}_\gamma \star \mathrm{Coll}(\kappa_\gamma, < \kappa_{\gamma+1}) \star \mathcal{P}^\gamma$ where $\mathcal{P}_\gamma$ is the Easton support iteration of $\mathrm{Coll}(\kappa_\beta, < \kappa_{\beta+1})$ for $\beta < \gamma$ and $\mathcal{P}^\gamma$ is the Easton support iteration of $\mathrm{Coll}(\kappa_\beta, < \kappa_{\beta+1})$ for $\beta > \gamma$. It is easily seen that $\mathcal{P}_\gamma$ has cardinality $\leq \kappa_\gamma$ and that $\mathcal{P}^\gamma$ is $\kappa_{\gamma+1}$ directed closed. Also by standard arguments the resulting model satisfies GCH.

In the resulting model each successor of a regular cardinal is of the form $\kappa_{\gamma+1}$. We show that in the resulting model $\kappa_{\gamma+1}$ is cwH weakly compact. We do it by showing that in $V^{\mathcal{P}}$ for $\lambda < \kappa_{\gamma+1}$ regular we have $\mathsf{REF}(\kappa_{\gamma+1})$. Note that in $V^{\mathcal{P}}$ the predecessor of $\kappa_{\gamma+1}$ is κ_γ and its successor is $\kappa_{\gamma+2}$. So in order to simplify notation we put $\kappa = \kappa_{\gamma+1}$, $\kappa^- = \kappa_\gamma$ and $\kappa^* = \kappa_{\gamma+2}$. Note that in $V^{\mathcal{P}}$ a cardinal $\delta < \kappa$ is satisfies $\delta \leq \kappa^-$.

Let $\dot{S}$ be a $\mathcal{P}$ name for a subset of $\wp_\delta(\mathbf{H}_\kappa) \cap \mathrm{IA}_\kappa$ which is stationary in $\wp_\delta(\mathbf{H}_\kappa)$ and let $\dot{\mathcal{A}}$ be a $\mathcal{P}$ name for an algebra with countably many operations on $\mathbf{H}_\kappa$. Since $V^{\mathcal{P}} \models |\mathbf{H}_\kappa| = \kappa$ and since $\mathcal{P} = \mathcal{P}_{\gamma+1} \star \mathrm{Coll}(\kappa, < \kappa^*) \star \mathcal{P}^{\gamma+1}$ we can assume that $\dot{S}$ and $\dot{\mathcal{A}}$ are $\mathcal{P} = \mathcal{P}_{\gamma+1} \star \mathrm{Coll}(\kappa, < \kappa^+)$ names. Denote $\mathcal{P}_{\gamma+1} \star \mathrm{Coll}(\kappa, < \kappa^*)$ by $\mathcal{P}^*$. Also let S and $\mathcal{A}$ be the realizations in $V^{\mathcal{P}^*}$ of $\dot{S}$ and $\dot{\mathcal{A}}$ respectively.

In V, κ is supercompact, so let $j : V \to M$ be a $2^{\kappa_{\gamma+2}}$-supercompact embedding with critical point κ. A finer representation of $\mathcal{P}^*$ is $\mathcal{P}_\gamma \star \mathrm{Coll}(\kappa^-, < \kappa) \star \mathrm{Coll}(\kappa, < \kappa^*)$. A standard argument (cf., e.g., [8]) shows that in $V_1 = V^{\mathcal{P}^*}$ there is a κ^- closed forcing $\mathcal{Q}$ such that in $V_1^{\mathcal{Q}}$ there is an embedding $j^* : V_1 \to M_1$. Note that j^* is the identity on $\mathbf{H}_\kappa \subseteq H^{M_1}_{j^*(\kappa)}$

and that $S = j^*(S) \cap \wp_\lambda(H^{M_1}_{j^*(\kappa)})$. Also $\mathbf{H}^{V_1}_\kappa$ is a subalgebra of $j^*(\mathcal{A})$. Since $S \subseteq \wp_\delta(\mathbf{H}_\kappa) \cap \mathrm{IA}_\kappa$ and since $\mathcal{Q}$ is λ closed we get that

$$V_1^{\mathcal{Q}} \models S \text{ is a stationary subset of } \wp_\delta(\mathbf{H}_\kappa),$$

so the same holds in $M_1 \subseteq V_1^{\mathcal{Q}}$. In M_1 the cardinality of $\mathbf{H}^{V_1}_\kappa$ is $\kappa^- < j^*(\kappa)$. Thus $\mathbf{H}^{V_1}_\kappa$ witnesses that in M_1 there is a set $A \in j^*(\wp_\kappa(\mathbf{H}_\kappa))$ which is a subalgebra of $j^*(\mathcal{A})$ and such that $S = j^*(S) \cap \wp_\lambda(A)$ is stationary in $\wp_\lambda(A)$.

By elementarity of j^* we get that in V_1 there is $A \in \wp_\kappa(\mathbf{H}_\kappa)$ closed under the operations of the algebra $\mathcal{A}$ such that $S \cap \wp_\lambda(A)$ is stationary in $\wp_\lambda(A)$. This shows that $\mathsf{REF}(\kappa, \lambda)$ holds in V_1, which by Lemma 2.3 proves Theorem 1.9. Q.E.D.

3 Some cases of incompactness

In this section we prove Theorem 1.11.

Proof of Theorem 1.11. Let $\kappa < \lambda$ be regular cardinals and let $\tau = \lambda^{+\kappa+1}$. We shall describe a space $X \in \Lambda_{\lambda^+}$ which is τ-cwH but not τ^+-cwH. The proof is a variation of [9, Theorem 1.4].

Theorem 3.1. Let $\kappa < \lambda$ be two regular cardinals. Let $\tau = \lambda^{+\kappa+1}$. Then there exists a stationary subset S of τ of points of cofinality λ and a sequence of sets $\langle A_\alpha \mid \alpha \in S\rangle$ such that for $\alpha \in S$, A_α is a subset of $\lambda^{+\kappa}$ of order type κ such that for every non-stationary subset of S, T, one can find $\langle D_\alpha \mid \alpha \in T\rangle$ and $\langle B_\alpha \mid \alpha \in T\rangle$ such that for $\alpha \in T$, D_α is a closed unbounded subset of α and B_α is a subset of A_α satisfying $|A_\alpha - B_\alpha| < \kappa$ and the sets $\langle B_\alpha \times D_\alpha \mid \alpha \in T\rangle$ are mutually disjoint. [9]

Let $S, \langle A_\alpha \mid \alpha \in S\rangle$ be as guaranteed by Theorem 3.1. Denote $\tau^- = \lambda^{+\kappa}$. We shall define the space X which will witness the proof of Theorem 1.11. Let $X = S \cup (\tau^- \times \tau)$. The topology of X is defined by taking the discrete topology on $\tau^- \times \tau$ and defining a neighborhood of $\alpha \in S$ to be $\alpha \in Y \subseteq X$ such that for some D a closed unbounded subset of α, D and for some $B \subseteq A_\alpha$ such that $|A_\alpha - B| < \kappa$ we have $B \times D \subseteq Y$. It is easily seen that this defines a topology on X and that $\lambda(X) = \lambda^+$. Note that for $\alpha \in S$ the cofinality of α is λ and $|A_\alpha| = \kappa$. So if D is a club in α of order type λ then $\{\alpha\} \cup A_\alpha \times D$ is an open neighborhood of α of cardinality λ.

S as a subset of X is a closed discrete set of points, S can not separated, because if $\langle W_\alpha \mid \alpha \in S\rangle$ is a sequence of open sets separating S (in particular $\alpha \in W_\alpha$) then for every $\alpha \in S$ there is $\varrho_\alpha < \alpha$ and $\eta_\alpha < \tau^-$ such that $(\eta_\alpha, \varrho_\alpha) \in W_\alpha$ but S is stationary, there are two distinct $\alpha, \beta \in S$, $(\eta_\alpha, \varrho_\alpha) = (\eta_\beta, \varrho_\beta)$, contradicting the assumption that the W_α's are mutually disjoint.

On the other hand let $Y \subset X$ be a discrete subset of X such that $|Y| < \tau$. Let $T = Y \cap S$ and $Y^* = Y \cap (\tau^- \times \tau)$. Since T is of small cardinality, it is non stationary. By applying Theorem 3.1 to T, we obtain $\langle D_\alpha \mid \alpha \in T\rangle$ and $\langle B_\alpha \mid \alpha \in T\rangle$ satisfying the statement of Theorem 3.1. Since Y is discrete we can assume without loss of generality that for $\alpha \in T$, $B_\alpha \times D_\alpha$ is disjoint from Y. Now define for $y \in Y^*$, $W_y = \{y\}$ and for $y \in T$, $W_y = \{y\} \cup B_y \times D_y$ then $\{W_y \mid y \in Y\}$ is a family of open sets separating Y. Q.E.D.

4 Some more cases of compactness

In this section we show that Shelah's singular compactness (Theorem 1.10) as well as the consistency results of [9] (Theorem 1.16) apply to the the appropriate versions of cwH. Following [12] we say that for a χ majority of subsets of a set H a certain property holds if there is an algebra on H with χ many operations such that every subset closed under these operations has the given property. We also say that an increasing sequence of sets $\langle A_\alpha \mid \alpha < \lambda\rangle$ is continuous if for limit $\alpha < \lambda$, $A_\alpha = \bigcup_{\beta<\alpha} A_\beta$.

Suppose that X is a space with $\lambda(X) = \lambda$. We want to show that the general notion of *freeness* introduced by Shelah [10, 12] applies in this case. For very $x \in X$ let U_x be a neighborhood of x of cardinality $< \lambda$. We can easily define an algebra with λ many operations such that every subset of X, Y which is a subalgebra is full in the sense that for $x \in Y$, $U_x \subseteq Y$. For $Y \subseteq X$ let Y^* be the closure of Y under the operations of this algebra.

For a pair (A, B) of subset of X we say that it is a free pair ("A/B is free") if $\bar{B}^* \cap (A - B) = \varnothing$ and $A - B$ is separated. Note that if A/B is free then $A - B$ can be separated by a family of open sets which has empty intersection with B^*.

This notion of freeness satisfies the set of axioms introduced by Shelah ([12], cf. also [9]). These axioms have one cardinal parameter which in our case it is λ. We follow [12] in naming these axioms. Most of these axioms are obviously satisfied by our notion of freeness. In case it is less obvious we add a sketch of the argument following the statement of the axiom.

Axiom I** *If A/B is free and $A' \subset A$ then A'/B is free.*

Axiom II *A/B is free if and only if $A \cup B/B$ is free. Furthermore, A/A is free.*

Axiom III *If A/B, $C \subseteq B$ and B/C is free then A/C is free.* This is seen by separating $A \cap (B - C)$ by open sets disjoint from C^*. Without loss of generality we can assume that for $x \in A \cap (B - C)$ the open set covering x is a subset of U_x. Then separate $A - B$ by sets disjoint from B^*.

Axiom IV *If $(A_\alpha)_{\alpha<\mu}$ is an increasing and continuous sequence of subsets of X such that A_0/B if free and for all $\alpha < \mu$, $A_{\alpha+1}/A_\alpha \cup B$ is free* *then* $\bigcup_{\alpha<\mu} A_\alpha/B$ *is free.* The argument is similar to the argument for Axiom III by constructing inductively a separating family for $A_\alpha - B$.

Axiom VI *If* $A/B \cup C$ *is free, then for a* λ *majority of* $X \subseteq A \cup B$, $A \cap X/(B \cap X) \cup C$ *is free.* In our case this is true for every $X \subseteq A \cup B$.

Axiom VII *If* A/B *is free then for a* λ *majority of* $X \subseteq A$, $A/X \cup B$ *is free.* The argument is as follows: Fix a separation of $A - B$ such that the open set covering $x \in A - B$ is a subset of U_x. For $x \in (A \cup B)^*$ let $F(x)$ be the unique point $z \in (A - B)$ such that x belongs the open set of the separating family separating $A - B$ if it exists. If no such z exists let $F(x)$ any element of $A - B$. Say that $X \subseteq A \cup B$ is nice if $F(x) \in X$ for every $x \in X^*$. It is easily seen that the λ majority of X's are nice and that if X is nice then $A/(X \cup B)$ is free.

Axiom A *Suppose that* $A/(B \cup C)$ *is not free. Let* $\delta = \max(|A|, \lambda)$. *Then for a* δ *majority of* $X \subseteq B$, $A/(X \cup C)$ *is not free.* If $x \in (B \cup C)^* \cap (A - (B \cup C))$ then there are $D \subseteq B$, $E \subseteq C$ such that $|D|, |E| < \lambda$ and such that $x \in (D \cup E)^*$. Then $D \subseteq X$ then $A/X \cup C$ is not free. If $(B \cup C)^* \cap (A - (B \cup C)) = \varnothing$ the reason that $A/(B \cup C)$ is not free is that $A - (B \cup C)$ is not separated. But then for every $X \subset B$, $A - (X \cup C)$ is not separated. So $A/X \cup C$ is not free for every $X \subseteq B$.

Following [12] we say that A/B is κ free if for every $C \subseteq A$, $|C| < \kappa$ we have that C/B is free.

Shelah's singular compactness [12] claims that for every notion of freeness satisfying the above axioms and for κ singular κ freeness implies κ^+ freeness. So let $X \in \Lambda_\lambda$ be a space which is κ-cwH such that κ is singular and $\lambda < \kappa$. Let $Y \subset X$ be discrete closed set such that $|Y| = \kappa$. It is easily seen that $Y/\varnothing$ is κ free for our notion of freeness for the space X. So by Shelah's theorem $Y/\varnothing$ is free. This implies that Y is separated. So X is κ^+-cwH. So we proved Theorem 1.10.

The proof of Theorem 1.16 is similar. By [9] if one assumes the consistency of ω many supercompact cardinals then there is a model of set theory in which the following is true: For every notion of freeness which satisfies the above axioms for $\lambda < \aleph_{\omega^2}$, if A/B is $\aleph_{\omega^2}$ free then it is $\aleph_{\omega^2+2}$ free. In particular in the model we get that for every space $X \in \Lambda_\lambda$ ($\lambda < \aleph_{\omega^2}$), if X is $\aleph_{\omega^2+1}$ is cwH then it is $\aleph_{\omega^2+2}$-cwH.

Similarly in [9], again assuming the consistency of ω supercompact cardinals, one gets a model in which the following is true: Let κ be the first cardinal fixed point. Then for every notion of freeness which satisfies the above axioms for $\lambda < \kappa$, every pair (A, B) which is κ free is free. So every space $X \in \Lambda_{<\kappa}$ which is κ-cwH is cwH. So in the model κ is cwH compact.

5 Another model with compactness

In this section we prove Theorem 1.15. We are going to use a principle $\tilde{\Delta}_{\kappa,\lambda}$ which is a slight variation of a similar principle introduced in [9]. The principle is as follows:

For every $\aleph_0 \leq \mu < \kappa$, $S \subseteq \lambda$ which is stationary in λ and such that $\delta \in S$ implies $\mathrm{cf}(\delta) < \kappa$, and for every algebra $\mathcal{A}$ on λ with μ many operations, there is a subalgebra $\mathcal{B}$ whose order type (as a subset of λ) is a regular cardinal $\eta < \lambda$ such that $\mathcal{B} \cap S$ is stationary in $\sup(\mathcal{B})$

The difference between $\Delta_{\kappa,\lambda}$ introduced in [9] and $\tilde{\Delta}_{\kappa,\lambda}$ is that here we only require that the order type of $\mathcal{B}$ is less than $< \lambda$ where in [9] we require that the order type of $\mathcal{B}$ is less than κ.

The relevance of this principle is the following lemma:

Lemma 5.1. Assume that $\tilde{\Delta}_{\kappa,\lambda}$ holds where λ is regular and $\kappa < \lambda$. Then for every space $X \in \Lambda_{<\kappa}$ if X is λ-cwH then it is λ^+-cwH.

Proof. The proof is very similar to the proof of Lemma 2.3. Let X be as in the statement of the Lemma.As in the proof of Lemma 2.3 we can assume that $|X| = \lambda$, so we can assume $X = \lambda$. We pick for $x \in X$ an open neighborhood of x, U_x, of cardinality $< \kappa$. As above we have the notion of a subset of X being full and for $Y \subseteq X$ we have Y^*, the minimal full set containing Y. Let $Z \subseteq X$ be discrete closed subset. We consider the set

$$S = \{\alpha < \lambda \mid \alpha \text{ is full}, \bar{\alpha} \cap (Z - \alpha) \neq \varnothing\}.$$

As in the proof of Lemma 2.3 we have (using the fact that X is λ-cwH).

Claim 5.2. The set Z is separated iff S is not stationary in λ

Towards a contradiction, we assume that S is stationary.

Claim 5.3. The set $R = \{\alpha \in S \mid \mathrm{cf}(\alpha) < \kappa\}$ is stationary in λ.

Proof. For every $\alpha \in S$ pick $z_\alpha \in \bar{\alpha} \cap (Z - \alpha)$. Then z_α is in the closure of α. Since $\lambda(X) < \kappa$ there is a set $D_\alpha \subseteq \alpha$ such that $z_\alpha \in \bar{D_\alpha}$ where $|D_\alpha| < \kappa$. Let $T = \{\alpha \in S \mid D_\alpha$ is unbounded in $\alpha\}$. If T is stationary then our claim is verified. If T is not stationary then the function $K(\alpha) = \sup(D_\alpha)$ is pressing down on the stationary set $S - T$, so it is constant on a stationary subset of $S - T$. Let γ be the constant value. But then it is easily seen that every α with $\gamma < \alpha < \lambda$ is in S, which includes stationarily many α's of cofinality less than κ. Q.E.D.

Define an algebra $\mathcal{A}$ on λ with $\lambda(X)$ many operations such that a set Y closed under these operations is full, $Z \cap Y$ is cofinal in $\sup(Y)$ and for $\alpha \in Y$ if $\alpha \in S$ then there is $z \in Y \cap (Z - \alpha) \cap \bar{\alpha}$. Use $\tilde{\Delta}_{\kappa,\lambda}$ for the algebra

$\mathcal{A}$ and the stationary set R and get a subalgebra $\mathcal{B}$ whose order type is a regular cardinal $\eta < \lambda$ and such that $R \cap \mathcal{B}$ is stationary in $\sup(\mathcal{B})$. Then $\mathcal{B}$ is full. Let $\langle \alpha_\delta \mid \delta < \eta \rangle$ be a monotone enumeration of $\mathcal{B}$. Since R is stationary in $\sup(\mathcal{B})$ we get that $R^* = \{\delta < \eta \mid \alpha_\delta \in R\}$ is stationary in η.

Since X is λ-cwH, we have that $Z^* = Z \cap \mathcal{B}$ is separated. Let $\langle W_z \mid z \in Z^* \rangle$ be a family of mutually disjoint open family separating Z^*. Without loss of generality we can assume that $W_z \subseteq U_z$. As in previous proofs define for $x \in \mathcal{B}$, $F(x)$ be the unique $z \in Z^*$ such that $x \in W_z$ if it exists and any value in Z^* otherwise. The set of $\delta < \eta$ such that the set $E_\delta = \{\alpha_\varrho \mid \varrho < \delta\}$ is full and closed under F is closed unbounded in η. Hence there is $\delta \in R^*$ such that E_δ is full and closed under F. Let $\alpha = \alpha_\delta$. By the properties of $\mathcal{B}$ there is $z \in Z^* - \alpha$ such that $z \in \bar{\alpha}$. Since $\mathcal{B}$ is full, we have $W_z \subseteq U_z \subseteq \mathcal{B}$. Since $W_z \cap \alpha \neq \varnothing$ we can find $x \in W_z \cap \alpha$ but then $x \in \mathcal{B} \cap \alpha$. So $x = \alpha_\varrho$ for some $\varrho < \delta$, so $x \in E_\delta$. So $F(x) < \alpha$. This implies that $F(x) \neq z$ but $x \in W_z \cap W_{F(x)}$ which is a contradiction. Q.E.D.

We prove Theorem 1.15 by getting a model of $\tilde{\Delta}_{\aleph_\omega, \aleph_{\omega \cdot 2+1}}$. We assume that κ is a $(1, \omega+1)$ huge, so suppose that $j : V \to M$ is an elementary embedding with critical point κ and such that $M^{j(\kappa)^{+\omega+1}} \subseteq M$. Without loss of generality we can assume that our ground model satisfies GCH. We show that in an appropriate forcing extension of V, $\tilde{\Delta}_{\aleph_\omega, \aleph_{\omega \cdot 2+1}}$ holds. Denote $j(\kappa) = \mu$, $\kappa^{+\omega} = \varrho$, $\lambda = \mu^{\omega+1} = j(\varrho^+)$.

Lemma 5.4. Under the above notations $\tilde{\Delta}_{\varrho,\lambda}$ holds in V.

Proof. Let $\mathcal{A}$ be an algebra with $< \varrho$ many operations and let $S \subseteq \lambda$ be a stationary subset of λ such that for $\alpha \in S$, $\mathrm{cf}(\alpha) < \varrho$. Without loss of generality we can assume that the cofinality of the points of S is less than a fixed regular cardinal $\delta < \varrho$. Also we can assume that the algebra $\mathcal{A}$ has at most δ many operations. It follows that there is an algebra $\mathcal{B}$ with countably many operations such that any set X closed under the operations of $\mathcal{B}$ such that $\delta \subseteq X$ is also closed under the operations of $\mathcal{A}$. Consider the structure $\mathcal{C} = \langle \mathbf{H}_\lambda, \varepsilon, \trianglelefteq, S, \{\delta\}, \cdots \rangle$ where $\triangleleft$ is a well ordering of $\mathbf{H}_\lambda$. In addition $\mathcal{C}$ has the countably many operations of the algebra $\mathcal{B}$.

It is easy to see that $j``\mathbf{H}_\lambda$ is an elementary substructure of $j(\mathcal{C})$ whose set of ordinals $D = j``\lambda$ has order type $\lambda = j(\varrho^+)$. The map $j \restriction \varrho^+$ is the unique order preserving map from the order type of D onto D. Also $j^{-1}(j(S)) = S$ is stationary in λ. Also for $n \in \omega$, $D \cap j(\mu^{+n})$ has order type μ^{+n}. Furthermore $j``\mathbf{H}_\lambda \in M$, and hence by elementarity of j we get

Claim 5.5. There is an elementary substructure of $\mathcal{C}$, E, where the order type of the set of its ordinals, D, is ϱ^+, for every $n \in \omega$ the order type of $D \cap \mu^{+n}$ is κ^{+n} and if F is the unique order preserving map from ϱ^+ onto D then $S^* = F^{-1}(S)$ is stationary in ϱ^+.

Fix E, D, F be as in the claim. Then $\mathcal{C}$ has definable Skolem functions. Let E^* be the Skolem hull of $E \cup \delta$ in $\mathcal{C}$. Obviously $\delta \subseteq E^*$.

Claim 5.6. The order type of the set of ordinals of E^*, D^*, is ϱ^+.

Proof. We first show that D is cofinal in D^*. Every element of D^* is of the form $g(\vec{d}, \vec{\eta})$ where g is a Skolem function for $\mathcal{C}$, $\vec{d}$ is a finite sequence of members of D and $\vec{\eta}$ is a finite sequence of ordinals in δ. Since $\delta < \lambda$, the set $\{g(\vec{d}, \vec{\nu}) \mid \vec{\nu} \in \delta^{<\omega}\}$ is bounded in λ. The bound is definable from $\vec{d}$ in $\mathcal{C}$ so there is such a bound in E, so it is in D. Hence it is enough to show that for $\alpha \in D$, $|D^* \cap \alpha| \leq \varrho$. In $\mathbf{H}_\lambda$ there is an injection, h, of α into the predecessor of λ which is $\sup(\langle \mu^{+n} \mid n \in \omega \rangle)$. We can assume $h \in E$. By E being elementary substructure of $\mathcal{C}$ we get that $D^* \cap \alpha = \bigcup_{n<\omega} h^{-1}(D^* \cap \mu^{+n})$.

So it is enough to show that for $n \in \omega$, $|D^* \cap \mu^{+n}| < \varrho$. Any member of $D^* \cap \mu^{+n}$ is of the form $g(\vec{d}, \vec{\eta})$ where g is a Skolem function for $\mathcal{C}$, $\vec{d}$ a finite sequence from D and $\vec{\eta}$ a finite sequence from δ. Without of generality we can assume that g gets values only in μ^{+n}. So the function defined on finite sequences from δ by $h(\vec{\nu}) = g(\vec{d}, \vec{\nu})$ is in $\mathbf{H}_\lambda$ and it is definable in $\mathcal{C}$ from $\vec{d}$. So $h \in E$. Since $\mathbf{H}_\lambda \models |(\mu^{+n})^\delta| < \mu^{+\omega}$, we have $|E \cap (\mu^{+n})^\delta| < \varrho$.

$D^* \cap \mu^{+n}$ is obtained by applying all the functions in $E \cap (\mu^{+n})^\delta$ to all the values in δ. Since $\delta < \varrho$ we get that the cardinality of $D^* \cap \mu^{+n}$ is less than ϱ. And we proved our claim. Q.E.D.

Using the last claim, let G be an order preserving map from ϱ^+ onto D^*. There is a closed unbounded subset of ϱ^+, L such that for $\beta \in L$, $G(\beta) = F(\beta)$. Note that since the cofinality of every member of S is less or equal to δ and since $\delta \subseteq E^*$, we get that for $\alpha \in S \cap E^*$, $E^* \cap \alpha$ is cofinal in α.

Claim 5.7. The set $S \cap E^*$ is stationary in $\sup(E^*)$.

Proof. Assume otherwise. Let C be closed unbounded in $\sup(D^*)$ disjoint from $S \cap E^*$. For $\alpha \in S \cap E^*$ let $t(\alpha) < \alpha$ be a bound for $C \cap \alpha$. Since D^* is unbounded in α we can assume that $t(\alpha) \in D^*$. So $G^{-1}(t(\alpha)) < G^{-1}(\alpha)$. For $\beta \in S^* \cap L$ we get $w(\beta) = G^{-1}(t(F(\beta))) < \beta$. Since $S^* \cap L$ is stationary in ϱ^+ we get that w is constant on an unbounded subset of $S^* \cap L$. Let ξ be this constant. Then $G(\xi) = t(\alpha)$ for unboundedly many α's in D^*, which is clearly a contradiction. Q.E.D.

The set D^* is a subset of λ which has order type $\varrho^+ < \lambda$; it is closed under the operations of $\mathcal{A}$ and $S \cap D^*$ is stationary in $\sup(E^*)$. So it is a witness for the truth of $\tilde{\Delta}_{\varrho,\lambda}$. Q.E.D.

Checking the proof of Lemma 5.4 we see that we get very specific information about the elementary substructure of the algebra on λ.

Lemma 5.8. Under the above assumptions about κ, μ, λ and ϱ we get that for every structure $\mathcal{A}$ on $\mathbf{H}_\lambda$ with less than ϱ many relations and operations, for every stationary subset $S \subseteq \lambda$ such that for $\alpha \in S$, $\mathrm{cf}(\alpha) < \varrho$ and for every ordinal $\delta < \varrho$ there is an elementary substructure $\mathcal{B} \subseteq \mathcal{A}$ such that the order type of the ordinals of $\mathcal{B}$ is ϱ^+, $S \cap \mathcal{B}$ is stationary in $\sup(\mathcal{B} \cap \lambda)$ and $\delta \subseteq \mathcal{B}$.

We now force over V such that in the resulting model $\aleph_1 = \kappa, \varrho = \aleph_\omega$, $\mu = \aleph_{\omega+3}$, $\lambda = \aleph_{\omega\cdot 2+1}$ and $\tilde{\Delta}_{\varrho,\lambda}$ still holds. The forcing we use is $\mathcal{P} \star \mathcal{Q}$ where $\mathcal{P} = \mathrm{Coll}(\omega, < \kappa)$ and $\mathcal{Q} = \mathrm{Coll}(\varrho^{++}, < \mu)$. It is immediate that in the extensions the cardinal structure is as we specified above. Let $V_1 = V^{\mathcal{P}\star\mathcal{Q}}$. We verify that $V_1 \models \tilde{\Delta}_{\varrho,\lambda}$.

Let $\dot{A} \in V$ be a $\mathcal{P}\star\mathcal{Q}$ name forced to denote an algebra on λ with $\delta < \varrho$ many operations and let $\dot{S}$ be a $\mathcal{P}\star\mathcal{Q}$ name forced to denote a stationary subset of λ all of its members having cofinality less than ϱ. Since our forcing has cardinality less than λ, every stationary subset of λ in V_1 contains a stationary subset of λ which in V, we can assume that $S = \dot{S}$ is a set in the ground model V. Our forcing is an element of $\mathbf{H}_\lambda$, so we can assume that the name $\dot{A}$ is a subset of $\mathbf{H}_\lambda$. In V consider the structure

$$\mathcal{C} = \langle \mathbf{H}_\lambda, \varepsilon, \lhd, S, \dot{A}, \mathcal{P}\star\mathcal{Q}, \Vdash \rangle.$$

As before $\lhd$ is a well ordering of $\mathbf{H}_\lambda$ in order type λ and $\Vdash$ is the forcing relation for $\mathcal{P}\star\mathcal{Q}$.

By Lemma 5.8, let $\mathcal{B}$ be an elementary substructure of $\mathcal{C}$ such that the order type of the ordinals of $\mathcal{B}$ is ϱ^+, $S \cap \mathcal{B}$ is stationary in $\sup(\mathcal{B} \cap \lambda)$ and $\delta \subseteq \mathcal{B}$. Let $G \subseteq \mathcal{P}\star\mathcal{Q}$ be the generic filter over V such that $V_1 = V[G]$ and let $\mathcal{B}[G]$ be the set of the realizations of the $\mathcal{P}\star\mathcal{Q}$ names in $\mathcal{B}$ according to G. The following claim is easy (cf., e.g., [7, Lemma 2.4]).

Claim 5.9. The structure $\mathcal{B}[G]$ is an elementary substructure of $\langle \mathbf{H}_\lambda^{V_1}, \varepsilon, \mathcal{A} \rangle$ where $\mathcal{A}$ is the realization of the name $\dot{A}$ according to G. In particular $\mathcal{B}[G] \cap \lambda$ is a subalgebra of $\mathcal{A}$.

Lemma 5.10. 1. $\sup(\mathcal{B}[G] \cap \lambda) = \sup(\mathcal{B} \cap \lambda)$.

2. The order type of $\mathcal{B}[G] \cap \lambda$ is ϱ^+.

Theorem 1.15 follows from Lemmas 5.6 and 5.10 since $\mathcal{B}[G] \cap \lambda$ is a witness in V_1 to the truth of $\tilde{\Delta}_{\aleph_\omega, \aleph_{\omega\cdot 2+1}}$: It is a subalgebra of $\mathcal{A}$ with the right order type. In V, $S \cap \mathcal{B}$ is stationary in $\gamma = \sup(\mathcal{B}[G] \cap \lambda) = \sup(\mathcal{B} \cap \lambda)$. The cofinality of γ is ϱ^+. The forcing $\mathcal{P}$ has cardinality $\kappa < \varrho^+$, so it does not kill the stationarity of $S \cap \gamma$ in γ. The forcing $\mathcal{Q}$ is ϱ^{++} closed, so it does add any sequences of ordinals of length ϱ^+, thus it also does not kill the stationarity of $S \cap \gamma$ in γ. So we get that V_1 satisfies $S \cap \mathcal{B} \subseteq S \cap \mathcal{B}[G]$ is stationary in $\gamma = \sup(\mathcal{B}[G] \cap \lambda) = \sup(\mathcal{B} \cap \lambda)$.

Proof of Lemma 5.10. For the first claim assume that $\alpha \in \mathcal{B}[G] \cap \lambda$. Then α is the G-realization of some name τ which without loss of generality can be assumed to be forced to denote an element of λ. Since our forcing has cardinality $\mu < \lambda$ there is an ordinal $\eta < \lambda$ such that every condition forces that τ is less than η. By elementarity of $\mathcal{B}$ we can assume that $\eta \in \mathcal{B}$. Since $\alpha < \eta$ we get $\alpha < \sup(\mathcal{B} \cap \lambda)$.

For the second claim, using the first claim, it is enough to show that for every $\alpha \in \mathcal{B}$, $\mathcal{B}[G] \cap \alpha$ has cardinality at most ϱ. As in the proof of Claim 5.6 we pick an injection $h \in \mathcal{B}$ of α into $\sup(\langle \mu^{+n} \mid n \in \omega \rangle)$. By elementarity

$$\mathcal{B}[G] = h^{-1}(\bigcup_{n<\omega} \mathcal{B}[G] \cap \mu^{+n}).$$

So it is enough to show that for $n \in \omega$, $|\mathcal{B}[G] \cap \mu^{+n}| \leq \varrho$.

Fix $n < \omega$ and let β be an ordinal $\mathcal{B}[G] \cap \mu^{+n}$. The ordinal β is the G realization of some name τ in $\mathcal{B}$. Without loss of generality τ is forced by every condition to be an ordinal in μ^{+n}. (We say that τ is a μ^{+n} name.) The cardinality of the forcing notion is μ so the cardinality of the set of all such names (up to equivalence of names) is at most $2^{\mu \times \mu^{+n}}$ which is at most μ^{+n+1}. The set $\mathcal{B} \cap \mu^{+n+1}$ has cardinality at most ϱ, so we have at most ϱ many μ^{+n+1} names in $\mathcal{B}$, which implies that $|\mathcal{B}[G] \cap \mu^{+n}| \leq \varrho$. Q.E.D.

References

[1] J. E. Baumgartner. A new class of order types. *Annals of Mathematical Logic*, 9:187–222, 1976.

[2] R. Bing. Metrization of topological spaces. *Canadian Journal of Mathematics*, 3:175–186, 1951.

[3] P. Eklof. On the existence of κ-free abelian groups. *Proceedings of the American Mathematical Society*, 47:65–72, 1975.

[4] W. G. Fleissner. On λ-collectionwise Hausdorff spaces. *Topology Proceedings*, 2(2):445–456, 1977.

[5] W. G. Fleissner and S. Shelah. Collectionwise Hausdorff incompactness at singulars. *Topology and its Applications*, 31:101–107, 1989.

[6] M. D. Foreman, M. Magidor, and S. Shelah. Martin's Maximum, saturated ideals and non-regular ultrafilters, part I. *Annals of Mathematics*, 127:1–47, 1988.

[7] J.-P. Levinski, M. Magidor, and S. Shelah. Chang's conjecture for $\aleph_\omega$. *Israel Journal of Mathematics*, 69:161–172, 1990.

[8] M. Magidor. Reflecting stationary sets. *Journal of Symbolic Logic*, 47:755–771, 1982.

[9] M. Magidor and S. Shelah. When does almost free imply free (for groups, transversals etc.). *Journal of the American Mathematical Society*, 7:769–830, 1994.

[10] S. Shelah. A compactness theorem for singular cardinals, free algebras, whitehead problem and transversals. *Israel Journal of Mathematics*, 21:319–349, 1975.

[11] S. Shelah. Remarks on λ-collectionwise Hausdorff spaces. *Topology Proceedings*, 2(2):583–592, 1977.

[12] S. Shelah. On successors of singular cardinals. In M. Boffa, D. van Dalen, and K. MacAloon, editors, *Logic Colloquium '78, Proceedings of the Colloquium held in Mons, August 24–September 1, 1978*, volume 97 of *Studies in Logic and the Foundations of Mathematics*, pages 357–380. North Holland, 1979.

Cardinal sequences for superatomic Boolean algebras

Juan Carlos Martínez*

Facultat de Matemàtiques, Universitat de Barcelona, Barcelona, Spain

Abstract

This expository paper contains fundamental results on cardinal sequences for superatomic Boolean algebras as well as a list of open problems on the subject.

1 Introduction

A Boolean algebra B is *superatomic* (sBA) if and only if every homomorphic image of B is atomic. It is well-known that a Boolean algebra B is superatomic if and only if its Stone space $\mathrm{St}(B)$ is scattered, i.e., every non-empty subspace of $\mathrm{St}(B)$ has some isolated points. A useful tool in the study of scattered spaces is the Cantor-Bendixson derivative $A^{(\alpha)}$ of a subset A of $\mathrm{St}(B)$, which is defined by induction on α as follows: $A^{(0)} = A$; if $\alpha = \beta + 1$, $A^{(\alpha)} = \{x \in \mathrm{St}(B) : x$ is an accumulation point of $A^{(\beta)}\}$; and if α is a limit ordinal, $A^{(\alpha)} = \bigcap\{A^{(\beta)} : \beta < \alpha\}$. It is easy to check that $\mathrm{St}(B)$ is scattered if and only if there is an ordinal α such that $\mathrm{St}(B)^{(\alpha)} = \varnothing$.

The Cantor-Bendixson process for topological spaces can be transferred to the context of Boolean algebras, obtaining in this way an increasing sequence of ideals. If B is a Boolean algebra and α is an ordinal, we define *the ideal* I_α as follows: $I_0 = \{0\}$; if $\alpha = \beta + 1$, $I_\alpha =$ the ideal generated by $I_\beta \cup \{b \in B : b/I_\beta$ is an atom in $B/I_\beta\}$; and if α is a limit, $I_\alpha = \bigcup\{I_\beta : \beta < \alpha\}$. Then B is superatomic if and only if there is an ordinal α such that $B = I_\alpha$.

For basic facts on superatomic Boolean algebras, we refer the reader to [15] and [22]. In particular, if B is a superatomic Boolean algebra, we have $|B| = |\mathrm{St}(B)|$. Also, it is known that the notion of a superatomic Boolean algebra is absolute in the set-theoretic sense.

Assume that B is a superatomic Boolean algebra. The *height* of B, in symbols $\mathrm{ht}(B)$, is the least ordinal α such that $B = I_\alpha$. This ordinal α is always a successor ordinal. Then we define the *reduced height* of B, in symbols $\mathrm{ht}^-(B)$, as the least ordinal α such that $B = I_{\alpha+1}$. It is well-known that if $\mathrm{ht}^-(B) = \alpha$, then $I_{\alpha+1} \setminus I_\alpha$ is a finite set. For each $\alpha < \mathrm{ht}(B)$ we write $\mathrm{wd}_\alpha(B) = |I_{\alpha+1} \setminus I_\alpha|$, the number of atoms in B/I_α. It is easy to

*Supported by the Spanish Ministry of Education DGI grant MTM2011-26840 and by the Catalan DURSI grant 2009SGR00187.

see that the atoms in B/I_α correspond canonically to the isolated points of $\mathrm{St}(B)^\alpha$. Then we define the *cardinal sequence* of B by $\mathrm{CS}(B) = \langle \mathrm{wd}_\alpha(B) : \alpha < \mathrm{ht}^-(B)\rangle$. Note that all the cardinals in the sequence $\mathrm{CS}(B)$ are infinite.

For example, it is easy to check that if κ is an infinite cardinal, $\alpha < \kappa^+$ is a non-zero ordinal and B is the clopen algebra of the ordinal $\kappa^\alpha + 1$ endowed with the order topology, then $\mathrm{ht}(B) = \alpha + 1$, $\mathrm{wd}_\beta(B) = \kappa$ for $\beta < \alpha$ and $\mathrm{wd}_\alpha(B) = 1$. If $s = \langle \kappa_\alpha : \alpha < \beta\rangle$ is a sequence of infinite cardinals, we say that s is the *cardinal sequence of a superatomic Boolean algebra* if and only if there is a sBA B such that $\mathrm{CS}(B) = s$. In this paper, we are concerned with the following general problem:

Main Question. What are the cardinal sequences of superatomic Boolean algebras?

The following proposition contains important restrictions about the cardinal sequence which a superatomic Boolean algebra can have.

Proposition 1.1. Suppose that B is a superatomic Boolean algebra with $\mathrm{CS}(B) = \langle \kappa_\alpha : \alpha < \beta\rangle$. Then the following conditions hold:

(a) $|\beta| \leq 2^{\kappa_0}$ and $\kappa_\alpha \leq 2^{\kappa_0}$ for each $\alpha < \beta$.

(b) If $\alpha + 1 < \beta$ then $\kappa_{\alpha+1} \leq \kappa_\alpha^\omega$.

(c) If $\delta < \beta$ is a limit ordinal of cofinality λ and C is a sequence of order-type λ converging to δ, then $\kappa_\delta \leq \prod\{\kappa_\alpha : \alpha \in C\}$.

For a proof of Proposition 1.1, we refer the reader to [1, § 1] or to [11, Lemmas 1-3].

Also, the following limitation has been noticed by Soukup.

Proposition 1.2. Assume that $s = \langle \kappa_\alpha : \alpha < \beta\rangle$ is a sequence of infinite cardinals such that there are a limit ordinal $\alpha < \beta$ and a cardinal $\lambda \leq \kappa_\alpha$ such that $\lambda < \mathrm{cf}(\alpha)$ and there is a strictly increasing sequence of ordinals $\langle \alpha_\xi : \xi < \mathrm{cf}(\alpha)\rangle$ whose supremum is α in such a way that $\kappa_{\alpha_\xi} < \lambda$ for each $\xi < \mathrm{cf}(\alpha)$. Then there is no sBA whose cardinal sequence is s.

Proof. Suppose on the contrary that there is a sBA B such that $\mathrm{CS}(B) = s$. Let $\vartheta = \mathrm{cf}(\alpha)$. Let Y be a set of size λ formed by representatives of different atoms in B/I_α. Note that if $\{b, b'\} \in [Y]^2$, it follows that $b \wedge b' \in I_\gamma$ for some $\gamma < \alpha$. Then we define the function $F : [Y]^2 \to \{\alpha_\xi : \xi < \vartheta\}$ as follows. If $\{b, b'\} \in [Y]^2$ we let $F\{b, b'\}$ be the the least ordinal $\gamma < \vartheta$ such that $b \wedge b' \in I_{\alpha_\gamma}$. As $|Y| = \lambda$ and $\lambda < \vartheta$, there is an $\eta < \vartheta$ such that for every $\{b, b'\} \in [Y]^2$, we have $F\{b, b'\} < \eta$. It follows that B/I_{α_η} contains at least λ atoms, which contradicts the assumption that $\kappa_{\alpha_\xi} < \lambda$ for each $\xi < \vartheta$.
Q.E.D.

On the other hand, there is a standard mechanism for building in a direct way sBAs from certain well-founded partial orderings, which is mainly used in forcing constructions of sBAs (cf. [1, § 6]).

Our set-theoretic terminology is standard. Terms not defined here can be found in [6] or in [13].

2 Cardinal sequences of length $< \omega_2$

By using Proposition 1.1, it is easy to prove by transfinite induction that if B is a sBA with $\mathrm{CS}(B) = \langle \kappa_\alpha : \alpha < \delta \rangle$, we have that $\kappa_\beta \leq \kappa_\alpha^{|\beta \setminus \alpha| + \omega}$ whenever $\alpha < \beta < \delta$. Then the following characterization of the cardinal sequences for superatomic Boolean algebras of length $\leq \omega_1$ was shown by Juhász and Weiss (cf. [11]).

Theorem 2.1. Suppose that $s = \langle \kappa_\alpha : \alpha < \delta \rangle$ is a sequence of infinite cardinals where $\delta \leq \omega_1$. Then the following two conditions are equivalent:

(A) There is a sBA B such that $\mathrm{CS}(B) = s$.

(B) $\kappa_\beta \leq \kappa_\alpha^\omega$ for every $\alpha < \beta < \delta$.

Previously, the equivalence of conditions (A) and (B) for cardinal sequences of countable length had been shown by La Grange (cf. [15, § 17]). The following substantial result proved by Baumgartner in [2] shows that the situation is completely different when we consider cardinal sequences of length $\omega_1 + 1$.

Theorem 2.2. In the Mitchell model there is no sBA B such that $\mathrm{CS}(B) = \langle \kappa_\alpha : \alpha \leq \omega_1 \rangle$ where $\kappa_\alpha \leq \omega_1$ for every $\alpha < \omega_1$ and $\kappa_{\omega_1} = \omega_2$.

Also, it was proved by Just in [12] that if we add any number of Cohen reals to a model of CH, then in the resulting generic extension there is no sBA B with $\mathrm{CS}(B) = \langle \kappa_\alpha : \alpha \leq \omega_1 \rangle$ where $\kappa_\alpha \leq \omega_1$ for every $\alpha < \omega_1$, $\kappa_\alpha = \omega$ for some $\alpha < \omega_1$ and $\kappa_{\omega_1} = \omega_2$.

Therefore, it is not possible to extend Theorem 2.1 to cardinal sequences of length $\omega_1 + 1$. However, in [11] the following result was shown which provides us with a general sufficient condition under which a cardinal sequence of length $< \omega_2$ is the cardinal sequence of a sBA.

Theorem 2.3. Let $s = \langle \kappa_\alpha : \alpha < \delta \rangle$ be a sequence of infinite cardinals with $\delta < \omega_2$ satisfying the following conditions:

1. $\kappa_\beta \leq \kappa_\alpha^\omega$ for every $\alpha < \beta < \delta$.
2. $\kappa_\alpha \leq \omega_1$ for every $\alpha < \delta$ with $\mathrm{cf}(\alpha) = \omega_1$.

Then there is a sBA B with $\mathrm{CS}(B) = s$.

Note that the Baumgartner-Just impediment mentioned above is the reason for condition 2 in Theorem 2.3.

Also, an important characterization under GCH of the cardinal sequences for sBAs of length $< \omega_2$ was obtained by Juhász, Soukup and Weiss in [9]. If α is a non-zero ordinal and λ is an infinite cardinal, we denote by $C_\lambda(\alpha)$ the class of cardinal sequences s of length α of sBAs such that $s(0) = \lambda$ and $s(\beta) \geq \lambda$ for every $\beta < \alpha$. The following proposition was shown in [9, Theorem 2.1] in order to prove the GCH characterization. If s, t are cardinal sequences, we denote by $s^\frown t$ the concatenation of s and t.

Theorem 2.4. Let s be a sequence of infinite cardinals and let α be the length of s. Then s is the cardinal sequence of a sBA if and only if there are a natural number n, a decreasing sequence $\lambda_0 > \lambda_1 > \ldots > \lambda_{n-1}$ of infinite cardinals and there are ordinals $\alpha_0, \ldots, \alpha_{n-1}$ such that $\alpha = \alpha_0 + \ldots + \alpha_{n-1}$ and $s = s_0^\frown s_1^\frown \ldots ^\frown s_{n-1}$ with $s_i \in C_{\lambda_i}(\alpha_i)$ for each $i < n$.

Then a description of the classes $C_\lambda(\alpha)$ under GCH is given in [9, Theorem 4.1] for every ordinal $\alpha < \omega_2$ and every infinite cardinal λ, and hence we obtain from Theorem 2.4 the desired characterization under GCH for cardinal sequences of length $< \omega_2$. However, it is unknown whether a GCH characterization can be obtained for longer sequences, for example for sequences of length ω_2. We refer the reader to [25] for a list of results and open questions on cardinal sequences in relation to GCH.

3 Cardinal sequences of length $< \omega_3$

We define the *width* of a sBA B by $\mathrm{wd}(B) = \sup\{\mathrm{wd}_\alpha(B) : \alpha < ht^-(B)\}$. If α is a non-zero ordinal and κ is an infinite cardinal, we say that a sBA B is a (κ, α)-*Boolean algebra* if and only if $\mathrm{ht}(B) = \alpha + 1$ and $\mathrm{wd}(B) \leq \kappa$. And we say that B is a *thin-tall* Boolean algebra if and only if $\mathrm{wd}(B) < |\mathrm{ht}(B)|$.

It was shown in [18] that if GCH holds and $s = \langle \kappa_\alpha : \alpha < \beta \rangle$ is a sequence of infinite cardinals such that $\kappa_\alpha \geq |\beta|$ for every $\alpha < \beta$, then in some cardinal-preserving generic extension there is a sBA B with $\mathrm{CS}(B) = s$. The situation becomes, however, more complicated when we want to construct superatomic Boolean algebras B with $\mathrm{CS}(B) = \langle \kappa_\alpha : \alpha < \beta \rangle$ where $\beta > \omega_1$ and $\kappa_\alpha < |\beta|$ for some $\alpha < \beta$, in particular when we want to construct thin-tall Boolean algebras.

The following result, which was first proved in [10], is an immediate consequence of Theorem 2.3.

Theorem 3.1. For every ordinal $\alpha < \omega_2$ there is an (ω, α)-Boolean algebra.

It was even proved in [4] that for every ordinal $\alpha < \omega_2$ there are 2^{ω_1} (as many as possible) pairwise non-isomorphic (ω, α)-Boolean algebras. Note that we can not extend Theorem 3.1 to cardinal sequences of length ω_2, because it follows from Proposition 1.1(a) that under CH there is no (ω, ω_2)-Boolean algebra. Also, it follows from [8, Theorem 2.1] that if we add any number of Cohen reals to a model of CH, in the resulting generic extension we have that if B is a sBA with $\mathrm{CS}(B) = \langle \kappa_\alpha : \alpha < \beta \rangle$ then $|\{\alpha \in \beta : \kappa_\alpha = \omega\}| \leq \omega_1$. However, the following result was proved by Baumgartner and Shelah in [2].

Theorem 3.2. It is relatively consistent with ZFC that there is an (ω, ω_2)-Boolean algebra.

In order to prove Theorem 3.2, the notion of a Δ-function was introduced, an important combinatorial tool due to Shelah that has been also used in other forcing constructions of sBAs. We say that a function $f : [\omega_2]^2 \to [\omega_2]^{\leq\omega}$ is a *Δ-function* if and only if $f\{\alpha, \beta\} \subseteq \min\{\alpha, \beta\}$ for every $\{\alpha, \beta\} \in [\omega_2]^2$ and for every uncountable set D of finite subsets of ω_2 there are $a, b \in D$ with $a \neq b$ such that for every $\alpha \in a \setminus b$, $\beta \in b \setminus a$ and $\tau \in a \cap b$ the following conditions hold:

1. $\tau < \alpha, \beta$ implies $\tau \in f\{\alpha, \beta\}$,
2. $\tau < \alpha$ implies $f\{\beta, \tau\} \subseteq f\{\alpha, \beta\}$,
3. $\tau < \beta$ implies $f\{\alpha, \tau\} \subseteq f\{\alpha, \beta\}$.

The forcing construction of an (ω, ω_2)-Boolean algebra carried out in [2] is achieved in two steps. First, a Δ-function is adjoined by means of an ω_1-closed and ω_2-c.c. forcing notion. Then by using the Δ-function, a c.c.c. forcing notion is defined which adds the required (ω, ω_2)-Boolean algebra. Recently, an alternative proof of Theorem 3.2 has been given in [27] by using Neeman's method of generalized side conditions instead of a Δ-function.

The following problems are long-standing open questions.

Problem 3.3. Is there in ZFC an (ω_1, ω_2)-Boolean algebra?

Problem 3.4. Is it relatively consistent with ZFC that there is an (ω_1, ω_3)-Boolean algebra?

However, some extensions of the theorem by Baumgartner and Shelah have been obtained. The following result was first proved in [17].

Theorem 3.5. It is relatively consistent with ZFC that there is an (ω, α)-Boolean algebra for every ordinal $\alpha < \omega_3$.

In order to prove Theorem 3.5, a c.c.c. notion of forcing to add an (ω, α)-Boolean algebra is defined by using the notion of a tree of intervals, a combinatorial tool that permits us to use different Δ-functions at ordinals

$\gamma \leq \alpha$ of cofinality ω_2 in a coherent way. Generalizations of Theorem 3.5 have been given in [5] and [24]. In [24], it was shown that from a natural c.c.c. notion of forcing that adjoins an (ω, ω_2)-Boolean algebra, we can define a natural c.c.c. notion of forcing that adjoins an (ω, α)-Boolean algebra for any $\alpha < \omega_3$. And in [5], the following result was shown.

Theorem 3.6. It is relatively consistent with ZFC that there are PCF structures of height α for every ordinal $\alpha < \omega_3$.

A PCF structure is a well-founded partial ordering defined on an infinite successor ordinal that satisfies certain conditions that reflect fundamental properties of the PCF operator defined on $\{\aleph_n : n \geq 1\}$. Every PCF structure T has the form $T = \bigcup\{T_\alpha : \alpha < \gamma\}$ where γ is a non-zero ordinal and $|T_\alpha| \leq |\alpha|$ for every infinite ordinal $\alpha < \gamma$. If $T = \bigcup\{T_\alpha : \alpha < \gamma\}$ is a PCF structure, then T is associated with a superatomic Boolean algebra B such that $\mathrm{ht}(B) = \gamma$ and $\mathrm{wd}_\alpha(B) = |T_\alpha|$ for every $\alpha < \gamma$. PCF structures are important in the proof of Shelah's fundamental theorem on cardinal arithmetic, saying that $2^{\aleph_\omega} < \aleph_{\omega_4}$ if $\aleph_\omega$ is a strong limit cardinal, because this fundamental theorem follows from the fact that there is no PCF structure of size $\geq \omega_4$ (cf. [3] or [23]). If in ZFC there is no PCF structure of size ω_3, then we could improve Shelah's bound from $\aleph_{\omega_4}$ to $\aleph_{\omega_3}$. And if it is relatively consistent with ZFC that there is a PCF structure of size ω_3, then in order to improve the bound from $\aleph_{\omega_4}$ to $\aleph_{\omega_3}$ we should use methods different from those used by Shelah in his theorem. To decide this question, we should first solve the following open problem.

Problem 3.7. Is it relatively consistent with ZFC that there is an (ω, ω_3)-Boolean algebra?

Also, in [26] the following generalization of the Baumgartner-Shelah theorem was shown, in which wide sBAs of height $\omega_2 + 1$ are constructed by forcing.

Theorem 3.8. If GCH holds and $\lambda \geq \omega_2$ is a regular cardinal, then in some cardinal-preserving generic extension $2^\omega = \lambda$ and every sequence $\langle \kappa_\alpha : \alpha < \omega_2 \rangle$ of infinite cardinals with $\kappa_\alpha \leq \lambda$ for each $\alpha < \omega_2$ is the cardinal sequence of some sBA.

In order to prove Theorem 3.8, the notion of a $\Delta(\omega_2 \times \lambda)$-function was used, a generalization of Shelah's notion of Δ-function, in which the domain of the function is $[\omega_2 \times \lambda]^2$ and its range is a subset of $[\omega_2]^{<\omega}$. The forcing construction in Theorem 3.8 is then carried out in three steps. In the first step, an (ω_1, λ)-semimorass in the sense of Koszmider with certain additional properties is adjoined by means of an ω_1-closed and ω_2-c.c. notion of forcing. Using this strong semimorass, in the second step a $\Delta(\omega_2 \times \lambda)$-function is adjoined to the first extension. The notion of forcing defined in this second step is ω_2-c.c. and it is shown that it preserves ω_1 by using a proper forcing argument. Finally, by using the $\Delta(\omega_2 \times \lambda)$-function, a c.c.c. notion of forcing is defined in order to obtain a cardinal-preserving generic extension satisfying the required property.

Note that it follows from Proposition 1.2 that there is no sBA B with $\mathrm{CS}(B) = \langle \kappa_\alpha : \alpha \leq \omega_2 \rangle$ where $\kappa_\alpha = \omega$ for $\alpha < \omega_2$ and $\kappa_{\omega_2} \geq \omega_1$. More generally, if B is a sBA with $\mathrm{CS}(B) = \langle \kappa_\alpha : \alpha < \gamma \rangle$ where $\omega_2 < \gamma < \omega_3$, then for every ordinal $\alpha < \gamma$ with $\mathrm{cf}(\alpha) = \omega_2$ we have that if there is a strictly increasing sequence of ordinals $\langle \alpha_\xi : \xi < \omega_2 \rangle$ converging to α in such a way that $\kappa_{\alpha_\xi} = \omega$ for every $\xi < \omega_2$, then $\kappa_\alpha = \omega$. So, Theorem 3.8 can not be extended from cardinal sequences of length ω_2 to cardinal sequences of length $< \omega_3$. So it is natural to raise the following question.

Problem 3.9. Assume that GCH holds, $\lambda \geq \omega_2$ is an infinite cardinal and γ is an ordinal with $\omega_2 < \gamma < \omega_3$. Is there a cardinal-preserving generic extension where for every sequence $s = \langle \kappa_\alpha : \alpha < \gamma \rangle$ of infinite cardinals with $\kappa_\alpha \leq \lambda$ for each $\alpha < \gamma$ and $\kappa_\alpha = \omega$ whenever $\mathrm{cf}(\alpha) = \omega_2$, s is the cardinal sequence of some sBA?

We hope to give a positive answer to Problem 3.9 in a future paper.

4 Long cardinal sequences

For thin-tall spaces of height $> \omega_3$, the following result was proved in [14].

Theorem 4.1. The axiom **V=L** implies the existence of a (κ, κ^+)-Boolean algebra for every regular cardinal κ.

In order to obtain Theorem 4.1, it was shown in [14] that for every regular cardinal κ, the existence of a simplified $(\kappa, 1)$-morass implies the existence of a (κ, κ^+)-Boolean algebra. Briefly, the idea of the proof for this fact is as follows. Assume that $\langle \langle \vartheta_\xi : \xi \leq \kappa \rangle, \langle \mathcal{F}_{\mu\xi} : \mu < \xi \leq \kappa \rangle \rangle$ is a simplified $(\kappa, 1)$-morass. For $\xi \leq \kappa$, a sBA $B_\xi \subseteq P((\vartheta_\xi + 1) \times \kappa)$ with $\mathrm{ht}(B_\xi) = \vartheta_\xi + 1$ is constructed such that, for every $\xi < \kappa$, we have that $|B_\xi| < \kappa$ and if $\alpha < \beta < \vartheta_\xi$ and b is the representative of an atom in B_ξ / I_β then there is an $\eta > \xi$ such that $(\alpha, \eta) \in b$. If $\xi < \eta \leq \kappa$ and $f \in \mathcal{F}_{\xi\eta}$, we can project the Boolean algebra B_ξ on $P((f[\vartheta_\xi] + 1) \times \kappa)$ obtaining a sBA B^f_ξ isomorphic to B_ξ. The approximation B_0 is easy to construct. If $\xi = \mu + 1$ and $\mathcal{F}_{\mu\xi} = \{f, g\}$, we can construct B_ξ from a certain amalgamation of B^f_μ and B^g_μ. And if ξ is a limit, B_ξ is constructed from the Boolean algebras B^f_μ for $\mu < \xi$ and $f \in \mathcal{F}_{\mu\xi}$. Then B_κ is the required Boolean algebra. On the other hand, the following problem appears to be open.

Problem 4.2. Does GCH imply the existence of a (κ, κ^+)-Boolean algebra for every regular cardinal κ?

In relation to Problem 4.2, we have the following result shown by Juhász, Soukup and Weiss, which is a consequence of [9, Theorem 3.9].

Theorem 4.3. If κ is an uncountable cardinal such that $\kappa^{<\kappa} = \kappa$, then for every ordinal $\gamma < \kappa^+$ with $\mathrm{cf}(\gamma) = \kappa$ there is a sBA B with $\mathrm{CS}(B) = \langle \kappa_\alpha : \alpha < \kappa^+ \rangle$ where $\kappa_\alpha = \kappa$ for $\alpha < \gamma$ and $\kappa_\alpha = \kappa^+$ for $\gamma \leq \alpha < \kappa^+$.

Theorem 4.3 is an extension of the main result obtained in [7], where it was shown that in ZFC there is a sBA with just ω_1 atoms and height $\omega_2 + 1$. Also, the following result was proved in [16].

Theorem 4.4. If GCH holds, κ is a regular cardinal and $\alpha < \kappa^{++}$ is an ordinal, then in some cardinal-preserving generic extension GCH holds and there is a (κ, α)-Boolean algebra.

However, we do not know whether the construction of these Boolean algebras can be carried out directly in the constructible universe.

Problem 4.5. Does **V**=**L** imply the existence of a (κ, α)-Boolean algebra for every regular cardinal κ and every ordinal $\alpha < \kappa^{++}$?

Furthermore, a generalization of Theorem 4.4 was obtained in [19], where a characterization was given under GCH of the sequences s of regular cardinals of any length such that s is the cardinal sequence of a sBA in some cardinal-preserving and GCH-preserving generic extension of the ground model.

On the other hand, no consistency result is known about the existence of thin-tall Boolean algebras whose width is a singular cardinal. In particular, the following problem is open.

Problem 4.6. Is it relatively consistent with ZFC that there exists an $(\aleph_\omega, \aleph_\omega^+)$-Boolean algebra?

Further results on long cardinal sequences have been obtained for the class of the thin-thick Boolean algebras (cf. [14, 20, 21]). If γ is an infinite ordinal and B is a sBA, we say that B is a *γ-thin-thick* Boolean algebra if and only if $\mathrm{CS}(B) = \langle \kappa_\alpha : \alpha \leq \gamma \rangle$ where $\kappa_\alpha \leq |\gamma|$ for each $\alpha < \gamma$ and $\kappa_\gamma \geq |\gamma|^+$. As a consequence of the main result of [20], we obtain that if κ, λ are any specific infinite cardinals such that κ is regular and $\kappa < \lambda$, then it is relatively consistent with ZFC that there is a sBA B with $\mathrm{CS}(B) = \langle \kappa_\alpha : \alpha \leq \kappa^+ \rangle$ where $\kappa_\alpha = \kappa$ for $\alpha < \kappa^+$ and $\kappa_{\kappa^+} = \lambda$. Also, by using standard arguments, we can obtain from Theorem 2.2 that the consistency of the existence of an inaccessible cardinal is equivalent to the consistency of the non-existence of an ω_1-thin-thick Boolean algebra (cf. [22, § 5]). However, it is not known whether Theorem 2.2 can be generalized to larger cardinals.

Problem 4.7. If $\kappa \geq \omega_2$ is a regular cardinal, does the consistency of the existence of an inaccessible cardinal imply the consistency of the non-existence of a κ-thin-thick Boolean algebra?

5 Cardinal sequences and large cardinals

By using the refinement of Prikry forcing due to Magidor (cf. [6, Chapter 36]) and the fact that the notion of superatomic Boolean algebra is absolute, we obtain the following corollary of Theorem 4.3.

Theorem 5.1. If it is consistent that there is a measurable cardinal, then it is consistent that there is a sBA B with $\mathrm{CS}(B) = \langle \kappa_\alpha : \alpha < \aleph_\omega^+ \rangle$ where $\kappa_\alpha = \aleph_\omega$ for $\alpha < \aleph_\omega$ and $\kappa_\alpha = \aleph_\omega^+$ for $\aleph_\omega \leq \alpha < \aleph_\omega^+$.

We do not know whether the sBA in Theorem 5.1 can be constructed by using large cardinals smaller than a measurable cardinal or even whether it can be constructed without using inaccessible cardinals.

Problem 5.2. Is the consistency of the existence of a measurable cardinal necessary in the statement of Theorem 5.1?

The proof of the following result is an adaptation of a classical argument of Vopěnka on cardinal arithmetic (cf. [13, § 5]).

Theorem 5.3. Assume that κ is a measurable cardinal, U is a normal ultrafilter over κ and $\{\lambda \in \kappa : \lambda$ is a cardinal and there is a (λ, λ^+)-Boolean algebra$\} \in U$. Then there is a (κ, κ^+)-Boolean algebra.

Proof. Suppose that $X = \{\lambda \in \kappa : \lambda$ is a cardinal and there is a (λ, λ^+)-Boolean algebra$\}$ and $Y = \{\alpha \in \kappa : \alpha$ is an infinite ordinal and there is a $(|\alpha|, |\alpha|^+)$- Boolean algebra$\}$. Note that $X \subseteq Y$, and so $Y \in U$. We denote by M the transitive class isomorphic with $\mathrm{Ult}(V, U)$. As $[id]_U = \kappa$ and $Y \in U$, we deduce that in M there is a (κ, κ^+)-Boolean algebra B. Now since $M \subseteq V$, $(\kappa)^M = \kappa$, $(\kappa^+)^M = \kappa^+$ and the notion of sBA is absolute, we infer that B is a (κ, κ^+)-Boolean algebra in V. Q.E.D.

The following result is an immediate consequence of Theorem 5.3.

Theorem 5.4. Assume that κ is a measurable cardinal. If for every infinite cardinal $\lambda < \kappa$ there is a (λ, λ^+)-Boolean algebra, then there is a (κ, κ^+)-Boolean algebra.

Now, it is natural to raise the following question.

Problem 5.5. Does Theorem 5.4 hold with "measurable" replaced by "Ramsey" or "weakly compact"?

Theorem 5.6. Assume that κ is a strong cardinal and there is a (λ, λ^+)-Boolean algebra for every infinite cardinal $\lambda < \kappa$. Then there is a (λ, λ^+)-Boolean algebra for every infinite cardinal λ.

Proof. Assume that $\lambda \geq \kappa$ is a cardinal. Let ϑ be a cardinal such that $\max\{\lambda, \kappa \times \omega\} < \vartheta$. As κ is ϑ-strong, there is an elementary embedding $j : V \to M$ such that $\text{crit}(j) = \kappa$, $\vartheta < j(\kappa)$ and $V_\vartheta \subseteq M$. Since in V there is a (ϱ, ϱ^+)-Boolean algebra for every infinite cardinal $\varrho < \kappa$, it follows that in M there is a (ϱ, ϱ^+)-Boolean algebra for every infinite cardinal $\varrho < j(\kappa)$, and so in M there is a (λ, λ^+)-Boolean algebra B. But since the notion of sBA is absolute, we infer that B is a sBA in V. Also as $V_\vartheta \subseteq M$, we have $(\lambda)^M = \lambda$ and $(\lambda^+)^M = \lambda^+$, and hence B is a (λ, λ^+)-Boolean algebra in V. Q.E.D.

However, we do not know the answer to the following question.

Problem 5.7. Assume that κ is a strongly compact cardinal. Does the existence of a (λ, λ^+)-Boolean algebra for every infinite cardinal $\lambda < \kappa$ imply the existence of a (λ, λ^+)-Boolean algebra for every infinite cardinal λ ?

References

[1] J. Bagaria. Thin-tall spaces and cardinal sequences. In E. Pearl, editor, *Open problems in topology. II*, pages 115–124. Elsevier, 2007.

[2] J. E. Baumgartner and S. Shelah. Remarks on superatomic Boolean algebras. *Annals of Pure and Applied Logic*, 33(2):109–129, 1987.

[3] M. R. Burke and M. Magidor. Shelah's pcf theory and its applications. *Annals of Pure and Applied Logic*, 50(3):207–254, 1990.

[4] A. Dow and P. Simon. Thin-tall Boolean algebras and their automorphism groups. *Algebra Universalis*, 29(2):211–226, 1992.

[5] K. Er-rhaimini and B. Veličković. PCF structures of height less than ω_3. *Journal of Symbolic Logic*, 75(4):1231–1248, 2010.

[6] T. J. Jech. *Set theory.* Springer Monographs in Mathematics. Springer, third millenium edition, 2003.

[7] I. Juhász, S. Shelah, L. Soukup, and Z. Szentmiklóssy. A tall space with a small bottom. *Proceedings of the American Mathematical Society*, 131(6):1907–1916, 2003.

[8] I. Juhász, S. Shelah, L. Soukup, and Z. Szentmiklóssy. Cardinal sequences and Cohen real extensions. *Fundamenta Mathematicae*, 181(1):75–88, 2004.

[9] I. Juhász, L. Soukup, and W. Weiss. Cardinal sequences of length $< \omega_2$ under GCH. *Fundamenta Mathematicae*, 189(1):35–52, 2006.

[10] I. Juhász and W. Weiss. On thin-tall scattered spaces. *Colloquium Mathematicum*, 40(1):63–68, 1978/79.

[11] I. Juhász and W. Weiss. Cardinal sequences. *Annals of Pure and Applied Logic*, 144(1-3):96–106, 2006.

[12] W. Just. Two consistency results concerning thin-tall Boolean algebras. *Algebra Universalis*, 20(2):135–142, 1985.

[13] A. Kanamori. *The higher infinite, Large cardinals in set theory from their beginnings*. Springer Monographs in Mathematics. Springer, 2nd edition, 2003.

[14] P. Koepke and J. C. Martínez. Superatomic Boolean algebras constructed from morasses. *Journal of Symbolic Logic*, 60(3):940–951, 1995.

[15] S. Koppelberg. General theory of Boolean algebras. In J. D. Monk and R. Bonnet, editors, *Handbook of Boolean algebras. Vol. 1*, pages 741–773. North-Holland, 1989.

[16] J. C. Martínez. A forcing construction of thin-tall Boolean algebras. *Fundamenta Mathematicae*, 159(2):99–113, 1999.

[17] J. C. Martínez. A consistency result on thin-very tall Boolean algebras. *Israel Journal of Mathematics*, 123:273–284, 2001.

[18] J. C. Martínez. A consistency result on cardinal sequences of scattered Boolean spaces. *Mathematical Logic Quarterly*, 51(6):586–590, 2005.

[19] J. C. Martínez and L. Soukup. Cardinal sequences of LCS spaces under GCH. *Annals of Pure and Applied Logic*, 161(9):1180–1193, 2010.

[20] J. C. Martínez and L. Soukup. Superatomic Boolean algebras constructed from strongly unbounded functions. *Mathematical Logic Quarterly*, 57(5):456–469, 2011.

[21] J. Roitman. Height and width of superatomic Boolean algebras. *Proceedings of the American Mathematical Society*, 94(1):9–14, 1985.

[22] J. Roitman. Superatomic Boolean algebras. In J. D. Monk and R. Bonnet, editors, *Handbook of Boolean algebras. Vol. 3*, pages 719–740. North-Holland, 1989.

[23] S. Shelah. *Cardinal arithmetic*, volume 29 of *Oxford Logic Guides*. Clarendon Press Oxford University Press, New York, 1994. Oxford Science Publications.

[24] L. Soukup. A lifting theorem on forcing LCS spaces. In E. Győri, G. O. H. Katona, and L. Lovász, editors, *More sets, graphs and numbers, A salute to Vera Sós and András Hajnal*, volume 15 of *Bolyai Society Mathematical Studies*, pages 341–358. Springer-Verlag, 2006.

[25] L. Soukup. Cardinal sequences and universal spaces. In E. Pearl, editor, *Open problems in topology. II*, pages 743–746. Elsevier, 2007.

[26] L. Soukup. Wide scattered spaces and morasses. *Topology and its Applications*, 158(5):697–707, 2011.

[27] B. Veličković and G. Venturi. Proper forcing remastered. Preprint.

One-dimensional Σ-presentations of structures over $\mathbf{HF}(\mathbb{R})$

Andrey S. Morozov*

Sobolev Institute of Mathematics, Russian Academy of Sciences, Novosibirsk, Russia
Department of Mathematics, Novosibirsk State University, Novosibirsk, Russia

Abstract

We introduce and study the notion of dimension for Σ-definable structures over $\mathbf{HF}(\mathbb{R})$, the hereditarily finite superstructure over $\mathbb{R}$, and prove a general result which implies that the following structures fail to have one-dimensional Σ-presentations over $\mathbf{HF}(\mathbb{R})$ with any parameters: any free system of uncountable rank whose signature contains at least one at least binary operation symbol, the direct product of two Boolean algebras of cardinality more than ω, the Boolean algebra $\wp(\omega)$ of all subsets of ω, the symmetric group $\mathrm{Sym}(\omega)$ of a countable set.

1 Introduction

The study of computable structures, i.e., of the structures whose diagrams are computably enumerable, was started in the middle of the last century by Ershov, Crossley, Fröhlich, Malcev, Nerode, Rabin, Shepherdson, Vaught, and others. The importance of the study of such structures is based on the observation that these structures are exactly the structures that could be studied with the use of a computer. The study of computable structures still remains an actively developing area on the border of model theory and recursion theory; it became so broad that nowadays it is almost impossible to give a more or less complete survey of its results and ideas. Later on, this area came to involve issues related to various kinds of generalized computability, mostly to relative computability and Turing degrees. Actually, an interesting series of questions still attracting many researches is the study of the complexities of various algebraic and model-theoretic constructions.

Along with new ideas in generalized computability and as reflection and application of these ideas, new generalizations of the definition of computable structures have appeared. One of these approaches is related to a specific kind of computability, namely to Σ-definability over admissible sets. Ershov [3] has formulated a definition of structures Σ-definable over admissible sets, which seems to be a natural analog of computable structures in this situation. The author considers the hereditarily finite superstructure

*The author thanks the anonymous referee for valuable remarks.

Stefan Geschke, Benedikt Löwe, Philipp Schlicht (*eds.*).
Infinity, computability, and metamathematics: Festschrift celebrating the 60th birthdays of Peter Koepke and Philip Welch. College Publications, London, 2014. Tributes, Volume 23.

$\mathbf{HF}(\mathbb{R})$ over the ordered field of reals $\mathbb{R}$ as the most important and interesting candidate amongst concrete admissible sets to start the study of generalizations of computable models. One of the reasons for this is the idea that a Σ-definable function over this set could be considered as an opportunity to develop a program computing our function in a hypothetical powerful programming system like C++ provided that we had an opportunity to use an *exact realization of real numbers* (instead of the existing implementation of reals as approximations by rationals) and we could take exact solutions of algebraic equations and use them in further computations.

In this paper, we introduce and study the notion of dimension for structures Σ-definable over $\mathbf{HF}(\mathbb{R})$ and prove that the following structures fail to have one-dimensional Σ-presentations over $\mathbf{HF}(\mathbb{R})$ with any parameters: any free system of uncountable rank whose signature contains at least one binary operation, any direct product of two Boolean algebras of cardinality greater than ω, the Boolean algebra $\wp(\omega)$ of all subsets of ω, the symmetric group $\mathrm{Sym}(\omega)$ of a countable set. These facts will be obtained from a general proposition.

Now we pass to definitions. The basic definitions related to admissible sets can be found in [1] or [2].

In this paper we shall use the notion of hereditarily finite superstructure, i.e., of admissible sets of the kind $\mathbf{HF}(\mathfrak{M})$. Here we give an informal definition of it. Roughly speaking, to construct $\mathbf{HF}(\mathfrak{M})$ we start with a structure $\mathfrak{M}$ of finite predicate signature, whose elements are considered as *urelements*,, i.e., they differ from the empty set $\varnothing$ but nevertheless they cannot contain elements; then we add finite sets of them, finite sets of what we have already constructed, etc. In other words, the structure $\mathbf{HF}(\mathfrak{M})$ is formed from elements of $\mathfrak{M}$ and sets that could be written down as finite strings using elements of $\mathfrak{M}$, $\varnothing$, $\{$, $\}$ only. Typical representatives of the elements of $\mathbf{HF}(\mathfrak{M})$ are $\varnothing$, m, $(m \in \mathfrak{M})$, $\{m, \varnothing, \{\varnothing\}\}$, etc. This admissible set could be viewed from a computer scientist's point of view: we may think of $\mathfrak{M}$ as of an a priori given basic data type and we may consider the elements of $\mathbf{HF}(\mathfrak{M})$ as potential interpretations of elements of data types that could be constructed from the basic data types. For instance, ordered pairs $\langle a, b\rangle$ could be viewed as it is usually done in classical set theory: $\langle a, b\rangle = \{\{a\}, \{a, b\}\}$, etc. Thus, actually $\mathbf{HF}(\mathfrak{M})$ contains all tuples of urelements, tuples of tuples, etc., and many other things.

The structure $\mathbf{HF}(\mathfrak{M})$ is a classical example of admissible set. The set of all ordinals that are contained in it coincides with ω, the set of all natural numbers.

It is well known that the definition of computable enumerability (c.e.) is enough to define all the remaining most important notions related to computability: a set is computable if and only if it is c.e. together with its complement, a function is computable if and only if its graph is c.e. In admissible sets, the analog of computable enumerability is given by the notion of Σ-definable sets, i.e., the sets definable by means of Σ-formulas with parameters. To define the class of Σ-formulas, we first define the class of Δ_0-formulas as the smallest class containing atomic formulas which is closed under propositional connectives and bounded quantifiers, i.e., quantifiers of the kind $\exists x \in y$ and $\forall x \in y$; then we define the class of Σ-formulas as the smallest class containing Δ_0-formulas which is closed under positive propositional connectives $\wedge$, $\vee$, bounded quantifiers, and unbounded existential quantifiers $\exists x$. The sets that are Σ-definable together with their complements are called Δ-sets and are considered as computable, the functions whose graphs are Σ-definable are also considered as computable. It is well known fact that any computable function on ω is Σ-definable without parameters.

One can code formulas of some formal languages by elements of admissible sets; more exactly, we can consider some elements of admissible sets as formulas. There exists a coding in which the class of Σ-formulas is Δ-definable. It can be proved that there exists a Σ-definable predicate $\mathrm{Sat}(x, y)$ which is true if and only if x is a formula, y is a valuation of all its free variables, and x is satisfied on y [1, 2].

If we want to stress that some object is defined by Σ-definitions with parameters $\bar{p}$, we shall use the expression $\Sigma_{\bar{p}}$*-definable*. If we do not indicate parameters explicitly and allow the use of any parameters, we shall refer to this object as to Σ*-definable*.

If $\varphi(x_1, \ldots, x_k, \bar{z})$ is a formula, $\mathfrak{A}$ is a structure, and $\bar{p} \in \mathfrak{A}$, the set $\{\langle a_1, \ldots, a_k\rangle \mid \mathfrak{A} \models \varphi(a_1, \ldots, a_k, \bar{p})\}$ will be denoted as $\varphi^{\mathfrak{A}}[x_1, \ldots, x_k, \bar{p}]$. If $A \cup B = C$ and $A \cap B = \varnothing$, we shall write this fact as $A \oplus B = C$.

If we take one of the equivalent variants of classical definitions of computable structure and replace the concepts related to computability with their analogs in admissible sets, we arrive at the following definition. In the last item of this definition, we make an unessential change, namely, we use the condition $\mathfrak{B}/E = \mathfrak{A}$ instead of $\mathfrak{B}/E \cong \mathfrak{A}$ used in the original text. The reason will be clear from the remark that follows this definition.

Definition 1.1 (Ershov, [3]). We say that a structure

$$\mathfrak{A} = \left\langle A; P_0^{m_0}, \ldots, P_{k-1}^{m_{k-1}} \right\rangle$$

is Σ-definable over $\mathbf{HF}(\mathfrak{M})$ if there exist a $\bar{p} \in \mathbf{HF}(\mathfrak{M})$ and Σ-formulas $V(x, \bar{z})$, $E^+(x, y, \bar{z})$, $E^-(x, y, \bar{z})$, $P_0^+(x_1, \ldots, x_{m_0}, \bar{z})$, $P_0^-(x_1, \ldots, x_{m_0}, \bar{z})$, $\ldots$, $P_{k-1}^+(x_1, \ldots, x_{m_{k-1}}, \bar{z})$, $P_{k-1}^-(x_1, \ldots, x_{m_{k-1}}, \bar{z})$ such that

1. $(E^+)^{\mathbf{HF}(\mathfrak{M})}[x, y, \bar{p}] \oplus (E^-)^{\mathbf{HF}(\mathfrak{M})}[x, y, \bar{p}] = \left(V^{\mathbf{HF}(\mathfrak{M})}[x, \bar{p}]\right)^2$;

2. for all $i < k$,

$$\left(P_i^+\right)^{\mathbf{HF}(\mathfrak{M})}[\bar{x},\bar{p}] \oplus \left(P_i^-\right)^{\mathbf{HF}(\mathfrak{M})}[\bar{x},\bar{p}] = \left(V^{\mathbf{HF}(\mathfrak{M})}[x,\bar{p}]\right)^{m_i};$$

3. the set $E = (E^+)^{\mathbf{HF}(\mathfrak{M})}[x,y,\bar{p}]$ is a congruence on the structure

$$\mathfrak{B} = \left\langle V^{\mathbf{HF}(\mathfrak{M})}[\bar{x},\bar{p}]; \left(P_0^+\right)^{\mathbf{HF}(\mathfrak{M})}[\bar{x},\bar{p}], \ldots, \left(P_{k-1}^+\right)^{\mathbf{HF}(\mathfrak{M})}[\bar{x},\bar{p}]\right\rangle$$

and $\mathfrak{B}/E = \mathfrak{A}$.

In our study, we shall need to distinguish between the structures of the kind $\mathfrak{B}/E$ (and call them Σ-definable or $\Sigma_{\bar{p}}$-definable) and structures isomorphic to them, which will be called Σ-presentable or $\Sigma_{\bar{p}}$-presentable. If a Σ-definable structure $\mathfrak{A}_0$ is isomorphic to a structure $\mathfrak{A}_1$ then $\mathfrak{A}_0$ is called a Σ-presentation of $\mathfrak{A}_1$.

The formulas V, E^+, E^-, P_i^+, P_i^-, $i < k$ defining a structure will be called its *Σ-definition (or $\Sigma_{\bar{p}}$-definition, respectively).*

We say that a definition of a structure is a *definition with trivial equivalence* if E^+ is the identity relation. In this case we shall identify $V^{\mathfrak{M}}[x,\bar{p}]$ and $V^{\mathfrak{M}}[x,\bar{p}]/_E$.

One can easily prove that in case of $\mathbf{HF}(\mathfrak{M})$ every Σ-formula with parameters $\bar{a} \in \mathbf{HF}(\mathfrak{M})$ is equivalent to some Σ-formula with parameters $\bar{p}$ such that $\bar{p} \subseteq \mathrm{sp}(\bar{a})$, where $\mathrm{sp}(\bar{a})$ is the support of $\bar{a}$, i.e., the set of urelements which take part in the construction of $\bar{a}$.

Recall the definition of algebraically independent tuples. Let $\bar{p} = p_1, \ldots, p_k$ and $\bar{\alpha} = \alpha_1, \ldots, \alpha_s$ be some tuples of reals We say that $\bar{\alpha}$ is algebraically independent over $\bar{p}$, if for any polynomial

$$f(x_1, \ldots, x_s, y_1, \ldots, y_k) \in \mathbb{Q}[x_1, \ldots, x_s, y_1, \ldots, y_k]$$

the condition $f(\bar{\alpha}, \bar{p}) = 0$ implies $\forall \bar{x}\, (f(\bar{x}, \bar{p}) = 0)$.

The following statement, which could be easily proved in many different ways, plays an important role in the study of Σ-definability over $\mathbf{HF}(\mathbb{R})$.

Theorem 1.2 (Algebraic generalization principle, AGP, [7]). Let $\varphi(\bar{x}, \bar{p})$ be an infinite disjunction of first order formulas which is satisfied on a tuple of reals $\bar{\alpha} = \alpha_1, \ldots, \alpha_n \in R^n$ which is algebraically independent over $\bar{p}$. Then this formula is satisfied in some neighborhood of $\bar{\alpha}$. The same holds for Σ-formulas and for usual formulas with parameters $\bar{p}$ and free variables $\bar{x}$.

In particular, this principle could be derived from the quantifier elimination for $\mathbb{R}$ (cf., e.g., [12] or [4, Section 3.3]) and the following more general result, which one could easily prove by induction:

Theorem 1.3. Assume that a structure $\mathfrak{M}$ is o-minimal and the set of its algebraic elements is dense in its order. Suppose that $\bar{p} \in \mathfrak{M}$ and a tuple $\bar{\alpha} = \alpha_0, \dots, \alpha_{k-1}$ has the property that for each $i < k$, α_i is not definable from $\{\alpha_j \mid j \neq i, j < k\} \cup \{\bar{p}\}$ and that $\varphi(\bar{\alpha}, \bar{p})$. Then there exist open intervals $I_1, \dots, I_k$ such that $\bar{\alpha} \in I_1 \times \cdots \times I_k$ and the formula $\varphi(\bar{x}, \bar{p})$ is satisfied by any tuple $\bar{\beta} \in I_1 \times \cdots \times I_k$.

We shall use set-theoretical terms (*s-terms*) that generate all elements of $\mathbf{HF}(\mathbb{R})$ from its urelements. We define the set of all s-terms as the smallest set containing $\varnothing$, all variables, and together with each $t_0, \dots, t_k$ it also contains $\{t_0, \dots, t_k\}$. Fix some Gödel numbering of the set of all s-terms.

Earlier, some results were obtained mostly related to Σ-definable structures with trivial equivalence whose basic set (i.e., $V^{\mathbf{HF}(\mathbb{R})}[x, \bar{p}]$) is a subset of $\mathbb{R}$ (cf. [8, 5, 6, 7]). In some of these papers, these presentations were also called one-dimensional. It will be clear from the results of this paper, that the notion of one-dimensional presentation introduced in this paper below (partial case of Definition 2.2) actually means the same (cf. Corollary 2.6), and there is no need to change the formulations of the earlier results.

The first interesting question in the study of Σ-presentable structures was, whether we can find a Σ-definition of a structure classically isomorphic to $\mathbb{R}$, the ordered field of reals, which would have some extra algorithmic properties in the sense of Σ-definability (for more details and discussion, cf. [7]). The answer appeared to be negative:

Theorem 1.4 ([7]). Let $\bar{p}$ be a tuple of parameters from $\mathbb{R}$ and assume

1. that the sets $\widetilde{R} \subseteq R$ and $<^* \subseteq \widetilde{R} \times \widetilde{R}$ are Σ-definable over $\mathbf{HF}(\mathbb{R})$ with parameters $\bar{p}$,
2. that the operations $+^*, \times^* : \widetilde{R} \to \widetilde{R}$ are Σ-definable operations over $\mathbf{HF}(\mathbb{R})$ with parameters $\bar{p}$, and
3. that γ is an isomorphism from $\langle \widetilde{R}; +^*, \times^*, <^* \rangle$ onto $\langle R; +, \times, < \rangle$ (this isomorphism is unique, if it exists).

Then the mapping γ is Σ-definable over $\mathbf{HF}(\mathbb{R})$ with parameters $\bar{p}$.

Actually it follows that all ordered fields of reals which are Σ-definable over $\mathbf{HF}(\mathbb{R})$ with the basic set within $\mathbb{R}$ are actually the same, they are 'computably' isomorphic. It follows that there is no hope to create something one-dimensional which would be isomorphic to $\mathbb{R}$ in which you could define some functions like exp or logarithm by means of Σ-formulas, without any use of infinite series.

The author has studied Σ-presentations of linear orderings isomorphic to $\langle \mathbb{R}; < \rangle$ [5]. It was proved that there exist 2^ω pairwise non-Σ-isomorphic Σ-definable orders classically isomorphic to $\langle \mathbb{R}; < \rangle$, and even if we restrict ourselves to some tuple of parameters $\bar{p}$ then the class of $\Sigma_{\bar{p}}$-presentations of such orders will have no good parameterizations, i.e., it is in some sense unobservable. It appears that there exist $\Sigma_{\bar{p}}$-presentations of this ordering that have no nontrivial $\Sigma_{\bar{p}}$-embeddings, moreover, even the class of such presentations is infinite up to $\Sigma_{\bar{p}}$-isomorphism and has no good parameterizations [6].

It is also known that the semigroup of all functions from ω to ω has no Σ-presentations over $\mathbf{HF}(\mathbb{R})$ [8] with trivial equivalence, this result is also proved even without extra restriction on the dimension of presentations.

Korovina and the present author studied countable structures Σ-presentable over $\mathbf{HF}(\mathbb{R})$ [9]. It was established that each countable structure has such a presentation with at most one parameter; that every countable structure Σ-presentable over $\mathbf{HF}(\mathbb{R})$ without parameters with at most countable classes of the equivalence E, has a computable copy; and that in general case countable structures Σ-presentable without parameters over $\mathbf{HF}(\mathbb{R})$ could have arbitrary high hyperarithmetical complexity.

2 Dimensions of elements of $\mathbf{HF}(\mathbb{R})$

In the classical computability concept over ω, all c.e. sets of the same cardinality are in some sense isomorphic, i.e., for any c.e. $A, B \subseteq \omega$, $|A| = |B|$ implies that there exists a computable bijection f from A onto B. This statement fails to be true for admissible sets, e.g., if S is a countable structure of empty signature then there is no Σ-definable bijection over $\mathbf{HF}(S)$ from S onto the set of all its ordinals $\omega \subseteq \mathbf{HF}(S)$; it follows easily from the fact that, if a formula $\varphi(x, y, \bar{p})$ defines such an isomorphism, then for any automorphism ϑ of the structure $\mathbf{HF}(S)$ generated by a permutation on S, it must also be true that $\varphi(\vartheta(x), \vartheta(y), \bar{p})$.

That is why the presentability of structures essentially depends on the sets which we would like to have as their universes. Here we start the study of definability over sets of different dimensions.

It is well-known that for each set of reals and for each $\bar{p} \in \mathbb{R}$, all the cardinalities of its maximal algebraically independent subsets over $\bar{p}$ are the same. By this, the following definition is correct:

Definition 2.1. Let $a \in \mathbf{HF}(\mathbb{R})$. We define $\dim_{\bar{p}}(a)$, the dimension of a over $\bar{p}$, as the cardinality of any maximal algebraically independent over $\bar{p}$ subset of $\mathrm{sp}(a)$. We let

$$D_m(\bar{p}) = \{x \in \mathbf{HF}(\mathbb{R}) \mid \dim_{\bar{p}}(x) \leqslant m\}.$$

Definition 2.2. Let $\bar{p} \in \mathbb{R}$ and $m \in \omega$. A $\Sigma_{\bar{p}}$-definable structure with trivial equivalence is called m-dimensional over $\bar{p}$, if its universe is a subset of $D_m(\bar{p})$.

Theorem 2.3. For each m there exists a Σ-formula $\psi_m(x, y)$ such that for any tuple $\bar{p} \in \mathbb{R}$ holds

$$D_m(\bar{p}) = \{a \in \mathbf{HF}(\mathbb{R}) \mid \mathbf{HF}(\mathbb{R}) \models \psi_m(a, \bar{p})\}.$$

Proof. Let $a_0 \ldots a_{k-1} \in \mathbb{R}$. It can be easily seen that $\{a_0 \ldots a_{k-1}\} \in D_m(\bar{p})$ if and only if

> there exist $\bar{t} \in \mathbb{R}^m$ and quantifier-free formulas $\varphi_i(\bar{x}, y, \bar{z})$, $i < k$, such that for all $i < k$, $\exists^{=1} y \varphi_i(\bar{t}, y, \bar{p})$ and $\varphi_i(\bar{t}, a_i, \bar{p})$.

For any quantifier-free formula $\varphi(\bar{t}, y, \bar{p})$, let $\widehat{\varphi}(\bar{t}, \bar{p})$ be a quantifier-free formula equivalent to $\exists^{=1} y \varphi(\bar{t}, y, \bar{p})$. Inasmuch as quantifier elimination for $\mathbb{R}$ is effective, one can assume that $\widehat{\varphi}$ is produced from φ by means of some fixed effective procedure. Thus, an arbitrary $x \in \mathbf{HF}(\mathbb{R})$ belongs to $D_m(\bar{p})$ if and only if there exist $k < \omega$, an s-term $\tau(w_0, \ldots, w_{k-1})$, quantifier-free formulas $\varphi_i(\bar{x}, y, \bar{z})$, $i < k$, and $\bar{t} = t_0, \ldots, t_{m-1} \in \mathbb{R}$ such that x is the only element satisfying the following Σ_p-condition:

$$\exists y_0 \ldots y_{k-1} \left[x = \tau(y_0 \ldots y_{k-1}) \wedge \bigwedge_{i<k} \left(\widehat{\varphi}_i(\bar{t}, \bar{p}) \wedge \varphi_i(\bar{t}, y_i, \bar{p}) \right) \right],$$

which is a Σ-formula with free variable x and parameters $\bar{p}$. Note now that the set of all possible Σ-formulas as above is computably enumerable, so we can organize an effective sequence $\vartheta_j(\bar{u}, x, \bar{z})$, $j < \omega$ of such formulas so that any condition as above is equivalent to an appropriate formula $\vartheta_j(\bar{u}, x, \bar{p})$, $j < \omega$. Thus, we have:

$$x \in D_m(\bar{p}) \Leftrightarrow \exists \bar{t}\, \exists j\, \mathrm{Sat}\big(\vartheta_j(\bar{u}, x, \bar{z}), \bar{t}, x, \bar{p}\big),$$

which yields the required formula. Q.E.D.

Proposition 2.4. The sets $\mathbf{HF}(\mathbb{R}) \setminus D_m(\bar{p})$ and $D_{m+1}(\bar{p}) \setminus D_m(\bar{p})$ are not $\Sigma_{\bar{p}}$-definable, for all $m < \omega$.

Proof. Take an arbitrary $m < \omega$ and assume that $\bar{\alpha} = \alpha_0 \ldots \alpha_m$ is an arbitrary m-tuple independent over $\bar{p}$. If some of the sets $\mathbf{HF}(\mathbb{R}) \setminus D_m(\bar{p})$, $D_{m+1}(\bar{p}) \setminus D_m(\bar{p})$ were Σ_p-definable, by the AGP then all tuples which are close enough to α would also belong to the same set, which is a contradiction. Q.E.D.

Theorem 2.5. Assume that $\bar{p} \in \mathbb{R}$ and $m \in \omega$. Then there exists an injective $\Sigma_{\bar{p}}$-mapping from $D_m(\bar{p})$ into $\mathbb{R}^m \subseteq \mathbf{HF}(\mathbb{R})$.

Proof. Recall that a function f is said to uniformize a relation $Q \subseteq A \times B$, if $f \subseteq Q$ and $\mathrm{dom}(f) = \mathrm{dom}(Q)$. Stukachev has proved that the admissible set $\mathbf{HF}(\mathbb{R})$ satisfies the uniformization principle, i.e., any binary Σ-predicate over $\mathbf{HF}(\mathbb{R})$ can be uniformized by means of a Σ-function (a general statement that implies this fact first appeared in [10] but it was incorrect; it was corrected in [11]; although, one should mention that the reasonings from [10] already prove the uniformization for $\mathbf{HF}(\mathbb{R})$). Examining Stukachev's proof, one can ascertain that one can always find a uniformizing Σ-function f which could be defined with the same parameters as the initial relation Q.

Our proof will use this result by Stukachev. First we define a $\Sigma_{\bar{p}}$-relation $F_m \subseteq D_m(\bar{p}) \times \mathbb{R}^m$ so that for all $x \in D_m(\bar{p})$ there is an $y \in \mathbb{R}^m$ such that $F_m(x, y)$, and if $x_0 \neq x_1$ then $x_0 F_m \cap x_1 F_m = \varnothing$. After this, it remains to uniformize the predicate F_m and the resulting function will be the required mapping.

Below we shall use the following notations: $\ulcorner A \urcorner$ will denote the Gödel number of an object A; for a real r, $\mathrm{int}(r)$ will denote the integer part of $|r|$, $\mathrm{fr}(r) = |r| - \mathrm{fr}(r)$, the fractional part of $|r|$, $\mathrm{sg}(r)$ equals to 1 if $r > 0$, 0 if $r = 0$, and -1 if $r < 0$ i.e., $r = (-1)^{\mathrm{sg}(r)}(\mathrm{int}(r) + \mathrm{fr}(r))$. Define the relation F_m as follows:

> $F_m(x, t)$ means that there are $k < \omega$, an s-term $\tau(w_0, \ldots, w_{k-1})$, quantifier-free formulas $\varphi_i(u_0, \ldots, u_{m-1}, y, \bar{z})$, $i < k$, $w_0, \ldots, w_{k-1} \in \mathbb{R}$, and $\bar{a} = a_0, \ldots, a_{m-1} \in \mathbb{R}$ such that
>
> $$x = \tau(w_0, \ldots, w_{k-1}) \wedge \bigwedge_{i<k} (\widehat{\varphi}_i(\bar{a}, \bar{p}) \wedge \varphi_i(\bar{a}, w_i, \bar{p}))$$
>
> and
>
> $$t = \langle \ulcorner \tau, \varphi_0, \ldots, \varphi_{k-1}, \mathrm{sg}(a_0), \mathrm{int}(a_0) \urcorner + \mathrm{fr}(a_0), a_1, \ldots, a_{k-1} \rangle.$$

One can easily ascertain that F_m is a $\Sigma_{\bar{p}}$-predicate; and given a $t \in \mathbb{R}$ one can restore the x such that $F_m(x, t)$, if such x exists. (We still need the uniformization because given an $x \in D_m(\bar{p})$, one can generally find several t with the property $F_m(x, t)$, due to the fact that terms τ are not uniquely defined from x.) Thus, F_m is the required relation. Q.E.D.

Corollary 2.6. A structure has an m-dimensional $\Sigma_{\bar{p}}$-presentation if and only if it has a $\Sigma_{\bar{p}}$-presentation whose universe is a subset of $\mathbb{R}^m$.

Proof. To find such a presentation, fix a $\Sigma_{\bar{p}}$-injection from $D_m(\bar{p})$ into $\mathbb{R}^m$ and replace all the $\Sigma_{\bar{p}}$-formulas $\varphi(x_0, x_1, \ldots)$ taking part in the definition of an initial structure with the formulas $\varphi(f^{-1}(x_0), f^{-1}(x_1), \ldots)$, which will be also equivalent to $\Sigma_{\bar{p}}$-formulas as well. Q.E.D.

Theorem 2.7. There exists an injective Σ-mapping from the superstructure $\mathbf{HF}(\mathbb{R})$ into $\bigcup_{m<\omega} \mathbb{R}^m$.

Proof. First note that $\mathbf{HF}(\mathbb{R}) = \bigcup_{m<\omega} D_m(\varnothing)$. It follows from the proof of Theorem 2.5 that the Σ-formulas defining the relation $F_m(x, y)$ could be found effectively from m. This means that in this case the predicate $F = \bigcup_{m<\omega} F_m$ is Σ-definable. Then any Σ-uniformization for F will be a Σ-mapping as required. Q.E.D.

Corollary 2.8. Any structure $\Sigma_{\bar{p}}$-definable over $\mathbf{HF}(\mathbb{R})$ with trivial equivalence has a $\Sigma_{\bar{p}}$-presentation with trivial equivalence whose basic set is a subset of $\bigcup_{m<\omega} \mathbb{R}^m$.

3 Presentations of some structures

In what follows, $\exists^{=1} \langle xy \rangle\, \varphi$ will mean that there exists a unique pair $\langle x, y \rangle$ satisfying φ.

Theorem 3.1. Assume that $\mathfrak{M}$ is a structure of finite signature for which there exists an $\exists$-formula $\varphi(x, y, z, \bar{a})$ with parameters $\bar{a}$ and uncountable sets $A, B \subseteq |\mathfrak{M}|$ such that

1. $\mathfrak{M} \models \forall x \in A \forall y \in B \exists^{=1} z \varphi(x, y, z, \bar{a})$;
2. $\mathfrak{M} \models \forall z \big(\exists x \in A \exists y \in B \varphi(x, y, z, \bar{a}) \to \exists^{=1} \langle xy \rangle\, \varphi(x, y, z, \bar{a})\big)$.

Then $\mathfrak{M}$ has no one-dimensional presentations over $\mathbf{HF}(\mathbb{R})$.

Proof. Assume that γ is an isomorphism from $\mathfrak{M}$ onto its one-dimensional presentation over $\mathbf{HF}(\mathbb{R})$ whose basic set is a subset of $\mathbb{R}$.

Without loss of generality, we may assume that atomic subformulas of φ do not contain compositions of operation symbols in their terms. Using Σ-presentations of basic relations and operations of $\mathfrak{M}$ by means of Σ-formulas and restricting x, y, z and the existential quantifiers of the initial formula φ by the set $\gamma(\mathfrak{M})$, which is a Σ-subset of $\mathbf{HF}(\mathbb{R})$, we arrive at a Σ-formula $\varphi'(x, y, z, \bar{a}')$ with parameters $\bar{a}'$ from $\mathbb{R}$ such that for all $x, y, z \in |\mathfrak{M}|$ holds

$$\mathfrak{M} \models \varphi(x, y, z, \bar{a}) \Leftrightarrow \mathbf{HF}(\mathbb{R}) \models \varphi'(\gamma(x), \gamma(y), \gamma(z), \bar{a}').$$

Since the sets A and B are uncountable and γ is a bijection, we can fix some $a_0 \in A$ and $b_0 \in B$ so that the tuple $\langle \alpha_0, \beta_0 \rangle = \langle \gamma(a_0), \gamma(b_0) \rangle \in \mathbb{R}^2$ will be independent over the parameters $\bar{a}'$. Then

$$\exists^{=1} z \varphi'(\alpha_0, \beta_0, z, \bar{a}').$$

Decompose the formula $\varphi'(x, y, z, \bar{a}')$ into an infinite disjunction of quantifier-free formulas and fix a quantifier-free member $\vartheta(x, y, z, \bar{a}')$ in this decomposition that satisfies $\mathbb{R} \models \exists z \vartheta(\alpha_0, \beta_0, z, \bar{a}')$. Of course, it is also true that

$$\mathbb{R} \models \exists^{=1} z \vartheta(\alpha_0, \beta_0, z, \bar{a}'). \tag{1}$$

Since α_0 and β_0 form a tuple independent over the parameters, by the AGP the property (1) is also true near the point $\langle \alpha_0, \beta_0 \rangle$. As it was noted in [7], by the AGP this mapping is continuous near this point.

Further on, note that the unique real $z = z_0$ satisfying $z \in \gamma(\mathfrak{M}) \wedge \vartheta(\alpha_0, \beta_0, z, \bar{a}')$ is independent over the parameters $\bar{a}'$. Otherwise we could express it from $\bar{a}'$ and then use condition 2. of the theorem to express α_0 and β_0 from $\bar{a}'$. Since $\vartheta(\alpha_0, \beta_0, z_0)$, we also have $\exists^{=1} \langle xy \rangle \vartheta(x, y, z_0, \bar{a}')$. By the AGP, $\vartheta(x, y, z, \bar{a}')$ also defines an injection from some neighborhood of z_0 to $\mathbb{R}^2$.

Thus, the mapping $\langle x, y \rangle \mapsto z$, where z is the unique real with the property $\vartheta(x, y, z, \bar{a}')$ is a continuous injection from some product of two nonempty open intervals into $\mathbb{R}$. It is well known that it is impossible, and thus we have a contradiction. Q.E.D.

Theorem 3.2. There is no one-dimensional Σ-presentation for the following structures:

1. any free system of uncountable rank whose signature contains at least one at least binary operation;
2. the direct product of two Boolean algebras of cardinality more than ω;
3. the Boolean algebra $\wp(\omega)$ of all subsets of ω;
4. the symmetric group of a countable set.

Proof. 1. Let X be a free basis of cardinality 2^ω and let $x_0 \in X$. The formula $\varphi(x, y, z, x_0) \leftrightharpoons f(x, y, x_0, \dots, x_0) = z$, where f is an at least binary operation symbol and the sets $A = B = X$ satisfy the conditions of Theorem 3.1, which proves part 1.

2. Let B be a Boolean algebra. By the condition, there exists an $a \in B$ such that the sets $A = \{x \in B \mid x \leqslant a\}$ and $B = \{x \in B \mid x \leqslant \bar{a}\}$ are uncountable. The formula $\varphi(x, y, z, a) \leftrightharpoons x \leqslant a \wedge y \leqslant \bar{a} \wedge z = x \cup y$ together with the above mentioned sets A and B satisfies the conditions of Theorem 3.1, which proves part 2.

3. Follows from part 2.

4. We are going to prove that the presentability of the group $\mathrm{Sym}(\omega)$ will imply the presentability of $\wp(\omega)$.

We shall use standard abbreviation for conjugation $f^g = gfg^{-1}$ and commutators $[f, g] = fgf^{-1}g^{-1}$. If f is a permutation on ω, we define its support as $\mathrm{Supp}(f) = \{x \mid f(x) \neq x\}$. Fix a family $(a_i)_{i<\omega}$ of natural numbers so that $a_i \neq a_j$, for all $i, j < \omega$, $i \neq j$, and the set $\omega \setminus A$ is infinite, where $A = \{a_i \mid i < \omega\}$. Define the permutations

$$\begin{aligned} g_0 &= \prod_{i<\omega}(a_{2i}, a_{2i+1}) \text{ and} \\ g_1 &= \prod_{i<\omega}(a_{2i+1}, a_{2i+2}). \end{aligned}$$

Let w be any permutation on the natural numbers such that $w^2 = 1$ and w maps A onto $\omega \setminus A$ and $\omega \setminus A$ onto A. Let $\widehat{g}_0 = g_0^w$, $\widehat{g}_1 = g_1^w$, and $a = g_0 = \prod_{i<\omega}(a_{2i}, a_{2i+1})$. Note that a is an infinite product of 2-cycles whose support is an infinite and co-infinite subset of ω. In what follows, we are going to identify the sets $X \subseteq \omega$ with permutations $\prod_{i \in X}(a_{2_i}, a_{2i+1})$, which could be thought of as parts of the permutation a.

Lemma 3.1. The following three conditions are equivalent:

1. x is the identity on A;
2. x is the identity on $\mathrm{Supp}(g_0) \cup \mathrm{Supp}(g_1)$;
3. $[x, g_0] = [x, g_1] = 1$.

Proof. We give an informal argument. Note that for any $f, g \in \mathrm{Sym}(\omega)$, $f^g(g(x)) = g(f(x))$, i.e., each time f takes x to y, f^g takes $g(x)$ to $g(y)$ and vice versa. This means that the conjugation of f by means of g acts just as if we replaced in the picture of the action of f each x with $g(x)$. Now we use this observation together with the properties of g_0 and g_1 to prove our statement.

The implications 1⇒2⇒3 are obvious. It remains to prove 3⇒1. Since $g_0^2 = g_1^2 = 1$, the condition $[x, g_0] = 1 \wedge [x, g_1] = 1$ is equivalent to the condition $g_0^x = g_0 \wedge g_1^x = g_1$, i.e., x must preserve pictures of actions of g_0 and g_1. Just by drawing the pictures of actions of g_0 and g_1 on ω, one can easily ascertain that x must be identity on A. Q.E.D.

Note that a similar result is also true for conjugates of g_0 and g_1, i.e., for pairs of the kind g_0^u, g_1^u, $u \in \mathrm{Sym}(\omega)$: any permutation commuting with g_0^u and g_1^u must be the identity on the union of the supports of g_0^u and g_1^u.

Lemma 3.2. The following properties can be expressed in $\mathrm{Sym}(\omega)$ by means of $\exists$-formulas with parameters $a, g_0, g_1, \widehat{g}_0, \widehat{g}_1$:

1. $\mathrm{Apart}(x,y)$: the supports of x and y are disjoint and their complements are infinite.

2. $x \otimes y = z$: supports of x and y are disjoint, both have infinite complements, and $xy = z$.

3. $U(x)$: there exists an $I \subseteq \omega$ such that $x = \prod_{i\in I}(a_{2i}, a_{2i+1})$.

4. $x \sqsubseteq y$: there exist $I \subseteq J \subseteq \omega$ such that $x = \prod_{i\in I}(a_{2i}, a_{2i+1})$ and $y = \prod_{i\in J}(a_{2i}, a_{2i+1})$.

5. $x \sqcap y = z$: there exist $I, J \subseteq \omega$ such that $x = \prod_{i\in I}(a_{2i}, a_{2i+1})$, $y = \prod_{i\in J}(a_{2i}, a_{2i+1})$, and $z = \prod_{i\in I\cap J}(a_{2i}, a_{2i+1})$.

6. $\mathrm{C}(x) = y$: there exist $I, J \subseteq \omega$ such that $I \cup J = \omega$, $I \cap J = \varnothing$, and $x = \prod_{i\in I}(a_{2i}, a_{2i+1})$ and $y = \prod_{i\in J}(a_{2i}, a_{2i+1})$.

7. $x \sqcup y = z$: there exist $I, J \subseteq \omega$ such that $x = \prod_{i\in I}(a_{2i}, a_{2i+1})$, $y = \prod_{i\in J}(a_{2i}, a_{2i+1})$, and $z = \prod_{i\in I\cup J}(a_{2i}, a_{2i+1})$.

Proof. Define

$$\begin{aligned}
\mathrm{Apart}(x,y) &\leftrightharpoons \exists z([y, g_0^z] = [y, g_1^z] = 1 \wedge [x, \widehat{g}_0^z] = [x, \widehat{g}_1^z] = 1);\\
x \otimes y = z &\leftrightharpoons \mathrm{Apart}(x,y) \wedge xy = z\\
U(x) &\leftrightharpoons \exists y(x \otimes y = a)\\
x \sqsubseteq y &\leftrightharpoons U(x) \wedge U(y) \wedge \exists z(x \otimes z = y)\\
x \sqcap y = z &\leftrightharpoons \exists u \exists v(\mathrm{Apart}(u,v) \wedge x = u \otimes z \wedge y = v \otimes z)\\
\mathrm{C}(x) = y &\leftrightharpoons (x \otimes y = a)\\
x \sqcup y = z &\leftrightharpoons z = \mathrm{C}(\mathrm{C}(x) \sqcap \mathrm{C}(y))
\end{aligned}$$

We leave it to the reader to check this. Q.E.D.

One can easily ascertain that the mapping $X \subseteq \omega \mapsto \prod_{i\in X}(a_{2i}, a_{2i+1})$ is an isomorphism from $\langle U^{\mathrm{Sym}(\omega)}; \sqcap, \sqcup, \mathrm{id}_\omega, a\rangle$ onto $\langle \wp(\omega); \cap, \cup, \varnothing, \omega\rangle$.

Assume that there exists a one-dimensional $\Sigma_{\bar{p}}$-presentation of $\mathrm{Sym}(\omega)$, and the reals $r_0, r_1, \widehat{r}_0, \widehat{r}_1, s$ are images of the elements in this presentation of $g_0, g_1, \widehat{g}_0, \widehat{g}_1, a$ respectively. Using the $\exists$-formulas above, one can transform this one-dimensional $\Sigma_{\bar{p}}$-presentation of $\mathrm{Sym}(\omega)$ into a $\Sigma_{\bar{p}, r_0, r_1, \widehat{r}_0, \widehat{r}_1, s}$-presentation of the Boolean algebra $\wp(\omega)$, which contradicts part 3 of our theorem. Q.E.D.

The case of $\mathrm{Sym}(\omega)$ can also be handled with the use of Theorem 3.1, e.g., we can use the formula

$$([x, g_0] = [x, g_1] = 1) \wedge ([y, \widehat{g}_0] = [y, \widehat{g}_1] = 1) \wedge (xy = z),$$

where g_0, g_1, $\widehat{g}_0$, $\widehat{g}_1$ are as in the proof of part 4; this formula expresses the fact that z is the product of permutations x and y whose supports are contained in two fixed infinite disjoint sets. We intentionally give a reduction of the presentability of $\wp(\omega)$ to presentability of $\mathrm{Sym}(\omega)$ hoping that this fact will help in further studies.

Note that for any finite signature, the free algebra of rank 2^ω of this signature has a Σ-presentation with trivial equivalence.

The author believes that it should not be very difficult to investigate Σ-presentability in the general case for structures like the ones considered in this paper and plans to do so in the near future.

References

[1] J. Barwise. *Admissible sets and structures, An approach to definability theory.* Perspectives in Mathematical Logic. Springer-Verlag, 1975.

[2] Y. L. Ershov. *Definability and computability.* Siberian School of Algebra and Logic. Consultants Bureau, 1996.

[3] Y. L. Ershov. Σ-definability of algebraic structures. In Y. L. Ershov, S. S. Goncharov, A. Nerode, J. B. Remmel, and V. W. Marek, editors, *Handbook of recursive mathematics. Vol. 1: Recursive Model Theory,* volume 138 of *Studies in Logic and the Foundations of Mathematics,* pages 235–260. North-Holland, 1998.

[4] D. Marker. *Model theory, An Introduction,* volume 217 of *Graduate Texts in Mathematics.* Springer-Verlag, 2002.

[5] A. S. Morozov. On Σ-presentations of the real order. To appear in *Algebra and Logic.*

[6] A. S. Morozov. Σ-rigid presentations of the real order. To appear in *Siberian Journal of Mathematics.*

[7] A. S. Morozov. On some presentations of the real number field. *Algebra and Logic,* 51(1):66–88, 2012.

[8] A. S. Morozov. Nonpresentability of the semigroup ω^ω over $\mathbf{HF}(\mathbb{R})$. *Siberian Journal of Mathematics,* 55(1):125–131, 2014.

[9] A. S. Morozov and M. V. Korovina. Σ-definability of countable structures over real numbers, complex numbers, and quaternions. *Algebra and Logic,* 47(3):193–209, 2008.

[10] A. I. Stukachev. The uniformization property in hereditary finite superstructures. *Siberian Advances in Mathematics*, 7(1):123–131, 1997.

[11] A. I. Stukachev. Effective model theory via the Σ-definability approach. In N. Greenberg, J. D. Hamkins, D. Hirschfeldt, and R. Miller, editors, *Effective Mathematics of the Uncountable*, volume 41 of *Lecture Notes in Logic*, pages 164–197. Association for Symbolic Logic, 2013.

[12] A. Tarski. *A decision method for elementary algebra and geometry.* University of California Press, 2nd edition, 1951.

Remarkable cardinals

Ralf Schindler*

Institut für Mathematische Logik und Grundlagenforschung, Westfälische Wilhelms-Universität Münster, Münster, Germany

1 Introduction

To a large extent, the scientific work of both Peter Koepke and Philip Welch has always been inspired by Ronald Jensen's fine structure theory and his Covering Lemma for $\mathbf{L}$, the constructible universe. Cf., e.g., their joint papers [5] and [6].

In this paper we aim to play with the theme that large cardinals compatible with "$\mathbf{V}=\mathbf{L}$," specifically: remarkable cardinals, allow us to create situations below $\aleph_2$ which above $\aleph_2$ can only occur if Jensen's Covering fails.

Large cardinals are a central tool in set theory. A cardinal κ is called supercompact iff for every λ there is an elementary embedding $j\colon V \to M$ such that M is transitive, κ is the critical point of j, $j(\kappa) > \lambda$, and ${}^{\lambda}M \subset M$. The following elegant characterization of supercompact cardinals is due to Magidor, cf. [7].

Lemma 1.1. Let κ be a cardinal. Then κ is supercompact iff for every cardinal $\lambda > \kappa$ there is some $X \prec \mathbf{H}_\lambda$ such that $\overline{\overline{X}} < \kappa$, $X \cap \kappa \in \kappa$, and there is some cardinal $\bar{\lambda}$ such that X condenses to $\mathbf{H}_{\bar{\lambda}}$, i.e., $X \cong \mathbf{H}_{\bar{\lambda}}$.

It is well-known that supercompact cardinals cannot exist in $\mathbf{L}$, the smallest inner model of set theory. The situation changes if we don't require X as in Lemma 1.1 to exist in $\mathbf{V}$, but rather in the extension of $\mathbf{V}$ obtained by Lévy collapsing κ to become $\aleph_1$:

Definition 1.2. Let κ be a cardinal. Then κ is called *remarkable* iff in $\mathbf{V}^{\mathrm{Col}(\omega,<\kappa)}$, for every cardinal $\lambda > \kappa$ there is some $X \prec (\mathbf{H}_\lambda)^{\mathbf{V}}$ such that $\overline{\overline{X}} = \aleph_0$, $X \cap \kappa \in \kappa$, and there is some $\mathbf{V}$-cardinal $\bar{\lambda}$ such that X condenses to $(\mathbf{H}_{\bar{\lambda}})^{\mathbf{V}}$, i.e., $X \cong (\mathbf{H}_{\bar{\lambda}})^{\mathbf{V}}$.

Notice that if κ is remarkable, then in $\mathbf{V}^{\mathrm{Col}(\omega,<\kappa)}$, for every cardinal $\lambda > \kappa$ the set

$$S_\lambda = \{X \prec (\mathbf{H}_\lambda)^{\mathbf{V}} \colon \overline{\overline{X}} = \aleph_0 \wedge X \cap \kappa \in \kappa \wedge \exists \bar{\lambda} \in \mathrm{Card}^{\mathbf{V}}\, X \cong (\mathbf{H}_{\bar{\lambda}})^{\mathbf{V}}\} \tag{1}$$

is in fact *stationary* in $[(\mathbf{H}_\lambda)^{\mathbf{V}}]^\omega$. To see this, let us fix a cardinal $\lambda > \kappa$, and let us work in $V[G]$, where G is $\mathrm{Col}(\omega, <\kappa)$-generic over $\mathbf{V}$. Set $\lambda^* = (2^{<\lambda})^+$, and pick $Y \in S_{\lambda^*}$. Notice that $Y[G] \prec (\mathbf{H}_{\lambda^*})^{\mathbf{V}}[G]$; moreover,

*Dedicated to Peter Koepke and Philip Welch on the occasion of their 60th birthdays.

Stefan Geschke, Benedikt Löwe, Philipp Schlicht (*eds.*).
Infinity, computability, and metamathematics: Festschrift celebrating the 60th birthdays of Peter Koepke and Philip Welch. College Publications, London, 2014. Tributes, Volume 23.

$Y[G] \cap (\mathbf{H}_{\lambda^*})^{\mathbf{V}} = Y$, as $\mathrm{Col}(\omega, < \kappa)$ has the κ-c.c. If S_λ is not stationary, then there is some $\mathbf{V}$-cardinal $\lambda' \in Y$ with $2^{<\lambda'} < \lambda^*$ and there is also some club $C \in \wp([(\mathbf{H}_{\lambda'})^{\mathbf{V}}]^\omega) \cap Y[G]$ such that $S_{\lambda'} \cap C = \emptyset$. But then $Y \cap (\mathbf{H}_{\lambda'})^{\mathbf{V}} \in S_{\lambda'} \cap C$. Contradiction!

Remarkable cardinals were introduced by the author in [12], cf. also [11] and [9]. There is a characterization of remarkable cardinals which does not mention forcing and which is in fact taken in [12, Definition 1.1] to be the official definition of remarkability; [12, Lemma 1.6] shows the equivalence of Definition 1.2 and [12, Definition 1.1]. It is shown in [12] that remarkable cardinals relativize down to $\mathbf{L}$, cf. [12, Lemma 1.7]; let us sketch the argument, for the reader's convenience.

Let κ be remarkable, and let $\lambda > \kappa$ be a cardinal. In $\mathbf{V}^{\mathrm{Col}(\omega,<\kappa)}$, there is then some $\mathbf{V}$-cardinal $\bar{\lambda} < \kappa$ and some elementary embedding $\pi\colon (\mathbf{H}_{\bar{\lambda}})^{\mathbf{V}} \to (\mathbf{H}_\lambda)^{\mathbf{V}}$ with critical point $\pi^{-1}(\kappa)$. We have that $\pi \restriction \mathbf{L}_{\bar{\lambda}}\colon \mathbf{L}_{\bar{\lambda}} \to \mathbf{L}_\lambda$ is elementary. But $\{\mathbf{L}_{\bar{\lambda}}, \mathbf{L}_\lambda\} \subset L$, and $\mathbf{L}_{\bar{\lambda}}$ is countable in $\mathbf{L}^{\mathrm{Col}(\omega,<\kappa)}$, so that by absoluteness[1] between $\mathbf{L}^{\mathrm{Col}(\omega,<\kappa)}$ and $\mathbf{V}^{\mathrm{Col}(\omega,<\kappa)}$ there is some elementary embedding $\sigma\colon \mathbf{L}_{\bar{\lambda}} \to \mathbf{L}_\lambda$ with $\sigma \in \mathbf{L}^{\mathrm{Col}(\omega,<\kappa)}$. We have verified that κ is remarkable in $\mathbf{L}$.

It is shown in [12, Lemmas 1.2 and 1.4] that consistency-wise remarkable cardinals lie strictly between ineffable and ω-Erdős cardinals. Gitman and Welch, cf. [3, Theorems 4.8 and 4.11], produce better bounds for their strength by showing that they lie strictly between "1-iterable" and "2-iterable" cardinals; one should think about this result in terms of inner model theory, as follows.

Because remarkable cardinals don't yield $0^\#$ (as they relativize down to $\mathbf{L}$), they cannot be used to prove the existence of an ω_1-iterable premouse. (Cf., e.g., [10, Chapter 10] on the relevant notion of a premouse.) However, they yield the existence of premice which are 1-iterable, i.e., whose ultrapower is well-founded, but they don't yield the existence of a premouse which is 2-iterable, i.e., whose second ultrapower is also well-founded.

Remarkable cardinals were not introduced because the author of [12] was full of mischief. Rather, they appear quite naturally; let us state a theorem.

Theorem 1.3. The following theories are equiconsistent:

(i) ZFC + proper forcing cannot change the theory of $\mathbf{L}(\mathbb{R})$.
(ii) ZFC + semi-proper forcing cannot change the theory of $\mathbf{L}(\mathbb{R})$.
(iii) Third order number theory + Harrington's principle ("there is a real x such that every x-admissible ordinal is an $\mathbf{L}$-cardinal").
(iv) ZFC + there is a remarkable cardinal.

[1] There is a tree T in $\mathbf{L}^{\mathrm{Col}(\omega,<\kappa)}$ of height ω searching for some such σ; T is ill-founded in $\mathbf{V}^{\mathrm{Col}(\omega,<\kappa)}$, hence in $\mathbf{L}^{\mathrm{Col}(\omega,<\kappa)}$. Cf., e.g., [12, Lemma 0.1].

We refer the reader to [12, Theorem 2.4 and Lemma 3.5], [9, Theorem 1.1], and [2, Theorem 3.2] for a proof of Theorem 1.3. One may also formulate an anti-coding theorem, cf. [12, Definition 2.5 and Corollary 3.6], which may be shown to be equi-consistent with a remarkable cardinal.

In the next section of this paper, we shall consider the combinatorial heart of the situation represented by (i),(ii), and (iii) of Theorem 1.3; $(\aleph_1)^{\mathbf{V}}$ will have to be a remarkable cardinal in $\mathbf{L}$.

Another use of remarkable cardinals is:

Theorem 1.4. The following theories are equiconsistent:

(i) ZFC + for every cardinal $\lambda > \aleph_2$ the set

$$\{X \prec \mathbf{H}_\lambda \colon \overline{\overline{X}} = \aleph_1 \wedge X \cap \omega_2 \in \omega_2 \wedge \operatorname{otp}(X \cap \lambda) \in \mathrm{Card}^{\mathbf{L}}\}$$

is stationary.

(ii) ZFC + there is a remarkable cardinal.

This theorem is produced as [8, Theorem 4]. We shall discuss the situation of this theorem in the second next section; (i) of Theorem 1.4 is realized by having $(\aleph_2)^{\mathbf{V}}$ be a remarkable cardinal in $\mathbf{L}$.

In the last section of this paper we shall produce a new result, Theorem 4.1, making use of *two* remarkable cardinals. This will involve taking another look at the notion of *subcomplete forcing*, cf. [4].

2 Why $\aleph_2$ is a threshold—turning a remarkable cardinal into $\aleph_1$

Let us write for an $\mathbf{L}$-cardinal $\lambda > (\aleph_1)^{\mathbf{V}}$,[2]

$$S_\lambda(\mathbf{L}) = \{X \prec \mathbf{L}_\lambda \colon \overline{\overline{X}} = \aleph_0 \wedge X \cap \omega_1 \in \omega_1 \wedge \exists \bar{\lambda} \in \mathrm{Card}^{\mathbf{L}}\, X \cong \mathbf{L}_{\bar{\lambda}}\}. \quad (2)$$

Hence $(\aleph_1)^{\mathbf{V}}$ is remarkable in $\mathbf{L}$ iff $S_\lambda(\mathbf{L})$ is stationary in $\mathbf{L}^{\mathrm{Col}(\omega, <(\aleph_1)^{\mathbf{V}})}$ for every $\mathbf{L}$-cardinal $\lambda > (\aleph_1)^{\mathbf{V}}$. An absoluteness argument as above immediately gives the following sufficient criterion.

Lemma 2.1. Suppose that in $\mathbf{V}$, $S_\lambda(\mathbf{L})$ is stationary for every $\mathbf{L}$-cardinal $\lambda > (\aleph_1)^{\mathbf{V}}$. Then $(\aleph_1)^{\mathbf{V}}$ is remarkable in $\mathbf{L}$.

Let us address the question if $S_\lambda(\mathbf{L})$ can contain a club. The situation for $\lambda \leq (\aleph_2)^{\mathbf{V}}$ differs significantly from the situation for $\lambda > (\aleph_2)^{\mathbf{V}}$.

[2] We here understand that in a model of set theory, $S_\lambda(\mathbf{L})$ is to denote the set given by the right hand side of (2) as computed in that model. The same remark applies to the principles $S^\kappa_\lambda(\mathbf{L})$ and $\tilde{S}_\lambda(\mathbf{L})$ which will be defined later.

Lemma 2.2. Let $\lambda > (\aleph_2)^{\mathbf{V}}$ be an $\mathbf{L}$-cardinal. Then $S_\lambda(\mathbf{L})$ contains a club iff $0^\#$ exists.

Proof. If $0^\#$ exists, we get that if $\pi\colon \mathbf{L}_{\bar\lambda}[0^\#] \to \mathbf{L}_\lambda[0^\#]$ is elementary with $\lambda > \aleph_0$ being an $\mathbf{L}$-cardinal, then $\bar\lambda$ must be an $\mathbf{L}$-cardinal also.

Let us now assume that $\lambda > (\aleph_2)^{\mathbf{V}}$ is an $\mathbf{L}$-cardinal and $0^\#$ does not exist. We want to show that $S_\lambda(\mathbf{L})$ does not contain a club. Let us write $\tau = ((\aleph_2)^{\mathbf{V}})^{+\mathbf{L}} \leq \lambda$. We in fact show that $S_\tau(\mathbf{L})$ does not contain a club.[3] We are going to use Kueker's lemma.

Let $(\mathbf{L}_\tau; \in, \ldots) \in V$ be any model in a countable language with universe $\mathbf{L}_\tau$. By the proof of the Jensen Covering Lemma, cf., e.g., [10, Section 11.2], we may pick some elementary embedding

$$\pi\colon \bar{H} \to (\mathbf{H}_{\omega_4})^{\mathbf{V}},$$

where $\bar{H}$ is transitive and of size $\aleph_1$ with $(\mathbf{L}_\tau; \in, \ldots) \in \operatorname{ran}(\pi)$, such that if $\delta = \pi^{-1}((\aleph_2)^{\mathbf{V}})$ is the critical point of π, $\eta = \bar{H} \cap \mathrm{OR}$, and $\bar\tau = \pi^{-1}(\tau)$, then for every $\mu > \eta$ with $\wp(\delta) \cap \mathbf{L}_\mu \subset \mathbf{L}_\eta$,

$$\operatorname{ult}(\mathbf{L}_\mu; E_{\pi \restriction \mathbf{L}_{\bar\tau}}) \text{ is well–founded.} \tag{3}$$

(Cf. [10, Claim 11.58].) If $\bar\tau$ is an $\mathbf{L}$-cardinal, then $\wp(\delta) \cap L \subset \mathbf{L}_\eta$, so that (3), applied with $\mu = \infty$, yields a non–trivial elementary embedding from $\mathbf{L}$ to $\mathbf{L}$, i.e., the existence of $0^\#$.

Therefore, $\bar\tau$ is not an $\mathbf{L}$-cardinal. Let

$$\sigma\colon H \to (\mathbf{H}_{\omega_2})^{\mathbf{V}}$$

be an elementary embedding such that H is countable and transitive and $\{\bar\tau, \pi^{-1}((\mathbf{L}_\tau; \in, \ldots))\} \subset \operatorname{ran}(\sigma)$. Then $\sigma^{-1}(\bar\tau) = (\pi \circ \sigma)^{-1}(\tau) = \operatorname{otp}(\operatorname{ran}(\pi \circ \sigma) \cap \tau)$ is not an $\mathbf{L}$-cardinal, but[4]

$$\operatorname{ran}(\pi \circ \sigma) \cap \mathbf{L}_\tau \prec (\mathbf{L}_\tau; \in, \ldots).$$

As $(\mathbf{L}_\tau; \in, \ldots)$ was arbitrary, $S_\tau(\mathbf{L})$ does not contain a club. Q.E.D.

If $S_{(\aleph_2)^{\mathbf{V}}}(\mathbf{L})$ contains a club, then $(\aleph_2)^{\mathbf{V}}$ must be inaccessible in $\mathbf{L}$. This is because if $(\mathbf{L}_{(\aleph_2)^{\mathbf{V}}}; \in, \ldots) \in V$ is a model in a countable language with universe $\mathbf{L}_{(\aleph_2)^{\mathbf{V}}}$ and if $(\aleph_2)^{\mathbf{V}} = \varrho^{+\mathbf{L}}$, then there is some η, $\varrho < \eta < (\aleph_2)^{\mathbf{V}}$, such that $\mathbf{L}_\eta \prec (\mathbf{L}_{(\aleph_2)^{\mathbf{V}}}; \in, \ldots)$. But then if $X \prec (\mathbf{H}_{\omega_3})^{\mathbf{V}}$ is such that $\overline{\overline{X}} = \aleph_0$ and $\{\eta, (\mathbf{L}_{(\aleph_2)^{\mathbf{V}}}; \in, \ldots)\} \subset X$, then $X \cap \mathbf{L}_\eta \prec (\mathbf{L}_{(\aleph_2)^{\mathbf{V}}}; \in, \ldots)$, but $\operatorname{otp}(X \cap \eta)$ is not an $\mathbf{L}$-cardinal. The following is now easy to verify.

[3] If $\bar\lambda \leq \lambda$ and $S_{\bar\lambda}(\mathbf{L})$ does not contain a club, then $S_\lambda(\mathbf{L})$ does not contain a club either.

[4] Notice that $(\operatorname{ran}(\sigma) \cap \mathbf{L}_{\bar\tau}) \cup \{\mathbf{L}_{\bar\tau}\} \subset \operatorname{dom}(\pi)$, so that $\pi \circ \sigma \restriction (\mathbf{L}_{\sigma^{-1}(\bar\tau)} \cup \{\mathbf{L}_{\sigma^{-1}(\bar\tau)}\})$ makes sense.

Lemma 2.3. The set $S_{(\aleph_2)^{\mathbf{V}}}(\mathbf{L})$ contains a club iff [$(\aleph_2)^{\mathbf{V}}$ is inaccessible in $\mathbf{L}$ and $S_\lambda(\mathbf{L})$ contains a club for every $\mathbf{L}$-cardinal λ such that $(\aleph_1)^{\mathbf{V}} < \lambda < (\aleph_2)^{\mathbf{V}}$].

In the light of Lemmas 2.1 and 2.3, the large cardinal hypothesis of the next lemma is optimal.

Lemma 2.4. It is consistent, relative to the existence of an inaccessible cardinal κ such that $\mathbf{V}_\kappa \models$ "there is a remarkable cardinal," that $S_{(\aleph_2)^{\mathbf{V}}}(\mathbf{L})$ contains a club. In particular, the fact that $S_{(\aleph_2)^{\mathbf{V}}}(\mathbf{L})$ contains a club does not yield the existence of $0^{\#}$.

Proof. Let κ be inaccessible in $\mathbf{L}$, and let $\mathbf{L}_\kappa \models$ "μ is remarkable." We aim to produce a generic extension of $\mathbf{L}$ in which $\kappa = \aleph_2$ and $S_\kappa(\mathbf{L})$ contains a club.

Let G be $\mathrm{Col}(\omega, <\mu)$-generic over $\mathbf{L}$, and let H be $\mathrm{Col}(\mu, <\kappa)$-generic over $\mathbf{L}[G]$. In $\mathbf{L}[G]$, $S_\lambda(\mathbf{L})$ is stationary for every cardinal $\lambda > \mu$, $\lambda < \kappa$. As $\mathrm{Col}(\mu, <\kappa)$ is ω-closed, in $V[G, H]$, $S_\lambda(\mathbf{L})$ is still stationary for every $\mathbf{L}$-cardinal $\lambda > \mu$, $\lambda < \kappa$. For the record, $\aleph_1 = \mu$ and $\aleph_2 = \kappa$ in $\mathbf{L}[G, H]$.

Now let, for an $\mathbf{L}$-cardinal λ with $\mu < \lambda < \kappa$, $\mathbb{Q}_\lambda$ be defined in $\mathbf{L}[G, H]$ to be the set of all strictly increasing and continuous sequences $(X_i : i \leq \alpha)$ of length some $\alpha < \mu$ consisting of elements of $S_\lambda(\mathbf{L})$. We order $\mathbb{Q}_\lambda$ by end-extension. Hence $\mathbb{Q}_\lambda$ shoots a club through $S_\lambda(\mathbf{L})$.

Let us work in $\mathbf{L}[G, H]$ until further notice. We let $\mathbb{Q}$ be the countable support product of all $\mathbb{Q}_\lambda$ for $\mathbf{L}$-cardinals λ strictly between μ and κ. If $p \in \mathbb{Q}$, then we may write

$$p = \{(X_i^\lambda(p) : i \leq \alpha_\lambda(p)) : \lambda \in \mathrm{supp}(p)\}.$$

We claim that $\mathbb{Q}$ is ω-distributive. To this end, let $\vec{D} = (D_n : n < \omega)$ be a sequence of open dense sets, and let $p \in \mathbb{Q}$. Let us pick some $Y \prec \mathbf{H}_{\omega_3}$ such that $\mu \cup \{p, \vec{D}, \mathbb{Q}\} \subset Y$, $Y \cap \kappa \in \kappa$, and Y has size $\aleph_1$. Writing $\gamma = Y \cap \kappa$, γ must then be an $\mathbf{L}$-cardinal. Exploiting the fact that $S_\gamma(\mathbf{L})$ is stationary, we may pick some countable $X \prec \mathbf{H}_{\omega_3}$ such that $\{p, \vec{D}, \mathbb{Q}, Y, \gamma\} \subset X$ and $X \cap \mathbf{L}_\gamma \in S_\gamma(\mathbf{L})$. We have that

$$\{p, \vec{D}, \mathbb{Q}\} \subset X \cap Y \prec Y \prec \mathbf{H}_{\omega_3}.$$

We may therefore build a descending sequence $(p_n : n < \omega)$ of conditions $p_n \in \mathbb{Q}$ such that $p_0 = p$, $\{p_n : n < \omega\} \subset X \cap Y$, $p_{n+1} \in D_n$, and for every $\mathbf{L}$-cardinal $\lambda \in X \cap \gamma$ and every $x \in X \cap \mathbf{L}_\gamma$ there is some $n < \omega$ such that $\lambda \in \mathrm{supp}(p_n)$ and $x \in X_i^\lambda(p_n)$ for some $i \leq \alpha_\lambda(p_n)$. Let us write $\alpha = X \cap \omega_1$ and

$$q = \{(X_i^\lambda : i \leq \alpha) : \lambda \in X \cap \gamma \text{ is an } \mathbf{L}\text{-cardinal}\},$$

where for every **L**-cardinal $\lambda \in X \cap \gamma$, if $i < \alpha$, then $X_i^\lambda = X_i^\lambda(p_n)$ for some (all) sufficiently large $n < \omega$, and $X_\alpha^\lambda = X \cap \mathbf{L}_\lambda$. It is then straightforward to verify that $q \in \mathbb{Q}$, $q \leq_{\mathbb{Q}} p$, and $q \in D_n$ for all $n < \omega$. We have seen that $\mathbb{Q}$ is ω-distributive.

Using CH, $\overline{\overline{\mathbb{Q}_\lambda}} = \aleph_1$ for every $\lambda < \kappa$. An easy application of the Δ-system Lemma then gives that $\mathbb{Q}$ has the $\aleph_2$-chain condition.

Let us now step outside of $\mathbf{L}[G, H]$, and let K be $\mathbb{Q}$-generic over $\mathbf{L}[G, H]$. In $\mathbf{L}[G, H, K]$, $\mu = \aleph_1$, $\kappa = \aleph_2$, and if λ is an **L**-cardinal strictly between $\aleph_1$ and $\aleph_2$, then $S_\lambda(\mathbf{L})$ contains a club. Q.E.D.

A weaker version of Lemma 2.4 is presented in [1], and the full version of Lemma 2.4 is implicit in [2]. Theorem 1.4 is shown in [11, 9, 2] by further exploiting the arguments of this section.

3 Uncountable substructures—turning a remarkable cardinal into $\aleph_2$

The set $S_\lambda(\mathbf{L})$, as defined in (2), consists of countable substructures of $\mathbf{L}_\lambda$. As in [8], we now aim to talk about uncountable such substructures.

Let us write, for an uncountable regular cardinal κ and an **L**-cardinal $\lambda > \kappa$,

$$S_\lambda^\kappa(\mathbf{L}) = \{X \prec \mathbf{L}_\lambda : \overline{\overline{X}} < \kappa \wedge X \cap \kappa \in \kappa \wedge \exists \bar{\lambda} \in \mathrm{Card}^{\mathbf{L}}\, X \cong \mathbf{L}_{\bar{\lambda}}\}. \tag{4}$$

Obviously, $S_\lambda^{(\aleph_1)^{\mathbf{V}}}(\mathbf{L}) = S_\lambda(\mathbf{L})$. The following is easy to verify, cf. [8] and also the proof of Lemma 3.3 below.

Lemma 3.1. Let κ be regular, $\kappa \geq (\aleph_3)^{\mathbf{V}}$, and let $\lambda > \kappa$ be an **L**-cardinal. The following are equivalent:

1. $S_\lambda^\kappa(\mathbf{L})$ is stationary.
2. $S_\lambda^\kappa(\mathbf{L})$ contains a club.
3. $0^\#$ exists.

We shall thus now focus on $S_\lambda^{(\aleph_2)^{\mathbf{V}}}(\mathbf{L})$, which we shall now denote by $\tilde{S}_\lambda(\mathbf{L})$. Compare this with the set from (i) of Theorem 1.4.

Lemma 3.2. Suppose that for every **L**-cardinal $\lambda > (\aleph_2)^{\mathbf{V}}$, $\tilde{S}_\lambda(\mathbf{L}) \neq \varnothing$. Then $(\aleph_2)^{\mathbf{V}}$ is remarkable in **L**.

Proof. Write $\kappa = (\aleph_2)^{\mathbf{V}}$. Let $\lambda > \kappa$ be an **L**-cardinal, and let $X \in \tilde{S}_\lambda(\mathbf{L})$. In particular, $X \cong \mathbf{L}_{\bar{\lambda}}$ for some **L**-cardinal $\bar{\lambda} < \kappa$. We have that $\mathbf{L}_{\bar{\lambda}}$ is countable in $\mathbf{L}^{\mathrm{Col}(\omega,<\kappa)}$, so that by absoluteness between $\mathbf{L}^{\mathrm{Col}(\omega,<\kappa)}$ and $\mathbf{V}^{\mathrm{Col}(\omega,<\kappa)}$ there is some $\sigma : \mathbf{L}_{\bar{\lambda}} \to \mathbf{L}_\lambda$ in $\mathbf{L}^{\mathrm{Col}(\omega,<\kappa)}$. The set $\mathrm{ran}(\sigma)$ is then in S_λ as defined in $\mathbf{L}^{\mathrm{Col}(\omega,<\kappa)}$, cf. (1) on p. 299 and Definition 1.2. We have verified that κ is remarkable in **L**. Q.E.D.

Lemma 3.3. Set $\lambda = ((\aleph_2)^{\mathbf{V}})^{+\mathbf{L}}$. If $\tilde{S}_\lambda(\mathbf{L})$ contains a club, then $0^\#$ exists.

Proof. This immediately follows from the first part of the proof of "$\Longrightarrow$" of Lemma 2.2. Q.E.D.

Lemma 3.4. It is consistent, relative to the existence of a remarkable cardinal, that for every $\mathbf{L}$-cardinal $\lambda > (\aleph_2)^{\mathbf{V}}$, $\tilde{S}_\lambda(\mathbf{L})$ be stationary. In particular, the fact that for every $\mathbf{L}$-cardinal $\lambda > (\aleph_2)^{\mathbf{V}}$ the set $\tilde{S}_\lambda(\mathbf{L})$ is stationary does not yield the existence of $0^\#$.

Proof. We give a streamlined version of the proof of [8, Theorem 4]. We shall make use of Jensen's theory of subcomplete forcings, cf. [4] where also an iteration theorem is shown for RCS (revised countable support) iterations of subcomplete forcings.

Let us assume that $\mathbf{L} \models$ "κ is a remarkable cardinal," but there is no $\mathbf{L}$-inaccessible cardinal μ such that $\mathbf{L}_\mu \models$ "there is a remarkable cardinal." Let us perform an RCS iteration of legth κ over $\mathbf{L}$, as follows.

Let $\mathbb{N}$ denote Namba forcing. By [4], $\mathbb{N}$ is subcomplete. At stage $i < \kappa$ of the iteration, we let $\mu(i)$ denote the least $\mathbf{L}$-inaccessible above the current $\aleph_2$, and we force with

$$\mathrm{Col}(\omega_2, \mu(i)) * \mathbb{N} * \mathrm{Col}(\omega_1, (\mu(i))^{+\mathbf{L}}),$$

as defined in the current model. At limit stages $i \leq \kappa$, we take the revised limit. Let us denote by $\mathbb{P}$ the resulting RCS iteration of length κ.

By [4], $\mathbb{P}$ is subcomplete, so that in particular forcing with $\mathbb{P}$ preserves ω_1. Also, $\mathbb{P}$ has the κ-c.c., so that κ will be turned into $\aleph_2$.

Let G be $\mathbb{P}$-generic over $\mathbf{L}$. We claim that in $\mathbf{L}[G]$, $\tilde{S}_\lambda(\mathbf{L})$ is stationary for every ($\mathbf{L}$-)cardinal $\lambda > \kappa$.

Let us fix a regular cardinal $\lambda > \kappa$. As κ is remarkable in $\mathbf{L}$, inside $\mathbf{L}^{\mathrm{Col}(\omega,<\kappa)}$ we may pick some

$$\pi \colon \mathbf{L}_{\bar{\lambda}} \to \mathbf{L}_\lambda \tag{5}$$

with critical point $\pi^{-1}(\kappa)$ such that $\bar{\lambda} < \kappa$ is a cardinal in $\mathbf{L}$; as λ was chosen to be regular, $\bar{\lambda}$ will also be regular in $\mathbf{L}$. Let us write $\bar{\kappa} = \pi^{-1}(\kappa)$. We must have that $\bar{\kappa}$ is inaccessible in $\mathbf{L}$, so that the initial segment of $\mathbb{P}$ given by the first $\bar{\kappa}$ steps of the iteration has the $\bar{\kappa}$-c.c. Hence $\bar{\kappa} = (\aleph_2)^{V[G\restriction\bar{\kappa}]}$. Moreover, as $\mathbb{P}$ has the κ-c.c., $\mathrm{ran}(\pi)[G] \cap \mathbf{L}_\lambda = \mathbf{L}_\lambda$, so that inside $\mathbf{L}[G]^{\mathrm{Col}(\omega,<\kappa)}$, we may actually extend π to an embedding

$$\tilde{\pi} \colon \mathbf{L}_{\bar{\lambda}}[G\restriction\bar{\kappa}] \to \mathbf{L}_\lambda[G].$$

By our smallness hypothesis on $\mathbf{L}$, $\mu(\bar{\kappa}) > \bar{\lambda}$. The first two components of the $\bar{\kappa}$th stage of the iteration $\mathbb{P}$ will therefore add a surjection $f \colon \bar{\kappa} \to \mathbf{L}_{\bar{\lambda}}$

such that $f\restriction\xi \in \mathbf{L}[G\restriction\bar\kappa]$ for all $\xi < \bar\kappa$ and some (Namba) sequence $(\kappa_n : n < \omega)$ which is cofinal in $\bar\kappa$. Writing $X_n = f''\kappa_n$, we therefore have $X_n \in \mathbf{L}[G\restriction\bar\kappa]$ for every $n < \omega$ and $\mathbf{L}_{\bar\lambda} = \bigcup_{n<\omega} X_n$. By replacing X_n with the Skolem hull of X_n inside $\mathbf{L}_{\bar\lambda}$, we may in addition assume that $X_n \prec \mathbf{L}_{\bar\lambda}$ for every $n < \omega$. Of course $(X_n : n < \omega) \in \mathbf{L}[G]$.

Let $n < \omega$. As $\bar\lambda$ is regular in $\mathbf{L}[G\restriction\bar\kappa]$, $f\restriction\kappa_n \in \mathbf{L}_{\bar\lambda}[G\restriction\bar\kappa]$. Then $\pi\restriction X_n \in \mathbf{L}[G]$, as it may be computed in $\mathbf{L}[G]$ via $(\pi\restriction X_n)((f\restriction\kappa_n)(\xi)) = \tilde\pi(f\restriction\kappa_n)(\xi)$ for every $\xi < \kappa_n$. Also, $(\pi\restriction n : n < \omega) \in \mathbf{L}[G]^{\mathrm{Col}(\omega,<\kappa)}$.

Now let $T \in \mathbf{L}[G]$ be the tree of attempts to find a map as in (5), more precisely, let T be the set of all $\sigma \in \mathbf{L}[G]$ for which there is some $n < \omega$ such that $\sigma : X_n \to \mathbf{L}_\lambda$ is elementary, ordered by end-extension. As $(\pi\restriction n : n < \omega) \in \mathbf{L}[G]^{\mathrm{Col}(\omega,<\kappa)}$ and $\pi\restriction X_n \in \mathbf{L}[G]$ for every $n < \omega$, T is ill-founded in $\mathbf{L}[G]^{\mathrm{Col}(\omega,<\kappa)}$ and hence in $\mathbf{L}[G]$. There is therefore in $\mathbf{L}[G]$ a system $(\sigma_n : n < \omega)$ such that for $n < \omega$, $\sigma_n : X_n \to \mathbf{L}_\lambda$ is elementary, and if $n \leq m$, then $\sigma_m \supset \sigma_n$. Therefore, in $\mathbf{L}[G]$ we get an elementary embedding

$$\sigma : \mathbf{L}_{\bar\lambda} \to \mathbf{L}_\lambda,$$

where for each $x \in \mathbf{L}_{\bar\lambda}$,

$$\sigma(x) = \sigma_n(x)$$

for some (all) sufficently large $n < \omega$. But then $\mathrm{ran}(\sigma) \in \tilde S_\lambda(\mathbf{L}) \cap \mathbf{L}[G]$.

We have verified that $\tilde S_\lambda(\mathbf{L})$ is stationary in $\mathbf{L}[G]$. Q.E.D.

4 Two remarkable cardinals

We now aim to explore the argument for Lemma 3.4 further by trying to arrange that for each $\mathbf{L}$-cardinal $\lambda > (\aleph_2)^{\mathbf{V}}$, $S_\lambda(\mathbf{L})$ and $\tilde S_\lambda(\mathbf{L})$ are simultaneouly stationary, without $0^\#$. It is worth mentioning that if $\tilde S_\lambda(\mathbf{L})$ is true for every $\mathbf{L}$-cardinal $\lambda > \aleph_2$ and $S_{\bar\lambda}(\mathbf{L})$ is true for every $\mathbf{L}$-cardinal $\bar\lambda$ between $\aleph_1$ and $\aleph_2$, then in fact $S_\lambda(\mathbf{L})$ is true for every $\mathbf{L}$-cardinal λ. This corresponds to the fact that if $\mathbf{V}_\kappa \models$ "μ is remarkable," and κ is remarkable, then μ is remarkable.

By Lemmas 2.1 and 3.2, if both $S_\lambda(\mathbf{L})$ and $\tilde S_\lambda(\mathbf{L})$ are stationary for every $\mathbf{L}$-cardinal λ, then $(\omega_1)^{\mathbf{V}}$ and $(\omega_2)^{\mathbf{V}}$ are both remarkable in $\mathbf{L}$. We shall therefore now work with *two* remarkable cardinals.

Theorem 4.1. It is consistent, relative to the existence of two remarkable cardinals, that for every $\mathbf{L}$-cardinal $\lambda > (\aleph_2)^{\mathbf{V}}$, both $S_\lambda(\mathbf{L})$ and $\tilde S_\lambda(\mathbf{L})$ are stationary, and that $S_{(\aleph_2)^{\mathbf{V}}}(\mathbf{L})$ contains a club.

Proof. Let us assume that in $\mathbf{L}$ there are two remarkable cardinals, $\mu < \kappa$. Let G be $\mathrm{Col}(\omega, <\mu)$-generic over $\mathbf{L}$, so that in $\mathbf{L}[G]$, $S_\lambda(\mathbf{L})$ is stationary for every $\mathbf{L}$-cardinal $\lambda > \mu$. As the remarkability of κ is indestructible by small forcing, κ is still remarkable in $\mathbf{L}[G]$.

Now let $\mathbb{P}$ be the forcing defined in the proof of Lemma 3.4, but defined over $\mathbf{L}[G]$ instead of over $\mathbf{L}$. We may and shall assume that there is no $\mathbf{L}$-inaccessible cardinal $\varrho > \kappa$. Let H be $\mathbb{P}$-generic over $\mathbf{L}[G]$. We shall then have that in $\mathbf{L}[G, H]$, $\aleph_1 = \mu$, $\aleph_2 = \kappa$, and $\tilde{S}_\lambda(\mathbf{L})$ is stationary for every $\mathbf{L}$-cardinal $\lambda > \aleph_2$.

We claim that in $\mathbf{L}[G, H]$, $S_\lambda(\mathbf{L})$ is stationary for every $\mathbf{L}$-cardinal $\lambda > \aleph_1$. We shall make use of the fact that $\mathbb{P}$ is subcomplete. Suppose that there is $p \in \mathbb{P}$ and $\tau \in \mathbf{L}[G]^{\mathbb{P}}$ such that

$$p \Vdash \tau \text{ is a model with universe } \mathbf{L}_{\check{\lambda}},$$

and also

$$p \Vdash \text{ if } X \prec \tau, \overline{\overline{X}} = \aleph_0, \text{ and } X \cap \omega_1 \in \omega_1, \text{ then } \operatorname{otp}(X \cap \check{\lambda}) \notin \mathrm{Card}^{\mathbf{L}}. \quad (6)$$

Let $\vartheta \gg \max(\lambda, \kappa)$ be a cardinal. Working in $\mathbf{L}[G]$, where $S_{(2^{<\vartheta})^+}(\mathbf{L})$ is stationary, we may pick some

$$Y \prec (\mathbf{H}_{(2^{<\vartheta})^+})^{\mathbf{L}[G]}$$

such that Y is countable, $Y \cap \mu \in \mu$, $\{\mathbb{P}, p, \tau, \lambda\} \subset Y$, and $\operatorname{otp}(Y \cap (2^{<\vartheta})^+) \in \mathrm{Card}^{\mathbf{L}}$. Let

$$\sigma \colon H' \cong Y,$$

where H' is transitive, and write $H = \sigma^{-1}(\mathbf{H}_\vartheta)$, $\bar{\mathbb{P}}, \bar{p}, \bar{\tau}, \bar{\lambda} = \sigma^{-1}(\mathbb{P}, p, \tau, \lambda)$. Because $\mathbb{P}$ is subcomplete (we in fact only need that $\mathbb{P}$ is subproper, cf. [4]), there are k, K, and σ' such that $\bar{p} \in k$, k is $\bar{\mathbb{P}}$-generic over H, K is $\mathbb{P}$-generic over $\mathbf{V}$, $\sigma' \in \mathbf{L}[G, K]$,

$$\sigma' \colon H[k] \to \mathbf{H}_\vartheta[K]$$

is an elementary embedding, and $\sigma'(\bar{\mathbb{P}}, \bar{p}, \bar{\tau}, \bar{\lambda}) = \mathbb{P}, p, \tau, \lambda$. Setting $X = \mathbf{L}_\lambda \cap \operatorname{ran}(\sigma')$, $X \prec \tau^H$ and $\bar{\lambda} = \operatorname{otp}(X \cap \lambda)$ is an $\mathbf{L}$-cardinal. As $p \in H$, this contradicts (6).

Now let $\mathbb{Q}$ be the forcing from the proof of Lemma 2.4 for simultaneously shooting clubs through all $S_\lambda(\mathbf{L})$, where λ is an $\mathbf{L}$-cardinal between μ and κ, and let I be $\mathbb{Q}$-generic over $\mathbf{L}[G, H]$. In $\mathbf{L}[G, H, I]$, $S_\kappa(\mathbf{L})$ will contain a club. It is easy to verify that the proof that $\mathbb{Q}$ is ω-distributive also gives that $\mathbb{Q}$ preserves the stationarity of every $S_\vartheta(\mathbf{L})$, ϑ any $\mathbf{L}$-cardinal above κ. (In the proof of Lemma 2.4, pick $X \subset \mathbf{H}_\vartheta$ with $X \cap \mathbf{L}_\vartheta \in S_\vartheta(\mathbf{L})$, rather than $X \prec \mathbf{H}_{\omega_3}$ and $X \cap \mathbf{L}_\gamma \in S_\gamma(\mathbf{L})$.) Moreover, $\mathbb{Q}$ has the κ-c.c., which implies that every $\tilde{S}_\lambda$, λ any $\mathbf{L}$-cardinal above κ, remains stationary in $\mathbf{L}[G, H, I]$. Therefore, $\mathbf{L}[G, H, I]$ is a model as desired. Q.E.D.

References

[1] Y. Cheng. *Analysis of Martin–Harrington Theorem in Higher Order Arithmetic.* PhD thesis, National University of Singapore, 2012.

[2] Y. Cheng and R. Schindler. Harrington's principle in higher order arithmetic, 2014. Submitted.

[3] V. Gitman and P. D. Welch. Ramsey-like cardinals II. *Journal of Symbolic Logic*, 76(2):541–560, 2011.

[4] R. B. Jensen. Iteration theorems for subcomplete and related forcings. Handwritten notes.

[5] P. Koepke and P. Welch. On the strength of mutual stationarity. In J. Bagaria and S. Todorcevic, editors, *Set theory, Centre de Recerca Matemàtica, Barcelona, 2003–2004*, Trends in Mathematics, pages 309–320. Birkhäuser Verlag, 2006.

[6] P. Koepke and P. D. Welch. Global square and mutual stationarity at the $\aleph_n$. *Annals of Pure and Applied Logic*, 162(10):787–806, 2011.

[7] M. Magidor. On the role of supercompact and extendible cardinals in logic. *Israel Journal of Mathematics*, 10:147–157, 1971.

[8] T. Räsch and R. Schindler. A new condensation principle. *Archive for Mathematical Logic*, 44(2):159–166, 2005.

[9] R. Schindler. Semi-proper forcing, remarkable cardinals, and bounded Martin's maximum. *Mathematical Logic Quarterly*, 50(6):527–532, 2004.

[10] R. Schindler. *Set theory, Exploring independence and truth.* Springer-Verlag, 2014. To appear.

[11] R.-D. Schindler. Proper forcing and remarkable cardinals. *Bulletin of Symbolic Logic*, 6(2):176–184, 2000.

[12] R.-D. Schindler. Proper forcing and remarkable cardinals. II. *Journal of Symbolic Logic*, 66(3):1481–1492, 2001.

The weak ultimate L conjecture

W. Hugh Woodin

Mathematics and Philosophy Departments, Harvard University, Cambridge MA, United States of America

1 Introduction

We introduce a test question, in the form of the conjecture in the title, for the solution of the inner model problem at the level of one supercompact cardinal. Our goal it to isolate a reasonable feature that one expects of such a solution and which is sufficient to prove the **HOD** conjecture of [5]. The discussion at the beginning of § 5 gives a very brief and technical summary of the primary issues.

We also use these ideas to obtain a number of consequences of a variation of the axiom **V**=**UltL** from [7]. These consequences include **V**=**HOD** and that **V** is not a generic extension of any inner model $N \subsetneq \mathbf{V}$. In fact we obtain that **V** is the minimum universe of the *generic multiverse*—this is defined in [6]. We also prove that (this strengthened version of) **V**=**UltL** implies the Ω conjecture.

Finally we collect and review a number of the relevant results from [5]; this is simply in an attempt to make this a material a bit more accessible, though [10] also serves that purpose. Here though we focus just on the notions of weak extender models and the **HOD** conjecture, developing the necessary elements of the theory essentially from just elementary notions. Another reason for reviewing this development is that the new arguments are really just variations of those previous arguments and it seems worthwhile to package everything together in a single account.

2 Preliminaries

The following definition is from [1].

Definition 2.1. A set of reals $A \subseteq \mathbb{R}$ is *universally Baire* if for all topological spaces, Ω, and for all continuous functions $\pi : \Omega \to \mathbb{R}$, the preimage of A under π, $\pi^{-1}[A]$, has the property of Baire in Ω.

The following theorem, from [1], gives the fundamental connection between universally Baire sets, determinacy, and large cardinals.

Theorem 2.2. Suppose there is a proper class of Woodin cardinals and that $A \subseteq \mathbb{R}$ is universally Baire. Then

(1) every set $B \in \wp(\mathbb{R}) \cap \mathbf{L}(A, \mathbb{R})$ is universally Baire and

(2) $\mathbf{L}(A, \mathbb{R}) \models \mathsf{AD}^+$.

Stefan Geschke, Benedikt Löwe, Philipp Schlicht (*eds.*).
Infinity, computability, and metamathematics: Festschrift celebrating the 60th birthdays of Peter Koepke and Philip Welch. College Publications, London, 2014. Tributes, Volume 23.

The next theorem shows that in the presence of a supercompact cardinal, the inner model $\mathbf{L}(\Gamma_\infty)$ can be *sealed* in a very strong sense where Γ_∞ is the collection of all universally Baire sets $A \subseteq \mathbb{R}$. Cf. [2] for a proof.

Theorem 2.3 (Sealing Theorem). Suppose there is a proper class of Woodin cardinals, δ is a supercompact cardinal, and that $G \subset \mathrm{Coll}(\omega, \mathbf{V}_{\delta+2})$ is $\mathbf{V}$-generic. Suppose $\mathbf{V}[G][H]$ is a set generic extension of $\mathbf{V}[G]$. Then the following hold where $\Gamma_\infty^{\mathbf{V}[G]}$ is the set of universally Baire sets as defined in $\mathbf{V}[G]$ and $\Gamma_\infty^{\mathbf{V}[G][H]}$ is the set of universally Baire sets as defined in $\mathbf{V}[G][H]$.

(1) $\Gamma_\infty^{\mathbf{V}[G]} = \mathbf{L}(\Gamma_\infty^{\mathbf{V}[G]}) \cap (\wp(\mathbb{R}))^{\mathbf{V}[G]}$.

(2) $\Gamma_\infty^{\mathbf{V}[G][H]} = \mathbf{L}(\Gamma_\infty^{\mathbf{V}[G][H]}) \cap (\wp(\mathbb{R}))^{\mathbf{V}[G][H]}$.

(3) There is an elementary embedding $j : \mathbf{L}(\Gamma_\infty^{\mathbf{V}[G]}) \to \mathbf{L}(\Gamma_\infty^{\mathbf{V}[G][H]})$.

The simplest test question for inner model theory at the level a supercompact cardinal is whether the existence of a supercompact cardinal implies that $\Gamma_\infty = \mathbf{L}(\Gamma_\infty) \cap \mathbf{L}(\wp(\mathbb{R}))$ where Γ_∞ is the set of all universally Baire sets, or even if the conclusion of Theorem 2.3 holds for all pairs of extensions of $\mathbf{V}$, $\mathbf{V}[G] \subset \mathbf{V}[G][H]$. But positive answers to these test questions do not obviously yield many other applications and so we shall define a more useful test question. The *weak ultimate* $\mathbf{L}$ *conjecture* is simply the conjecture that there is a positive solution to this question.

We shall need some more definitions from the theory of AD^+.

Definition 2.4 ($\mathsf{ZF} + \mathsf{AD}^+$). We denote by Θ the supremum of the set of $\alpha \in \mathrm{Ord}$ such that there is a surjection $\pi : \mathbb{R} \to \alpha$. The sequence $\langle \Theta_\alpha : \alpha \leq \Omega \rangle$ is called the *Solovay sequence* and is defined by induction on α as follows:

1. For $\alpha = 0$, the supremum of the set of $\xi \in \mathrm{Ord}$ such that there is a surjection $\pi : \mathbb{R} \to \xi$ such that π is $\mathbf{OD}$ is called Θ_0.
2. For $\alpha = \beta + 1$, Θ_α is the supremum of the set of $\xi \in \mathrm{Ord}$ such that there is a surjection $\pi : \wp(\Theta_\beta) \to \xi$ such that π is $\mathbf{OD}$.
3. For a nonzero limit ordinal α, $\Theta_\alpha = \sup\{\Theta_\beta \mid \beta < \alpha\}$.
4. Finally, $\Theta = \Theta_\Omega$.

The following lemma is a relatively standard consequence of the basic theory of AD^+.

Lemma 2.5 ($\mathsf{ZF} + \mathsf{DC}_\mathbb{R} + \mathsf{AD}^+$). Assume $\mathbf{V}$=$\mathbf{L}(\wp(\mathbb{R}))$ and that $\Theta_2 < \Theta$. Then there is a partial order $\mathbb{P} \in \mathbf{HOD} \cap \mathbf{V}_{\Theta_2}$ such that $\mathbf{HOD}^\mathbb{P} \cap \mathbf{V}_\Theta \models$ "Every $\mathbf{OD}$ set $A \subset \mathbb{R}$ is universally Baire."

We shall exploit this lemma using the following theorem from [5].

Theorem 2.6. Suppose there is a proper class of Woodin cardinals and that every **OD** subset of $\mathbb{R}$ is universally Baire. Then **HOD** $\models$ "The Ω conjecture".

3 Weak extender models

We review some of the basic notions from [5] and we include proofs of some of the relevant lemmas.

Definition 3.1. Suppose that N is a transitive class, $\mathrm{Ord} \subset N$, and $N \models$ ZFC. Then N is a *weak extender model for the supercompactness of* δ if for each $\lambda > \delta$ there is a δ-complete normal fine ultrafilter U on $\wp_\delta(\lambda)$ such that $N \cap \wp_\delta(\lambda) \in U$ and $U \cap N \in N$.

The motivation for the definition of a weak extender model for supercompactness is to isolate a property that the inner models produced by the successful solution to the Inner Model Problem at the level of supercompact cardinals should have. The assumption of course is that these inner models will (or can be) "backgrounded".

The proof below of Solovay's Lemma is really the key to much of that which we shall do.

Lemma 3.2 (Solovay)**.** Suppose that $\delta < \lambda$ are uncountable regular cardinals and that U is a δ-complete normal fine ultrafilter on $\wp_\delta(\lambda)$. Then there exists a set $A \in U$ such that for all $\sigma, \tau \in A$, $\sigma = \tau$ if and only if $\sup(\sigma) = \sup(\tau)$.

Proof. Remarkably, the set A does not depend on U. Let $\langle S_\alpha : \alpha < \lambda \rangle$ be a partition of $\{\alpha < \lambda \mid \mathrm{cf}(\alpha) = \omega\}$ into pairwise disjoint stationary sets. For each $\eta < \lambda$ such that $\delta > \mathrm{cf}(\eta) > \omega$, let Z_η be the set of $\alpha < \eta$ such that $S_\alpha \cap \eta$ is stationary in η (in the sense that for all closed cofinal sets $C \subseteq \eta$, $S_\alpha \cap C \neq \varnothing$). Let $A = \{Z_\eta \mid \eta < \lambda, \omega < \mathrm{cf}(\eta) < \delta, \text{ and } \sup(Z_\eta) = \eta\}$. It suffices to show that $A \in U$. Let $j : V \to M \cong \mathbf{V}^{\wp_\delta(\lambda)}/U$ be the ultrapower embedding given by U. Thus $\mathrm{crit}(j) = \delta$ and $j(\delta) > \lambda$, $\{j(\alpha) \mid \alpha < \lambda\} \in M$, and for all $Y \subseteq \wp_\delta(\lambda)$, $Y \in U$ if and only if $\{j(\alpha) \mid \alpha < \lambda\} \in j(Y)$. Let $\eta = \sup\{j(\alpha) \mid \alpha < \lambda\}$. Since $\{j(\alpha) \mid \alpha < \lambda\}$ is ω-closed, for each $\alpha < \lambda$, $M \models$ "$j(S_\alpha) \cap \eta$ is stationary in η". Since $\{j(\alpha) \mid \alpha < \lambda\} \in M$, it follows that $\{j(\alpha) \mid \alpha < \lambda\} \in j(A)$ and so $A \in U$. Q.E.D.

Lemma 3.4 shows that if N is a weak extender model for the supercompactness of δ, then fairly strong covering properties hold for N above δ. We first prove a preliminary lemma.

Lemma 3.3. Suppose that N is a weak extender model for the supercompactness of δ, $\gamma \geq \delta$, and γ is a regular cardinal in N. Then $(\mathrm{cf}(\gamma))^{\mathbf{V}} = |\gamma|^{\mathbf{V}}$.

Proof. Let U be a normal fine δ-complete ultrafilter on $\wp_\delta(\gamma)$ such that

(a) $\left(N \cap \wp_\delta(\gamma)\right) \in U$,

(b) $U \cap N \in N$.

By Lemma 3.2 applied within N, there exists a set $A \in U \cap N$ such that for all $\sigma, \tau \in A$ if $\sup(\sigma) = \sup(\tau)$ then $\sigma = \tau$.

Assume toward a contradiction that $\mathrm{cf}(\gamma) < |\gamma|$. Since $N \cap \wp_\delta(\gamma) \in U$ and since γ is regular in N, $\mathrm{cf}(\gamma) \geq \delta$.

For each closed unbounded set $C \subset \gamma$, $\{\sigma \in \wp_\delta(\gamma) \mid \sup(\sigma) \in C\} \in U$. Let $C \subseteq \gamma$ be a closed unbounded set with ordertype $(\mathrm{cf}(\gamma))^{\mathbf{V}}$ and let $A_C = \{\sigma \in A \mid \sup(\sigma) \in C\}$. Thus $A_C \in U$ and so since U is fine, $\gamma = \cup\{\sigma \mid \sigma \in A_C\}$. Finally, $|A_C| \leq \gamma \cdot |C|^{\mathbf{V}} = \delta \cdot (\mathrm{cf}(\gamma))^{\mathbf{V}}$. Therefore $|\gamma|^{\mathbf{V}} = (\mathrm{cf}(\gamma))^{\mathbf{V}}$. Q.E.D.

Lemma 3.4. Suppose that N is a weak extender model for the supercompactness of δ, $\gamma \geq \delta$, and γ is a singular cardinal in $\mathbf{V}$. Then the following hold:

(1) γ is a singular cardinal in N and

(2) $\gamma^+ = (\gamma^+)^N$.

Proof. We first prove (1). Assume toward a contradiction that γ is a regular cardinal in N. Then by Lemma 3.3, $|\gamma|^{\mathbf{V}} = (\mathrm{cf}(\gamma))^{\mathbf{V}}$ and so since γ is a cardinal in $\mathbf{V}$, $\gamma = \mathrm{cf}(\gamma)$ which contradicts that γ is singular.

We finish by proving (2). Let $\kappa = (\gamma^+)^N$. Thus κ is a regular cardinal in N and so again by Lemma 3.3, $|\kappa|^{\mathbf{V}} = (\mathrm{cf}(\kappa))^{\mathbf{V}}$. But $\gamma \leq |\kappa|^{\mathbf{V}}$ and so $\gamma \leq (\mathrm{cf}(\kappa))^{\mathbf{V}}$. Therefore since γ is a singular cardinal, $\gamma < (\mathrm{cf}(\kappa))^{\mathbf{V}}$, which implies that $\kappa = \gamma^+$. Q.E.D.

We are headed toward a useful reformulation of the definition that N be a weak extender model for the supercompactness of δ. The proof of that reformulation requires the following simple lemma.

Lemma 3.5. Suppose that N is transitive class containing Ord, $N \models \mathsf{ZFC}$, $\gamma > \delta$, and that $\gamma = |\mathbf{V}_\gamma|$. Suppose that U is a δ-complete normal fine ultrafilter on $\wp_\delta(\gamma)$ such that $N \cap \wp_\delta(\gamma) \in U$ and such that $U \cap N \in N$. Let $j : V \to M \cong \mathbf{V}^{\wp_\delta(\gamma)}/U$ be the ultrapower embedding. Then $N \cap \mathbf{V}_\gamma = j(N \cap \mathbf{V}_\delta) \cap \mathbf{V}_\gamma$.

Proof. Since $N \cap \wp_\delta(\gamma) \in U$, it follows that $j[\gamma] \in j(N)$ and so since $\gamma = |N_\gamma|^N$, $N \cap \mathbf{V}_\gamma \subset j(N) \cap \mathbf{V}_\gamma$. We must show that $N \cap \mathbf{V}_\gamma = j(N) \cap \mathbf{V}_\gamma$. Fix a bijection $e : \gamma \to N \cap \mathbf{V}_\gamma$ such that $e \in N$. Since $\gamma = |\mathbf{V}_\gamma|$, $\gamma = |N_\gamma|^N$ and so e exists.

Let A be the set of $\sigma \in \wp_\delta(\gamma)$ such that $e[\sigma] \prec N \cap \mathbf{V}_\gamma$ and the transitive collapse of $e[\sigma]$ is $N \cap \mathbf{V}_\alpha$ where α is the ordertype of σ. The key claim is that

$$A \in U. \tag{1}$$

Working in N we have that $U \cap N$ is a δ-complete normal fine ultrafilter on $\wp_\delta(\gamma)$ and that $e : \gamma \to N_\gamma$ is a bijection. Let $j_N : N \to N_U \cong N^{\wp^N_\delta(\gamma)}/U$ be the ultrapower embedding as computed in N and let $\sigma = j_N[\gamma]$. We have that $j_N(e)[\sigma] \prec j_N(N \cap \mathbf{V}_\gamma)$, that $N \cap \mathbf{V}_\gamma$ is the transitive collapse of $j_N(e)[\sigma]$, and that $N_U \cap \mathbf{V}_\gamma = N \cap \mathbf{V}_\gamma$. Therefore $A \cap N \in U$ and so $A \in U$. This proves (1).

Thus $j[\gamma] \in j(A)$ and so by applying j to the definition of A, the transitive collapse of $j(e)[\gamma]$ must be $j(N)_\gamma$. But the transitive collapse of $j(e)[\gamma]$ is N_γ and therefore $N \cap \mathbf{V}_\gamma = j(N) \cap \mathbf{V}_\gamma$. This proves the lemma. Q.E.D.

With $N = \mathbf{V}$ the next lemma is in essence Magidor's Lemma on the reformulation of the supercompactness of δ.

Lemma 3.6. Suppose that N is a weak extender model for the supercompactness of δ, $\gamma > \delta$, and that $a \in \mathbf{V}_\gamma$. Then there exist $\bar{\delta} < \bar{\gamma} < \delta$, $\bar{a} \in \mathbf{V}_{\bar{\gamma}}$, and an elementary embedding, $\pi : \mathbf{V}_{\bar{\gamma}+1} \to \mathbf{V}_{\gamma+1}$ such that

(1) $\bar{\delta} = \mathrm{crit}(j)$ and $\pi(\bar{\delta}) = \delta$,

(2) $\pi(\langle \bar{a}, \bar{\gamma}\rangle) = \langle a, \gamma\rangle$,

(3) $\pi(N \cap \mathbf{V}_{\bar{\gamma}}) = N \cap \mathbf{V}_\gamma$ and $\pi|(N \cap \mathbf{V}_{\bar{\gamma}+1}) \in N$.

Proof. Fix $\kappa > \gamma$ such that $\kappa = |\mathbf{V}_\kappa|$. Let $j : V \to M$ be an elementary embedding with critical point δ such that $M^{\mathbf{V}_{\kappa+1}} \subseteq M$, such that $\kappa < j(\delta)$, and such that $\mathbf{V}_\kappa \cap N = \mathbf{V}_\kappa \cap j(N \cap \mathbf{V}_\delta)$ and $j[\kappa] \in j(N)$. The elementary embedding exists by Lemma 3.5.

Let $\pi = j|\mathbf{V}_{\gamma+1}$. Thus $\pi : M_{\gamma+1} \to M_{j(\gamma)+1}$ is an elementary embedding and $\pi \in M$. Further for all $\alpha < \kappa$, $\pi|(\mathbf{V}_\alpha \cap N) \in j(N)$ since $j[\kappa] \in j(N)$ and since $\kappa = |\mathbf{V}_\kappa|$. Thus π witnesses that the lemma holds in M at $j(\delta)$ for $(j(\gamma), j(a))$. Therefore the lemma holds in $\mathbf{V}$ at δ for (γ, a). Q.E.D.

We now come to the key theorem for weak extender models for supercompactness. There is a more general version of Theorem 3.7 for extenders, cf. [5, Theorem 142]. The proof of that theorem is just an elaboration of the proof here.

Theorem 3.7 (The Universality Theorem)**.** Suppose that N is a weak extender model for the supercompactness of δ, $\gamma > \delta$ is a cardinal in N, and that $j : H(\gamma^+)^N \to M$ is an elementary embedding such that $M \subset N$ and such that $\delta \leq \mathrm{crit}(j)$. Then $j \in N$.

Proof. Fix $\kappa > \gamma$ such that $|\mathbf{V}_\kappa| = \kappa$ and such that $j \in \mathbf{V}_\kappa$. By Lemma 3.6, with (κ, j) as (γ, a) in the hypothesis, there exist $\bar{\delta} < \bar{\gamma} < \bar{\kappa} < \delta$, a transitive set $\bar{M} \in N$, an elementary embedding, $\bar{j} : (H(\bar{\gamma}^+))^N \to \bar{M}$ with $\bar{j} \in \mathbf{V}_{\bar{\kappa}}$, and an elementary embedding $\pi : \mathbf{V}_{\bar{\kappa}+1} \to \mathbf{V}_{\kappa+1}$, such that the following hold: $\mathrm{crit}(\pi) = \bar{\delta}$, $\pi(\bar{\delta}) = \delta$, $\pi(\bar{M}) = M$, $\pi(\bar{j}) = j$, $\pi(N \cap \mathbf{V}_{\bar{\kappa}}) = N \cap \mathbf{V}_\kappa$, and $\pi|(N \cap \mathbf{V}_{\bar{\kappa}+1}) \in N$.

We prove that $\bar{j} \in N$. Since $\pi(\bar{j}) = j$ and since $\pi|(N \cap \mathbf{V}_{\bar{\kappa}+1}) \in N$, this implies that $j \in N$.

The key points are

$$\pi|(H(\bar{\gamma}^+))^N \in N \text{ and} \tag{2}$$

$$\pi|(H(\bar{\gamma}^+))^N \in H(\gamma^+)^N \tag{3}$$

since $(H(\bar{\gamma}^+))^N$ is closed under γ-sequences in N. Let $\pi^* = j\left(\pi|(H(\bar{\gamma}^+))^N\right)$. Since $\bar{\gamma} < \delta \leq \mathrm{crit}(j)$, π^* is an elementary embedding, $\pi^* : (H(\bar{\gamma}^+))^N \to M$. The last key point is that $\pi^* \in M$ and so $\pi^* \in N$.

We can now compute $\bar{j}$ from π^* and $\pi|(N \cap \mathbf{V}_{\bar{\kappa}+1})$ as follows. For all $a \in (H(\bar{\gamma}^+))^N$, for all $b \in \bar{M}$: $b = \bar{j}(a)$ if and only if $\pi(b) = \pi^*(a)$. Here we are using that

$$\pi(\bar{j}(a)) = j(\pi(a)) = j\left(\pi|(H(\bar{\gamma}^+))^N\right)(j(a)) = j\left(\pi|(H(\bar{\gamma}^+))^N\right)(a) = \pi^*(a).$$

This proves that $\bar{j} \in N$. Q.E.D.

The following lemma from [5] shows that the requirement in Theorem 3.7 that $\mathrm{crit}(j) \geq \delta$ is necessary. The indicated weak extender model N is easily specified. Let U be a κ-complete normal ultrafilter on κ where $\kappa < \delta$. Let $j_0 : V \to N_0 \cong \mathbf{V}^\kappa/U$ be the associated ultrapower embedding. Let $j : V \to N$ be the ω-th iterate of j_0. Then N is a weak extender model for the supercompactness of δ and $j_0(N) = N$ and so $j_0|N$ gives a nontrivial elementary embedding $j_N : N \to N$. The details together with more general versions of this claim can be found in [5].

Lemma 3.8. Suppose δ is a supercompact cardinal. Then there exists a weak extender model N for the supercompactness of δ such that there is a non-trivial elementary embedding, $j : N \to N$.

We briefly describe the basic obstruction to developing a nontrivial theory of weak extender models for large cardinal notions beyond the level of superstrong cardinals. The issues arise well within the finite levels of supercompactness and so certainly arise in the hierarchy of weak extender models for the existence of cardinals κ which are κ^{+n} supercompact where $n < \omega$. The following definition is from [5].

Definition 3.9. A sequence $N = \langle N_\alpha : \alpha \in \mathrm{Ord}\rangle$ is *weakly Σ_2-definable* if there is a formula $\varphi(x)$ such that

1. for all $\beta < \eta_1 < \eta_2 < \eta_3$, if $(N_\varphi)^{\mathbf{V}_{\eta_1}}|\beta = (N_\varphi)^{\mathbf{V}_{\eta_3}}|\beta$ then
$$(N_\varphi)^{\mathbf{V}_{\eta_1}}|\beta = (N_\varphi)^{\mathbf{V}_{\eta_2}}|\beta = (N_\varphi)^{\mathbf{V}_{\eta_3}}|\beta;$$
2. for all $\beta \in \mathrm{Ord}$, $N|\beta = (N_\varphi)^{\mathbf{V}_\eta}|\beta$ for all sufficiently large η,

where for all γ, $(N_\varphi)^{\mathbf{V}_\gamma} = \{a \in \mathbf{V}_\gamma \mid \mathbf{V}_\gamma \models \varphi[a]\}$.

We can require without loss of generality that φ is chosen such that for all γ, $(N_\varphi)^{\mathbf{V}_\gamma}$ is a function with domain ξ where ξ is the largest limit ordinal $\xi \leq \gamma$.

Definition 3.10. Suppose that $N \subset \mathbf{V}$ is an inner model and $N \models \mathsf{ZFC}$. Then N is *weakly Σ_2-definable* if the sequence $\langle N \cap \mathbf{V}_\alpha : \alpha \in \mathrm{Ord}\rangle$ is weakly Σ_2-definable.

Remark 3.11. If $N \subset \mathbf{V}$ is a class which is Σ_2-definable then the sequence $\langle N \cap \mathbf{V}_\alpha : \alpha \in \mathrm{Ord}\rangle$ is weakly Σ_2-definable. Therefore inner models N which are Σ_2-definable are weakly Σ_2-definable and as a special case **HOD**, being Σ_2-definable, is weakly Σ_2-definable.

This implies of course that the sequence $\langle \mathbf{HOD} \cap \mathbf{V}_\alpha : \alpha \in \mathrm{Ord}\rangle$ is weakly Σ_2-definable. More generally, for each $\alpha \in \mathrm{Ord}$, let T_α be the Σ_2-theory of $\mathbf{V}$ with parameters from $\mathbf{V}_\alpha$. Then the sequence $\langle T_\alpha : \alpha \in \mathrm{Ord}\rangle$ is weakly Σ_2-definable (and here we correct the incorrect comment in [5]).

Remark 3.12. The increasing enumeration $\langle \delta_\alpha : \alpha \in \mathrm{Ord}\rangle$ of all supercompact cardinals is weakly Σ_2-definable.

Suppose that $N = \langle N_\alpha : \alpha \in \mathrm{Ord}\rangle$ is weakly Σ_2-definable and $\mathbf{V}_\delta \prec_{\Sigma_2} \mathbf{V}$. Then $(N)^{\mathbf{V}_\delta}$ denotes $(N_\varphi)^{\mathbf{V}_\delta}$ where φ is a formula which witnesses that N is weakly Σ_2-definable. This is well-defined in the sense that it does not depend on the choice of the formula φ which witnesses that N is weakly Σ_2-definable. With this notation we prove the following lemma.

Lemma 3.13. Suppose that $N = \langle N_\alpha : \alpha \in \mathrm{Ord}\rangle$ is weakly Σ_2-definable and δ is a strong cardinal. Then $N \cap \mathbf{V}_\delta = (N)^{\mathbf{V}_\delta}$.

Proof. Let $\varphi(x)$ be a formula which witnesses that N is weakly Σ_2-definable.

Assume toward a contradiction that $N|\delta \neq (N)^{\mathbf{V}_\delta}$. Then there exists $\eta > \delta$ and $\beta < \delta$ such that $N|\beta = (N_\varphi)^{\mathbf{V}_\eta}|\beta \neq (N_\varphi)^{\mathbf{V}_\delta}|\beta$. Since δ is a strong cardinal, $\mathbf{V}_\delta \prec_{\Sigma_2} \mathbf{V}$ and so there exists $\beta < \eta_0 < \delta$ such that $N|\beta = (N_\varphi)^{\mathbf{V}_{\eta_0}}|\beta$. But then $\beta < \eta_0 < \delta < \eta$, $(N_\varphi)^{\mathbf{V}_{\eta_0}}|\beta = (N_\varphi)^{\mathbf{V}_\eta}$, and $(N_\varphi)^{\mathbf{V}_{\eta_0}}|\beta \neq (N_\varphi)^{\mathbf{V}_\delta}|\beta$, which together form a contradiction. Q.E.D.

The following theorem is a sharper version of the corresponding [5, Theorem 168] and the proof is a minor variation of the proof there. Both theorems significantly constrain the construction of extender models and this begins just past the level of superstrong cardinals. Recall that a cardinal δ is an *extendible cardinal* if for each $\alpha > \delta$, there is an elementary embedding, $j : \mathbf{V}_{\alpha+1} \to \mathbf{V}_{j(\alpha)+1}$ such that $\mathrm{crit}(j) = \delta$ and such that $j(\delta) > \alpha$.

Theorem 3.14 also arguably shows that constructing backgrounded weak extender models for δ is δ^{+n} supercompact for all $n < \omega$ is a critical test case for the extension of the results of [4] to levels beyond superstrong, this is the subject of [8]. The concern of course is that the contraints we have outlined, along with the universality theorems for weak extender models of for the supercompactness of δ, are simply the precursors of an anti inner model theorem at the level of supercompactness.

Theorem 3.14. Suppose that δ is an extendible cardinal. Then there is a class-generic extension $\mathbf{V}[G]$ of $\mathbf{V}$ in which the following hold:

(1) $\mathbf{V}[G] = (\mathbf{HOD})^{\mathbf{V}[G]}$.

(2) δ is an extendible cardinal.

(3) Suppose $\mathbb{E} \subset \mathrm{Ord}$ is such that the following hold.

   (a) $\mathbf{L}[\mathbb{E}]$ is weakly Σ_2-definable,

   (b) Let $X \subset \delta$ be the set of all $\kappa < \delta$ such that there is an elementary embedding, $j : \mathbf{V}[G]_{\kappa+\gamma+1} \to \mathbf{V}[G]_{\delta+j(\gamma)+1}$ with $\mathrm{crit}(j) = \kappa$ where γ is the least inaccessible cardinal above κ. Then there exists $Y \subset X$ such that $Y \cap \xi \in \mathbf{L}[\mathbb{E}]$ for all $\xi < \delta$ and such that $\sup(Y) = \sup(X) = \delta$.

   Then $\mathbf{V}[G]_\delta = \mathbf{L}_\delta[\mathbb{E}]$.

4 The HOD conjecture

We are headed toward the **HOD** Dichotomy Theorem which shows that if there is an extendible cardinal then either **HOD** is very "close" to **V** or **HOD** is very "far" from **V**, at least above any extendible cardinal. This theorem is naturally viewed as an abstract generalization of Jensen's Covering Lemma.

Definition 4.1. Suppose that κ is an uncountable regular cardinal. Then κ is *ω-strongly measurable in* **HOD** if there exists $\lambda < \kappa$ such that the following hold:

1. the ordinal λ is a cardinal in **HOD** such that $(2^\lambda)^{\mathbf{HOD}} < \kappa$ and

2. there is no partition $\langle S_\alpha : \alpha < \lambda \rangle \in \mathbf{HOD}$ of the set $\{\alpha < \kappa \mid (\mathrm{cf}(\alpha))^{\mathbf{V}} = \omega\}$ such that for each $\alpha < \lambda$, S_α is a stationary subset of κ.

Lemma 4.2. Suppose that κ is an uncountable regular cardinal which is ω-strongly measurable in **HOD**. Then there exist $\gamma < \kappa$ and a partition $\langle A_\alpha : \alpha < \gamma \rangle \in \mathbf{HOD}$ of the set $\{\alpha < \kappa \mid (\mathrm{cf}(\alpha))^{\mathbf{V}} = \omega\}$ into sets stationary in $\mathbf{V}$ such that for each $\alpha < \gamma$ and for each $S \subset A_\alpha$, if $S \in \mathbf{HOD}$ then either S is not stationary in $\mathbf{V}$ or $A_\alpha \backslash S$ is not stationary in $\mathbf{V}$.

Proof. Let $S_\omega = \{\alpha < \kappa \mid \mathrm{cf}(\alpha) = \omega\}$ and let $\mathcal{I}$ be the ideal of non-stationary subsets of S_ω. Thus $S_\omega \in \mathbf{HOD}$ and $\mathcal{I} \cap \mathbf{HOD} \in \mathbf{HOD}$. Let $\lambda < \kappa$ witness that κ is ω-strongly measurable in **HOD**.

There are two cases. First suppose that the Boolean algebra, $\wp(S_\omega) \cap \mathbf{HOD}/\mathcal{I}$ is atomic. Let γ be the cardinality as computed in **HOD** of the set of atoms. If $\gamma \geq \lambda$ then there is a partition $\langle T_\alpha : \alpha < \lambda \rangle \in \mathbf{HOD}$ of S_ω into sets each of which is stationary in $\mathbf{V}$ and this contradicts the choice of λ. Therefore $\gamma < \lambda$ and so there is a partition $\langle A_\alpha : \alpha < \gamma \rangle \in \mathbf{HOD}$ of S_ω into stationary sets in $\mathbf{V}$ which witnesses the lemma.

Thus we can reduce to the second case which is the case that the Boolean algebra, $\wp(S_\omega) \cap \mathbf{HOD}/\mathcal{I}$ is not atomic and we show this case is vacuous. Fix a stationary set $A \subset S_\omega$ such that $A \in \mathbf{HOD}$ and such that the Boolean algebra, $\wp(A) \cap \mathbf{HOD}/\mathcal{I}$ has no atoms. The ideal $\mathcal{I} \cap \mathbf{HOD}$ is κ-complete in **HOD** and $(2^\lambda)^{\mathbf{HOD}} < \kappa$. Therefore working in **HOD** there is an order preserving map $\pi : \mathbb{P} \to (\wp(A) \cap \mathbf{HOD}, \subseteq)$ where $\mathbb{P} \in \mathbf{HOD}$ is the partial order of all binary sequences of length at most $\lambda + 1$ (ordered by reverse extension, so the $s \in \mathbb{P}$ with $\mathrm{dom}(s) = \lambda$ are atoms) such that for all $s, t \in \mathbb{P}$:

1. if $s \leq t$ then $\pi(s) \subseteq \pi(t)$ and if $\mathrm{dom}(t)$ is a limit ordinal then $\pi(t) = \cap\{\pi(u) \mid u < t\}$,

2. if s, t are incompatible then $\pi(s) \cap \pi(t) = \varnothing$,

3. $\pi(s) = \pi(s^\frown 0) \cup \pi(s^\frown 1)$ and if $\pi(s) \notin \mathcal{I}$ then $\pi(s^\frown 0) \notin \mathcal{I}$ and $\pi(s^\frown 1) \notin \mathcal{I}$.

4. $A = \cup\{\pi(s) \mid \mathrm{dom}(s) = \lambda\}$.

Since $\mathcal{I}$ is a κ-complete ideal in **HOD** and since $(2^\lambda)^{\mathbf{HOD}} < \kappa$, there must exist $s \in \mathbb{P}$ such that $\mathrm{dom}(s) = \lambda$ and such that $\pi(s) \in \wp(A) \backslash \mathcal{I}$. For each $\alpha < \lambda$, let $T_\alpha = \pi(s_\alpha)$ where $s_\alpha = (s|(\alpha+1))^\frown 0$ if $s(\alpha+1) = 1$ and $s_\alpha = (s|(\alpha+1))^\frown 1$ otherwise.

Thus $\langle T_\alpha : \alpha < \lambda \rangle \in \mathbf{HOD}$ is a sequence of pairwise disjoint subsets of S_ω each of which is stationary in $\mathbf{V}$. This contradicts the choice of λ. Q.E.D.

Corollary 4.3. Suppose that κ is an uncountable regular cardinal which is ω-strongly measurable in **HOD**. Then κ is a measurable cardinal in **HOD**.

Thus the two possibilities in the conclusion of the **HOD** Dichotomy Theorem below, are in fact mutually exclusive.

Theorem 4.4 (The **HOD** Dichotomy Theorem)**.** Suppose that δ is an extendible cardinal. Then one of the following hold:

(1) Every regular cardinal $\gamma \geq \delta$ is ω-strongly measurable in **HOD**.

(2) For each singular cardinal $\gamma > \delta$, γ is a singular cardinal in **HOD** and $\gamma^+ = (\gamma^+)^{\mathbf{HOD}}$.

Proof. We assume (1) fails and prove that **HOD** is a weak extender model for the supercompactness of δ. This implies (2) by Lemma 3.4.

Fix a regular cardinal $\kappa \geq \delta$ which is not ω-strongly measurable in **HOD**. Let I be the class of all regular cardinals γ such that there exists $\eta > \gamma$ such that $\eta = |\mathbf{V}_\eta|$ and

$$\mathbf{V}_\eta \models \text{"}\gamma \text{ is not } \omega\text{-strongly measurable in } \mathbf{HOD}\text{"}. \tag{4}$$

Note that if $\gamma \in I$ then γ is not ω-strongly measurable in **HOD**. We verify this. Assume toward a contradiction that γ is ω-strongly measurable in **HOD** and let η witness that $\gamma \in I$. Since γ is ω-strongly measurable in **HOD** there must exist $\lambda < \gamma$ such that $(2^\lambda)^{\mathbf{HOD}} < \gamma$ and there is no partition $\langle S_\alpha : \alpha < \lambda \rangle \in \mathbf{HOD}$ of $\{\xi < \gamma \mid \mathrm{cf}(\xi) = \omega\}$ into stationary sets. But $(\mathbf{HOD})^{\mathbf{V}_\eta} \subseteq \mathbf{HOD}$ and so 2^λ as computed in $(\mathbf{HOD})^{\mathbf{V}_\eta}$ is necessarily smaller than γ and this contradicts 4. This verifies that no $\gamma \in I$ can be ω-strongly measurable in **HOD**.

Note that κ is not ω-strongly measurable in $(\mathbf{HOD})^{\mathbf{V}_\eta}$ for all sufficiently large η such that $\eta = |\mathbf{V}_\eta|$ (since κ is not ω-strongly measurable in **HOD**). Therefore $\kappa \in I$. Since δ is extendible and since $\kappa \geq \delta$, $I \cap \delta$ is cofinal in δ. Finally again since δ is extendible it follows that I is a proper class. The point here is that the property of *not* being ω-strongly measurable in **HOD** is a Σ_2-property.

Fix a strongly inaccessible cardinal $\lambda > \delta$. Since I is a proper class there exists a regular cardinal $\gamma \in I$ such that $2^\lambda < \gamma$. Let $S^\gamma_\omega = \{\alpha < \lambda \mid \mathrm{cf}(\alpha) = \omega\}$. Therefore since γ is not ω-strongly measurable in **HOD** there exists a partition $\langle S_\alpha : \alpha < \lambda \rangle \in \mathbf{HOD}$ of S^γ_ω such that for all $\alpha < \lambda$, S_α is a stationary set in $\mathbf{V}$.

Fix $\eta > \gamma$ such that $\mathbf{V}_\eta \prec_{\Sigma_2} \mathbf{V}$. Thus $\mathbf{HOD} \cap \mathbf{V}_\eta = (\mathbf{HOD})^{\mathbf{V}_\eta}$ and so $\langle S_\alpha : \alpha < \lambda \rangle \in (\mathbf{HOD})^{\mathbf{V}_\eta}$. Let $j : \mathbf{V}_{\eta+1} \to \mathbf{V}_{j(\eta)+1}$ be an elementary embedding such that $\mathrm{crit}(j) = \delta$ and such that $j(\delta) > \eta$. Let $\langle T_\beta : \beta < j(\lambda) \rangle \in j(\langle S_\alpha : \alpha < \lambda \rangle)$. Thus $\langle T_\beta : \beta < j(\lambda) \rangle \in (\mathbf{HOD})^{\mathbf{V}_{j(\eta)}} \subset \mathbf{HOD}$. Let

$Z = \{\beta < j(\lambda) \mid T_\beta \cap C \neq \varnothing$ for all closed cofinal sets $C \subset \sup(j[\gamma])\}$. Thus $Z \in (\mathbf{HOD})^{\mathbf{V}_{j(\eta)}}$. We claim that $Z = j[\lambda]$.

For each closed cofinal set $D \subset \sup(j[\gamma])$ there exists a closed cofinal set $D^* \subset \gamma$ such that $j[D^* \cap S^\gamma_\omega] \subseteq D$. This is because j is continuous on S^γ_ω. This implies that $j(\alpha) \in Z$ for each $\alpha < \lambda$ since S_α is stationary in S^λ_ω. Thus $j[\lambda] \subseteq Z$. Now suppose that $\beta \in Z$ and let $T^* = \{\xi \in S^\gamma_\omega \mid j(\xi) \in T_\beta\}$. Again since j is continuous on S^γ_ω, T^* is a stationary subset of S^γ_ω and in particular $T^* \neq \varnothing$. Therefore there exists $\alpha_0 < \lambda$ such that $S_{\alpha_0} \cap T^* \neq \varnothing$. But then $T_\beta \cap T_{j(\alpha_0)} \neq \varnothing$ and so $\beta = j(\alpha_0)$. This proves $Z = j[\lambda]$ as claimed.

Finally there is a bijection $e : \lambda \to \mathbf{HOD} \cap \mathbf{V}_\lambda$ such that $e \in (\mathbf{HOD})^{\mathbf{V}_\eta}$ (since such a bijection exists in $\mathbf{HOD}$ and since $\mathbf{V}_\eta \prec_{\Sigma_2} \mathbf{V}$).

Therefore $j|(\mathbf{HOD} \cap \mathbf{V}_\lambda) \in (\mathbf{HOD})^{\mathbf{V}_{j(\eta)}}$. For each cardinal υ with $\delta < \upsilon < \lambda$, let U_υ be the δ-complete normal fine ultrafilter on $\wp_\delta(\upsilon)$ given by j. Thus since $j|(\mathbf{HOD} \cap \mathbf{V}_\lambda) \in (\mathbf{HOD})^{\mathbf{V}_{j(\eta)}}$ and since λ is strongly inaccessible in $\mathbf{HOD}$, for each $\delta < \upsilon < \lambda$, we have that $\mathbf{HOD} \cap \wp_\delta(\upsilon) \in U_\upsilon$ and $U_\upsilon \cap \mathbf{HOD} \in \mathbf{HOD}$. Since λ can be chosen to be arbitrarily large this proves that $\mathbf{HOD}$ is a weak extender model for the supercompactness of δ. Q.E.D.

In fact the dichotomy is much stronger as indicated by the next theorem, the proof of which is essentially identical to that of Theorem 4.4. For each regular uncountable cardinal λ, let $\mathcal{I}^\lambda_{\mathrm{NS}}$ be the ideal of non-stationary subsets of λ, and for each infinite regular cardinal $\kappa < \lambda$, let $S^\lambda_\kappa = \{\alpha < \lambda \mid \mathrm{cf}(\alpha) = \kappa\}$

Theorem 4.5. Suppose that δ is an extendible cardinal. Then one of the following holds:

(1) Suppose that $\kappa < \lambda$ are regular cardinals with $\omega \leq \kappa < \delta < \lambda$. Then $\wp(S^\lambda_\kappa) \cap \mathbf{HOD}/\mathcal{I}^\lambda_{\mathrm{NS}}$ is atomic with fewer than δ-many atoms.

(2) For each singular cardinal $\gamma > \delta$, γ is a singular cardinal in $\mathbf{HOD}$ and $\gamma^+ = (\gamma^+)^{\mathbf{HOD}}$.

Proof. Suppose that (1) fails. Fix a pair of regular cardinals (κ_0, λ_0) such that $\omega \leq \kappa_0 < \delta < \lambda_0$ and such that (1) fails for the pair (κ_0, λ_0). Arguing as in the proof of Lemma 4.2, there exists a partition $\langle S_\alpha : \alpha < \delta\rangle \in \mathbf{HOD}$ of $S^{\lambda_0}_{\kappa_0}$ into sets which are stationary in $\mathbf{V}$. Therefore since δ is an extendible cardinal, for each $\gamma \in \mathrm{Ord}$ there exists a regular cardinal $\lambda > \gamma$ such that there is a partition $\langle T_\alpha : \alpha < \gamma\rangle \in \mathbf{HOD}$ of $S^\lambda_{\kappa_0}$ into sets which are stationary in $\mathbf{V}$. Now arguing as in the proof of Theorem 4.4, it follows that $\mathbf{HOD}$ is a weak extender model for the supercompactness of δ and so by Lemma 3.4, (2) must hold. Q.E.D.

Note that (1) implies that every regular cardinal λ above δ is κ-strongly measurable in **HOD** (defined in the obvious fashion) for *all* possible (infinite) cofinalities, $\kappa < \delta$. Thus the dichotomy of possibilities for the relationship between **V** and **HOD** (above an extendible cardinal) is analogous to the relationship between **V** and **L** given by Jensen's Covering Lemma. Nevertheless we conjecture there is a key difference.

Definition 4.6. We call the following statement *the* **HOD** *conjecture*: (ZFC) There is a proper class of cardinals γ which are not ω-strongly measurable in **HOD**.

The following theorem explains the motivation for the **HOD** conjecture showing that the **HOD** conjecture (in the presence of an extendible cardinal) is simply the conjecture that there is an extender based inner model theory at the level of one supercompact cardinal.

Theorem 4.7. Suppose that δ is an extendible cardinal. Then the following are equivalent:

(1) There is a regular cardinal $\gamma \geq \delta$ which is not ω-strongly measurable in **HOD**.

(2) There is a cardinal $\gamma \geq \delta$ such that $\gamma^+ = (\gamma^+)^{\mathbf{HOD}}$.

(3) The **HOD** conjecture holds.

(4) There is a weak extender model N for the supercompactness of δ such that $N \subseteq$ **HOD**.

(5) **HOD** is a weak extender model for the supercompactness of δ.

Proof. The proof of Theorem 4.4 explicitly shows that (1) implies (5). The equivalence of (1)–(5) is immediate from this. Q.E.D.

5 The weak ultimate L conjecture

The natural scenario for proving that if there is an extendible cardinal then the **HOD** conjecture must hold, is to prove that if δ is an extendible cardinal then there must exist a weak extender model N for the supercompactness of δ such that $N \subseteq$ **HOD**. However the backgrounded construction of weak extender models may require iteration hypotheses which do not in general hold. The point is that while it seems likely the iteration trees induced on **V** by fine-structural iteration trees on background premice may have unique wellfounded branches, the situation for more general iteration trees, including those iteration trees on **V** induced by fine-structural iteration trees on background bicephali is less clear. But it is exactly through the analysis of backgrounded bicephali that one in general proves the definability of the construction, [4] and [8]. This concern is amplified by the emerging picture from [9] that general iteration hypotheses may actually be false.

If the **HOD** conjecture holds and δ is an extendible cardinal then by Theorem 4.7, **HOD** is a weak extender model for for the supercompactness of δ and so the definability of backgrounded constructions is not an issue since there is a rich class of extenders E in $\mathbf{V}$ such that $E \cap \mathbf{HOD} \in \mathbf{HOD}$. But this does not help with the indicated scenario for proving the **HOD** conjecture.

Conjecture 5.1. We call the following statement the *weak ultimate* **L** *conjecture*: (ZFC) Suppose that δ is an extendible cardinal. Then for all sufficiently large strongly inaccessible $\kappa > \delta$, there exists a transitive set $M \models \mathsf{ZFC}$ with $\mathrm{Ord}^M = \kappa$ and an elementary embedding $\pi : M \to N$ such that the following hold for some $\delta_M \in M$:

(1) $\mathbf{V}_{\kappa+1} \models$ "M is a weak extender model for the supercompactness of δ_M".

(2) For each $\lambda \in M$, if λ is a regular cardinal of M and if $\pi(\lambda) \neq \sup(\pi[\lambda])$, then $\lambda > \delta$ and there exists $\delta < \gamma \leq \lambda$ such that γ is λ-supercompact.

(3) $N \in \mathbf{HOD}^{\mathbf{L}(\Gamma_\infty^{\mathbf{V}[G]})}$ where $G \subset \mathrm{Coll}(\omega, \kappa)$ is $\mathbf{V}$-generic and $\Gamma_\infty^{\mathbf{V}[G]}$ is the collection of universally Baire sets in $\mathbf{V}[G]$.

Suppose that M is a transitive class, $M \models \mathsf{ZFC}$, and that $n < \omega$. Define M to be a weak extender model for the δ^{+n}-supercompactness of δ if there exists a δ-complete normal fine ultrafilter U on $\wp_\delta(\lambda)$ such that $M \cap \wp_\delta(\lambda) \in U$ and $U \cap M \in M$, where $\lambda = (\delta^{+n})^M$.

The following theorem is a reformulation of the main theorem from [8] and this provides the principal motivation for the weak ultimate **L** conjecture. The (ω_1+1)-Iteration Hypothesis is an *iteration hypothesis* defined in [8], it is a weakening of the Strong (ω_1+1)-Iteration Hypothesis of [5] and more closely related to the (ω_1+1)-Iteration Hypothesis of [3].

Theorem 5.2 ((ω_1+1)-Iteration Hypothesis)**.** Suppose that δ is an extendible cardinal. Then there exists a strongly inaccessible cardinal $\kappa > \delta$, a transitive set $M \models \mathsf{ZFC}$ with $\mathrm{Ord}^M = \kappa$, and an elementary embedding $\pi : M \to N$ such that for some $\delta_M \in M$, $\delta_M > \delta$, for all $n < \omega$,

$$\mathbf{V}_{\kappa+1} \models \text{“}M \text{ is a weak extender model for the } \delta_M^{+n}\text{-supercompactness of } \delta_M\text{”},$$

and such that the following hold:

(1) For each $\lambda \in M$, if λ is a regular cardinal of M and if $\pi(\lambda) \neq \sup(\pi[\lambda])$, then $\lambda > \delta$ and there exists $\delta < \gamma \leq \lambda$ such that γ is λ-supercompact.

(2) $N \in \mathbf{HOD}^{\mathbf{L}(\Gamma_\infty^{\mathbf{V}[G]})}$ where $G \subset \mathrm{Coll}(\omega, \kappa)$ is $\mathbf{V}$-generic and $\Gamma_\infty^{\mathbf{V}[G]}$ is the collection of universally Baire sets in $\mathbf{V}[G]$.

The following theorem is a key theorem since it suggests a route to actually proving that if there is an extendible cardinal then the **HOD** conjecture must hold. At this stage, as illustrated by the **HOD** Dichotomy Theorem, it is in the context of such large cardinal assumptions that the **HOD** conjecture is most interesting

Theorem 5.3. Suppose there is an extendible cardinal and that the weak ultimate **L** conjecture holds. Then the **HOD** conjecture holds.

Proof. We just need a single instance (M, π) of the conclusion of weak ultimate **L** conjecture and moreover we only need that $N \in \mathbf{HOD}$. Fix such an instance (M, π) and let $\gamma = \delta_M^{+\omega}$.

By Lemma 3.4, γ is a singular cardinal in M and $\gamma^+ = (\gamma^+)^M$. For each $\alpha < \gamma^+$, $(\mathrm{cf}(\alpha))^M < \gamma$ and so there must exist $\lambda < \gamma$ such that the set $S = \{\alpha < \gamma^+ \mid \mathrm{cf}(\alpha) = \omega \text{ and } (\mathrm{cf}(\alpha))^M = \lambda\}$ is a stationary subset of γ^+. By (2) of the conclusion of weak ultimate **L** conjecture (and this is the only use of (2)), π is continuous on S.

The key point is that since $\gamma^+ = (\gamma^+)^M$, there must exist a partition $\langle S_\alpha : \alpha < \gamma^+\rangle \in M$ of S such that $|\{\alpha < \gamma^+ \mid S_\alpha \text{ is stationary in } \gamma^+\}| = \gamma^+$. Let $\langle T_\beta : \beta < \pi(\gamma^+)\rangle = \pi(\langle S_\alpha : \alpha < \gamma^+\rangle)$. We come to a key claim.

There exists $X \subset \pi[\gamma^+]$ such that $X \in \mathbf{HOD}$ and such that $|X| = \gamma^+$. (5)

Define

$$X = \{\beta < \pi(\gamma^+) \mid T_\beta \cap D \neq \varnothing \text{ for all closed cofinal sets } D \subset \sup(\pi[\gamma^+])\}.$$

Note that since $N \in \mathbf{HOD}$, $X \in \mathbf{HOD}$. For each closed cofinal set $D \subset \sup(\pi[\gamma^+])$ there exists a closed cofinal set $D^* \subset \gamma^+$ such that $\pi[D^* \cap S] \subseteq D$. This is because π is continuous on S. Therefore for each $\alpha < \gamma^+$, if S_α is stationary in S then $\pi(\alpha) \in X$ and so $|X| \geq \gamma^+$. Now suppose that $\beta \in X$ and let $T^* = \{\xi \in S \mid \pi(\xi) \in T_\beta\}$. Again since π is continuous on S, T^* is a stationary subset of S, and in particular $T^* \neq \varnothing$. Therefore there exists $\alpha < \gamma^+$ such that $T^* \cap S_\alpha \neq \varnothing$ and this implies that $T_\beta \cap T_{\pi(\alpha)} \neq \varnothing$. The sets $\langle T_\xi : \xi < \pi(\gamma^+)\rangle$ are pairwise disjoint and so $\beta = \pi(\alpha)$. This proves $X \subseteq \pi[\gamma^+]$ and this proves (5).

Let $<_M \in M$ be a wellordering of $(H(\gamma^+))^M$ and let M^* be the transitive collapse of $Z \prec \left(\pi\left((H(\gamma^+))^M\right), \pi(<_M)\right)$ where Z is the set of all $a \in$

$\pi\left((H(\gamma^+))^M\right)$ such that a is definable in $\left(\pi\left((H(\gamma^+))^M\right), \pi(<_M)\right)$ with parameters from X. Note that since $Z \subset \pi[M]$ and since $X \subset Z$, necessarily $\mathrm{Ord}^{M^*} = \gamma^+$. Since $(N, X) \in \mathbf{HOD}$, it follows that $M^* \in \mathbf{HOD}$ and so $\gamma^+ = (\gamma^+)^{\mathbf{HOD}}$. Finally since $\gamma > \delta$ and since δ is an extendible cardinal, the **HOD** conjecture must hold by Theorem 4.7. Q.E.D.

6 The axiom V=UltL

We fix some notation to simplify various statements.

Definition 6.1. Assume there is a proper class of Woodin cardinals. Then Γ_∞ is the set of all universally Baire sets, and we write $\Gamma \lhd \Gamma_\infty$ if the following hold: $\Gamma \subsetneq \Gamma_\infty$, $\Gamma = \wp(\mathbb{R}) \cap \mathbf{L}(\Gamma, \mathbb{R})$, and $\mathbf{L}(\Gamma, \mathbb{R}) \models \neg \mathsf{AD}_{\mathbb{R}}$.

We note the following from the basic theory of AD^+.

Lemma 6.2. Suppose there is a proper class of Woodin cardinals and that $\Gamma \lhd \Gamma_\infty$. Then there is a largest Suslin cardinal in $\mathbf{L}(\Gamma, \mathbb{R})$.

The following is a strengthening the axiom **V=UltL** defined on [7, p. 321]. There the indicated reflection is restricted to Σ_3 sentences, Γ is restricted to be of the form $\wp(\mathbb{R}) \cap \mathbf{L}(A, \mathbb{R})$ for some $A \in \Gamma_\infty$, and the large cardinal hypothesis is weaker. In [9] we shall work with a version much closer to the orginal version, the only change is in the large cardinal hypothesis where as here we add that there exist a proper class of strong cardinals. But working with that version requires elements from the theory of homogeneously Suslin sets and related matters.

Definition 6.3. We say that **V=UltL** if

1. there is a proper class of Woodin cardinals,
2. there is a proper class of strong cardinals, and
3. for each Σ_4-sentence φ, if φ holds in **V** there exists $\Gamma \lhd \Gamma_\infty$ such that $\mathbf{HOD}^{\mathbf{L}(\Gamma,\mathbb{R})} \cap \mathbf{V}_\Theta \models \varphi$ where $\Theta = \Theta^{\mathbf{L}(\Gamma,\mathbb{R})}$.

One can show that if **V=UltL** then the weak ultimate **L** conjecture holds. More importantly, the ultimate **L** conjecture, which we define at the end of this section, implies the weak ultimate **L** conjecture.

Remark 6.4. One could formulate **V=UltL** as a scheme and require that for each sentence φ if φ holds in **V** then there exists $\Gamma \lhd \Gamma_\infty$ such that $\mathbf{HOD}^{\mathbf{L}(\Gamma,\mathbb{R})} \cap \mathbf{V}_\Theta \models \varphi$ where $\Theta = \Theta^{\mathbf{L}(\Gamma,\mathbb{R})}$.

The following theorem from the basic theory of AD^+ motivates in part the formulation of the axiom **V=UltL**. We shall also need to appeal to this theorem in the analysis of this axiom.

Theorem 6.5. Suppose there is a proper class of Woodin cardinals and that $\Gamma \lhd \Gamma_\infty$. Let δ be the largest Suslin cardinal of $\mathbf{L}(\Gamma, \mathbb{R})$. Then:

(1) $\Theta^{\mathbf{L}(\Gamma,\mathbb{R})}$ is a Woodin cardinal in $\mathbf{HOD}^{\mathbf{L}(\Gamma,\mathbb{R})}$.

(2) δ is a strong cardinal in $\mathbf{HOD}^{\mathbf{L}(\Gamma,\mathbb{R})} \cap \mathbf{V}_\Theta$ where $\Theta = \Theta^{\mathbf{L}(\Gamma,\mathbb{R})}$.

The analysis of the axiom, "**V**=**UltL**", uses the following theorem and the proof of this theorem will appear in [9] along with the detailed versions of the proofs which we sketch here. Theorem 6.6 and Theorem 6.11 are in essence theorems in the context of AD^+ reformulated as theorems about $\mathbf{L}(\Gamma, \mathbb{R})$ where $\Gamma \lhd \Gamma_\infty$. This is also true of the previous theorem, Theorem 6.5.

Recall from Definition 2.4, that Ω denotes the length of the Solovay sequence.

Theorem 6.6. Suppose there is a proper class of Woodin cardinals and that $\Gamma \lhd \Gamma_\infty$. Let δ be the largest Suslin cardinal of $\mathbf{L}(\Gamma, \mathbb{R})$. Then the following are equivalent:

(1) δ is a limit of Woodin cardinals in $\mathbf{HOD}^{\mathbf{L}(\Gamma,\mathbb{R})}$.

(2) $\delta < \Omega^{\mathbf{L}(\Gamma,\mathbb{R})}$ and $\delta = (\Theta_\delta)^{\mathbf{L}(\Gamma,\mathbb{R})}$.

Remark 6.7. Assuming AD^+ and that $\mathbf{V}=\mathbf{L}(\wp(\mathbb{R}))$, it is known that for a cone of $x \in \mathbb{R}$, Θ_0 is the least Woodin cardinal in $\mathbf{HOD}_{\{x\}}$. This strongly suggests that necessarily Θ_0 is the least Woodin cardinal in **HOD**.

The previous theorem actually establishes a weak version of this conjecture showing that Θ_0 cannot be a limit of Woodin cardinals in **HOD**. The proof of this claim uses the following additional elements of the theory of AD^+ in the context that $\mathbf{V}=\mathbf{L}(\wp(\mathbb{R}))$:

1. $\underset{\sim}{\delta}^2_1$ is the largest Suslin cardinal below Θ_0 and every Σ^2_1-set admits a Σ^2_1-scale.

2. Let T be the tree of a Σ^2_1-scale on a universal Σ^2_1-set. Then $\mathbf{HOD}^{\mathbf{L}(T,\mathbb{R})}|\Theta^{\mathbf{L}(T,\mathbb{R})} = \mathbf{HOD}|\Theta^{\mathbf{L}(T,\mathbb{R})}$ and $\mathbf{L}_{\underset{\sim}{\delta}^2_1}[T](\mathbb{R}) \prec_{\Sigma_1} \mathbf{L}(\wp(\mathbb{R}))$.

The point here is that if Θ_0 is a limit of Woodin cardinals in **HOD** then $\underset{\sim}{\delta}^2_1$ must be a limit of Woodin cardinals in **HOD** since $\underset{\sim}{\delta}^2_1$ is $(<\Theta_0)$-strong in **HOD**. Therefore by (1) and (2) above, in $\mathbf{L}(T, \mathbb{R})$, $\underset{\sim}{\delta}^2_1$ must be a limit of Woodin cardinals in **HOD** and this is a contradiction since $(\Theta_0)^{\mathbf{L}(T,\mathbb{R})} = \Theta^{\mathbf{L}(T,\mathbb{R})}$ and since $\underset{\sim}{\delta}^2_1$ is the largest Suslin cardinal in $\mathbf{L}(T, \mathbb{R})$. The details will be given in [9].

We define the notion of an $\mathsf{AD}_{\mathbb{R}}$-*cardinal* which is the key to our proof that if the axiom **V=UltL** holds then the Ω conjecture holds. The context is ZFC together with the existence of a proper class of Woodin cardinals.

Definition 6.8. Suppose there is a proper class of Woodin cardinals. Then δ is an $\mathsf{AD}_{\mathbb{R}}$-*cardinal* if whenever $G \subset \mathrm{Coll}(\omega, \delta)$ is $\mathbf{V}$-generic then in $\mathbf{V}[G]$ there is a universally Baire set $A \subset \mathbb{R}^{\mathbf{V}[G]}$ such that for all γ, if $H \subset \mathrm{Coll}(\omega, \gamma)$ is $\mathbf{V}[G]$-generic then in $\mathbf{V}[G][H]$ for all universally Baire sets $B \subset \mathbb{R}^{\mathbf{V}[G][H]}$, $(\Theta_\omega)^{\mathbf{L}(\mathbb{R}^{\mathbf{V}[G][H]}, B, A_H)} < \Theta^{\mathbf{L}(\mathbb{R}^{\mathbf{V}[G][H]}, A_H)}$ where A_H is the interpretation of A in $\mathbf{V}[G][H]$.

Remark 6.9. Suppose there is a proper class of Woodin cardinals and that δ is an $\mathsf{AD}_{\mathbb{R}}$-cardinal. Suppose $G \subset \mathrm{Coll}(\omega, \delta)$ is $\mathbf{V}$-generic. Then in $\mathbf{V}[G]$, 0 is an $\mathsf{AD}_{\mathbb{R}}$-cardinal.

We need the following two theorems. The first of these is from the theory of derived models and the second is a relatively standard theorem from the theory of AD^+.

Theorem 6.10. Suppose there is a proper class of Woodin cardinals and that δ is a limit of strong cardinals. Then there is an $\mathsf{AD}_{\mathbb{R}}$-cardinal κ such that $\kappa \leq \delta$.

Theorem 6.11. Suppose there is a proper class of Woodin cardinals, $\Gamma \lhd \Gamma_\infty$, and that $(\Theta_\omega)^{\mathbf{L}(\Gamma,\mathbb{R})} < \Theta^{\mathbf{L}(\Gamma,\mathbb{R})}$. Suppose that $(\Theta_\omega)^{\mathbf{L}(\Gamma,\mathbb{R})} < \delta$ and that δ is a inaccessible limit of Woodin cardinals in $\mathbf{HOD}^{\mathbf{L}(\Gamma,\mathbb{R})}$. Then $(\Theta_\omega)^{\mathbf{L}(\Gamma,\mathbb{R})}$ is the least $\mathsf{AD}_{\mathbb{R}}$-cardinal in $\mathbf{HOD}^{\mathbf{L}(\Gamma,\mathbb{R})} \cap \mathbf{V}_\delta$.

Remark 6.12. Suppose there is a proper class of Woodin cardinals, $\Gamma \lhd \Gamma_\infty$, and that $(\Theta_\omega)^{\mathbf{L}(\Gamma,\mathbb{R})} < \Theta^{\mathbf{L}(\Gamma,\mathbb{R})}$. It should be the case that $(\Theta_\omega)^{\mathbf{L}(\Gamma,\mathbb{R})}$ is the ω-th Woodin cardinal of $\mathbf{HOD}^{\mathbf{L}(\Gamma,\mathbb{R})}$. This holds is in all the models where the **HOD**-analysis can be carried out.

In general, the best known result is that there exists $s \in \mathrm{Ord}^\omega \cap \mathbf{L}(\Gamma, \mathbb{R})$ such that $(\Theta_\omega)^{\mathbf{L}(\Gamma,\mathbb{R})}$ is the ω-th Woodin cardinal of $(\mathbf{HOD}_{\{s,x\}})^{\mathbf{L}(\Gamma,\mathbb{R})}$ for all $x \in \mathbb{R}$.

We shall use the following theorem to show that if **V=UltL** then the Ω conjecture holds.

Theorem 6.13 (V=UltL). There is a partial order $\mathbb{P}$ such that if $G \subset \mathbb{P}$ is $\mathbf{V}$-generic then in $\mathbf{V}[G]$ every **OD** subset of $\mathbb{R}$ is universally Baire.

Proof. Assume toward a contradiction that no such partial order exists. By Theorem 6.10, there is an $\mathsf{AD}_{\mathbb{R}}$-cardinal. Let δ be the least $\mathsf{AD}_{\mathbb{R}}$-cardinal. We claim that there must exist $\eta_0 \in \mathrm{Ord}$ such that for all $\eta > \eta_0$ such that η is an inaccessible limit of Woodin cardinals the following hold in $\mathbf{V}_\eta$:

1. There is a $\mathsf{AD}_{\mathbb{R}}$-cardinal.

2. Let κ be the least $\mathsf{AD}_{\mathbb{R}}$-cardinal. Then for all partial orders $\mathbb{P} \in \mathbf{V}_\kappa$, if $G \subset \mathbb{P}$ is $\mathbf{V}$-generic then in $\mathbf{V}[G]$ there is an $\mathbf{OD}$ set $A \subseteq \mathbb{R}$ which is not universally Baire.

This is essentially immediate since for all $\eta > \delta$, if η is an inaccessible limit of Woodin cardinals then δ must be the least $\mathsf{AD}_{\mathbb{R}}$-cardinal of $\mathbf{V}_\eta$. Also for any partial order $\mathbb{P}$, if $G \subset \mathbb{P}$ is $\mathbf{V}$-generic then for all $A \subset (\wp(\mathbb{R}))^{\mathbf{V}[G]}$, if A is not universally Baire in $\mathbf{V}[G]$ then for all sufficiently large η such that $\eta = |\mathbf{V}_\eta|$, A is not universally Baire in $\mathbf{V}_\eta[G]$. Finally A is $\mathbf{OD}$ in $\mathbf{V}[G]$ if and only if A is $\mathbf{OD}$ in $\mathbf{V}_\eta[G]$ for some $\eta \in \mathrm{Ord}$.

The indicated property of η_0 is a Π_2-property and so the existence of η_0 is expressible as a Σ_3-sentence, φ_0. Since $\mathbf{V}{=}\mathbf{UltL}$, there must exist $\Gamma \lhd \Gamma_\infty$ such that $M \models \varphi_0$ and $M \models$ "There is a proper class of Woodin cardinals", where $M = \mathbf{HOD}^{\mathbf{L}(\Gamma,\mathbb{R})} \cap \mathbf{V}_\Theta$ and $\Theta = \Theta^{\mathbf{L}(\Gamma,\mathbb{R})}$. Here we are reflecting the conjunction of φ_0 and the Π_3 sentence which asserts there is a proper class of Woodin cardinals.

Fix Γ and M as above. Let κ_M be the least $\mathsf{AD}_{\mathbb{R}}$-cardinal of M and let δ_Γ be the largest Suslin cardinal of $\mathbf{L}(\Gamma, \mathbb{R})$. By Theorem 6.5, δ_Γ is a strong cardinal in M and so δ_Γ must be a limit of Woodin cardinals in M. This implies by Theorem 6.6 that $\delta_\Gamma = (\Theta_{\delta_\Gamma})^{\mathbf{L}(\Gamma,\mathbb{R})}$. Therefore by Theorem 6.11, $\kappa_M = (\Theta_\omega)^{\mathbf{L}(\Gamma,\mathbb{R})}$ and in particular, $(\Theta_2)^{\mathbf{L}(\Gamma,\mathbb{R})} < \kappa_M$. Finally by Lemma 2.5, there is a partial order $\mathbb{P} \in M \cap \mathbf{V}_{\kappa_M}$ such that if $G \subset \mathbb{P}$ is M-generic then $M[G] \models$ "Every $\mathbf{OD}$-set $A \subset \mathbb{R}$ is universally Baire". This contradicts that $M \models \varphi_0$. Q.E.D.

The second consequence of $\mathbf{V}{=}\mathbf{UltL}$ we shall need is given in the next theorem which will be proved in [9]. The proof requires much deeper elements of the theory of AD^+.

Theorem 6.14 ($\mathbf{V}{=}\mathbf{UltL}$). Suppose κ is a cardinal and that G is $\mathbf{V}$-generic for $\mathrm{Coll}(\omega, \kappa)$. Then there exists a (possibly trivial) elementary embedding $\pi : (H(\kappa^+))^{\mathbf{V}} \to N$ such that $(\pi, N) \in \mathbf{V}$ and such that $N \in \mathbf{HOD}^{\mathbf{L}(\Gamma_\infty^{\mathbf{V}[G]})}$ where $\Gamma_\infty^{\mathbf{V}[G]}$ is the collection of universally Baire sets in $\mathbf{V}[G]$.

The following theorem summarizes the key consequences of the axiom $\mathbf{V}{=}\mathbf{UltL}$ which we shall prove. The Generic-Multiverse is the generic-multiverse generated by $\mathbf{V}$ and this is defined in [6].

Theorem 6.15 ($\mathbf{V}{=}\mathbf{UltL}$). (1) CH holds.

(2) $\mathbf{V}{=}\mathbf{HOD}$.

(3) $\mathbf{V}$ is the minimum universe of the Generic-Multiverse.

(4) The Ω conjecture holds.

The following theorem from the theory of AD^+ immediately gives that if **V**=**UltL** then CH holds.

Theorem 6.16. Suppose there is a proper class of Woodin cardinals and that $\Gamma \lhd \Gamma_\infty$. Then $\mathbf{HOD}^{\mathbf{L}(\Gamma,\mathbb{R})} \models \mathsf{CH}$.

The conclusions (2)–(4) of Theorem 6.15 follow from Theorem 6.14 as we shall show below. We first prove the following corollary of Theorem 6.14 which easily gives (2) and (3) of Theorem 6.15.

Theorem 6.17 (V=UltL). Suppose $\mathbf{V}[G]$ is a set generic extension of **V**. Then $V \subseteq (\mathbf{HOD})^{\mathbf{V}[G]}$.

Proof. Fix a partial order $\mathbb{P} \in \mathbf{V}$ such that G is $\mathbf{V}$-generic for $\mathbb{P}$ and let $\delta = |\mathbb{P}|^{\mathbf{V}}$. We prove that for all regular cardinals $\kappa > \delta$, $(\wp(\kappa))^{\mathbf{V}} \subset (\mathbf{HOD})^{\mathbf{V}[G]}$ and this will show that $V \subseteq (\mathbf{HOD})^{\mathbf{V}[G]}$.

Fix a regular cardinal $\kappa > \delta$ and let $\langle S_\alpha : \alpha < \kappa\rangle \in \mathbf{V}$ be a partition of the set $S = \{\alpha < \kappa \mid (\mathrm{cf}(\alpha))^{\mathbf{V}} = \omega\}$ into stationary sets. Note that if $C \subseteq \kappa$ is a closed cofinal set with $C \in \mathbf{V}[G]$ then there must exist a closed cofinal set $D \subseteq C$ such that $D \in \mathbf{V}$. Therefore each S_α is a stationary set in $\mathbf{V}[G]$.

By Theorem 6.14 there exists an elementary embedding $\pi : (H(\kappa^+))^{\mathbf{V}} \to N$ such that $N \in (\mathbf{HOD})^{\mathbf{V}[G]}$ and such that $(\pi, N) \in V$. Let $\langle T_\beta : \beta < \pi(\kappa)\rangle = \pi(\langle S_\alpha : \alpha < \kappa\rangle)$. Working in $\mathbf{V}[G]$, define

$$Z = \{\beta < \pi(\kappa) \mid T_\beta \cap C \neq \varnothing \text{ for all closed cofinal sets } C \subset \sup(\pi[\kappa])\}.$$

Thus $Z \in (\mathbf{HOD})^{\mathbf{V}[G]}$ since $\langle T_\beta : \beta < \pi(\kappa)\rangle \in (\mathbf{HOD})^{\mathbf{V}[G]}$. Note that for all $\xi \in S$, $\pi(\xi) = \sup(\pi[\xi])$. Therefore since each set S_α is a stationary subset of κ in $\mathbf{V}[G]$, $\pi[\kappa] \subseteq Z$ and so necessarily, $Z = \pi[\kappa]$. The situation here is similar though simpler than in the proof of 5 within the proof of Theorem 5.3.

For each $X \in N$, let $X^* = \{\alpha < \kappa \mid \pi(\alpha) \in X\}$. Since $N \in (\mathbf{HOD})^{\mathbf{V}[G]}$ and since $\pi[\kappa] \in (\mathbf{HOD})^{\mathbf{V}[G]}$, $\{X^* \mid X \in N\} \subset (\mathbf{HOD})^{\mathbf{V}[G]}$. Finally we have $(\wp(\kappa))^{\mathbf{V}} \subset \mathrm{dom}(\pi)$ and so $(\wp(\kappa))^{\mathbf{V}} = \{X^* \mid X \in N\}$ which implies that $(\wp(\kappa))^{\mathbf{V}} \subset (\mathbf{HOD})^{\mathbf{V}[G]}$. Q.E.D.

Theorem 6.18 (V = UltL). V = HOD.

Proof. This is an immediate corollary of Theorem 6.17. Q.E.D.

Theorem 6.19 (V = UltL). V is the minimum universe of the Generic-Multiverse.

Proof. Suppose that $\mathbf{V}[G] = V_0[G_0]$, $G \subset \mathbb{P}$ is $\mathbf{V}$-generic for some partial order $\mathbb{P} \in \mathbf{V}$, and $G_0 \subset \mathbb{P}_0$ is V_0-generic for some partial order $\mathbb{P}_0 \in V_0$. We must prove that $V \subseteq V_0$.

Fix a cardinal $\delta \in \mathbf{V}$ such that $|\mathbb{P}|^{\mathbf{V}} < \delta$ and such that $|\mathbb{P}_0|^{V_0} < \delta$. The key points are that in $\mathbf{V}$, $\mathrm{RO}(\mathbb{P} \times \mathrm{Coll}(\omega, \delta)) \cong \mathrm{RO}(\mathrm{Coll}(\omega, \delta))$, and that in V_0, $\mathrm{RO}(\mathbb{P}_0 \times \mathrm{Coll}(\omega, \delta)) \cong \mathrm{RO}(\mathrm{Coll}(\omega, \delta))$. Suppose $g \subset \mathrm{Coll}(\omega, \delta)$ is $\mathbf{V}[G]$-generic. Therefore by the homogeneity of $\mathrm{Coll}(\omega, \delta)$ and the isomorphisms above,

$$(\mathbf{HOD})^{V[g]} = (\mathbf{HOD})^{\mathbf{V}[G][g]} = (\mathbf{HOD})^{V_0[G_0][g]} = (\mathbf{HOD})^{V_0[g]} \subseteq V_0.$$

By Theorem 6.17, $V \subseteq (\mathbf{HOD})^{V[g]}$ and so $V \subseteq (\mathbf{HOD})^{V_0[g]} \subseteq V_0$. Q.E.D.

Theorem 6.20 (V=UltL). The Ω conjecture holds.

Proof. By Theorem 6.13, there is a partial order $\mathbb{P}$ such that if $G \subset \mathbb{P}$ is $\mathbf{V}$-generic then in $\mathbf{V}[G]$, every $\mathbf{OD}$ subset of $\mathbb{R}$ is universally Baire. By Theorem 2.6, $(\mathbf{HOD})^{\mathbf{V}[G]} \models$ "The Ω conjecture" and by Theorem 6.17, $V \subseteq (\mathbf{HOD})^{\mathbf{V}[G]}$. Therefore $(\mathbf{HOD})^{\mathbf{V}[G]}$ must be a generic extension of $\mathbf{V}$. Finally the Ω conjecture is absolute between set-generic extensions and so the Ω conjecture holds in $\mathbf{V}$. Q.E.D.

We end with the following conjecture from [5]. This conjecture if true would show that there is no analog of Scott's Theorem for **V**=**UltL** and would also prove the weak ultimate **L**conjecture. Suitable extender models are defined in [5]. The essential additional requirement that a weak extender model N for the supercompactness of δ must satisfy to be a suitable extender model, is that there exist a sequence of extenders $\langle E_\alpha : \alpha < \delta \rangle$ such that $\langle E_\alpha \cap N : \alpha < \delta \rangle \in N$ and such that $\langle E_\alpha \cap N : \alpha < \delta \rangle$ witnesses in N that δ is a Woodin cardinal, [5]. This guarantees that every bounded subset of δ is set generic over $N \cap \mathbf{V}_\delta$ which is really all this additional requirement is used for. This can fail if N is just a weak extender model for the supercompactness of δ and the example is the same as indicated in the discussion just before Lemma 3.8. Cf. [5, Lemma 150] for the details.

Conjecture 6.21. We call the following statement the *ultimate* **L** *conjecture*: (ZFC) Suppose that δ is an extendible cardinal. Then there exists a transitive class N such that

(1) N is a suitable extender model for the supercompactness of δ.

(2) $N \subseteq \mathbf{HOD}$,

(3) $N \models$ "**V**=**UltL**".

References

[1] Q. Feng, M. Magidor, and H. Woodin. Universally Baire sets of reals. In H. Judah, W. Just, and H. Woodin, editors, *Set theory of the continuum, Papers from the workshop held in Berkeley, California, October 16–20, 1989*, volume 26 of *Mathematical Sciences Research Institute Publications*, pages 203–242. Springer-Verlag, 1992.

[2] P. B. Larson. *The stationary tower, Notes on a course by W. Hugh Woodin*, volume 32 of *University Lecture Series*. American Mathematical Society, 2004.

[3] D. A. Martin and J. R. Steel. Iteration trees. *Journal of the American Mathematical Society*, 7(1):1–73, 1994.

[4] W. J. Mitchell and J. R. Steel. *Fine structure and iteration trees*, volume 3 of *Lecture Notes in Logic*. Springer-Verlag, 1994.

[5] W. H. Woodin. Suitable extender models I. *Journal of Mathematical Logic*, 10(1-2):101–339, 2010.

[6] W. H. Woodin. The continuum hypothesis, the generic-multiverse of sets, and the Ω conjecture. In J. Kennedy and R. Kossak, editors, *Set theory, arithmetic, and foundations of mathematics: theorems, philosophies*, volume 36 of *Lecture Notes in Logic*, pages 13–42. Association for Symbolic Logic, 2011.

[7] W. H. Woodin. Suitable extender models II: beyond ω-huge. *Journal of Mathematical Logic*, 11(2):115–436, 2011.

[8] W. H. Woodin. The fine structure of suitable extender models I, 2013. typoscript.

[9] W. H. Woodin. The fine structure of suitable extender models II, 2014. In preparation.

[10] W. H. Woodin, J. Davis, and D. Rodriquez. The **HOD** dichotomy. In J. Cummings and E. Schimmerling, editors, *Appalachian Set Theory 2006–2012*, volume 406 of *London Mathematical Society Lecture Notes Series*, pages 397–419. Cambridge University Press, 2012.